Scheduling Theory Single-Stage Systems

Mathematics and Its Applications

Volume 284

Scheduling Theory. Single-Stage Systems

by

V. S. Tanaev
V. S. Gordon
and
Y. M. Shafransky

Institute of Engineering Cybernetics,
Byelorussian Academy of Sciences,
Minsk, Byelorussia

SPRINGER-SCIENCE+BUSINESS MEDIA, B.V.

A C.I.P. Catalogue record for this book is available from the Library of Congress

ISBN 978-94-010-4520-9   ISBN 978-94-011-1190-4 (eBook)
DOI 10.1007/978-94-011-1190-4

This is the updated and revised translation
of the original Russian work *Scheduling Theory. Single-Stage Systems.*
Translated by the authors.
Published by Nauka, Moscow © 1984

Printed on acid-free paper

CONTENTS

Preface	vii
Introduction	1
Chapter 1	
Elements of Graph Theory and Computational Complexity of Algorithms	40
1. Sets, Orders, Graphs	40
2. Balanced 2-3-Trees	50
3. Polynomial Reducibility of Discrete Problems. Complexity of Algorithms	58
4. Bibliography and Review	68
Chapter 2	
Polynomially Solvable Problems	69
1. Preemption	70
2. Deadline-Feasible Schedules	78
3. Single Machine. Maximal Cost	93
4. Single Machine. Total Cost	103
5. Identical Machines. Maximal Completion Time. Equal Processing Times	122
6. Identical Machines. Maximal Completion Time. Preemption	134
7. Identical Machines. Due Dates. Equal Processing Times	147
8. Identical Machines. Maximal Lateness.	158
9. Uniform and Unrelated Parallel Machines. Total and Maximal Cost	163
10. Bibliography and Review	175

Chapter 3
Priority-Generating Functions. Ordered Sets of Jobs 186
1. Priority-Generating Functions 187
2. Elimination Conditions 196
3. Tree-like Order 202
4. Series-Parallel Order 213
5. General Case 224
6. Convergence Conditions 233
7. 1-Priority-Generating Functions 246
8. Bibliography and Review 249

Chapter 4
NP-Hard Problems 253
1. Reducibility of the Partition Problem 253
2. Reducibility of the 3-Partition Problem 260
3. Reducibility of the Vertex Covering Problem 280
4. Reducibility of the Clique Problem 288
5. Reducibility of the Linear Arrangement Problem 297
6. Bibliographic Notes 303

Appendix
Approximation Algorithms 312

References 324

Additional References 354

Index 365

Also of Interest 371

PREFACE

Scheduling theory is an important branch of operations research. Problems studied within the framework of that theory have numerous applications in various fields of human activity. As an independent discipline scheduling theory appeared in the middle of the fifties, and has attracted the attention of researchers in many countries. In the Soviet Union, research in this direction has been mainly related to production scheduling, especially to the development of automated systems for production control.

In 1975 *Nauka* ("Science") *Publishers*, Moscow, issued two books providing systematic descriptions of scheduling theory. The first one was the Russian translation of the classical book *Theory of Scheduling* by American mathematicians R. W. Conway, W. L. Maxwell and L. W. Miller. The other one was the book *Introduction to Scheduling Theory* by Soviet mathematicians V. S. Tanaev and V. V. Shkurba. These books well complement each other. Both books well represent major results known by that time, contain an exhaustive bibliography on the subject. Thus, the books, as well as the Russian translation of *Computer and Job-Shop Scheduling Theory* edited by E. G. Coffman, Jr., (Nauka, 1984) have contributed to the development of scheduling theory in the Soviet Union.

Many different models, the large number of new results make it difficult for the researchers who work in related fields to follow the fast development of scheduling theory and to master new methods and approaches quickly.

Bibliography on scheduling theory includes more than 1,500 titles. Unfortunately, many of papers and some of the books originally published in Russian are practically unknown for the Western specialists.

In the early eighties a group of Byelorussian mathematicians made an attempt to give an up-to-date description of standard scheduling theory. As a result, two books appeared: *Scheduling Theory. Single-Stage Systems* by V. S. Tanaev, V. S. Gordon and Y. M. Shafransky (Nauka, 1984) and *Scheduling Theory. Multi-Stage Systems* by V. S. Tanaev, Y. N. Sotskov and V. A. Strusevich (Nauka, 1989). These two books cover two different major problem areas of scheduling theory and can be considered as a two-volume monograph that provides a systematic and comprehensive exposition of the subject.

The authors are grateful to *Kluwer Academic Publishers* for creating the opportunity to publish the English translations of these two books. We are indebted to M. Hazewinkel, J. K. Lenstra, A. H. G. Rinnooy Kan, D. B. Shmoys and W. Szwarc for their supporting the idea of translating the books into English.

The first of the books proposed to the reader is devoted to the problems of finding optimal schedules for systems consisting either of a single machine or of several parallel machines. The book describes in detail the most important statements and algorithms which contain typical scheduling ideas and approaches. Some propositions are accompanied only with schematic proofs. Besides that, each chapter of the book presents a bibliographic review containing all necessary references. Some major results are grouped into three tables given in Introduction, thus creating a visual guideline.

In the process of preparing this book for publication a number of small errors and misprints were observed. These have been revised without special mention. To present the results not reflected in the Russian edition, a list of additional references has been included and corresponding amendments have been made to the bibliographic sections and to the tables given in Introduction. The references to the additional list are marked in the text by "*". The list mainly contains the papers and books that appeared after 1983. This translation also includes a specially written Appendix that presents a review of approximation algorithms.

It should be noted that Russian and English scheduling terminologies are not quite stable and may differ from each other. There are also some notational differences. However, those are not significant and will not create difficulties for the reader.

It has been a pleasure to cooperate with Dr. D. J. Larner and his colleagues from Kluwer Academic Publishers. We are also grateful to V. A. Strusevich for his assistance in preparing this translation.

We hope this book will be of interest for different groups of readers working in applied mathematics, production planning, flexible manufacturing systems and related areas, and will contribute to the further development of scheduling theory as well as to expanding spheres of its possible applications.

V. S. Tanaev

V. S. Gordon

Y. M. Shafransky

INTRODUCTION

Scheduling theory studies the problems of optimal *distribution* and *sequencing* of the jobs of a finite set to be processed on either a deterministic single machine or in a multi-machine system under different assumptions on the nature of this processing.

Machine tools, railway lines, classrooms, computers, etc., may be treated as "machines". Workpieces, trains, student teams, computer programs, etc., may be interpreted as "jobs". Since the nature of "machines" and "jobs" is, in fact, immaterial, those can be numbered by the integers 1, 2,..., M and by 1, 2,..., n, respectively. In what follows, we formulate scheduling problems in terms of the jobs of a set $N = \{1, 2,..., n\}$ to be processed in a system consisting of M machines 1, 2,..., M.

As a rule, each job $i \in N$ is given a set $Q^{(i)} \subseteq \{1, 2,..., M\}$ of machines such that each of the machines in this set either may or must process this job. If each job i is allowed to be processed on any machine $L \in Q^{(i)}$, then the processing system is called *a single-stage* system (consisting either of one machine or of several *parallel* machines).

In *multi-stage* systems, the processing of job i involves l_i stages. Each job $i \in N$ at a stage j, $1 \le j \le l_i$, is associated with some set $Q_j^{(i)} \subseteq Q^{(i)}$ of machines, so that job i at stage j may be processed on a machine $L \in Q_j^{(i)}$, but on at most one machine at a time. In any case, it is assumed that any machine can process at most one job at a time.

If $l_i = l \ge 2$, $Q_j^{(i)} = Q_j$, $i = 1, 2,..., n$, $Q_{j_1} \cap Q_{j_2} = \emptyset$, $1 \le j_1 \ne j_2 \le l$, then a processing system is *a flow-type system with parallel machines*. For a *job shop* system, we have $|Q_j^{(i)}| = 1$, $i = 1, 2,..., n$, $j = 1, 2,..., l_i$. In a *flow shop* system (without parallel machines), the machines are normally numbered so that each job is first processed on machine 1, then on machine 2, and so on, until it is processed on machine $l = M$. Of some interest are *open shop* systems and *mixed shop* systems with non-fixed processing routes of all or some jobs.

This book *concentrates on single-stage processing systems* in which:

(i) $Q^{(i)} = \{1, 2,..., M\}$, $i = 1, 2,..., n$, i.e. a machine can process any job of set N;

(ii) each job can be processed on at most one machine at a time, and each machine can process at most one job at a time.

For each $i \in N$, the release date $d_i \geq 0$ is given (a time at which job i becomes available for processing).

The processing time $t_{iL} > 0$ of a job $i \in N$ on a machine L, $1 \leq L \leq M$, is known in advance. If $t_{iL} = a_L t_i$, $i = 1, 2,..., n$, $L = 1, 2,..., M$, machine L is said to have a processing *speed* equal to $1/a_L$. If $a_L = 1$, $L = 1, 2,..., M$, the machines are called *identical*.

Depending on the nature of the processing system, *preemption* in the processing of a job may or may not be allowed. Allowing preemption implies that the processing of a job may be interrupted and resumed at a later time on any of the machines. Preemptions may be allowed either at some specific times or at arbitrary times. As a rule, it is assumed that preemption does not involve additional expenses, and their number is finite.

Processing the jobs can be described by a family $s = \{s_1(t), s_2(t),..., s_M(t)\}$ of piecewise-constant left-semicontinuous functions $s_L = s_L(t)$, $L = 1, 2,..., M$, each being defined over the interval $0 \leq t < \infty$ and assuming the values $0, 1,..., n$. If $s_L(t') = i \neq 0$, then at time t' machine L processes job i. If $s_L(t') = 0$, then at time t' machine L is idle.

Since a job cannot be processed on two or more machines at a time, the condition $s_L(t') = i \neq 0$ implies that $s_H(t') \neq i$ for all $1 \leq H \neq L \leq M$. Since d_i is the release date for a job i, $i = 1, 2,..., n$, it follows that $s_L(t) \neq i$, $L = 1, 2,..., M$, for all $t < d_i$.

If t'_{iL} is the total length of time intervals where the function $s_L(t)$ has the value i, then the relations $\sum_{L=1}^{M} (t'_{iL}/t_{iL}) = 1$, $i = 1, 2,..., n$, hold. For example, if the machines are identical, then the total length of all time intervals in which all functions $s_L(t)$, $L = 1, 2,..., M$, have the same value i must be equal to t_i.

A family s of functions with the described properties is called *a schedule* for processing the jobs of set N in a system consisting of M parallel machines.

Figure I.1 presents the diagram of a schedule $s(t)$ for single-machine processing of the jobs of set $N = \{1, 2, 3, 4\}$. Here $d_1 = 0$, $d_2 = 2$, $d_3 = d_4 = 3$, $t_1 = 4$, $t_2 = 1$, $t_3 = t_4 = 2$.

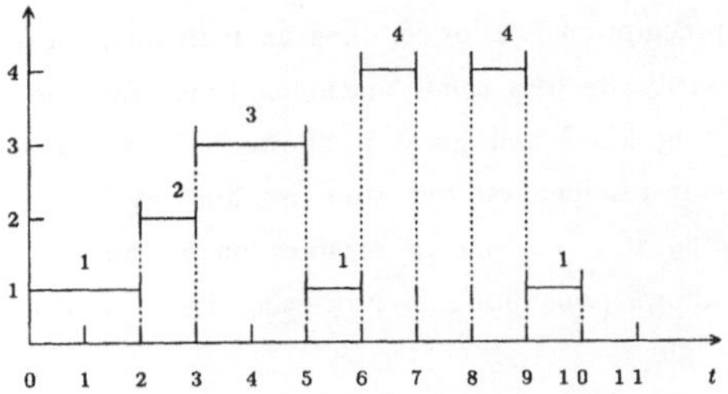

Fig. I.1

If a system consists of two or more machines, the diagrams of functions $s_L(t)$ are normally combined as in Fig. I.2. Here $M = 3$, the machines are identical, $N = \{1, 2, 3, 4, 5\}$; $d_1 = d_2 = d_3 = 0$, $d_4 = 1$, $d_5 = 2$; $t_1 = 1$, $t_2 = t_3 = 3$, $t_4 = t_5 = 2$. Machine 1 processes job 1 in the time interval (0, 1] and job 5 in the interval (3, 4]. Machine 2 processes job 2 in the time intervals (0, 1] and (3, 5], and job 4 in the interval (1, 3]. Machine 3 processes job 3 in (0, 3] and job 5 in (4, 5].

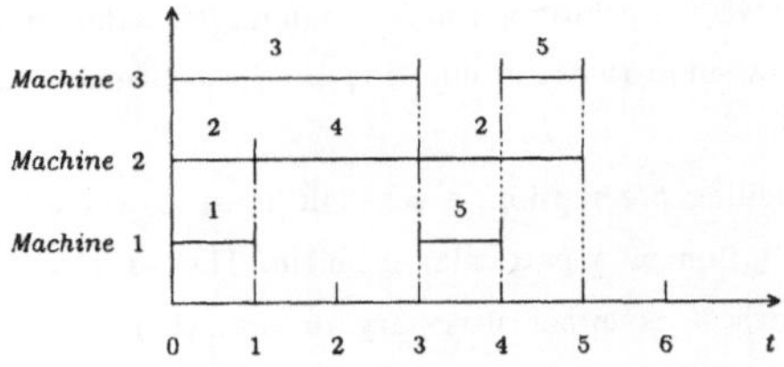

Fig. I.2

A schedule $s = \{s_1(t), s_2(t),..., s_M(t)\}$ is said to be *preemptive* if there exist an i, $1 \le i \le n$, both L and H, $1 \le L \ne H \le M$, and times t' and t'', $0 \le t' < t < t'' < \infty$, such that at least one of the following conditions holds:

(1) $s_L(t') = s_L(t'') = i$, but $s_L(t) \ne i$;

(2) $s_L(t') = s_H(t'') = i$.

Here, if $s_L(t'+\delta) \ne i$ for a sufficiently small $\delta > 0$, then the processing of job i on machine L is interrupted at time t' and may be resumed on another machine at the same time.

The non-preemptive processing of jobs satisfies the following condition. A job is processed on at most one machine at a time. If the processing of some job i on a machine L starts at time t_i^0, then the job is processed only on machine L and is completed at time $\overline{t}_i = t_i^0+t_{iL}$. It is obvious that, in this case, the schedule is completely determined by distributing the jobs over the machines and assigning the starting time t_i^0 to each job i. If in job processing preemption is allowed, then an individual job can be processed "part by part", not necessarily on the same machine. Thus, for the schedule in Fig. I.2, preemption in processing job 2 and job 5 is allowed. Processing job 2 on machine 2 is interrupted at $t = 1$ is resumed on the same machine at $t = 3$. Processing job 5 on machine 1 is interrupted at $t = 4$, and is resumed on machine 3 at the same time. The schedule in Fig. I.1 allows preemption in processing jobs 1 and 4. In the time interval (7, 8] the machine is idle.

In practical applications, the numbers d_i and t_{iL} are rational, and may be considered to be integers by choosing an appropriate scale. In this case, we can restrict our consideration to a class of schedules in which preemption occurs only at integer times. It is assumed that, for each job, the starting and the resumption times are also integers. Such schedules are specified by an M-dimensional vector with components $0, 1, \dots, n$ determined for each unit length time interval. If, for some unit time interval, the Lth component of the vector is $i \neq 0$, then in this interval machine L processes job i. Otherwise, machine L is idle.

If preemption is allowed at arbitrary times, assuming that the number of preemptions is finite, it is natural to assume that the duration of the continuous processing of a job is also finite.

In addition to forbidding preemption, a schedule must satisfy other requirements which follow from the formulation of a particular problem. Thus, for each job i, *a due date* D_i may be given, by which it is either necessary or desirable to complete processing this job. A schedule in which all jobs meet their due dates is called *feasible with respect to the due dates*. In a general case, such a schedule need not exist.

Situations in which some restrictions are introduced on the possible job processing sequence are also quite common. If, according to the problem formulation, the processing of a job j may start only after another job i is completed, then a schedule s must satisfy the condition: if $s_L(t') = i$ for some $1 \leq L \leq M$ and some $t' > 0$, then $s_H(t) \neq j$ for all $1 \leq H \leq M$ and $t \leq t'$. Situations of this type are usually described by specifying some precedence relation $\rightarrow$ over the set N of jobs such that the notation $i \rightarrow j$ implies that the processing of job i must be completed before the processing of job j can start. In this case, the schedule is said to be *feasible with respect to precedence relation defined over* N.

The processing of jobs may involve the consumption or usage of some additional *resources*. A typical situation of this type can be described as follows. There are q types of resources which are used in job processing. At time t, there are $R_k(t)$, $k = 1, 2,\ldots, q$, units of resource of type k available. The processing of job i at time t requires $r_{ik}(t)$, $i = 1, 2,\ldots, n$, $k = 1, 2,\ldots, q$, units of resource of type k. If at time t only the jobs $i_1, i_2,\ldots, i_l$ are processed, then the inequality $\sum_{j=1}^{l} r_{i_j k}(t) \le R_k(t)$ must hold for all k, $1 \le k \le q$. A schedule s in which the above resource constraints are satisfied at any time $t \ge 0$ is called *feasible with respect to resources.*

Schedules which meet restrictions connected with machine setups, job grouping, etc., are also of practical interest. In such situations, a schedule is feasible if it satisfies all requirements which follow from the formulation of a particular problem.

It should be noted that constructing a feasible schedule or even checking whether such a schedule exists is frequently a far from trivial problem. At the same time, in many situations constructing feasible schedules does not involve any special difficulties, and then the problem of choosing the best (in a certain sense) schedule arises.

In scheduling theory, *the quality* of a schedule is normally estimated in the following way. A schedule s is associated with the vector $\bar{t}(s) = (\bar{t}_1(s), \bar{t}_2(s),\ldots, \bar{t}_n(s))$ of the job completion times. Here $\bar{t}_i(s)$ denotes the largest value of t such that there exists a $L \in \{1, 2,\ldots, M\}$ for which $s_L(t) = i$. A real-valued function $F(x) = F(x_1, x_2,\ldots, x_n)$ is specified, non-decreasing with respect to each of its n arguments. The quality of schedule s is characterized by the value of this function evaluated at $x = \bar{t}(s)$. Among any two schedules, that with a smaller value of $F(x)$ is considered to be the better one. The schedule with the smallest value of $F(x)$ (among all feasible schedules) is called an *optimal* schedule.

A function $F(x)$ is normally determined by associating each job i with some non-decreasing function, called a *cost* function $\varphi_i(t)$, which specifies a "penalty" to be "paid" for having this job completed by time t. The quality of a schedule is characterized by *the total* or *the maximal cost* that must be paid for processing the jobs according to a schedule s, i.e. $F_\Sigma(s) = \sum_{i \in N} \varphi_i(\bar{t}_i(s))$ or $F_{max}(s) = \max\{\varphi_i(\bar{t}_i(s)) \mid i \in N\}$.

In particular, if $\varphi_i(t) = t$, $i = 1, 2,\ldots, n$, then $F_{max}(s) = \max\{\bar{t}_i(s) \mid i \in N\}$ is the *makespan* (or the maximal completion time). In this case, $F_{max}(s)$ is denoted by $\bar{t}_{max}(s)$, and a schedule s with the smallest value of $\bar{t}_{max}(s)$ is called a *time-optimal* schedule.

If $\varphi_i(t) = t - D_i$, then $F_{max}(s)$ is denoted by $L_{max}(s)$. We have $L_{max}(s) = \max\{L_i(s) \mid i \in N\}$, where $L_i(s) = \bar{t}_i(s) - D_i$ is the *lateness* of job i with respect to the due date D_i.

If $\varphi_i(t) = \max\{0, t - D_i\}$, then $F_{max}(s)$ is denoted by $z_{max}(s)$. We have $z_{max}(s) = \max\{z_i(s) \mid i \in N\}$, where $z_i(s) = \max\{0, \bar{t}_i(s) - D_i\}$ is the *tardiness* of job i with respect to the due date D_i. In this case, $F_\Sigma(s) = \sum_{i \in N} z_i(s)$ is *the total tardiness.*

If $\varphi_i(t) = \text{sign}(\max\{0, t-D_i\})$, then $F_\Sigma(s) = \sum_{i \in N} u_i(s)$, where $u_i(s) = \text{sign}(z_i(s))$, is *the number of late* jobs (with respect to their due dates).

Each job i may also be given the number α_i representing the "weight" of the job, and we consider the *weighted* sum of job completion times $\sum_{i \in N} \alpha_i \bar{t}_i(s)$ (or the *weighted total flow time*), the weighted total tardiness $\sum_{i \in N} \alpha_i z_i(s)$, and the weighted number of late jobs $\sum_{i \in N} \alpha_i u_i(s)$.

The described optimality criteria reflect an intention to complete each job as soon as possible. Under these conditions, we may restrict our search to a class of schedules which do not allow unnecessary idle times. If preemption is either forbidden or allowed only at integer times, this class contains a finite number of schedules.

In fact, let s be a non-preemptive schedule in which the jobs are processed on a single machine according to the sequence $\pi = (i_1, i_2, \dots, i_n)$ where π is a permutation of the elements of set N. The starting time $t^0_{i_j}(s)$ of a job i_j satisfies the inequality $t^0_{i_j}(s) \geq \max\{\bar{t}_{i_{j-1}}(s), d_{i_j}\}$, while the completion time of this job is $\bar{t}_{i_j}(s) = t^0_{i_j}(s) + t_{i_j}$, $j = 1, 2, \dots, n$; $\bar{t}_{i_0}(s) = 0$. Consider a schedule s', in which the jobs are processed according to the same sequence π, and $t^0_{i_j}(s') = \max\{\bar{t}_{i_{j-1}}(s'), d_{i_j}\}$, $\bar{t}_{i_j}(s') = t^0_{i_j}(s') + t_{i_j}$, $j = 1, 2, \dots, n$, $\bar{t}_{i_0}(s') = 0$. It is easy to check that $\bar{t}_{i_j}(s') \leq \bar{t}_{i_j}(s)$, $j = 1, 2, \dots, n$, and, since function $F(s)$ is non-decreasing, it follows that $F(\bar{t}(s')) \leq F(\bar{t}(s))$. The schedule s' is uniquely specified by the permutation π, and, hence, the search for an optimal schedule can be restricted to the consideration of at most $n!$ schedules.

Similarly, let s be a non-preemptive schedule for processing the jobs on M parallel machines in which a machine L processes the jobs of set N_L according to the sequence $\pi^L = (i^L_1, i^L_2, \dots, i^L_{n_L})$, $L = 1, 2, \dots, M$. Here $N_1 \cup N_2 \cup \dots \cup N_M = N$, $N_H \cap N_R = \emptyset$, $1 \leq H \neq R \leq M$, and it is not necessary that $N_L \neq \emptyset$. Consider the schedule s', in which the starting time of a job i^L_j is $t^0_{i_j L} = \max\{\bar{t}_{i_{j-1} L}, d_{i_j L}\}$, and its completion time is $\bar{t}_{i_j L} = t^0_{i_j L} + \bar{t}_{i_j L}$, $j = 1, 2, \dots, n_L$, $\bar{t}_{i_0 L} = 0$, $L = 1, 2, \dots, M$. It is evident that $F(\bar{t}(s')) \leq F(\bar{t}(s))$. The schedule s' is uniquely specified by: (*i*) a partition of set N into subsets $N_1, N_2, \dots, N_M$ (some of them may be empty), and (*ii*) permutations of the elements of these sets. Therefore, the search for an optimal schedule can be restricted to considering at most $n!\binom{M+n-1}{n}$ schedules, where $\binom{p}{q}$ denotes a *binomial coefficient*. If the machines are identical, then the number of schedules under consideration is $M!$ times lower.

If interruptions in job processing are allowed only at integer time moments, then, as mentioned above, a schedule is uniquely specified by an M-dimensional vector with components 0, 1,..., n associated with each unit time interval. Here, it suffices to consider the time interval (called a planning interval) between $\min\{d_i \mid i \in N\}$ and $\max\{d_i \mid i \in N\} + \sum_{i \in N} \max\{t_{iL} \mid L = 1, 2,..., M\}$. Denoting the length of the planning interval by T, we can conclude that the search for an optimal schedule can be restricted to considering at most $\binom{n+1}{M}T$ schedules.

If interruptions are allowed at arbitrary times, then, in general, an optimal schedule need not be found in a finite set of schedules. However, under certain conditions, a finite set of schedules containing at least one optimal schedule can be determined in this case as well.

Similar considerations can be given to various types of feasible schedules.

Therefore, as a rule, an optimal schedule can be found by enumerating a finite set of feasible variants. The main difficulty is that the number of such variants is usually extremely large (e.g., already 10! = 3 628 800), and this increases exponentially with the problem dimension. Research in scheduling theory concentrates on reducing that enumeration as much as possible, and on finding an optimal schedule requiring the least computational effort.

If the volume of calculations is limited by some polynomial of the length of the problem input, the problem is said to belong to the class of polynomially solvable problems. The corresponding algorithms are called polynomial-time ones. On the other hand, so-called NP-hard problems are known for which polynomial-time algorithms are unlikely to exist.

This book presents the state-of-the-art in research on single-stage scheduling systems. Chapter 1 contains some auxiliary information. In Section 1, some facts from combinatorial analysis and graph theory are given which will be useful for further consideration. Section 2 gives a description of a specific data representation using so-called 2-3-trees. Section 3 introduces the main concepts of computational complexity of combinatorial optimization problems and their solution algorithms.

Chapter 2 describes computationally effective algorithms for finding optimal schedules. Section 1 establishes sufficient conditions for the existence of optimal schedules without preemption at times different from d_i, $i = 1, 2,..., n$. Section 2 presents the necessary and sufficient conditions for the existence of schedules that are feasible with respect to the given due dates D_i, $i = 1, 2, \ldots, n$, and describes algorithms for finding such schedules. The problem of minimizing the maximal cost $F_{max}(s)$ for single-machine processing is considered in Section 3. Section 4 studies effective algorithms for finding optimal schedules for a number of problems of minimizing the total cost $F_{\Sigma}(s)$ for single-machine processing. Sections 5 and 6 consider the problem of finding time-optimal schedules for processing a partially ordered set of jobs in a system consisting of identical parallel machines. Section 7 describes algorithms for finding deadline-feasible schedules for processing a partially ordered set of jobs with equal processing times on parallel machines. The problems of minimizing the maximal lateness for identical parallel machines are presented in Section 8. In Section 9, the problems of minimizing the total and the maximal costs for unrelated parallel machines are discussed.

Chapter 3 is devoted to the problems of minimizing the so-called priority-generating functions over permutations of elements of an ordered finite set N. Many scheduling problems are naturally formulated in terms of minimizing priority-generating functions. Such examples are given in Section 1. This section also introduces the concept of a priority-generating function. Section 2 describes transformations of graph G, which is the reduction graph of a precedence relation defined over set N. These transformations provide a basis for the algorithms for minimizing the priority-generating functions discussed in subsequent sections. Sections 3 and 4 study the cases when G is a tree-like graph and a series-parallel graph, respectively. The situation when G is an arbitrary graph is considered in Sections 5 and 6. Section 7 introduces the concept of the so-called 1-priority-generating function and discusses the methods for minimizing such functions.

In Chapter 4, a number of scheduling problems are proved to be NP-hard. Most of these problems are shown to be NP-hard in the strong sense.

Each chapter is accompanied by a bibliographic review. The review given in Chapter 4 is supplemented with information on enumeration methods used for solving NP-hard problems. The interested reader can find some additional information on results in scheduling theory as well as on the methods for finding optimal and near-optimal schedules in a number of surveys [20, 24, 25, 37-39, 61, 65, 92, 95, 182, 208, 211, 243, 281, 314, 325, 340, 341, 347, 374, 377, 382, 384, 7*, 10*, 11*, 66*, 78*, 114*, 117*-119*] and monographs [12, 78, 89, 110, 115, 118, 120, 122, 126, 127, 143, 144, 158, 162, 185, 192, 193, 239, 345, 368, 38*]. An extensive list of references in scheduling theory is given in the classified bibliography [303].

In order to facilitate the search for information on any particular problem in which the reader could be interested, the tables provided below contain data on most of the problems discussed in this book. Only some problems mainly described in Chapter 3 are omitted.

Polynomially solvable problems are given in Table I.1; Table I.2 contains *NP*-hard problems. Table I.3 presents information on approximation algorithms for solving *NP*-hard problems discussed in Appendix.

The first five columns of each table give problem descriptions using appropriate notation. The last column contains references either to the corresponding sections of this book (Tables I.1 and I.2) or to the cited literature (Table I.3).

The first column gives the number of machines.

The second column describes two parameters: "processing time" and "release dates". The "processing time" parameter may have the following values:

"t_i" - corresponds to the situation in which all machines are identical;

"$a_H t_i$" - the processing system consists of machines of different speeds (uniform machines);

"t_{iH}" - the machines are unrelated parallel;

"$t_i = t$" - the machines are identical, the processing times for all jobs are the same (and equal to t);

"$t_{iH} = a_H$" - the machines operate at different speeds, the processing times of each job on a machine H are the same (and equal to a_H);

"$t_i \in \{c_1, c_2, \ldots, c_l\}$" - the machines are identical, job processing times may have only the values in the indicated set;

"$[t_{iH}]$" - the processing times are integers.

The "release date" parameter is either equal to "$d_i = 0$" or to "d_i" depending on whether the release dates are the same. If the release dates are integers, the notation "$[d_i]$" is used.

The third column contains the values of three parameters: "preemption", "precedence", and "resources".

The "preemption" parameter is equal either to "*Pr*" or to "[*Pr*]" depending on whether preemption is allowed at arbitrary or only at integer times. If none of these values is indicated, then preemption is forbidden.

Depending on the type of the reduction graph of precedence relation $\rightarrow$ defined over set N of jobs, the "precedence" parameter may have one of the following values:

"*G*" - the reduction graph of relation $\rightarrow$ is an arbitrary circuit-free graph;

"ω-*SP*" - the reduction graph is an ω-series-parallel graph;

"*SP*" - is a series-parallel graph;

"$\mathcal{T}$" - is a tree;

"$\mathcal{T}^+$" - is a forest of outtrees;

"$\mathcal{T}^-$" - is a forest of intrees;

"C" - each connected component of the reduction graph is a chain.

If none of these values is indicated, then the set N is not ordered.

The "resource" parameter has the value $Rs(q)$ only if there are resource constraints and the number of resource types is q.

In the fourth column, additional conditions are given. For example, the notation "$D_i = D$" implies that all due dates are the same (and equal to D); the notation "$t_i = GCD(d_i)$" implies that processing times are the same for all jobs and coincide with the greatest common divisor of the release dates d_i $i = 1, 2,..., n$; the notation "$r_{ik} \in \{c_1, c_2,..., c_l\}$" says that r_{ik} may have only the values in the indicated set. The notation "$(d_i\uparrow, t_i\uparrow, D_i\uparrow, \alpha_i\downarrow)$" implies that the jobs of set N can be numbered in such a way that $d_i \leq d_{i+1}$, $t_i \leq t_{i+1}$, $D_i \leq D_{i+1}$, $\alpha_i \geq \alpha_{i+1}$, $i = 1, 2,..., n-1$. The notation "$\varphi_i\uparrow$" has a similar meaning, and here $\varphi_i \leq \varphi_{i+1}$ implies that $\varphi_i(t) \leq \varphi_{i+1}(t)$ for all t from the planning interval. The notation "$[D_i]$" indicates that due dates are integers. The notation "$M = M(N, D)$" implies that the number of machines M is a variable that is dependent on the set N of jobs and the common due date D.

Most of the problems presented in the tables involve minimizing a function whose form is indicated in the fifth column. Symbols F_{pg} and $F_{1\text{-}pg}$ denote priority-generating and 1-priority- generating functions, respectively. Some problems are to find a schedule that is feasible with respect to deadlines (the notation is "$\bar{t}_i \leq D_i$"). Some problems involve minimizing a certain function over a set of schedules that are feasible with respect to deadlines D_i. In this case, the function notation is supplemented with "$\bar{t}_i \leq D_i$". If the problem requires that inequalities $\bar{t}_i(s) \leq D_i$ must hold only for $i \in Q \subset N$, the previous notation accompanied by the condition "$i \in Q$".

The sixth column of Table I.1 gives estimates of the running times for solution algorithms (accurate up to a constant factor). Here the notation "*LP*" implies that the corresponding scheduling problem is reduced to a linear programming problem. The asterisk (*) in Table I.1 indicates problems in which allowing preemption does not reduce an optimal value of the objective function.

In Table I.2, the asterisk (*) marks *NP*-hard problems for which pseudopolynomial algorithms are known, and (**) indicates *NP*-hard problems for which pseudopolynomial algorithms are unknown, but *NP*-hardness in the strong sense is not established.

Table I.3 contains information on polynomial-time approximation algorithms presenting the estimate of the running time of an algorithm (if known) in column 6, and the performance guarantee (column 7). In column 6 of this table we use the notation "$P(\cdot, \cdot)$" to stress that the running time of an algorithm polynomially depends on the mentioned parameters. As a rule, column 7 provides the bound on the relative error of an obtained solution $\Delta = |F^0 - F^*|/|F^*|$, where F^0 is the value of the objective function for an approximate solution, and F^* is the optimal value.

In Tables I.1 and I.3, as well as elsewhere throughout the book, all logarithms are taken to the base 2 (unless stated otherwise).

In the tables, the following notation is used:

$t_{max} = \max\{t_{iH} \mid i \in N,\ H = 1, 2,\dots, M\}$;

$t_{min} = \min\{t_{iH} \mid i \in N,\ H = 1, 2,\dots, M\}$;

$t_\Sigma = \sum_{i \in N} t_i$ - for a single machine or identical parallel machines;

The values D_{max}, D_{min}, α_{max} etc., are defined analogously.

Estimates of the running time of algorithms given in the tables are valid, assuming that the precedence relation defined over the set of jobs (if the relation is not empty) is represented by its reduction graph. Note that transformation of an arbitrary circuit-free graph into its transitive closure or into the reduction graph requires at most $O(n^3)$ time [7], where n is the number of vertices of a graph.

As a rule, no special cases of the problems considered are included in Table I.1 unless simpler solution algorithms are known for them. A special case of a problem A is such a problem B that the set of all inputs of problem A contains all inputs of problem B as a subset. For example, the problem of minimizing a function $F(s)$ is a special case of the problem of minimizing $F(s)$ over the set of all schedules, satisfying the additional constraint $\bar{t}_i(s) \le D_i$, $i = 1, 2,\dots, n$. To see this, it suffices to take $D_i = W$, where W is a sufficiently large number.

Some polynomial-time solvable problems are not included in Table I.1 due to other reasons. It is easy to check that a schedule minimizing $L_{max}(s)$ simultaneously provides the minimum to functions $z_{max}(s)$ and $\max\{\varphi(\bar{t}_i(s) - D_i) \mid i \in N\}$, where $\varphi(x)$ is a non-decreasing function for $x > 0$. Therefore, if Table I.1 contains the problem with the objective function $L_{max}(s)$, then the problems with the objective functions $z_{max}(s)$ and $\max\{\varphi(\bar{t}_i(s) - D_i) \mid i \in N\}$ are omitted.

Table I.2 includes only "minimal" NP-hard problems, i.e., problems whose special cases are either polynomially solvable or have not been proved NP-hard. It is obvious that a problem with an NP-hard special case is NP-hard itself.

Table I.1

Number of machines	Processing times, release dates	Preemption, precedence and resource constraints	Additional conditions	Objective function	Running time	Section of the book
1	$t_i ; d_i=0$			$\bar{t}_i \le D_i$	$n \log n$	Ch. 2; 2.5(*)
1	$t_i ; d_i$	Pr		$\bar{t}_i \le D_i$	$n \log n$	Ch. 2; 2.5
1	$t_i ; d_i$		$(d_i\uparrow, D_i\uparrow)$	$\bar{t}_i \le D_i$	$n \log n$	Ch. 2; 2.5(*)
1	$t_i ; d_i$	Pr; G		$\bar{t}_i \le D_i$	n^2	Ch. 2; 3.7
1	$t_i=t ; d_i$	G		$\bar{t}_i \le D_i$	$n \log n$	Ch. 2; 10.1
1	$t_i ; d_i=0$	G		$\max \varphi_i(\bar{t}_i)$	n^2	Ch. 2; 3.2(*)
1	$t_i ; d_i=0$		$\varphi_i\uparrow$	$\max \varphi_i(\bar{t}_i)$	$n \log n$	Ch. 2; 3.3(*)
1	$t_i=1 ; d_i$			L_{max}	n	Ch. 2; 10.2
1	$t_i=1 ; d_i=0$	G		L_{max}	n	Ch. 2; 10.2
1	$t_i ; d_i=0$			L_{max}	$n \log n$	Ch. 2; 3.3(*)
1	$t_i ; d_i=0$	$\mathcal{T}$		L_{max}	$n \log n$	Ch. 3; 8,3(*)
1	$t_i ; d_i=0$	SP		L_{max}	$n \log n$	Ch. 3; 8,4(*)
1	$t_i ; d_i$		$D_i=D$	L_{max}	$n \log n$	Ch. 2; 3.4
1	$t_i ; d_i$	G	$D_i=D$	L_{max}	n^2	Ch. 2; 3.4
1	$t_i ; d_i=0$	SP	$\varphi(t_1+t_2)=\varphi(t_1)+\varphi(t_2)$; $\varphi(t)\ge 0$, $t>0$	$\max\{\varphi(\bar{t}_i)+\beta_i\}$	$n \log n$	Ch. 3; 8,4(*)
1	$t_i ; d_i=0$	SP	$\varphi(t_1+t_2)=\varphi(t_1)\varphi(t_2)$; $\varphi(t)\ge 1$, $t>0$	$\max\{\alpha_i\varphi(\bar{t}_i)\}$	$n \log n$	Ch. 3; 8,4(*)
1	$t_i=t ; d_i$	G		$\bar{t}_{max}$, $\bar{t}_i \le D_i$	$n \log n$	Ch. 2; 10.1
1	$t_i ; d_i$	Pr		L_{max}	$n \log n$	Ch. 2; 10.1
1	$t_i=t ; d_i$	G		L_{max}		Ch. 2; 10.1

(to be continued)

Table I.1

Number of machines	Processing times, release dates	Preemption, precedence and resource constraints	Additional conditions	Objective function	Running time	Section of the book
1	$t_i ; d_i$	$Pr;G$		$\max\varphi_i(\overline{t}_i)$	n^2	Ch.2;3.5,3.6
1	$[t_i];[d_i]$	G	$t_i=GCD(d_i)$	$\max\varphi_i(\overline{t}_i)$	n^2	Ch.2;3.8(*)
1	$t_i ; d_i$			$\max \varphi_i(\overline{t}_i-d_i)$	$n\log n$	Ch.2;3.8(*)
1	$t_i ; d_i=0$	SP	$t_i\in(-\infty,\infty)$	$\overline{t}_{max}$	$n\log n$	Ch.3;1,4
1	$t_i ; d_i=0$	ω-SP	$t_i\in(-\infty,\infty)$	$\overline{t}_{max}$	n^4	Ch.3;1,5,6
1	$t_i ; d_i=0$	$\mathcal{T}$		F_{pg}	$n\log n$	Ch.3;1,3(*)
1	$t_i ; d_i=0$	SP		F_{pg}	$n\log n$	Ch.3;1,4(*)
1	$t_i ; d_i=0$	ω-SP		F_{pg}	n^4	Ch.3;1,5,6(*)
1	$t_i ; d_i=0$			F_{1-pg}	$n\log n$	Ch.3;7(*)
1	$t_i ; d_i=0$			$\sum\varphi(\overline{t}_i)$	$n\log n$	Ch.3;7(*)
1	$t_i ; d_i=0$	SP		$\sum\alpha_i\overline{t}_i$	$n\log n$	Ch.3;1,4(*)
1	$t_i ; d_i=0$	ω-SP		$\sum\alpha_i\overline{t}_i$	n^4	Ch.3;1,5,6(*)
1	$t_i ; d_i=0$	SP		$\sum \alpha_i\times \exp(\gamma\overline{t}_i)$	$n\log n$	Ch.3;1,4(*)
1	$t_i ; d_i=0$	ω-SP		$\sum \alpha_i\times \exp(\gamma\overline{t}_i)$	n^4	Ch.3;1,5,6(*)
1	$t_i ; d_i$	Pr		$\sum\varphi(\overline{t}_i)$	$n\log n$	Ch.2;4.6
1	$t_i ; d_i=0$			$\sum\varphi(\overline{t}_i)$, $\overline{t}_i\le D_i$	n^2	Ch.2;10.1
1	$t_i=1;[d_i]$			$\sum\varphi_i(\overline{t}_i)$, $\overline{t}_i\le D_i$	n^3	Ch.2;4.5(*)
1	$t_i ; d_i=0$			$\sum u_i$	$n\log n$	Ch.2;4.3a(*)
1	$t_i=1; d_i=0$			$\sum u_i$	n	Ch.2;10.2

(to be continued)

Table I.1

Number of machines	Processing times, release dates	Preemption, precedence and resource constraints	Additional conditions	Objective function	Running time	Section of the book
1	$t_i\,;d_i$		$(d_i\uparrow, D_i\uparrow)$	$\sum u_i$	$n\log n$ n^2	Ch.2;10.2 Ch.2;4.3c(*)
1	$t_i\,;d_i=0$		$(t_i\uparrow, \alpha_i\downarrow)$	$\sum \alpha_i u_i$	$n\log n$	Ch.2;4.3b(*)
1	$t_i\,;d_i=0$		$(t_i\uparrow, \alpha_i\downarrow,$ $i\in N\backslash Q)$	$\sum \alpha_i u_i,$ $\bar{t}_i\le D_i,\ i\in Q$	$n\log n$	Ch.2;4.4(*)
1	$t_i\,;d_i$		$(d_i\uparrow, t_i\uparrow,$ $D_i\uparrow, \alpha_i\downarrow)$	$\sum \alpha_i u_i$	$n\log n$	Ch.2;4.3d(*)
1	$t_i\,;d_i$		$(d_i\uparrow, D_i\uparrow, \alpha_i\downarrow$ $t_i\le d_{i+1}-d_i)$	$\sum \alpha_i u_i$	$n\log n$	Ch.2;4.3e(*)
1	$t_i\,;d_i$		$(d_i=d_{i-1}+t,$ $D_i\uparrow, \alpha_i\uparrow,$ $2(n-i)t\le t_i\le$ $2(n-i)t+t)$	$\sum \alpha_i u_i$	$n\log n$	Ch.2;4.3f(*)
1	$t_i\,;d_i$		$(d_i\uparrow, D_i\uparrow,$ $i\in N\backslash Q)$	$\sum u_i,$ $\bar{t}_i\le D_i,\ i\in Q$	n^2	Ch.2;10.2
1	$t_i\,;d_i$		$(d_i\uparrow, D_i\uparrow,$ $t_i\uparrow, \alpha_i\downarrow)$	$\sum \alpha_i u_i,$ $\bar{t}_i\le D_i,\ i\in Q$	$n\log n$	Ch.2;10.2
1	$t_i\,;d_i$		$(d_i\uparrow, D_i\uparrow,$ $\alpha_i\downarrow,\ t_i\le d_{i+1}-$ $d_i)$	$\sum \alpha_i u_i,$ $\bar{t}_i\le D_i,\ i\in Q$	$n\log n$	Ch.2;10.2
1	$t_i=1\,;[d_i]$			$\sum z_i$	$n\log n$	Ch.2;10.2
2	$t_i=1\,;d_i=0$	G		$\bar{t}_i\le D_i$	n^2	Ch.2;7.3
2	$t_i=1\,;d_i$	G	$D_i=D$	$\bar{t}_i\le D_i$	n^2	Ch.2;7.3
2	$t_i=1\,;[d_i]$	G		$\bar{t}_i\le D_i$	n^3	Ch.2;10.1
2	$t_i=1\,;d_i=0$	$Rs(q)$	$D_i=D$	$\bar{t}_i\le D_i$	qn^2+ $n^{5/2}$	Ch.2;10.1
2	$t_{iH}=a_H;$	$Rs(1)$	$D_i=D$	$\bar{t}_i\le D_i$	$n\log n$	Ch.2;10.1

(to be continued)

Table I.1

Number of machi-nes	Processing times, release dates	Preemption, precedence and resource constraints	Additional conditions	Objective function	Running time	Section of the book
2	$a_H t_i$; $d_i=0$	Pr; G		$\bar{t}_i \le D_i$	n^2	Ch.2; 10.1
2	$a_H t_i$; d_i	Pr; G	$D_i=D$	$\bar{t}_i \le D_i$	n^2	Ch.2; 10.1
2	$a_H t_i$; d_i	Pr; G		$\bar{t}_i \le D_i$	n^3	Ch.2; 10.1
2	$t_i=1$; $d_i=0$	G		$\bar{t}_{max}$	n^2	Ch.2; 5.4, 5.5
2	$t_i=1$; $d_i=0$	G		$\bar{t}_{max}$	n	Ch.2; 10.2
2	$t_i \in \{1,2\}$; $d_i=0$	$\mathcal{T}^-$		$\bar{t}_{max}$	$n \log n$	Ch.2; 10.2
2	$t_i \in \{1,3\}$; $d_i=0$	$\mathcal{T}^-$		$\bar{t}_{max}$	$n^2 \log n$	Ch.2; 10.2
2	$t_i=1$; $d_i=0$	G		$\bar{t}_{max}$, $\bar{t}_i \le D_i$	$n^2 \log n$	Ch.2; 7.3
2	$t_i=1$; $[d_i]$	G		$\bar{t}_{max}$, $\bar{t}_i \le D_i$	$n^3 \log n$	Ch.2; 10.1, 7.3
2	$t_i=1$; $d_i=0$	G		L_{max}	n^2	Ch.2; 8.2
2	$t_i=1$; $[d_i]$	G		L_{max}	$n^3 \log n$	Ch.2; 10.1
2	$t_i=1$; d_i	G		$\bar{t}_{max}$	n^2	Ch.2; 8.4, 8.2
2	$t_i=1$; $d_i=0$	$Rs(q)$		$\bar{t}_{max}$	$qn^2+n^{5/2}$	Ch.2; 10.1
2	t_i; $d_i=0$	Pr; G		$\bar{t}_{max}$	n^2	Ch.2; 6.3-6.6
2	$t_{iH}=a_H$; $d_i=0$	$Rs(1)$		$\bar{t}_{max}$	$n \log n$	Ch.2; 10.1
2	$a_H t_i$; $d_i=0$	Pr; G		$\bar{t}_{max}$	n^2	Ch.2; 10.1
2	$a_H t_i$; $d_i=0$	Pr; G		L_{max}	n^2	Ch.2; 10.1
2	$a_H t_i$; d_i	Pr; G		$\bar{t}_{max}$	n^2	Ch.2; 10.1

(to be continued)

Table I.1

Number of machines	Processing times, release dates	Preemption, precedence and resource constraints	Additional conditions	Objective function	Running time	Section of the book
2	$a_H t_i$; $[d_i]$	$[Pr]$	$[t_i]$, $[a_H]$, $[D_i]$	L_{max}	$n^3 \min\{n^2/a_H, \log n - \log a_H + \log t_{max}\}$	Ch.2;10.1
2	$a_H t_i$; d_i	Pr; G		L_{max}	n^6	Ch.2;10.1
2	$a_H t_i$; $d_i=0$	Pr		$\sum u_i$	n^4	Ch.2;10.1
M	$t_i=1$; $[d_i]$			$\bar{t}_i \le D_i$	$n \log n$	Ch.2;10.1
M	$t_i=t$; d_i	·		$\bar{t}_i \le D_i$	$n^3 \log n$	Ch.2;10.1
M	$t_i=1$; $d_i=0$	$\mathcal{T}^+$	$D_i=D$	$\bar{t}_i \le D_i$	n	Ch.2;5.3
M	$t_i=1$; $d_i=0$	$\mathcal{T}^-$	$D_i=D$	$\bar{t}_i \le D_i$	n	Ch.2;5.2,5.3
M	$t_i=1$; $d_i=0$	$\mathcal{T}^-$		$\bar{t}_i \le D_i$	$n \log n$	Ch.2;7.2
M	$t_i=1$; d_i	$\mathcal{T}^+$	$D_i=D$	$\bar{t}_i \le D_i$	$n \log n$	Ch.2;7.2,7.3
M	t_i; $d_i=0$	Pr	$D_i=D$	$\bar{t}_i \le D_i$	n	Ch.2;2.6
M	t_i; $d_i=0$	Pr		$\bar{t}_i \le D_i$	$n \log n$	Ch.2;2.8
M	t_i; d_i	Pr	$D_i=D$	$\bar{t}_i \le D_i$	$n \log n$ nM	Ch.2;2.7,2.8 Ch.2;10.1
M	t_i; d_i	Pr		$\bar{t}_i \le D_i$	n^3	Ch.2;2.3
M	t_i; $d_i=0$	Pr; $\mathcal{T}^+$	$D_i=D$	$\bar{t}_i \le D_i$	n^2 $n \log M$	Ch.2;6.5-6.7 Ch.2;10.1
M	t_i; $d_i=0$	Pr; $\mathcal{T}^-$	$D_i=D$	$\bar{t}_i \le D_i$	n^2 $n \log M$	Ch.2;6.5-6.7 Ch.2;10,6.7
M	t_i; $d_i=0$	Pr; $\mathcal{T}^-$		$\bar{t}_i \le D_i$	n^2	Ch.2;10.1
M	t_i; d_i	Pr; $\mathcal{T}^+$	$D_i=D$	$\bar{t}_i \le D_i$	n^2	Ch.2;10.1
M	$t_{iH}=a_H$; $d_i=0$	$Rs(1)$	$r_i \in \{0,1\}$; $D_i=D$	$\bar{t}_i \le D_i$	n^3	Ch.2;10.1

(to be continued)

Table I.1

Number of machines	Processing times, release dates	Preemption, precedence and resource constraints	Additional conditions	Objective function	Running time	Section of the book
M	$a_H t_i; d_i=0$	Pr	$D_i=D$	$\bar{t}_i \le D_i$	$n+M\log M$	Ch.2;10.1
M	$a_H t_i; d_i$	Pr	$D_i=D$	$\bar{t}_i \le D_i$	$n\log n+Mn$	Ch.2;10.1
M	$a_H t_i; d_i=0$	Pr		$\bar{t}_i \le D_i$	$n\log n+Mn$	Ch.2;10.1
M	$a_H t_i; d_i$	Pr		$\bar{t}_i \le D_i$	$M^2n^4+n^5$	Ch.2;10.1
M	$t_{iH}; d_i=0$	Pr		$\bar{t}_i \le D_i$	LP	Ch.2;10.1
M	$t_{iH}; d_i$	Pr	$D_i=D$	$\bar{t}_i \le D_i$	LP	Ch.2;10.1
M	$t_i=1; [d_i]$		$[D_i]$	$\bar{t}_i \le D_i$	n	Ch.2;10.2
M	$t_i=1; d_i=0$	$\mathcal{T}^+$		$\bar{t}_{max}$	n	Ch.2;5.2,5.3
M	$t_i=1; d_i=0$	$\mathcal{T}^-$		$\bar{t}_{max}$	n	Ch.2;5.2,5.3
M	$t_i=1; [d_i]$			$\bar{t}_{max}$, $\bar{t}_i \le D_i$	$n\log n$	Ch.2;10.1
M	$t_i=t; d_i$			L_{max}	$n^3\log^2 n$	Ch.2;10.1
M	$t_i=1; d_i=0$			$\max\varphi_i(\bar{t}_i)$	n^3	Ch.2;9.4
M	$t_i=1; d_i=0$	$\mathcal{T}^-$		L_{max}	$n\log n$	Ch.2;8.2
M	$t_i=1; d_i$	$\mathcal{T}^+$		$\bar{t}_{max}$	$n\log n$	Ch.2;8.2,8.4
M	$t_i; d_i=0$	Pr		$\bar{t}_{max}$	n	Ch.2;6.2
M	$t_i; d_i$	Pr		$\bar{t}_{max}$	n^2 nM	Ch.2;10,8.4 Ch.2;10.1
M	$t_i; d_i=0$	Pr		L_{max}	n^2 nM	Ch.2;10.1 Ch.2;10,8.4
M	$[t_i]; [d_i]$	Pr	$[D_i]$	L_{max}	$n^3\max\{n^2, \log n+\log t_{min}\}$	Ch.2;8.3

(to be continued)

Table I.1

Number of machines	Processing times, release dates	Preemption, precedence and resource constraints	Additional conditions	Objective function	Running time	Section of the book
M	t_i; $d_i=0$	Pr; $\mathcal{T}^+$		$\bar{t}_{max}$	n^2 $n\log M$	Ch.2; 6.3-6.7 Ch.2; 10.1
M	t_i; $d_i=0$	Pr; $\mathcal{T}^-$		$\bar{t}_{max}$	n^2 $n\log M$	Ch.2; 6.3-6.7 Ch.2; 10, 6.7
M	t_i; $d_i=0$	Pr; $\mathcal{T}^-$		L_{max}	n^2	Ch.2; 10.1
M	t_i; d_i	Pr; $\mathcal{T}^+$		$\bar{t}_{max}$	n^2	Ch.2; 10, 8.4
M	$t_{iH}=a_H$; $d_i=0$			$\max\varphi_i(\bar{t}_i)$	n^3	Ch.2; 9.4
M	$t_{iH}=a_H$; $d_i=0$	$Rs(1)$	$r_i\in\{0,1\}$	$\bar{t}_{max}$	n^3	Ch.2; 10.1
M	$t_{iH}=a_H$; $d_i=0$			$\sum\alpha_i\bar{t}_i$	$n\log n$	Ch.2; 10.2
M	$t_{iH}=a_H$; $d_i=0$			$\sum\alpha_i u_i$	$n\log n$	Ch.2; 10.2
M	$t_{iH}=a_H$; $d_i=0$			$\sum z_i^p$	$n\log n$	Ch.2; 10.2
M	$t_{iH}=a_H$; $d_i=0$			$\sum\lvert L_i\rvert^p$	$n\log n$	Ch.2; 10.2
M	$t_{iH}=a_H$; $d_i=0$			$\sum\varphi(\bar{t}_i)$	$n+M\log M$	Ch.2; 10.2
M	$t_{iH}=a_H$; d_i			$\sum\bar{t}_i$	Mn^{2M+1}	Ch.2; 10.2
M	$t_{iH}=a_H$; $d_i=0$			$\max\ \varphi(\bar{t}_i)$	n^2	Ch.2; 10.2
M	$t_{iH}=a_H$; $d_i=0$			L_{max}	$n\log n$	Ch.2; 10.2
M	$t_{iH}=a_H$; d_i			$\bar{t}_{max}$	$n\log n$	Ch.2; 9.3
M	$t_{iH}=a_H$; $d_i=0$			$\max\ \alpha_i z_i$	$(\log n/m+\log\alpha_{max})\, n\log n$	Ch.2; 10.2
M	$t_{iH}=a_H$; d_i	Pr	t distinct machine speeds a_H	L_{max}	tn^2	Ch.2; 10.2
M	$a_H t_i$; $d_i=0$	Pr		$\bar{t}_{max}$	$n+M\log M$	Ch.2; 10.1
M	$a_H t_i$; d_i	Pr		$\bar{t}_{max}$	$Mn\log n+M^2n$	Ch.2; 10.1

(*to be continued*)

Table I.1

Number of machines	Processing times, release dates	Preemption, precedence and resource constraints	Additional conditions	Objective function	Running time	Section of the book
M	$a_H t_i$; $d_i=0$	Pr		L_{max}	$Mn\log n+M^2 n$	Ch.2;10.1
M	$a_H t_i$; $[d_i]$	Pr	$[a_H]$, $[t_i]$, $[D_i]$	L_{max}	$(n^2+\log(t_\Sigma+D_{max})-n\log a_H)\times(M^2 n^4+n^5)$	Ch.2;10.1
M	t_{iH}; d_i	Pr		$\bar{t}_{max}$	LP	Ch.2;9.6,9.7
M	t_{iH}; $d_i=0$	Pr		L_{max}	LP	Ch.2;9.7
M	$t_i=1$; $d_i=0$		$[D_i]$	$\sum\alpha_i u_i$	$n\log n$	Ch.2;9.5
M	t_i ; $d_i=0$			$\sum\bar{t}_i$	$n\log n$	Ch.2;9.3(*)
M	$t_{iH}=a_H$; $d_i=0$			$\sum\varphi_i(\bar{t}_i)$	n^3	Ch.2;9.4
M	$a_H t_i$; $d_i=0$			$\sum\bar{t}_i$	$n\log n$	Ch.2;10.1
M	$a_H t_i$; $d_i=0$	Pr		$\sum\bar{t}_i$	$n\log n+Mn$	Ch.2;10.1
M	$a_H t_i$; $d_i=0$	Pr	$(t_i\uparrow, \alpha_i\downarrow)$	$\sum\alpha_i\bar{t}_i$	$n\log n+Mn$	Ch.2;10.1
M	$a_H t_i$; $d_i=0$	Pr		$\sum u_i$	n^{3M-3}	Ch.2;10.1
M	t_{iH}; $d_i=0$			$\sum\bar{t}_i$	n^3	Ch.2;9.2

Table I.2

Number of machines	Processing times, release dates	Preemption, precedence and resource constraints	Additional conditions	Objective function	Section of the book
1	$t_i; d_i$			$\bar{t}_i \le D_i$	Ch.4;4.8,6
1	$t_i; d_i$			L_{max}	Ch.4;1.1, 1.5, 6
1	$t_i; d_i$			z_{max}	Ch.4;1.9,6
1	$t_i; d_i$			$\sum u_i$	Ch.4;1.9,6
1	$t_i; d_i=0$		$D_i' \ge D_i$	$\sum u_i, \bar{t}_i \le D_i'$	Ch.4;6
1	$t_i; d_i=0$		$D_i=D$	$\sum \alpha_i u_i$	Ch.4;1.1, 1.6(*)
1	$t_i; d_i=0$	Pr	$D_i=D$	$\sum \alpha_i u_i$	Ch.4;1.9(*)
1	$t_i; d_i$			$\sum \bar{t}_i$	Ch.4;2.1,2.5
1	$t_i; d_i$	Pr		$\sum \bar{t}_i, \bar{t}_i \le D_i$	Ch.4;6
1	$t_i; d_i$		$D_i=D$	$\sum z_i$	Ch.4;2.14
1	$t_i; d_i=0$			$\sum z_i$	Ch.4;6(*)
1	$t_i; d_i=0$	Pr		$\sum z_i$	Ch.4;6(*)
1	$t_i=1; d_i=0$	C		$\sum z_i$	Ch.4;6
1	$[t_i]; d_i=0$	$[Pr];C$		$\sum z_i$	Ch.4;6
1	$t_i=1; d_i=0$	$Pr;C$		$\sum z_i$	Ch.4;6
1	$t_i; d_i=0$			$\sum \min\{z_i, t_i\}$	Ch.4;6(*)
1	$t_i; d_i$	Pr		$\sum \alpha_i \bar{t}_i$	Ch.4;2.1,2.6
1	$t_i; d_i$	Pr	$D_i=D$	$\sum \alpha_i z_i$	Ch.4;2.14
1	$t_i=1; d_i$	C		$\sum \alpha_i \bar{t}_i$	Ch.4;2.1,2.7
1	$[t_i]; [d_i]$	$[Pr];C$		$\sum \alpha_i \bar{t}_i$	Ch.4;2.15
1	$t_i=1; d_i$	C	$D_i=D$	$\sum \alpha_i z_i$	Ch.4;2.14
1	$[t_i]; [d_i]$	$[Pr];C$	$D_i=D$	$\sum \alpha_i z_i$	Ch.4;2.15

(to be continued)

Table I.2

Number of machines	Processing times, release dates	Preemption, precedence and resource constraints	Additional conditions	Objective function	Section of the book
1	$t_i ; d_i=0$			$\sum\alpha_i z_i$	Ch. 4; 2.1, 2.8
1	$t_i ; d_i=0$	Pr		$\sum\alpha_i z_i$	Ch. 4; 2.14
1	$t_i ; d_i=0$		$D_i=D$	$\sum\alpha_i z_i$	Ch. 4; 6(**)
1	$t_i ; d_i=0$	Pr	$D_i=D$	$\sum\alpha_i z_i$	Ch. 4; 6(**)
1	$t_i ; d_i$	Pr		$\sum\alpha_i \bar{t}_i$	Ch. 4; 2.1, 2.6
1	$t_i ; d_i=0$			$\sum\alpha_i \bar{t}_i$, $\bar{t}_i \le D_i$	Ch. 4; 2.1, 2.9
1	$t_i=1 ; d_i=0$	C		$\sum\alpha_i \bar{t}_i$, $\bar{t}_i \le D_i$	Ch. 4; 2.1, 2.9
1	$[t_i] ; d_i=0$	$[Pr];C$		$\sum\alpha_i \bar{t}_i$, $\bar{t}_i \le D_i$	Ch. 4; 2.15
1	$t_i=1 ; d_i=0$	C		$\sum u_i$	Ch. 4; 2.1, 2.10
1	$[t_i] ; d_i=0$	$[Pr];C$		$\sum u_i$	Ch. 4; 2.15
1	$t_i=1 ; d_i=0$	$Pr;C$		$\sum u_i$	Ch. 4; 2.14
1	$t_i=1 ; d_i=0$	G		$\sum z_i$	Ch. 4; 4.1, 4.2
1	$[t_i] ; d_i=0$	$[Pr];G$		$\sum z_i$	Ch. 4; 4.7
1	$t_i=1 ; d_i=0$	$Pr;G$		$\sum z_i$	Ch. 4; 4.7
1	$t_i=1 ; d_i=0$	G	$\alpha_i \in \{\lambda, \lambda+1, \lambda+2\}$, $\lambda \in \{0, \pm1, \pm2, \ldots\}$	$\sum\alpha_i \bar{t}_i$	Ch. 4; 5.1, 5.2
1	$[t_i] ; d_i=0$	$[Pr];G$	$\alpha_i \in \{\lambda, \lambda+1, \lambda+2\}$, $\lambda \in \{0, \pm1, \pm2, \ldots\}$	$\sum\alpha_i \bar{t}_i$	Ch. 4; 5.5
1	$t_i=1 ; d_i=0$	$Pr;G$	$\alpha_i \in \{\lambda, \lambda+1, \lambda+2\}$, $\lambda \in \{0, \pm1, \pm2, \ldots\}$	$\sum\alpha_i \bar{t}_i$	Ch. 4; 5.5
1	$t_i=1 ; d_i=0$	G	$D_i=D$, $\alpha_i \in \{\lambda, \lambda+1, \lambda+2\}$,	$\sum\alpha_i z_i$	Ch. 4; 5.5
1	$[t_i] ; d_i=0$	$[Pr];G$	$D_i=D$, $\alpha_i \in \{\lambda, \lambda+1, \lambda+2\}$, $\lambda \in \{0, \pm1, \pm2, \ldots\}$	$\sum\alpha_i z_i$	Ch. 4; 5.5

(*to be continued*)

Table I.2

Number of machines	Processing times, release dates	Preemption, precedence and resource constraints	Additional conditions	Objective function	Section of the book
1	$t_i=1$; $d_i=0$	Pr; G	$D_i=D$, $\alpha_i\in\{\lambda, \lambda+1, \lambda+2\}$, $\lambda\in\{0, \pm1, \pm2, ...\}$	$\sum\alpha_i z_i$	Ch. 4; 5.5
1	$t_i\in\{1, 2\}$; $d_i=0$	G		$\sum\bar{t}_i$	Ch. 4; 5.1, 5.3
1	$t_i\in\{1, 2\}$; $d_i=0$	Pr; G		$\sum\bar{t}_i$	Ch. 4; 5.5
1	$t_i\in\{1, 2\}$; $d_i=0$	G	$D_i=D$	$\sum z_i$	Ch. 4; 5.1, 5.3
1	$t_i\in\{1, 2\}$; $d_i=0$	Pr; G	$D_i=D$	$\sum z_i$	Ch. 4; 5.5
1	$t_i\in\{0, 1\}$; $d_i=0$	G		$\sum\bar{t}_i$	Ch. 4; 5.1, 5.4
1	$[t_i]$; $d_i=0$	$[Pr]$; G		$\sum\bar{t}_i$	Ch. 4; 5.5
1	$t_i\in\{0, 1\}$; $d_i=0$	Pr; G		$\sum\bar{t}_i$	Ch. 4; 5.5
1	$t_i\in\{0, 1\}$; $d_i=0$	G	$D_i=D$	$\sum z_i$	Ch. 4; 5.5
1	$[t_i]$; $d_i=0$	$[Pr]$; G	$D_i=D$	$\sum z_i$	Ch. 4; 5.5
1	$t_i\in\{0, 1\}$; $d_i=0$	Pr; G	$D_i=D$	$\sum z_i$	Ch. 4; 5.5
1	$t_i=1$; $d_i=0$	G	$\alpha_i\in\{0, 1\}$	$\sum\alpha_i\bar{t}_i$	Ch. 4; 5.1, 5.4
1	$[t_i]$; $d_i=0$	$[Pr]$; G	$\alpha_i\in\{0, 1\}$	$\sum\alpha_i\bar{t}_i$	Ch. 4; 5.5
1	$t_i=1$; $d_i=0$	Pr; G	$\alpha_i\in\{0, 1\}$	$\sum\alpha_i\bar{t}_i$	Ch. 4; 5.5
1	$t_i=1$; $d_i=0$	G	$D_i=D$; $\alpha_i\in\{0, 1\}$	$\sum\alpha_i z_i$	Ch. 4; 5.5
1	$[t_i]$; $d_i=0$	$[Pr]$; G	$D_i=D$; $\alpha_i\in\{0, 1\}$	$\sum\alpha_i z_i$	Ch. 4; 5.5
1	$t_i=1$; $d_i=0$	Pr; G	$D_i=D$; $\alpha_i\in\{0, 1\}$	$\sum\alpha_i z_i$	Ch. 4; 5.5
1	t_i; $d_i=0$		$D_i=D<t_\Sigma$	$\sum\lvert\bar{t}_i-D_i\rvert$	Ch. 4; 6(**)
1	t_i; $d_i=0$		$D_i=D$	$\sum\alpha_i\lvert\bar{t}_i-D_i\rvert$	Ch. 4; 6(**)

(to be continued)

Table I.2

Number of machines	Processing times, release dates	Preemption, precedence and resource constraints	Additional conditions	Objective function	Section of the book
2	t_i; $d_i=0$		$D_i=D$	$\bar{t}_i \le D_i$	Ch.4; 4.8(*)
2	$t_{iH}=a_H$; $d_i=0$	$Rs(q)$	$R_k=1$, $r_{ik}\in\{0,1\}$; $D_i=D$	$\bar{t}_i \le D_i$	Ch.4; 6
2	t_i; $d_i=0$	C	$D_i=D$	$\bar{t}_i \le D_i$	Ch.4; 6
2	t_{iH}; $d_i=0$	Pr; $\mathcal{T}^-$	$D_i=D$	$\bar{t}_i \le D_i$	Ch.4; 6
2	t_{iH}; $d_i=0$	Pr; $\mathcal{T}^+$	$D_i=D$	$\bar{t}_i \le D_i$	Ch.4; 6
2	$t_i\in\{1,2\}$; $d_i=0$	G	$D_i=D$	$\bar{t}_i \le D_i$	Ch.4; 4.8
2	$t_i\in\{t^p \mid p\ge 0\}$, $t>1$; $d_i=0$	$\mathcal{T}^-$	$D_i=D$	$\bar{t}_i \le D_i$	Ch.4; 6(**)
2	$t_i\in\{t^p \mid p\ge 0\}$, $t>1$; $d_i=0$	$\mathcal{T}^+$	$D_i=D$	$\bar{t}_i \le D_i$	Ch.4; 6(**)
2	$t_i=1$; $d_i=0$	C; $Rs(1)$	$R_1=1$, $r_i\in\{0,1\}$; $D_i=D$	$\bar{t}_i \le D_i$	Ch.4; 4.8
2	$[t_i]$; $d_i=0$	$[Pr]$; C; $Rs(1)$	$R_1=1$, $r_i\in\{0,1\}$; $D_i=D$	$\bar{t}_i \le D_i$	Ch.4; 4.8
2	t_i; $d_i=0$			$\bar{t}_{max}$	Ch.4; 1.1, 1.2(*)
2	t_i; $d_i=0$	C		$\bar{t}_{max}$	Ch.4; 6
2	$t_i\in\{t^p \mid p\ge 0\}$, $t>1$; $d_i=0$	$\mathcal{T}^-$		$\bar{t}_{max}$	Ch.4; 6(**)
2	$t_i\in\{t^p \mid p\ge 0\}$, $t>1$; $d_i=0$	$\mathcal{T}^+$		$\bar{t}_{max}$	Ch.4; 6(**)
2	t_{iH}; $d_i=0$	Pr; $\mathcal{T}^-$		$\bar{t}_{max}$	Ch.4; 6
2	t_{iH}; $d_i=0$	Pr; $\mathcal{T}^+$		$\bar{t}_{max}$	Ch.4; 6
2	t_i; $d_i=0$		$D_i=D$	z_{max}	Ch.4; 1.9(*)
2	t_i; $d_i=0$	C	$D_i=D$	z_{max}	Ch.4; 6
2	t_i; $d_i=0$		$D_i=D$	L_{max}	Ch.4; 1.9(*)

(to be continued)

Table I.2

Number of machines	Processing times, release dates	Preemption, precedence and resource constraints	Additional conditions	Objective function	Section of the book
2	t_i ; $d_i=0$	C	$D_i=D$	L_{max}	Ch. 4; 6
2	$t_{iH}=a_H$; $d_i=0$	$Rs(q)$	$R_k=1$, $r_{ik}\in\{0,1\}$	$\bar{t}_{max}$	Ch. 4; 6
2	$t_{iH}=a_H$; $d_i=0$	$Rs(q)$	$R_k=1$, $r_{ik}\in\{0,1\}$ $D_i=D$	L_{max}	Ch. 4; 6
2	$t_{iH}=a_H$; $d_i=0$	$Rs(q)$	$R_k=1$, $r_{ik}\in\{0,1\}$; $D_i=D$	z_{max}	Ch. 4; 6
2	t_{iH}; $d_i=0$	Pr; $\mathcal{T}^-$	$D_i=D$	L_{max}	Ch. 4; 6
2	t_{iH}; $d_i=0$	Pr; $\mathcal{T}^-$	$D_i=D$	z_{max}	Ch. 4; 6
2	t_{iH}; $d_i=0$	Pr; $\mathcal{T}^+$	$D_i=D$	L_{max}	Ch. 4; 6
2	t_{iH}; $d_i=0$	Pr; $\mathcal{T}^+$	$D_i=D$	z_{max}	Ch. 4; 6
2	$t_i\in\{1,2\}$; $d_i=0$	G		$\bar{t}_{max}$	Ch. 4; 4.1, 4.3
2	$t_i\in\{1,2\}$; $d_i=0$	G	$D_i=D$	L_{max}	Ch. 4; 4.7
2	$t_i\in\{t^p \mid p\geq 0\}$, $t>1$; $d_i=0$	$\mathcal{T}^-$	$D_i=D$	L_{max}	Ch. 4; 6(**)
2	$t_i\in\{t^p \mid p\geq 0\}$, $t>1$; $d_i=0$	$\mathcal{T}^+$	$D_i=D$	L_{max}	Ch. 4; 6(**)
2	$t_i\in\{1,2\}$; $d_i=0$	G	$D_i=D$	z_{max}	Ch. 4; 4.7
2	$t_i\in\{t^p \mid p\geq 0\}$, $t>1$; $d_i=0$	$\mathcal{T}^-$	$D_i=D$	z_{max}	Ch. 4; 6(**)
2	$t_i\in\{t^p \mid p\geq 0\}$, $t>1$; $d_i=0$	$\mathcal{T}^+$	$D_i=D$	z_{max}	Ch. 4; 6(**)
2	$t_i=1$; $d_i=0$	C; $Rs(1)$	$R_1=1$, $r_i\in\{0,1\}$	$\bar{t}_{max}$	Ch. 4; 2.1, 2.12
2	[t_i]; $d_i=0$	[Pr]; C; $Rs(1)$	$R_1=1$, $r_i\in\{0,1\}$	$\bar{t}_{max}$	Ch. 4; 2.15
2	$t_i=1$; $d_i=0$	C; $Rs(1)$	$R_1=1$, $r_i\in\{0,1\}$ $D_i=D$	L_{max}	Ch. 4; 2.14

(to be continued)

Table I.2

Number of machi-nes	Processing times, release dates	Preemption, precedence and resource constraints	Additional conditions	Objective function	Section of the book
2	$[t_i]; d_i=0$	$[Pr]; C; Rs(1)$	$R_1=1, r_i \in \{0,1\}$ $D_i=D$	L_{max}	Ch.4; 2.15
2	$t_i=1; d_i=0$	$C; Rs(1)$	$R_1=1, r_i \in \{0,1\}$; $D_i=D$	z_{max}	Ch.4; 2.14
2	$[t_i]; d_i=0$	$[Pr]; C; Rs(1)$	$R_1=1, r_i \in \{0,1\}$; $D_i=D$	z_{max}	Ch.4; 2.15
2	$t_i; d_i=0$		$D_i=D$	$\sum z_i$	Ch.4; 1.9(*)
2	$t_i; d_i=0$	C	$D_i=D$	$\sum z_i$	Ch.4; 6
2	$t_i; d_i$	Pr	$D_i=D$	$\sum z_i$	Ch.4; 6
2	$t_i; d_i=0$		$D_i=D$	$\sum u_i$	Ch.4; 1.9(*)
2	$t_i; d_i=0$	C	$D_i=D$	$\sum u_i$	Ch.4; 6
2	$t_i; d_i$	Pr		$\sum u_i$	Ch.4; 6
2	$t_i; d_i=0$			$\sum \alpha_i \bar{t}_i$	Ch.4; 1.1, 1.4(*)
2	$t_i; d_i=0$	Pr		$\sum \alpha_i \bar{t}_i$	Ch.4; 1.9(*)
2	$t_i; d_i=0$	Pr	$D_i=D$	$\sum \alpha_i z_i$	Ch.4; 6(*)
2	$t_i; d_i=0$	C		$\sum \bar{t}_i$	Ch.4; 6
2	$t_i; d_i=0$	$Pr; \mathcal{T}^-$		$\sum \bar{t}_i$	Ch.4; 6
2	$t_i; d_i=0$	$Pr; \mathcal{T}^+$		$\sum \bar{t}_i$	Ch.4; 6
2	$t_i; d_i$	Pr		$\sum \bar{t}_i$	Ch.4; 6
2	$t_i; d_i=0$	$Pr; \mathcal{T}^-$	$D_i=D$	$\sum z_i$	Ch.4; 6
2	$t_i; d_i=0$	$Pr; \mathcal{T}^+$	$D_i=D$	$\sum z_i$	Ch.4; 6
2	$t_{iH}=a_H; d_i=0$	$Rs(q)$	$R_k=1, r_{ik} \in \{0,1\}$; $D_i=D$	$\sum z_i$	Ch.4; 6

(to be continued)

Table I.2

Number of machines	Processing times, release dates	Preemption, precedence and resource constraints	Additional conditions	Objective function	Section of the book
2	$t_{iH}=a_H; d_i=0$	$Rs(q)$	$R_k=1, r_{ik}\in\{0,1\}; D_i=D$	$\sum u_i$	Ch.4;6
2	$t_{iH}=a_H; d_i=0$	$Rs(q)$	$R_k=1, r_{ik}\in\{0,1\}$	$\sum \overline{t}_i$	Ch.4;6
2	$t_i\in\{1,2\}; d_i=0$	G	$D_i=D$	$\sum z_i$	Ch.4;4.7
2	$t_i\in\{t^p \mid p\geq 0\}, t>1; d_i=0$	$\mathcal{T}^-$	$D_i=D$	$\sum z_i$	Ch.4;6(**)
2	$t_i\in\{t^p \mid p\geq 0\}, t>1; d_i=0$	$\mathcal{T}^+$	$D_i=D$	$\sum z_i$	Ch.4;6(**)
2	$t_i\in\{1,2\}; d_i=0$	G	$D_i=D$	$\sum u_i$	Ch.4;4.7
2	$t_i\in\{t^p \mid p\geq 0\}, t>1; d_i=0$	$\mathcal{T}^-$	$D_i=D$	$\sum u_i$	Ch.4;6(**)
2	$t_i\in\{t^p \mid p\geq 0\}, t>1; d_i=0$	$\mathcal{T}^+$	$D_i=D$	$\sum u_i$	Ch.4;6(**)
2	$t_i\in\{1,2\}; d_i=0$	G		$\sum \overline{t}_i$	Ch.4;4.1,4.4
2	$t_i=1; d_i=0$	$C;Rs(1)$	$R_1=1, r_i\in\{0,1\}$ $D_i=D$	$\sum z_i$	Ch.4;2.14
2	$[t_i]; d_i=0$	$[Pr];C;Rs(1)$	$R_1=1, r_i\in\{0,1\}$ $D_i=D$	$\sum z_i$	Ch.4;2.15
2	$t_i=1; d_i=0$	$C;Rs(1)$	$R_1=1, r_i\in\{0,1\}$ $D_i=D$	$\sum u_i$	Ch.4;2.14
2	$[t_i]; d_i=0$	$[Pr];C;Rs(1)$	$R_1=1, r_i\in\{0,1\}$ $D_i=D$	$\sum u_i$	Ch.4;2.15
2	$t_i=1; d_i=0$	$C;Rs(1)$	$R_1=1, r_i\in\{0,1\}$	$\sum \overline{t}_i$	Ch.4;2.1,2.13
2	$[t_i]; d_i=0$	$[Pr];C;Rs(1)$	$R_1=1, r_i\in\{0,1\}$	$\sum \overline{t}_i$	Ch.4;2.15
2	$t_i; d_i=0$			$\overline{t}_{max}$ $\sum \overline{t}_i$	Ch.4;1.1, 1.3(*)

(*to be continued*)

Table I.2

Number of machines	Processing times, release dates	Preemption, precedence and resource constraints	Additional conditions	Objective function	Section of the book
2	t_i ; $d_i=0$		$D_i=D$	z_{max} $\sum z_i$	Ch. 4; 1.9(**)
2	t_i ; $d_i=0$		$D_i=D$	L_{max} $\sum L_i$	Ch. 4; 1.9(**)
3	$t_i=1$; $d_i=0$	$Rs(1)$	$D_i=D$	$\bar{t}_i \le D_i$	Ch. 4; 4.8
3	$[t_i]$; $d_i=0$	$[Pr]$; $Rs(1)$	$D_i=D$	$\bar{t}_i \le D_i$	Ch. 4; 4.8
3	$t_i=1$; $d_i=0$	$Rs(q)$	$R_k=1$, $r_{ik} \in \{0,1\}$; $D_i=D$	$\bar{t}_i \le D_i$	Ch. 4; 6
3	$[t_i]$; $d_i=0$	$[Pr]$; $Rs(q)$	$R_k=1$, $r_{ik} \in \{0,1\}$; $D_i=D$	$\bar{t}_i \le D_i$	Ch. 4; 6
3	$t_i=1$; $d_i=0$	$Rs(1)$		$\bar{t}_{max}$	Ch. 4; 2.1, 2.3
3	$[t_i]$; $d_i=0$	$[Pr]$; $Rs(1)$		$\bar{t}_{max}$	Ch. 4; 2.15
3	$t_i=1$; $d_i=0$	$Rs(q)$	$R_k=1$, $r_{ik} \in \{0,1\}$	$\bar{t}_{max}$	Ch. 4; 6
3	$[t_i]$; $d_i=0$	$[Pr]$; $Rs(q)$	$R_k=1$, $r_{ik} \in \{0,1\}$	$\bar{t}_{max}$	Ch. 4; 6
3	$t_i=1$; $d_i=0$	$Rs(1)$	$D_i=D$	L_{max}	Ch. 4; 2.14
3	$[t_i]$; $d_i=0$	$[Pr]$; $Rs(1)$	$D_i=D$	L_{max}	Ch. 4; 2.15
3	$t_i=1$; $d_i=0$	$Rs(q)$	$R_k=1$, $r_{ik} \in \{0,1\}$ $D_i=D$	L_{max}	Ch. 4; 6
3	$[t_i]$; $d_i=0$	$[Pr]$; $Rs(q)$	$R_k=1$, $r_{ik} \in \{0,1\}$ $D_i=D$	L_{max}	Ch. 4; 6
3	$t_i=1$; $d_i=0$	$Rs(1)$	$D_i=D$	z_{max}	Ch. 4; 2.14
3	$[t_i]$; $d_i=0$	$[Pr]$; $Rs(1)$	$D_i=D$	z_{max}	Ch. 4; 2.15
3	$t_i=1$; $d_i=0$	$Rs(q)$	$R_k=1$, $r_{ik} \in \{0,1\}$; $D_i=D$	z_{max}	Ch. 4; 6
3	$[t_i]$; $d_i=0$	$[Pr]$; $Rs(q)$	$R_k=1$, $r_{ik} \in \{0,1\}$; $D_i=D$	z_{max}	Ch. 4; 6
3	$t_i=1$; $d_i=0$	$Rs(1)$	$D_i=D$	$\sum z_i$	Ch. 4; 2.14

(to be continued)

Table I.2

Number of machines	Processing times, release dates	Preemption, precedence and resource constraints	Additional conditions	Objective function	Section of the book
3	$[t_i]; d_i=0$	$[Pr]; Rs(1)$	$D_i=D$	$\sum z_i$	Ch. 4; 2.15
3	$t_i=1; d_i=0$	$Rs(q)$	$R_k=1, r_{ik}\in\{0,1\}$; $D_i=D$	$\sum z_i$	Ch. 4; 6
3	$[t_i]; d_i=0$	$[Pr]; Rs(q)$	$R_k=1, r_{ik}\in\{0,1\}$; $D_i=D$	$\sum z_i$	Ch. 4; 6
3	$t_i=1; d_i=0$	$Rs(1)$	$D_i=D$	$\sum u_i$	Ch. 4; 2.14
3	$[t_i]; d_i=0$	$[Pr]; Rs(1)$	$D_i=D$	$\sum u_i$	Ch. 4; 2.15
3	$t_i=1; d_i=0$	$Rs(q)$	$R_k=1, r_{ik}\in\{0,1\}$; $D_i=D$	$\sum u_i$	Ch. 4; 6
3	$[t_i]; d_i=0$	$[Pr]; Rs(q)$	$R_k=1, r_{ik}\in\{0,1\}$; $D_i=D$	$\sum u_i$	Ch. 4; 6
3	$t_i=1; d_i=0$	$Rs(1)$		$\sum \bar{t}_i$	Ch. 4; 2.1, 2.4
3	$[t_i]; d_i=0$	$[Pr]; Rs(1)$		$\sum \bar{t}_i$	Ch. 4; 2.15
3	$t_i=1; d_i=0$	$Rs(q)$	$R_k=1, r_{ik}\in\{0,1\}$	$\sum \bar{t}_i$	Ch. 4; 6
3	$[t_i]; d_i=0$	$[Pr]; Rs(q)$	$R_k=1, r_{ik}\in\{0,1\}$	$\sum \bar{t}_i$	Ch. 4; 6
M	$t_i=1; d_i=0$	$\mathcal{T}^+$		$\bar{t}_i\le D_i$	Ch. 4; 4.8
M	$[t_i]; d_i=0$	$[Pr]; \mathcal{T}^+$		$\bar{t}_i\le D_i$	Ch. 4; 4.8
M	$t_i=1; d_i$	$\mathcal{T}^-$	$D_i=D$	$\bar{t}_i\le D_i$	Ch. 4; 6
M	$[t_i]; [d_i]$	$[Pr]; \mathcal{T}^-$	$D_i=D$	$\bar{t}_i\le D_i$	Ch. 4; 6
M	$t_i=1; d_i=0$	$\mathcal{T}$	$D_i=D$	$\bar{t}_i\le D_i$	Ch. 4; 6
M	$[t_i]; d_i=0$	$[Pr]; \mathcal{T}$	$D_i=D$	$\bar{t}_i\le D_i$	Ch. 4; 6
M	$t_i\in\{1,t\}$; $d_i=0$	$\mathcal{T}^-$	$D_i=D$	$\bar{t}_i\le D_i$	Ch. 4; 6
M	$t_i\in\{1,t\}$; $d_i=0$	$\mathcal{T}^+$	$D_i=D$	$\bar{t}_i\le D_i$	Ch. 4; 6

(to be continued)

Table I.2

Number of machines	Processing times, release dates	Preemption, precedence and resource constraints	Additional conditions	Objective function	Section of the book
M	$t_i=1$; $d_i=0$	Pr; G	$D_i=D$	$\bar{t}_i \le D_i$	Ch. 4; 6
M	t_i; $d_i=0$			$\bar{t}_{max}$	Ch. 4; 6
M	$t_i=1$; d_i	$\mathcal{T}^-$		$\bar{t}_{max}$	Ch. 4; 6
M	$[t_i]$; $[d_i]$	$[Pr]$; $\mathcal{T}^-$		$\bar{t}_{max}$	Ch. 4; 6
M	t_i; $d_i=0$		$D_i=D$	L_{max}	Ch. 4; 6
M	$t_i=1$; $d_i=0$	$\mathcal{T}^+$		L_{max}	Ch. 4; 3.1, 3.3
M	$[t_i]$; $d_i=0$	$[Pr]$; $\mathcal{T}^+$		L_{max}	Ch. 4; 3.4
M	$t_i=1$; d_i	$\mathcal{T}^-$	$D_i=D$	L_{max}	Ch. 4; 6
M	$[t_i]$; $[d_i]$	$[Pr]$; $\mathcal{T}^-$	$D_i=D$	L_{max}	Ch. 4; 6
M	$t_i=1$; $d_i=0$	$\mathcal{T}$	$D_i=D$	L_{max}	Ch. 4; 6
M	$[t_i]$; $d_i=0$	$[Pr]$; $\mathcal{T}$	$D_i=D$	L_{max}	Ch. 4; 6
M	$t_i \in \{1, t\}$; $d_i=0$	$\mathcal{T}^-$	$D_i=D$	L_{max}	Ch. 4; 6
M	$t_i \in \{1, t\}$; $d_i=0$	$\mathcal{T}^+$	$D_i=D$	L_{max}	Ch. 4; 6
M	t_i; $d_i=0$		$D_i=D$	z_{max}	Ch. 4; 6
M	$t_i=1$; $d_i=0$	$\mathcal{T}^+$		z_{max}	Ch. 4; 3.4
M	$[t_i]$; $d_i=0$	$[Pr]$; $\mathcal{T}^+$		z_{max}	Ch. 4; 3.4
M	$t_i=1$; $d_i=0$	$\mathcal{T}$	$D_i=D$	z_{max}	Ch. 4; 6
M	$[t_i]$; $d_i=0$	$[Pr]$; $\mathcal{T}$	$D_i=D$	z_{max}	Ch. 4; 6
M	$t_i \in \{1, t\}$; $d_i=0$	$\mathcal{T}^-$	$D_i=D$	z_{max}	Ch. 4; 6
M	$t_i \in \{1, t\}$; $d_i=0$	$\mathcal{T}^+$	$D_i=D$	z_{max}	Ch. 4; 6

(to be continued)

Table I.2

Number of machi-nes	Processing times, release dates	Preemption, precedence and resource constraints	Additional conditions	Objective function	Section of the book
M	$t_i=1$; d_i	$\mathcal{T}^-$	$D_i=D$	z_{max}	Ch. 4; 6
M	$[t_i]$; $[d_i]$	$[Pr]$;$\mathcal{T}^-$	$D_i=D$	z_{max}	Ch. 4; 6
M	$t_i=1$; $d_i=0$	$\mathcal{T}$		$\bar{t}_{max}$	Ch. 4; 6
M	$[t_i]$; $d_i=0$	$[Pr]$;$\mathcal{T}$		$\bar{t}_{max}$	Ch. 4; 6
M	$t_i\in\{1, t\}$; $d_i=0$	$\mathcal{T}^-$		$\bar{t}_{max}$	Ch. 4; 6
M	$t_i\in\{1, t\}$; $d_i=0$	$\mathcal{T}^+$		$\bar{t}_{max}$	Ch. 4; 6
M	$t_i=1$; $d_i=0$	Pr;G		$\bar{t}_{max}$	Ch. 4; 6
M	$t_i=1$; $d_i=0$	Pr;G	$D_i=D$	L_{max}	Ch. 4; 6
M	$t_i=1$; $d_i=0$	Pr;G	$D_i=D$	z_{max}	Ch. 4; 6
M	t_i; $d_i=0$		$D_i=D$	$\sum z_i$	Ch. 4; 6
M	$t_i=1$; $d_i=0$	$\mathcal{T}^+$		$\sum z_i$	Ch. 4; 3.4
M	$[t_i]$; $d_i=0$	$[Pr]$;$\mathcal{T}^+$		$\sum z_i$	Ch. 4; 3.4
M	$t_i=1$; d_i	$\mathcal{T}^-$	$D_i=D$	$\sum z_i$	Ch. 4; 6
M	$[t_i]$; $[d_i]$	$[Pr]$;$\mathcal{T}^-$	$D_i=D$	$\sum z_i$	Ch. 4; 6
M	t_i; $d_i=0$		$D_i=D$	$\sum u_i$	Ch. 4; 6
M	t_i; $d_i=0$	Pr		$\sum u_i$	Ch. 4; 6(**)
M	t_{iH}; $d_i=0$	Pr		$\sum u_i$	Ch. 4; 6
M	$t_i=1$; $d_i=0$	$\mathcal{T}^+$		$\sum u_i$	Ch. 4; 3.4
M	$[t_i]$; $d_i=0$	$[Pr]$;$\mathcal{T}^+$		$\sum u_i$	Ch. 4; 3.4
M	$t_i=1$; d_i	$\mathcal{T}^-$	$D_i=D$	$\sum u_i$	Ch. 4; 6
M	$[t_i]$; $[d_i]$	$[Pr]$;$\mathcal{T}^-$	$D_i=D$	$\sum u_i$	Ch. 4; 6
M	$t_i=1$; $d_i=0$	$\mathcal{T}$	$D_i=D$	$\sum z_i$	Ch. 4; 6

(*to be continued*)

Table I.2

Number of machines	Processing times, release dates	Preemption, precedence and resource constraints	Additional conditions	Objective function	Section of the book
M	$[t_i]; d_i=0$	$[Pr];\mathcal{T}$	$D_i=D$	$\sum z_i$	Ch. 4; 6
M	$t_i \in \{1, t\}$; $d_i=0$	$\mathcal{T}^-$	$D_i=D$	$\sum z_i$	Ch. 4; 6
M	$t_i \in \{1, t\}$; $d_i=0$	$\mathcal{T}^+$	$D_i=D$	$\sum z_i$	Ch. 4; 6
M	$t_i=1; d_i=0$	$Pr; G$	$D_i=D$	$\sum z_i$	Ch. 4; 6
M	$t_i=1; d_i=0$	$\mathcal{T}$	$D_i=D$	$\sum u_i$	Ch. 4; 6
M	$[t_i]; d_i=0$	$[Pr];\mathcal{T}$	$D_i=D$	$\sum u_i$	Ch. 4; 6
M	$t_i \in \{1, t\}$; $d_i=0$	$\mathcal{T}^-$	$D_i=D$	$\sum u_i$	Ch. 4; 6
M	$t_i \in \{1, t\}$; $d_i=0$	$\mathcal{T}^+$	$D_i=D$	$\sum u_i$	Ch. 4; 6
M	$t_i=1; d_i=0$	$Pr; G$	$D_i=D$	$\sum u_i$	Ch. 4; 6
M	$t_i=1; d_i=0$	G		$\sum \bar{t}_i$	Ch. 4; 4.1.4.6
M	$[t_i]; d_i=0$	G		$\sum \bar{t}_i$	Ch. 4; 4.7
M	$t_i; d_i=0$		machine speeds↓	$\sum \bar{t}_i$	Ch. 4; 6(**)
M	$t_i; d_i=0$		machine speeds↓ $D_i=D$	$\sum z_i$	Ch. 4; 6(**)
M	$t_i; d_i=0$			$\sum_{j=1}^{M}((\sum_{i \in N_j} \alpha_i) \times \sum_{i \in N_j} t_i)$	Ch. 4; 6(**)
M	$t_i; d_i=0$		$M=M(N, D)$	$M; \bar{t}_i \le D$	Ch. 4; 1.1, 1.7; 6

Table I.3

Number of machines	Processing times; release dates	Preemption; resource and precedence constraints	Additional conditions	Objective function	Running time	Performance guarantee	Section of the book
1	t_i; d_i			L_{max}	$n^2 \log n$	$(F^0-F^*)/(F^*+D_{max}) < \min\{1/2, t_{max}/t_\Sigma, 1-2t_{min}/t_\Sigma\}$	A.10
1	t_i; d_i			L_{max}	$n^2 \log n$	$(F^0-F^*)/(F^*+D_{max}) < 1/3$	A.10
2	t_i; $d_i=0$			$\overline{t}_{max}$	n	$\Delta \le 1/11$	A.2
2	$t_i=a_H$; $d_i=0$	G		$\overline{t}_{max}$	n^2	$\Delta \le 1-\min\{a_1, a_2\}/\max\{a_1, a_2\}$	A.5
2	t_{iH}; $d_i=0$			$\overline{t}_{max}$	n	$\Delta \le 1/2$	A.6
2	t_{iH}; $d_i=0$			$\overline{t}_{max}$	$n \log n$	$\Delta \le (\sqrt{5}-1)/2$	A.6
M	t_i; $d_i=0$			$\overline{t}_{max}$	n	$\Delta \le 1 - 1/M$	A.2
M	t_i; $d_i=0$			$\overline{t}_{max}$	$n \log n$	$\Delta \le 1/3 - 1/(3M)$	A.2
M	t_i; $d_i=0$			$\overline{t}_{max}$	$n \log n + kn \log M$	$\Delta \le \rho + 1/2^k$, $\rho = \begin{cases} 1/7, & M=2 \\ 2/13, & M=3 \\ 3/17, & M \in \{4, 5, 6, 7\} \\ 1/5, & M \ge 8 \end{cases}$	A.2
M	t_i; $d_i=0$			$\overline{t}_{max}$	$n \log n + kn \log M$	$\Delta \le 11/61 + 1/2^k$	A.2
M	t_i; $d_i=0$			$\overline{t}_{max}$	$(kn)^{k \log k}$	$\Delta \le 1/k + 1/2^k$	A.2
M	t_i; $d_i=0$			$\overline{t}_{max}$	$n \log n$	$\Delta \le 37/160$	A.2
M	t_i; $d_i=0$			$\overline{t}_{max}$	$n(M^4+\log n)$	$\Delta \le 35/192$	A.2
M	t_i; $d_i=0$			$\overline{t}_{max}$	$n \log n$	$F^0-F^* \le (1-1/M)t_{max}$	A.2

(*to be continued*)

Table I.3

Number of machines	Processing times; release dates	Preemption; resource and precedence constraints	Additional conditions	Objective function	Running time	Performance guarantee	Section of the book
M	t_i ; d_i			$\bar{t}_{max}$	$n\log n$	$F^0 - F^* \le (2-1/M)\, t_{max}$	A.2
M	t_i ; d_i			$\bar{t}_{max}$	$n\log n$	$\Delta < \min\{(2M-1)/M, (2M-1)\,t_{max}/t_\Sigma\}$	A.2
M	t_i ; $d_i=0$	G		$\bar{t}_{max}$	n^2	$\Delta \le 1-1/M$	A.3
M	t_i ; $d_i=0$	$\mathcal{T}^-$		$\bar{t}_{max}$	$n\log n$	$F^0 - F^* \le (1-1/M)\, t_{max}$	A.3
M	t_i ; $d_i=0$	$\mathcal{T}^-$		$\bar{t}_{max}$	$n\log n$	$\Delta \le 1-2/(M+1)$	A.3
M	t_i ; $d_i=0$	$\mathcal{T}^+$		$\bar{t}_{max}$	$n\log n$	$\Delta \le 1-2/(M+1)$	A.3
M	t_i ; $d_i=0$	C		$\bar{t}_{max}$	$n\log n$	$\Delta \le 2/3$	A.3
M	$t_i=1$; $d_i=0$	G		$\bar{t}_{max}$		$\Delta \le \begin{cases} 1/3, & M=2 \\ 1-1/M, & M\ge 3 \end{cases}$	A.3
M	$t_i \in \{1, t\}$; $d_i=0$	G		$\bar{t}_{max}$		$\Delta \le \begin{cases} 1/3, & t=2 \\ 1/2-1/(2t), & t\ge 3 \end{cases}$	A.3
M	$t_i=1$, $d_i=0$	G		$\bar{t}_{max}$	n^2	$\Delta \le 1-2/M$	A.3
M	t_i ; $d_i=0$	$Pr;G$		$\bar{t}_{max}$	n^2	$\Delta \le 1-2/M$	A.3
M	$t_i a_H$; $d_i=0$			$\bar{t}_{max}$	$n\log M$	$\Delta \le \begin{cases} (\sqrt{5}-1)/2, & M=2 \\ \sqrt{2M-2}/2, & M\ge 3 \end{cases}$	A.4
M	$t_i a_H$; $d_i=0$			$\bar{t}_{max}$		$\Delta \le a_{max}/a_{min} - 1/(a_{min}\Sigma(a_H)^{-1})$	A.4
M	$t_i a_H$; $d_i=0$			$\bar{t}_{max}$	$n\log M$	$\Delta \le \sqrt{M} - 1 + O(M^{1/4})$	A.4

(to be continued)

Table I.3

Number of machines	Processing times; release dates	Preemption; resource and precedence constraints	Additional conditions	Objective function	Running time	Performance guarantee	Section of the book
M	$t_i a_H$; $d_i=0$			$\bar{t}_{max}$	$n\log n$	$\Delta\le\begin{cases}(\sqrt{17}-3)/4, M=2\\ 1-2/(M+1), M\ge 3\end{cases}$	A.4
M	$t_i a_H$; $d_i=0$			$\bar{t}_{max}$	$n\log n$	$\Delta\le 1-2/(M+1)$	A.4
M	$t_i a_H$; $d_i=0$			$\bar{t}_{max}$	$n\log n$	$\Delta\le 7/12$	A.4
M	$t_i a_H$; $d_i=0$			$\bar{t}_{max}$	$n\log n+kn\log M$	$\Delta\le\begin{cases}(\sqrt{17}-3)/4+1/2^k, M=3\\ 1/2-1/(2M)+1/2^k, M\in\{4,5\}\\ 2/5+1/2^k, M\ge 6\end{cases}$	A.4
M	$t_i a_H$; $d_i=0$			$\bar{t}_{max}$	$n\log n+kn\log M$	$\Delta\le 1-1/M+1/2^k$	A.4
M	$t_i a_H$; $d_i=0$			$\bar{t}_{max}$	$n\log n+M$	$\Delta\le 1/2$	A.4
M	$t_i a_H$; $d_i=0$		$a_M<1$; $a_H=1$, $H\ne M$	$\bar{t}_{max}$	$n\log M$	$\Delta\le\begin{cases}(\sqrt{5}-1)/2, M=2\\ 2-4/(M+1), M\ge 3\end{cases}$	A.4
M	$t_i a_H$; $d_i=0$		$a_M<1$; $a_H=1$, $H\ne M$	$\bar{t}_{max}$	$n\log n$	$\Delta\le\begin{cases}(\sqrt{17}-3)/4, M=2\\ 1/2-1/(2M), M\ge 3\end{cases}$	A.4
M	$t_i a_H$; $d_i=0$		$a_M<1$; $a_H=1$, $H\ne M$	$\bar{t}_{max}$	$n\log n+kn\log M$	$\Delta\le\begin{cases}(\sqrt{6}-2)/2+1/2^k; M=2\\ (\sqrt{17}-3)/4+1/2^k, M\ge 3\end{cases}$	A.4
M	$t_i a_H$; $d_i=0$		$a_M<1$; $a_H=1$, $H\ne M$	$\bar{t}_{max}$	$n\log n+kn\log M$	$\Delta\le\begin{cases}(\sqrt{17}-3)/4+1/2^k, M=2\\ \sqrt{2}-1+1/2^k, M\ge 3\end{cases}$	A.4

(to be continued)

Table I.3

Number of machines	Processing times; release dates	Preemption; resource and precedence constraints	Additional conditions	Objective function	Running time	Performance guarantee	Section of the book
M	$t_i a_H$; $d_i=0$		$a_M>1$; $a_H=1$, $H\neq M$	$\bar{t}_{max}$	$n\log n+$ $kn\log M$	$\Delta\leq(\sqrt{17}-3)/4+1/2^k$	A.4
M	$t_i a_H$; $d_i=0$		$a_H=1$, $H\neq M$	$\bar{t}_{max}$	$n\log n$	$\Delta\leq\begin{cases}(Ma_M+1-3a_M)/(2a_M), a_M<1/2;\\ (2Ma_M+1-4a_M)/(2a_M+1),\\ a_M\in[1/2,1];\\ 1/a_M+1/(Ma_M+1-a_M), a_M>1\end{cases}$	A.4
M	$t_i a_H$; $d_i=0$	G		$\bar{t}_{max}$	n^2	$\Delta\leq 1-1/\sum_{H=1}^{M}(a_H)^{-1}$	A.5
M	$t_i a_H$; $d_i=0$	Pr; G		$\bar{t}_{max}$	n^2	$\Delta\leq\sqrt{3M/2}-1$	A.5
M	$t_i a_H$; $d_i=0$	Pr; G		$\bar{t}_{max}$		$\Delta\leq\sqrt{M}-1/2$	A.5
M	$t_i a_H$; $d_i=0$	Pr; G	$a_1\leq a_2$ $\leq\ldots\leq a_M$	$\bar{t}_{max}$	n^2	$\Delta\leq\begin{cases}\sum_{H=1}^{(M-1)/2}\max\{a_1/a_{2H-1}, a_2/a_{2H}\}\\ +a_1/a_M-1,\\ M\text{-odd};\\ \sum_{H=1}^{M/2}\max\{a_1/a_{2H-1}, a_2/a_{2H}\}\\ -1, M\text{-even}\end{cases}$	A.5
M	$t_i=a_H$; $d_i=0$	$Rs(1)$	$a_1\leq a_2$ $\leq\ldots\leq a_M$	$\bar{t}_{max}$	$n\log n$	$\Delta\leq\begin{cases}\sum_{H=1}^{(M-1)/2}\max\{a_1/a_{2H-1}, a_2/a_{2H}\}\\ +a_1/a_M-1,\\ M\text{-odd};\\ \sum_{H=1}^{M/2}\max\{a_1/a_{2H-1}, a_2/a_{2H}\}\\ -1, M\text{-even}\end{cases}$	A.5

(to be continued)

Table I.3

Number of machines	Processing times; release dates	Preemption; resource and precedence constraints	Additional conditions	Objective function	Running time	Performance guarantee	Section of the book
M	$t_{iH}; d_i=0$			$\bar{t}_{max}$	$Mn\log n$	$\Delta\le\sqrt{6M}+\sqrt{3}/\sqrt{8M}$	A.6
M	$t_{iH}; d_i=0$			$\bar{t}_{max}$	$Mn\log n$	$\Delta\le 2\sqrt{M}-1$	A.6
M	$t_{iH}; d_i=0$			$\bar{t}_{max}$	$M^M+Mn\log n$	$\Delta\le\sqrt{2M}+1/\sqrt{8M}$	A.6
M	$t_{iH}; d_i=0$			$\bar{t}_{max}$	Mn^2	$\Delta\le M-1$	A.6
M	$t_{iH}; d_i=0$			$\bar{t}_{max}$	$P(n,M)$	$\Delta<1$	A.6
M	$t_{iH}; d_i=0$	G		$\bar{t}_{max}$	$Mn+n^2$	$\Delta\le M-1$	A.6
M	$t_i; d_i=0$	$Rs(q)$		$\bar{t}_{max}$	$n\log n$	$\Delta\le\min\{(M-1)/2,$ $q+1-(2q+1)/M\}$	A.7
M	$t_i=1;$ $d_i=0$	$Rs(1)$		$\bar{t}_{max}$	$n\log n$	$\Delta\le 17/10-12/(5M)+$ $2/F^*$	A.7
M	$t_i=1;$ $d_i=0$	$Rs(1)$		$\bar{t}_{max}$	$n\log n$	$\Delta\le 1-2/M+1/F^*$	A.7
M	$t_i=1;$ $d_i=0$	$Rs(q)$		$\bar{t}_{max}$	$qn^2+n^{5/2}$	$\Delta\le\lceil M/2\rceil-1$	A.7
M	$t_i=1;$ $d_i=0$	$Rs(q)$		$\sum\bar{t}_i$	$qn+n^{5/2}$	$\Delta\le\lceil M/2\rceil-1$	A.7
M	$t_i; d_i=0$	$G; Rs(q)$		$\bar{t}_{max}$	n^2	$\Delta\le M-1$	A.8
M	$t_i; d_i=0$	$G; Rs(1)$		$\bar{t}_{max}$	n^2	$\Delta\le M-1$	A.8
M	$t_i=1;$ $d_i=0$	$G; Rs(q)$	$\max_j r_{ij}$ $=\sum_{j=1}^{q} r_{ij}$	$\bar{t}_{max}$		$\Delta\le\min\{M-1,\ q+1-$ $(q+1)/M\}$	A.8
M	$t_i=1;$ $d_i=0$	$G; Rs(q)$	$M\ge n$	$\bar{t}_{max}$	n^2	$\Delta\le q(1+F^*)/2$	A.9
M	$t_i=1;$ $d_i=0$	$G; Rs(q)$	$M\ge n$	$\bar{t}_{max}$	n^2	$\Delta\le 17q/10$	A.9

(to be continued)

Table I.3

Number of machines	Processing times; release dates	Preemption; resource and precedence constraints	Additional conditions	Objective function	Running time	Performance guarantee	Section of the book
M	$t_i=1$; $d_i=0$	$G; Rs(q)$	$M\geq n$; $\max_j r_{ij} = \sum_{j=1}^{q} r_{ij}$	$\bar{t}_{max}$		$\Delta\leq q$	A.9
M	$t_i=1$; $d_i=0$	$Rs(q)$	$M\geq n$;	$\bar{t}_{max}$	$n\log M$	$\Delta\leq q-3/10+5/(2F^*)$	A.9
M	$t_i=1$; $d_i=0$	$Rs(q)$	$M\geq n$;	$\bar{t}_{max}$	$nq+n\log n$	$\Delta\leq q-2/3$	A.9
M	$t_i=1$; $d_i=0$	$Rs(1)$	$M\geq n$;	$\bar{t}_{max}$	$n\log M$	$\Delta\leq 7/10+1/F^*$	A.9
M	$t_i=1$; $d_i=0$	$Rs(1)$	$M\geq n$;	$\bar{t}_{max}$	n^2	$\Delta\leq 1/3+1/F^*$	A.9
M	t_i; $d_i=0$	$Rs(q)$	$M\geq n$;	$\bar{t}_{max}$	$n\log M$	$\Delta\leq q$	A.9
M	t_i; $d_i=0$			L_{max}	$n\log n$	$F^0-F^*\leq(2M-1)t_{max}/M$	A.11
M	t_i; $d_i=0$			L_{max}	$\log n$	$(F^0-F^*)/(F^*+D_{max}) < 1-1/M$	A.11
M	t_i; d_i			L_{max}	$n\log n$	$(F^0-F^*)/(F^*+D_{max}) < \min\{4/3-1/(3M)-Mt_{min}/t_\Sigma,$ $1/3-1/(3M)-$ $-M(D_{max}-D_{min}/t_\Sigma\}$	A.11
M	$t_i=t$; $d_i=0$			L_{max}	$n\log n$	$F^0-F^*\leq t$	A.11
M	t_i; $d_i=0$			$\sum\alpha_i\bar{t}_i$	$n\log n$	$\Delta\leq(M-1)/(2M)$	A.12
M	t_i; $d_i=0$			$\sum\alpha_i\bar{t}_i$	$n\log n$	$\Delta\leq(\sqrt{2}-1)/2$	A.12
M	t_i; $d_i=0$			$\sum\bar{t}_i/t_{max}$	$n\log n$	$\Delta\leq(M-1)/(M+1)$	A.12

(*to be continued*)

Table I.3

Number of machines	Processing times; release dates	Preemption; resource and precedence constraints	Additional conditions	Objective function	Running time	Performance guarantee	Section of the book
M	t_i; $d_i=0$			$\sum T_H^2$	$n\log n$	$\Delta\le 1/24$	A.12
M	t_i; $d_i=0$		$M=M(N, D)$	M; $\bar{t}_i\le D$	$n\log M$	$\Delta\le 7/10+1/F^*$	A.1
M	t_i; $d_i=0$		$M=M(N, D)$	M; $\bar{t}_i\le D$	$n\log n$	$\Delta\le 2/9+4/F^*$	A.1
1	t_i; $d_i=0$			$\sum\alpha_i(1-u_i)\to\max$	n^2/ε	$\Delta\le\varepsilon$	A.13
1	t_i; $d_i=0$	$\mathcal{T}$	$D_i=D$	$\sum\alpha_i(1-u_i)\to\max$	n^2/ε	$\Delta\le\varepsilon$	A.13
1	t_i; $d_i=0$			$\sum\alpha_i u_i$	$n^2\log n+n^2/\varepsilon$	$\Delta\le\varepsilon$	A.13
1	t_i; d_i		$d_i<d_j\Rightarrow D_i\le D_j$	$\sum\alpha_i u_i$	$n^2\log n+n^2/\varepsilon$	$\Delta\le\varepsilon$	A.13
1	t_i; d_i			L_{max}	$n(1/\varepsilon)^{\rho}+n\log n$; $\rho=16/\varepsilon^2+8/\varepsilon$	$(F^0-F^*)/(F^*+D_{max})\le\varepsilon$	A.13
1	t_i; d_i			L_{max}	$2^{\rho}(n/\varepsilon)^{3+\rho}$; $\rho=4/\varepsilon$	$(F^0-F^*)/(F^*+D_{max})\le\varepsilon$	A.13
1	t_i; $d_i=0$			$\sum z_i$	n^7/ε	$\Delta\le\varepsilon$	A.13
1	t_i; $d_i=0$			$\sum z_i$	$n^6/\varepsilon+n^6\log n$	$\Delta\le\varepsilon$	A.13
1	t_i; $d_i=0$			$\sum\alpha_i\min\{t_i, z_i\}$	$n^3\log n+n^3/\varepsilon$	$\Delta\le\varepsilon$	A.13
1	t_i; $d_i=0$			$\sum\min\{t_i, z_i\}$	n^2/ε	$\Delta\le\varepsilon$	A.13
2	t_i; $d_i=0$			$\bar{t}_i\le D_i$	n/ε	$(\bar{t}_i(s^0)-D_i)/D_i\le\varepsilon$	A.13
2	t_i; $d_i=0$			$\bar{t}_{max}$	$\min\{n/\varepsilon, n+1/\varepsilon^2\}$	$\Delta\le\varepsilon$	A.13

(to be continued)

Table I.3

Number of machines	Processing times; release dates	Preemption; resource and precedence constraints	Additional conditions	Objective function	Running time	Performance guarantee	Section of the book
2	$t_i a_H$; $d_i=0$			$\overline{t}_{max}$	$\min\{n/\varepsilon, n+1/\varepsilon^3\}$	$\Delta\le\varepsilon$	A.13
2	t_{iH}; $d_i=0$			$\overline{t}_{max}$	n^2/ε	$\Delta\le\varepsilon$	A.13
2	t_i; $d_i=0$			$\sum\alpha_i\overline{t}_i$	n^2/ε	$\Delta\le\varepsilon$	A.13
2	t_i; $d_i=0$		$D_i=D$	$\sum z_i$	n^3/ε	$(F^0-F^*)/(F^*+D_{max})\le\varepsilon$	A.13
2	t_i; $d_i=0$			L_{max}	$\frac{n}{\varepsilon}(\log\frac{1}{\varepsilon}+n)$	$(F^0-F^*)/(F^*+D_{max})\le\varepsilon$	A.13
2	$t_i a_H$; $d_i=0$	Pr		$\sum\alpha_i u_i$	$P(n,1/\varepsilon)$	$\Delta\le\varepsilon$	A.13
M	t_i; $d_i=0$			$\overline{t}_{max}$	$n^{2M-1}/\varepsilon^{M-1}$	$\Delta\le\varepsilon$	A.13
M	t_i; $d_i=0$			$\overline{t}_{max}$; $T_M\le D$	n^M/ε^{M-1}	$\Delta\le\varepsilon$	A.13
M	t_i; $d_i=0$			L_{max}	$(\log\frac{1}{\varepsilon}+n)\times n^M/\varepsilon^{M-1}$	$(F^0-F^*)/(F^*+D_{max})\le\varepsilon$	A.13
M	t_i; $d_i=0$			$\overline{t}_{max}\times\sum\alpha_i\overline{t}_i$	n^M/ε^M	$\Delta\le\varepsilon$	A.13
M	t_i; $d_i=0$			$\sum T_H^2$	n^M/ε^M	$\Delta\le\varepsilon$	A.13
M	t_i; $d_i=0$			$\overline{t}_i\le D_i$	n^M/ε^{M-1}	$(\overline{t}_i(s^0)-D_i)/D_i\le\varepsilon$	A.13
M	$t_i a_H$; $d_i=0$			$\overline{t}_{max}$	n^{2M}/ε^{M-1}	$\Delta\le\varepsilon$	A.13
M	$t_i a_H$; $d_i=0$			$\overline{t}_{max}$	Mn^{3+10/ε^2}	$\Delta\le\varepsilon$	A.13
M	$t_i a_H$; $d_i=0$			$\sum\alpha_i\overline{t}_i$	$n^{2M-2}/\varepsilon^{M-1}$	$\Delta\le\varepsilon$	A.13
M	t_i; $d_i=0$			$\sum\alpha_i\overline{t}_i$	n^M/ε^M	$\Delta\le\varepsilon$	A.13
M	t_i; $d_i=0$			$\sum\alpha_i\overline{t}_i$	$n^{2M-1}/\varepsilon^{M-1}$	$\Delta\le\varepsilon$	A.13

CHAPTER 1

ELEMENTS OF GRAPH THEORY AND COMPUTATIONAL COMPLEXITY OF ALGORITHMS

This chapter is of auxiliary nature. It contains a number of facts from various areas of modern discrete mathematics. This information is widely used in further consideration.

Section 1 presents basic concepts of binary relations theory and graph theory. Various graph representations are discussed, and "effective" techniques for implementing some operations on graphs are described.

Section 2 considers a specific data structure, called balanced 2-3 trees. This structure is widely used in constructing fast algorithms for solving various problems discussed in Chapters 2 and 3.

The main concepts of the theory of the polynomial reducibility of discrete problems and the computational complexity of algorithms are introduced in Section 3. It should be noted that, unlike the first two sections, understanding Section 3 requires some preliminary background. The material in this section is used mainly in Chapter 4. To be able to follow the rest of the book it suffices to be aware of the concept of the running time of an algorithm.

1. Sets, Orders, Graphs

This section presents some facts from set theory and graph theory which are used in further considerations. We assume that the reader is familiar with such concepts as a set, a subset, union, intersection, difference of sets, etc.

1.1. In the following, only *finite sets* (i.e. the sets with a finite number of elements) are considered.

The Cartesian product of two non-empty sets X and Y (notation: $X \times Y$) is the set of all ordered pairs (x, y) such that $x \in X$, $y \in Y$. A subset $U \subseteq X \times Y$ is called *a binary relation between X and Y*. A subset $U \subseteq X \times X$ is called a *binary relation over X*. We write xUy if and only if $(x, y) \in U$. The binary relation U^{-1} is *the inverse* of U: $(x, y) \in U^{-1}$ if and only if $(y, x) \in U$.

A binary relation U defined over set X is:

(*i*) *Transitive* if for any x, y, z in X, such that xUy and yUz, the relation xUz holds.

(*ii*) *Reflexive* if for any $x \in X$ the relation xUx holds;

(*iii*) *Antireflexive* if the relation xUx does not hold for any $x \in X$;

(*iv*) *Symmetric* if for any x, y in X, such that xUy, the relation yUx holds;

(*v*) *Asymmetric* if for any x, y of X at least one of the relations xUy or yUx does not hold;

(*vi*) *Antisymmetric* if for any x, y of X such that if xUy and yUx hold simultaneously, it follows that $x = y$.

(*vii*) *Total* if for any x, y, in X, $x \neq y$, at least, one of the relations xUy and yUx holds.

A transitive relation defined over set X is called *a pseudo-order relation* (or *a pseudo-order*). In this case, set X is said to be *pseudo-ordered*.

A transitive and reflexive relation defined over set X is called *a quasi-order relation* (or *a quasi-order*). In this case, set X is said to be *quasi-ordered*.

A transitive and antireflexive relation defined over set X is called *a strict order relation* (or *a strict order*). In this case, set X is said to be *strictly ordered*.

A transitive, reflexive and antisymmetric relation defined over set X is called *a non-strict order relation* (or *a non-strict order*). In this case, set X is said to be *non-strictly ordered*.

A (pseudo-, quasi-, strictly, or non-strictly ordered) set X is called ***total*** if the binary relation defined over it is total.

A strictly ordered set X is said to be *linearly ordered* if the order is total. Otherwise, an ordered set X is called *partially ordered*.

Let X be a set of n-dimensional vectors $\mathbf{x} = (x_1, x_2, \ldots, x_n)$, where x_i are real numbers. We define the relation $\geq$ over set X as follows: for $\mathbf{x}, \mathbf{y} \in X$, $\mathbf{x} \geq \mathbf{y}$, if $x_i \geq y_i$, $i = 1, 2, \ldots, n$.

1.2. The pair consisting of a set X and a binary relation U defined over X is called a *directed graph* (notation: $G = (X, U)$). The elements of set X are called *vertices* of graph G, while the pairs $(x, y) \in U$ are called *arcs*. For an arc (x, y), the vertex x is its *beginning*, and the vertex y is its *end*. In this case, an arc (x, y) is said to *leave* the vertex x and *to enter* the vertex y. An arc (x, x) is called *a loop*.

If $\hat{U}$ is a set of non-ordered pairs of the elements of set X, then the pair $\hat{G} = (X, \hat{U})$ is called *a non-directed graph*. In this case, the pairs $(x, y) \in \hat{U}$ are called *edges* of graph $\hat{G}$.

A graph $\hat{G} = (X, \hat{U})$ is called *a complete graph* if $(x, y) \in \hat{U}$ for all $x, y \in X$, $x \neq y$.

Along with the notation $G = (X, U)$ for directed graphs and $\hat{G} = (X, \hat{U})$ for non-directed graphs, we use the notation $\tilde{G} = (X, \tilde{U})$ in the formulation of statements which hold for both directed and non-directed graphs.

If $(x, y) \in \tilde{U}$, the vertices x and y are said to be *adjacent*, and the arc (edge) (x, y) is said to be *incident* to the vertices x and y.

Two graphs $\tilde{G} = (X, \tilde{U})$ and $\tilde{G}' = (X', \tilde{U}')$ are called *isomorphic* if there exists a one-to-one mapping φ of the set X into the set X' such that $(x, y) \in \tilde{U}$ if and only if $(\varphi(x)), \varphi(y)) \in \tilde{U}'$, where $\varphi(x)$ and $\varphi(y)$ are the images of the elements x and y in mapping φ. In this case, mapping φ is called an *isomorphism* of graph $\tilde{G}$ onto graph $\tilde{G}'$.

A graph $\tilde{G}' = (X', \tilde{U}')$ is called *a subgraph* of a graph $\tilde{G} = (X, \tilde{U})$ if $X' \subseteq X$ and $(x, y) \in \tilde{U}'$ implies that $(x, y) \in \tilde{U}$. If, for any $x, y \in X'$, it follows from $(x, y) \in \tilde{U}$ that $(x, y) \in \tilde{U}'$, then $\tilde{G}'$ is called *an induced subgraph*.

A route in a graph $\tilde{G} = (X, \tilde{U})$ is a sequence of vertices $x_1, x_2, \ldots, x_r$ such that either $(x_k, x_{k+1}) \in \tilde{U}$ or $(x_{k+1}, x_k) \in \tilde{U}$, $k = 1, 2, \ldots, r-1$. In this case, the vertices x_1 and x_r are said to be connected by a route. *A path* in a directed graph $G = (X, U)$ is a sequence of arcs of the form $(x_1, x_2), (x_2, x_3), \ldots, (x_{r-1}, x_r)$, or, equivalently, a route $x_1, x_2, \ldots, x_r$ such that $(x_k, x_{k+1}) \in U$, $k = 1, 2, \ldots, r-1$. Here x_1 is the beginning and x_r is the end of the path. The number r is called the *length* of a path. In what follows, by a "path" is meant a *simple path*, i.e., a path in which all vertices are distinct. *A circuit* is a path where $x_1 = x_r$.

In a directed graph G, a vertex x is called *a predecessor* of a vertex y, if there is a path from x to y in G. In this case, vertex y is called *a successor* of vertex x. If G contains a circuit, then the same vertex x may be a predecessor and a successor of some vertex y at the same time. A vertex x of the directed graph $G = (X, U)$ is called *a direct predecessor* of a vertex y if $(x, y) \in U$ and G has no path from x to y without the arc (x, y). In this case, vertex y is called *a direct successor* of vertex x. Let $B_G(x)$ (or $A_G(x)$, respectively) denote the set of all predecessors (successors) of vertex x in graph

G. The set of all direct predecessors (direct successors) of vertex x is denoted by $B_G^0(x)$ (or by $A_G^0(x)$). Sometimes, if no confusion arises, the index G is omitted.

A connected component of a graph $\tilde{G} = (X, \tilde{U})$ is its induced subgraph such that, if it contains a vertex x, it does not contain a vertex which is not connected with x by a route. The connected components of a graph $\tilde{G}$ determine a partition of set X into subsets. The graph consisting of a single connected component is called *connected*.

The number of arcs (edges) incident to a vertex in a graph is *the degree* of a vertex. If a graph is directed, then the number of arcs leaving (entering) a vertex is called *the outdegree* (*the indegree*, respectively) of this vertex.

A vertex of a directed graph is called: (i) *initial*, if its indegree is zero; (ii) *terminal* or *a leaf*, if its outdegree is equal to zero; or (iii) *isolated*, if its degree is zero. The vertices which are not terminal are called *intermediate*. *The adjacency matrix* of a graph $\tilde{G} = (X, \tilde{U})$ is a square (0,1)-matrix $\|m_{ij}\|$ of order $|X|$ such that $m_{ij} = 1$ if and only if $(x_i, x_j) \in \tilde{U}$.

1.3. In what follows, we mainly consider *directed circuit-free graphs*.

The vertices of any circuit-free graph $G = (X, U)$ can be distributed *by ranks* (*levels*). The first rank includes all initial vertices. Eliminating the vertices of the first rank from the graph (together with the incident arcs) yields some subgraph. If this subgraph is not empty, assign the set of all its initial vertices to the second rank of the original graph. The procedure is repeated until each vertex of the original graph is given a rank. If the graph is given by its adjacency matrix, then distributing of its vertices by ranks can be implemented in at most $O(|X|^2)$ time.[1]

The height of a vertex of a circuit-free directed graph is the length of the longest path from x to a leaf. The height of a terminal vertex is 1.

A chain $C = (x_1, x_2,..., x_n)$ is a directed graph $G = (X, U)$ such that $X = \{x_1, x_2,..., x_n\}$, $U = \{(x_1, x_2), (x_2, x_3),..., (x_{n-1}, x_n)\}$. The vertex x_1 is *the beginning* and x_n is *the end* of chain C. In a chain C, a vertex x is said to be on the left of a vertex y if the path from x_1 to x is shorter than that from x_1 to y.

The chain $C' = (x_0, x_1, x_2,..., x_n)$ is said to be obtained from a chain $C = (x_1, x_2,..., x_n)$ by joining the vertex x_0 from the left. The operation of joining a vertex from the right is defined similarly. If $C_1 = (x_1, x_2,..., x_r)$, $C_2 = (y_1, y_2,..., y_s)$ are such

[1] Here and throughout the book $O(f(x))$ denotes a function $g(x)$ for which there exists a constant C such that $\limsup_{x\to\infty} \dfrac{g(x)}{f(x)} = C$.

chains that the sets of their vertices do not intersect, then (C_1, C_2) denotes the chain $C = (x_1, x_2, \ldots, x_r, y_1, y_2, \ldots, y_s)$.

A graph is called *an outtree* (denoted by $\mathcal{T}^+$) if it is connected, has a single initial vertex (called *a root*), and any other vertex has exactly one direct predecessor.

A subtree with a root x of an outtree $\mathcal{T}^+$ is a subgraph of the graph $\mathcal{T}^+$ induced by the vertex x and by all its successors. For a vertex x of an outtree $\mathcal{T}^+$, the subtrees with the roots that are the direct successors of vertex x are called *subtrees of vertex x*.

A graph is called *an intree* (denoted by $\mathcal{T}^-$) if the opposite orientation of all its arcs gives an outtree. A subtree with a root x of an intree $\mathcal{T}^-$, as well as subtrees of a vertex x of an intree $\mathcal{T}^-$ are defined analogously.

By definition, an isolated vertex is an outtree and an intree at the same time.

A graph $\mathcal{T}$ will be called a *tree-like* graph (or *a forest*) if each of its connected components is either an outtree or an intree).

An arc (x, y) of a graph is *transitive* if in this graph there is a path which goes from vertex x to vertex y and does not contain the arc (x, y). A graph G is *transitive* if, for any of its vertices x, y such that $x \in B_G(y)$, graph G contains arc (x, y). The transitive graph $\bar{G} = (X, \bar{U})$ is called *a transitive closure* of a graph $G = (X, U)$ if $U \subseteq \bar{U}$ and any arc $(x, y) \in \bar{U} \backslash U$ is transitive.

A graph $G = (X, U)$ is called *a parallel composition* of graphs $G_1 = (X_1, Y_1)$ and $G_2 = (X_2, Y_2)$ such that $X_1 \cap X_2 = \emptyset$, if $X = X_1 \cup X_2$ and $U = U_1 \cup U_2$. This is denoted by $G = G_1pG_2$.

A graph $G = (X, U)$ is a *series composition* of graphs $G_1 = (X_1, U_1)$ and $G_2 = (X_2, U_2)$ such that $X_1 \cap X_2 = \emptyset$, if $X = X_1 \cup X_2$ and $U = U_1 \cup U_2 \cup X_1' \times X_2'$, where X_1' is the set of all terminal vertices of the graph G_1 and X_2' is the set of all initial vertices of the graph G_2. This is denoted by $G = G_1sG_2$.

A graph G is said to be obtained by implementing *parallel* or *series composition* of graphs G_1 and G_2 if $G = G_1pG_2$ or $G = G_1sG_2$, respectively.

Let G^t denote the graph obtained from a graph G by the successive removal of all transitive arcs of G. A graph G is called *series-parallel* if the graph G^t can be obtained by successive implementation of series and parallel compositions of single-vertex graphs $G^{(i)} = (x_i, \emptyset)$, $x_i \in X$, $i = 1, 2, \ldots, |X|$. A single-vertex graph is series-parallel by definition. It can be easily seen that any *tree-like graph is series-parallel.*

A graph G is called *decomposable* if the graph G^t can be represented as a series or parallel composition of two graphs. If otherwise, G is called *non-decomposable*. Let the graphs $G_1, G_2, \ldots, G_m$ be such that the graph G can be obtained from them by successive implementation of $m-1$ operations of series and parallel composition. Then these graphs are called *decomposition components* of G, and a so-called *decomposition tree* of the graph G can be constructed which shows how G can be obtained from $G_1, G_2, \dots, G_m$ by successive implementation of composition operations.

A *decomposition tree* $T(G)$ of a graph G is a *binary* outtree (each intermediate vertex has exactly two direct successors) with m terminal vertices. The graphs $G_1, G_2, \dots, G_m$ are associated with the terminal vertices. The intermediate vertices called *operational*, and these are associated with the indices (s or p) of the operations of series or parallel composition, respectively. A decomposition tree $T(G)$ of a graph G is defined iteratively. Suppose that either $G = G_1'sG_2'$ or $G = G_1'pG_2'$, and the trees $T(G_1')$ and $T(G_2')$ have been constructed. Then construct a new vertex O to be the direct predecessor of the roots of the trees $T(G_1')$ and $T(G_2')$. The vertex O is given the index s (if $G = G_1'sG_2'$) or p (if $G = G_1'pG_2'$). The vertex O is now the root of the constructed tree $T(G)$. If either G_1' or G_2' is a non-decomposable graph, its decomposition tree is assumed to consist of a single vertex associated with the corresponding graph G_1' or G_2', respectively.

Since the operation of series composition is not commutative ($G_1'sG_2' \neq G_2'sG_1'$), the method of representing a tree $T(G)$ should be specified. We assume that tree $T(G)$ is embedded in the plane such that the vertices of one rank, and only these, are placed at the same horizontal level. The root of the tree $T(G_1')$ is assumed to be on the right of the root of the tree $T(G_2')$ with respect to the observer located at the operational vertex O.

Note that *any decomposition tree with m terminal vertices has exactly* $m-1$ *operational vertices*. This can be easily proved by induction with respect to m.

In what follows, we do not distinguish between the terminal vertices of a tree $T(G)$ and the corresponding decomposition components of graph G, since no confusion arises.

Let us consider *the procedure for reconstructing the graph G by its decomposition tree* $T(G)$. Find, in $T(G)$, an operation vertex O adjacent to two terminal vertices G_1' and G_2'. Remove the vertices G_1' and G_2' with the incident arcs from $T(G)$, and associate the vertex O either with $G_1'sG_2'$ or with $G_1'pG_2'$ depending on what operational index is assigned to the vertex O. The resulting decomposition tree $T'(G)$ of graph G has one terminal vertex less than the previous one. Repeat the described procedure until the decomposition tree is found that consists of a single vertex. The graph corresponding to this vertex is, in fact, the graph G.

In the following, a decomposition tree $T(G)$ of a graph G is not distinguished from a decomposition tree of the graph G^t. It is obvious that the graph reconstructed by the tree $T(G)$ can differ from the graph G by the transitive arcs. If $G = G^t$, then G is uniquely reconstructed by $T(G)$.

A decomposition tree $T(G)$ of a graph G is called *complete* if non-decomposable graphs correspond to all its terminal vertices.

The definition of a series-parallel graph implies that single-vertex graphs (any of which can be just considered as an element of set X) correspond to the terminal vertices of its complete decomposition tree. Note that the construction of the complete decomposition tree of a series-parallel graph G requires at most $O(|X|^2)$ time (see, e.g., [429, 430]).

Figure 1.1 gives an example of series-parallel graph G and its complete decomposition tree $T(G)$.

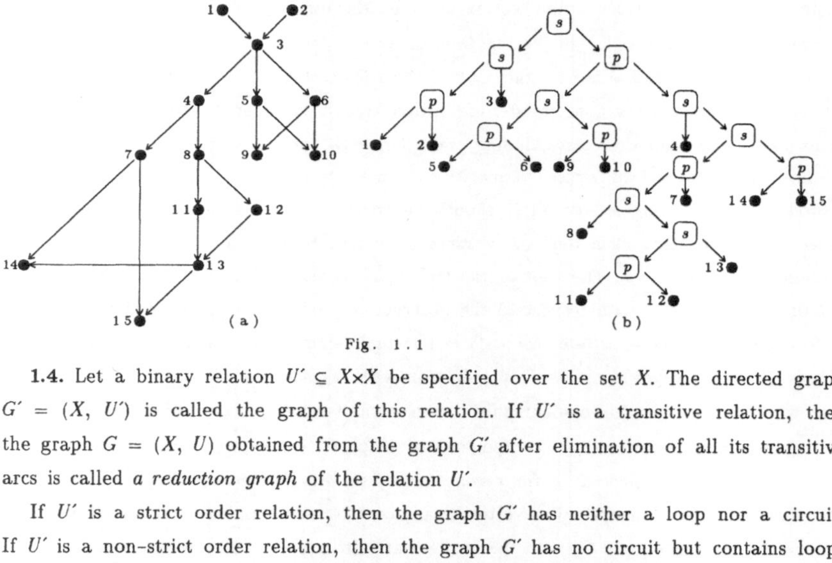

Fig. 1.1

1.4. Let a binary relation $U' \subseteq X \times X$ be specified over the set X. The directed graph $G' = (X, U')$ is called the graph of this relation. If U' is a transitive relation, then the graph $G = (X, U)$ obtained from the graph G' after elimination of all its transitive arcs is called *a reduction graph* of the relation U'.

If U' is a strict order relation, then the graph G' has neither a loop nor a circuit. If U' is a non-strict order relation, then the graph G' has no circuit but contains loops (x, x) for all $x \in X$. The graph G' of a quasi-order includes loops (x, x) for all $x \in X$ and may have circuits. The graph G' of a pseudo-order may contain circuits and loops (x, x) but not necessarily for all $x \in X$. In any case, the graph G' is transitively closed, i.e., for a path from a vertex x to a vertex y, it also contains the arc (x, y).

Let a total pseudo-order relation $\Longrightarrow$ be defined over set X. An element $x^0 \in X$ is called a *minimal* element of set X (with respect to $\Longrightarrow$), if the relation $x \Longrightarrow x^0$ holds for any $x \in X$. An element $x^0 \in X$ is a *maximal* element of X, if $x^0 \Longrightarrow x$ holds for any $x \in X$.

If G is the reduction graph of a total pseudo-order relation and x^0 is a minimal (or a maximal) element of set X with respect to $\Longrightarrow$, then, for any vertex $x \in X$, G contains a path from x to x^0 (or from x^0 to x). It is clear that the same element $x^0 \in X$ may be minimal and maximal at the same time. In particular, if G is a circuit, then any element $x \in X$ is both minimal and maximal.

Let a strict order relation $\longrightarrow$ be defined over set X, and $G = (X, U)$ be the reduction graph of this relation. It is obvious that, if $x \longrightarrow y$, then G contains a path from the vertex x to the vertex y. If $(x, y) \in U$, then we use notation $x \rightarrowtail y$. In this case, we have $x \longrightarrow y$ and no $z \in X$ exists such that $x \longrightarrow z$ and $z \longrightarrow y$. If none of the relations $x \longrightarrow y$ and $y \longrightarrow x$ holds (i.e., there is neither path from x to y nor from y to x in G), then we write $x \sim y$ and call the elements x and y *incomparable*.

In what follows, the notation $x \overset{G}{\longrightarrow} y$, $x \overset{G}{\rightarrowtail} y$, and $x \overset{G}{\sim} y$ is frequently used along with $x \longrightarrow y$, $x \rightarrowtail y$, and $x \sim y$, respectively. Here, the index G shows that the graph G is the reduction graph of the relation $\longrightarrow$.

It is clear that $y \in A_G(x)$ if and only if $x \longrightarrow y$, and $y \in B_G(x)$ if and only if $y \longrightarrow x$. Similarly, $y \in A_G^0(x)$ if and only if $x \overset{G}{\rightarrowtail} y$, and $y \in B_G^0(x)$ if and only if $y \overset{G}{\rightarrowtail} x$. We use the notation $E_G(x)$ to denote the set of all those $y \in X$ for which $x \overset{G}{\sim} y$. If no confusion arises, the index G is omitted.

If a graph G is given by its adjacency matrix, then finding the set $A_G(x)$ (or the set $B_G(x)$) requires at most $O(|X|^2)$ time. To see this, observe that to obtain the set $A_G(x)$ (or the set $B_G(x)$), it is sufficient to make at most $|X|$ steps. In the first step, direct successors (or direct predecessors) of the vertex x are to be found. In any subsequent step, all direct successors (or direct predecessors) of each vertex determined in the previous step are to be found. In order to find all direct successors (or direct predecessors) of a vertex, it is necessary to find all unit entries in the corresponding row (or column) of the adjacency matrix. Finding the unit entries requires $O(|X|)$ time.

An element $x^0 \in X$ is called *a minimal* element of set X (with respect to $\longrightarrow$) if there is no $x \in X$ such that $x^0 \longrightarrow x$. An element $x^0 \in X$ is *a maximal* element of set X if there is no $x \in X$ such that $x \longrightarrow x^0$. In the graph G, the terminal vertices correspond to the minimal elements, while the initial vertices correspond to the maximal elements. It is evident that the element corresponding to an isolated vertex of the graph G is both minimal and maximal. We denote the set of all minimal (maximal) elements of set X by X^- (or X^+, respectively).

In many situations, it is convenient to represent the reduction graph $G = (X, U)$ of a strict order relation $\longrightarrow$ by the lists of predecessors and/or successors of its vertices. In particular, this representation allows us to find the set of all minimal (maximal) elements of set X with respect to $\rightarrow$ in at most $O(|X|)$ time. Removing a certain minimal (maximal) element from X also requires at most $O(|X|)$ time.

If graph G is given by the list of predecessors, then its vertices are numbered by the integers 1, 2,..., $|X|$, and two one-dimensional arrays Q_B and S_B are constructed. The array Q_B contains $|X|$ elements, its kth element equal b_k shows how many direct predecessors the vertex k has. The array S_B consists of $\sum_{l=1}^{|X|} b_l$ elements, and its positions $\sum_{l=1}^{k-1} b_l+1, \sum_{l=1}^{k-1} b_l+2,..., \sum_{l=1}^{k} b_l$ contain the numbers of direct predecessors of the vertex k taken in an arbitrary order.

If a graph G is given by the list of successors, the arrays Q_A and S_A are constructed. The kth position of the array Q_A is equal to the number $a_k = |A_G^0(k)|$, while the positions $\sum_{l=1}^{k-1} a_l+1, \sum_{l=1}^{k-1} a_l+2,..., \sum_{l=1}^{k} a_l$ of the array S_A contain the numbers of the direct successors of the vertex k.

For finding the set of all minimal (maximal) elements of set X, it suffices to know array Q_A (or Q_B). An element k is minimal (maximal) if and only if the kth position of array Q_A (or of array Q_B) contains zero.

Let the elements of X (as well as vertices of G) be numbered by the integers 1, 2,..., $|X|$, and an element k be a minimal element of set X. To remove this element from X (maintaining the adopted representation form of the remaining subset), it suffices to know the arrays Q_B and S_B as well as the array Q_A. In this case, find the set of all direct predecessors of a kth element by scanning the positions $\sum_{l=1}^{k-1} b_l+1, \sum_{l=1}^{k-1} b_l+2,..., \sum_{l=1}^{k} b_l$ of the array S_B, and, for each found element j, decrease the number a_j located in the jth position of the array Q_A by 1. Mark the element k, for example, by placing the number (-1) in the kth position of the array Q_A. Removing a certain maximal element from set X can be implemented in a similar way; in this case, it suffices to know the arrays Q_B, Q_A, and S_A.

It is evident that removing a minimal (maximal) element from X followed by an appropriate correction of the array Q_A requires at most $O(|X|)$ time.

Note that, if graph G is an outtree (or an intree), it can be represented only by the array Q_A (or the array Q_B). To see this, suppose that the elements of set X are numbered in the following way. The root of a tree is given the number 1. Let G contain r_ν vertices of the νth rank. Then the vertices of the second rank are numbered by 2, 3,..., r_2+1. The vertices of the third rank are numbered by r_2+2, r_2+3,..., r_2+r_3+1, all successors of the vertex 2 being numbered first, followed by all successors of the vertex 3, and so on. The vertices of the other ranks are successively numbered in a similar way. In such a numbering, the direct successors of a vertex k (if it is not terminal) are the vertices with the numbers $\sum_{l=1}^{k-1} a_l+2$, $\sum_{l=1}^{k-1} a_l+3,\dots, \sum_{l=1}^{k} a_l+1$, and a direct predecessor of this vertex is a vertex with the number r such that $\sum_{l=1}^{r-1} a_l+2 \le k \le \sum_{l=1}^{r-1} a_l+1$.

For an intree, the numbering can be implemented similarly. It starts from the root followed by numbering all of vertices with the height equal to two, then the vertices with the height three are numbered, and so on. In this case, the direct predecessors of a vertex k (if it is not initial) are the vertices with the numbers $\sum_{l=1}^{k-1} b_l+2$, $\sum_{l=1}^{k-1} b_l+3,\dots$, $\sum_{l=1}^{k} b_l+1$, and its direct successor is a vertex with the number r that $\sum_{l=1}^{r-1} b_l+2 \le k \le \sum_{l=1}^{r} b_l+1$.

An outtree (an intree) can also be represented by a single array S_B (or S_A) because each vertex different form the root has exactly one direct predecessor (or direct successor). It is not necessary to use a special numbering of the vertices. Such representation, however, does not suit for finding a minimal (in the case of an intree) or a maximal (in the case of an outtree) element of set X.

1.5. Let $X = \{x_1, x_2,\dots, x_m\}$. *A permutation of the length* r of the elements of set X is an ordered sequence of r elements of this set. We suppose that $r \le m$ and there is no repetition in a permutation. If $r = m$, a permutation of the elements of set X is called *complete*. If $r < m$, a permutation is *partial*.

A symbolic expression for this construction is $\pi_r = (x_{i_1}, x_{i_2},\dots, x_{i_r})$ or $\pi_r = ([1], [2],\dots, [r])$, where x_{i_k} or $[k]$ is the element located at the kth position from the left in permutation π_r. If the nature of the elements of set X is immaterial, it is often more convenient to deal with the numbers of elements rather than with the elements themselves. In this case, $\pi_r = (i_1, i_2,\dots, i_r)$, where π_r is a permutation of the length r of elements of the set $\{1, 2,\dots, m\}$.

Sometimes, a permutation of the elements of set $X' \subseteq X$ is denoted by $\pi_{X'}$. The length of this permutation is equal to $|X'|$. If $\pi = \pi_{X'}$, then $\{\pi\}$ denotes the set X', i.e., $\{\pi\} = X'$. If $\pi' = (x_{i_1}, x_{i_2},\dots, x_{i_p})$, $\pi'' = (x_{j_1}, x_{j_2},\dots, x_{j_q})$ are permutations of the set X elements and $\{\pi'\} \cap \{\pi''\} = \emptyset$, then $\pi = (\pi', \pi'')$ denotes the permutation $(x_{i_1}, x_{i_2},\dots, x_{i_p}, x_{j_1}, x_{j_2},\dots, x_{j_q})$.

Let a strict order relation $\rightarrow$ be defined over a set X, and $G = (X, U)$ be the reduction graph of this relation. A permutation $\pi = (x_{i}, x_{i},\dots, x_{i})$ is called *feasible* with respect to the relation $\rightarrow$ or, equivalently, with respect to the graph G, if for all k and l, $1 \leq k$, $l \leq r$, the condition $x_{i_k} \rightarrow x_{i_l}$ implies that $k < l$.

2. Balanced 2-3-Trees

The material presented in this section can be used to develop effective algorithms for solving a wide range of discrete optimization problems including scheduling problems.

The data structure described below allows the implementation of a number of operations on a totally pseudo-ordered finite set X in at most $O(\log|X|)$ time. Such operations include, in particular, finding a minimal (or a maximal) element of set X with respect to a defined pseudo-order, deleting an element from X, and finding the union of subsets of set X.

2.1. An outtree is called a 2-3-*tree* if either two or three arcs leave from each of its non-terminal vertices. An outtree is *balanced* if all paths from the root to terminal vertices are of equal length. *The height of a tree* is the height of its root.

Let us estimate *the height of a balanced* 2-3-*tree*. Any 2-3-tree with m terminal vertices has at most $m-1$ intermediate vertices. In fact, the maximal number of intermediate vertices is attained if each intermediate vertex has exactly two leaving arcs. In this case, the number of intermediate vertices is equal to $m-1$, which can be easily verified by induction with respect to m.

The given definitions imply that there must be at least 2^{k-1} vertices of rank k in a balanced 2-3-tree. Let m be the number of the terminal vertices, and q denote the number of the ranks of a balanced 2-3-tree. Then $2m-1$ is the maximal number of the vertices and $2m-1 \geq \sum_{k=1}^{q} 2^{k-1} = 2^q-1$, which yields $q \leq \log 2m$. Therefore, the height of a balanced 2-3-tree does not exceed $1+\log m$.

2.2. Let a total pseudo-order relation $\Longrightarrow$ be specified over a set $X = \{x_1, x_2,..., x_m\}$. Let $(X', \Longrightarrow)$ denote a subset $X' \subset X$ such that the pseudo-order defined over set X is maintained over X'.

Let $T_b(X)$ denote a balanced 2-3-tree with m terminal vertices, each of which is in one-to-one correspondence with an element of set X. Thus, the terminal vertices of tree $T_b(X)$ may be assumed to be numbered by the integers from 1 to m. In what follows, we do not distinguish between the elements of set X and the terminal vertices of $T_b(X)$. The intermediate vertices of tree $T_b(X)$ are assumed to be numbered by the integers in the set $\{m+1,\ m+2,\ldots,\ 2m-1\}$.

An intermediate vertex v of tree $T_b(X)$ is connected by paths with some terminal vertices. The set of all such terminal vertices is denoted by X_v. Assign two labels v_{min} and v_{max} to vertex v, where v_{min} is the number of one of the minimal elements of the set $(X_v, \Longrightarrow)$, and v_{max} is the number of one of the maximal elements of this set.

A balanced 2-3-tree $T_b(X)$ can be conveniently represented by a table (see Table 2.1) consisting of five rows and at most $2m-1$ columns. The first row of the table contains the numbers of the vertices of 2-3-tree $T_b(X)$. The kth cell of the second row contains the number of the direct predecessor of vertex k. In the kth cell of the third and fourth rows, the labels k_{min} and k_{max}, respectively, are shown. The numbers of direct successors of vertex k (there are at most three such vertices) are written in the kth cell of the fifth row. Table 2.1 corresponds to the situation in which $m = 7$ (the procedure for constructing a balanced 2-3-tree is considered later).

Table 2.1

I	The number of a vertex	1	2	3	4	5	6	7	8	9	10	11
II	The number of direct predecessor	8	8	9	9	10	10	10	11	11	11	
III	The label v_{min}								1	4	7	7
IV	The label v_{max}								1	3	6	3
V	The number of direct successors								1 2	3 4	5 6 7	8 9 10

It is obvious that a balanced 2-3-tree can be specified by filling the first and second rows of the table. The fifth row is an auxiliary one, and this is used for labeling the vertices as well as for implementing some operations on 2-3-trees.

2.3. Given a set X, the following *procedure for constructing a tree* $T_b(X)$ can be applied. We split this procedure into several stages. The number of the stages is equal to the height of $T_b(X)$ minus 1. At the first stage, the first m cells of the second row of the table are filled in. Put the number $m+1$ in cells 1 and 2, fill cells 3 and 4 with the number $m+2$, and so on. If m is even, put the number $m+m/2$ in cell m. If m is odd, fill cell m (as well as cells $m-2$ and $m-1$) with the number $m+(m-1)/2$.

Let $\lfloor x \rfloor$ denote the largest integer not exceeding x. At the second stage, the cells from $m+1$ up to $m+\lfloor m/2 \rfloor$ are filled. Put the number $m+\lfloor m/2 \rfloor+1$ in cells $m+1$ and $m+2$, fill cells $m+3$ and $m+4$ with the number $m+\lfloor m/2 \rfloor+2$, and so on. If the number of cells filled at the second stage is odd, the last three cells contain the same number. At the last stage of this process, there are either two or three cells to be filled. At this stage, the number placed in the cells of the second row is the number of the root of the tree. As can be easily seen, the table obtained this way uniquely specifies a balanced 2-3-tree (with no labels).

Tables 2.2 and 2.3 give examples of filling the first two rows for $m = 11$ and $m = 8$, respectively. The dotted lines separate the stages. In the first example, the root of the tree is vertex 19, while in the second example, the root is vertex 15.

Table 2.2

I	1	2	3	4	5	6	7	8	9	10	11	12	13	14	15	16	17	18	19
II	12	12	13	13	14	14	15	15	16	16	16	17	17	18	18	18	19	19	

Table 2.3

I	1	2	3	4	5	6	7	8	9	10	11	12	13	14	15
II	9	9	10	10	11	11	12	12	13	13	14	14	15	15	

The fifth row of the table can be filled simultaneously with the second row: while a number l is placed in the kth cell of the second row, the number k is placed in the lth cell of the fifth row. The first m cells of the fifth row of the table representing a tree $T_b(X)$ are, obviously, empty.

The labeling of the intermediate vertices of a balanced 2-3-tree is implemented level by level starting from the level $q-1$ (here q is the height of the tree). At the $(q-l)$th level, $1 \le l \le q-1$, take the number of an arbitrary minimal (or maximal) element of the set $(X', \Longrightarrow)$ as the label v_{min} (respectively, v_{max}) for each vertex v. Here, for $l = 1$, we have $X' = X_v$, while for $l \ge 2$ set X' is the set of the elements whose numbers are the minimal (or maximal) labels of the direct successors of vertex v. Since $|X'| \le 3$, at most two comparisons are required for finding label v_{min} (or v_{max}). For finding all direct successors of a vertex v, the fifth row of the table can be used.

It is obvious that the vertices $m+1$, $m+2$,..., $m+\lfloor m/2 \rfloor$ belong to the $(q-1)$th level, the vertices $m+\lfloor m/2 \rfloor+1$, ..., $m+\lfloor m/2 \rfloor+\lfloor m/4 \rfloor$ belong to the $(q-2)$th level, etc.

As can easily be seen, the implementation of this procedure for constructing the tree $T_b(X)$ and labeling its vertices requires *at most O(m) time.*

Table 2.1 gives a balanced 2-3-tree for the set $X = \{1, 2, 3, 4, 5, 6, 7\}$ assuming that the relation $\Longrightarrow$ is defined over this set in the following way (here we write (x_i, x_j) instead of $x_i \Longrightarrow x_j$): (1, 2), (1, 4), (1, 5), (1, 6), (1, 7), (2, 1), (2, 4), (2, 5), (2, 6), (2, 7), (3, 1), (3, 2), (3, 4), (3, 5), (3, 6), (3, 7), (4, 5), (4, 6), (4, 7), (5, 7), (6, 5), (6, 7).

Note that, in general, a totally pseudo-ordered set may contain more than one minimal and more than one maximal element. This implies that the values of the labels v_{min} and v_{max} are not uniquely specified. For example, in the eighth cells of the third and fourth rows of Table 2.1 (or in any of them), the number 2 could be placed.

For finding either a minimal or a maximal element of set $(X, \Longrightarrow)$ represented by the tree $T_b(X)$, it suffices to check either the label v^0_{min} or v^0_{max} of the root v^0.

2.4. We now consider *the procedure for finding the union of two subsets of the X,* provided that each of them is represented by a balanced 2-3-tree. Let X_1, X_2 be non-empty subsets of a set X such that $X_1 \cap X_2 = \emptyset$, $|X_1| = m_1$, $|X_2| = m_2$, and $T_b(X_1)$ and $T_b(X_2)$ be balanced 2-3-trees of the heights q_1 and q_2 representing these sets. Without loss of generality, we may assume that $q_1 \geq q_2$. To find the union of sets X_1 and X_2 represented by the trees $T_b(X_1)$ and $T_b(X_2)$, we construct the tree $T_b(X_1 \cup X_2)$.

If $q_1 = q_2$, then for constructing $T_b(X_1 \cup X_2)$ it suffices to introduce a new vertex v^0 and make the roots of $T_b(X_1)$ and $T_b(X_2)$ direct successors of v^0. The constructed tree is a balanced 2-3-tree with the root v^0. It is not difficult to find the labels v^0_{min} and v^0_{max} by the corresponding labels of the roots of the trees $T_b(X_1)$ and $T_b(X_2)$.

If $q_1 > q_2$, then find in $T_b(X_1)$ a vertex v of the height q_2+1. If v has only two direct successors, make the root of the tree $T_b(X_2)$ its third successor. In the obtained balanced 2-3-tree, recalculate the labels of the vertex v and of all its predecessors (there are at most $\log m_1$ of them). If the chosen vertex v has three direct successors, take one of them, say, vertex v', and remove the arc (v, v') from $T_b(X_1)$. As a result, we obtain two balanced 2-3-trees: what is left from $T_b(X_1)$ (let us denote this tree by $T_b(X_1'))$ and the subtree with the root v'. Unite the latter tree and the tree $T_b(X_2)$ (their heights are equal) and denote the resulting tree by $T_b(X_2')$. The labels are not recalculated until the union of $T_b(X_1)$ and $T_b(X_2)$ is found. Note that the height of $T_b(X_2')$ is equal to q_2+1. Attempt to unite the trees $T_b(X_1')$ and $T_b(X_2')$ in the way described. If $q_1 > q_2+1$, then the direct predecessor of vertex v in $T_b(X_1')$ can be chosen as a vertex of the height q_2+2.

There may be at most $\log m_1$ such "attempts" to unite the trees. When $T_b(X_1 \cup X_2)$ is constructed, find the new labels of vertex v and those of all its predecessors, as well as the labels of the new vertices. The total number of labels to be recalculated does not exceed $O(\log m_1)$.

2.5. We now consider how to implement the described process of constructing $T_b(X_1 \cup X_2)$ followed by correcting the labels assuming that a balanced 2-3-tree is represented by the table. The implementation of this process in at most $O(\log m_1)$ time requires that both trees $T_b(X_1)$ and $T_b(X_2)$ are represented by a common table, and that the numbers of the roots of these trees are known.

Let $v^{(1)}$ and $v^{(2)}$ be the roots of the trees $T_b(X_1)$ and $T_b(X_2)$, respectively. We assume that both trees are represented by a common table having $2m-1$ columns. Without loss of generality, we may assume that the empty columns of the table are at the right-hand side, and m' is the number of the first of them.

The heights q_1 and q_2 of the trees $T_b(X_1)$ and $T_b(X_2)$ can be determined by finding a path from the root of a tree to some of its terminal vertices, which can easily be done using the fifth row of the table. Therefore, finding q_1 and q_2 requires at most $O(\log m_1)$ time.

If $q_1 = q_2$, then for constructing $T_b(X_1 \cup X_2)$ it suffices to place the number m' in cells $v^{(1)}$ and $v^{(2)}$ of the second row, to put the numbers $v^{(1)}$ and $v^{(2)}$ in cell m' of the fifth row, and then to fill the cells m' of the third and fourth rows as usual.

If $q_1 > q_2$, then using the fifth row (moving from the root of the tree $T_b(X_1)$), find a vertex v of height q_2+1. If in cell v of the fifth row there are two numbers, place the number $v^{(2)}$ in this cell as the third one. Then correct the labels of vertex v and those of all its predecessors. If the chosen cell of the fifth row contains three numbers, then remove one of them and place it along with the number $v^{(2)}$ in cell m' of the fifth row. Put the number m' in cell $v^{(2)}$ of the second row, and replace the number m' by v in the cell corresponding to the removed vertex. Keep the labels unchanged. Taking m' as the root of the second tree (the root of the first tree is $v^{(1)}$), attempt to unite these trees. The direct predecessor of vertex v (to be found in cell v of the second row) can be taken as a vertex of the first tree having the height q_2+2.

While constructing the tree $T_b(X_1 \cup X_2)$, store the numbers of the vertices whose labels are to be either defined or corrected. Note that such vertices are the vertex v, all its predecessors, as well as the new vertices added to the second tree. The process of constructing labels starts from the vertex with the minimal number.

It can be easily seen that the described method of implementing the procedure for finding the union of two balanced 2-3-trees followed by an appropriate correction of the labels requires at most $O(\log m_1)$ time.

2.6. Let us consider *the procedure for deleting an element* from a set X represented by a balanced 2-3-tree $T_b(X)$. Let $x_{i^0} \in X$ be the element to be deleted. The procedure for deleting x_{i^0} from X constructs a balanced 2-3-tree $T_b(X \backslash x_{i^0})$.

In the tree $T_b(X)$, let v_l^0 denote the predecessor of the vertex i^0 corresponding to the element x_{i^0} such that the path from v_l^0 to i^0 has the length $l+1$. If q is the height of the tree $T_b(X)$, then v_{q-1}^0 is the root of this tree.

If the vertex v_1^0 (i.e., the direct predecessor of the vertex i^0) has three direct successors, then for constructing the tree $T_b(X \backslash x_{i^0})$ it suffices to delete the vertex i^0 from the tree $T_b(X)$ (along with the arc (v_1^0, i^0)) and to correct the labels of the vertices $v_1^0, v_2^0, \ldots, v_{q-1}^0$.

If the tree $T_b(X)$ is given by the table, this can be implemented as follows. Remove the number i^0 from cell v_1^0 of the fifth row of the table and the number v_1^0 from the cell i^0 of the second row. Determine the new labels of the vertices $v_1^0, v_2^0, \ldots, v_{q-1}^0$. It is clear that, in this case, the deletion of an element from the set X requires at most $O(\log m)$ time, where $m = |X|$.

Suppose that v_1^0 has only two direct successors, i^0 and i'. Then the tree $T^{(1)}$, arising from $T_b(X)$ after the vertex i^0 has been deleted, is no longer a 2-3-tree (in this tree, the vertex v_1^0 has only one direct successor). Thus, additional transformations are required to obtain $T_b(X \backslash x_{i^0})$. In this case, correcting the labels starts only after these transformations are completed.

In the tree $T^{(1)}$, find a direct successor of the vertex v_2^0, say, vertex v'. If v' has three direct successors, make one of them (say, vertex i'') a direct successor of the vertex v_1^0 after the arc (v', i'') is removed. It is evident that the resulting tree is a balanced 2-3-tree. Correct the labels of the vertices $v', v_1^0, v_2^0, \ldots, v_{q-1}^0$.

If v' has two direct successors, make i' a direct successor of the vertex v' and delete the vertex v_1^0 from the tree. If, in the constructed tree $T^{(2)}$, the vertex v_2^0 has two direct successors, then $T^{(2)}$ is the desired tree $T_b(X \backslash x_{i^0})$, and we only have to correct the labels of the vertices $v', v_2^0, \ldots, v_{q-1}^0$. If otherwise, then $T^{(2)}$ is not a balanced 2-3-tree, and the vertex v_2^0 is the only its intermediate vertex having one direct successor. Transform $T^{(2)}$ in a similar way as for tree $T^{(1)}$. The only difference is that now the vertex v_3^0 acts as the vertex v_2^0, the vertex v_2^0 acts as v_1^0, and the vertex v' acts as the vertex i' (some direct successor v'' of the vertex v_3^0 plays the role of the vertex v'). If, in the tree $T^{(3)}$ obtained from $T^{(2)}$ by these transformations, the vertex v_3^0 has only one direct successor, transform $T^{(3)}$ in a similar way, and so on. It may turn out that a tree is obtained with the root having only one direct successor. In this case, the root is deleted and the vertex v_{q-2}^0 becomes the new root.

The implementation of this procedure for deleting an element requires storing the vertices such that their direct successors have been changed by the described transformations. There are at most q such vertices and their predecessors. Thus, the procedure for deleting an element can be implemented in at most $O(\log m)$ time.

2.7. A permutation $\pi = (x_{i_1}, x_{i_2},..., x_{i_m})$ of the elements of a totally pseudo-ordered set X is called *non-increasing* (or *non-decreasing*) with respect to $\Longrightarrow$ if, for any ν and μ, $\nu = 1, 2,..., m$, $\mu = 1, 2,..., m$, the condition $\nu < \mu$ implies that $x_{i_\nu} \Longrightarrow x_{i_\mu}$ (or $x_{i_\mu} \Longrightarrow x_{i_\nu}$).

We present *the procedure for constructing a monotone (either non-increasing or non-decreasing) with respect to* $\Longrightarrow$ *permutation* of the elements of set X. There is one-to-one correspondence between the permutations of the elements of set X and the permutations of their numbers. Therefore, we may talk about non-increasing or non-decreasing (with respect to a total pseudo-order defined over set X) permutations of the numbers of the elements of set X .

To find a non-increasing permutation $(i_1, i_2,..., i_m)$ of the numbers of the elements of set X it suffices to know a balanced 2-3-tree $T_b(X)$ in which each intermediate vertex v is given only one label v_{max}. Define i_1 to be equal to the number of the element of set X that is the label of the root of tree $T_b(X)$. Remove i_1 from tree $T_b(X)$ and, without transforming the resulting tree $T^{(1)}$ into a balanced 2-3-tree, find the new labels of its vertices. Define i_2 to be equal to the number of the element of set X that is the label of the root of tree $T^{(1)}$, and so on.

Since the height of the tree $T_b(X)$ does not exceed $1+\log m$, to find a non-increasing permutation of the elements of set X requires at most $O(m\log m)$ time.

A non-decreasing permutation of the numbers of the elements of set X can be found in a similar way using a balanced 2-3-tree in which each intermediate vertex v is given one label v_{min}.

2.8. It may be that solving a problem does not require finding a maximal (a minimal) element of a pseudo-ordered set X, but involves application one of the following procedures: given x', $x'' \in X$, find a maximal element of the set $X_1 = \{x \in X \mid x' \Longrightarrow x\}$ or a minimal element of the set $X_2 = \{x \in X \mid x \Longrightarrow x''\}$. To implement these procedures in $O(\log m)$ time, we need to modify the data structure under consideration.

A balanced 2-3-tree is called *ordered* if for any two of its vertices v and v' of the same rank either $v_{max} \Longleftarrow v'_{min}$ or $v_{min} \Longrightarrow v'_{max}$ holds. For constructing an ordered balanced 2-3-tree it suffices to find a non-decreasing (or non-increasing) permutation of the elements of set X and then to use the procedure described in Section 2.3. The latter procedure has to be implemented in such a way as if the elements of X were renumbered according to this permutation.

It is obvious that the construction of an ordered balanced 2-3-tree requires $O(\log m)$ time.

The search for a maximal element of the set X_1 using the ordered tree $T_b(X)$ is executed as follows (here we assume that the relation $x' \Longrightarrow x'$ does not hold, since otherwise, x' is the desired element). Find a direct successor of the root, say, vertex $v^{(1)}$, such that $v^{(1)}_{min} \Longleftarrow x'$ and $v^{(1)}_{max} \Longrightarrow x'$ (or $v^{(1)}_{max} = x'$). Then, find a direct successor of the vertex $v^{(1)}$, say, vertex $v^{(2)}$, which satisfies analogous conditions, and so on, until the desired element of set X_1 is found. At some step in the described procedure, it may turn out that the required vertex does not exist. In this case, among the vertices to be considered at this step, there exists a vertex v' such that $v'_{min} = x'$. If the number of vertices under consideration is two, and these are v' and v'', then v''_{max} is the desired element. If the number of the vertices under consideration is equal to three, and these are v', v'', and v''', then two cases are possible: (1) either $v''_{max} \Longleftarrow x'$ and $v'''_{max} \Longleftarrow v''_{min}$, or $v''_{max} \Longleftarrow x'$ and $v'''_{min} \Longrightarrow x'$; (2) either $v''_{max} \Longleftarrow x'$ and $v''_{max} \Longleftarrow v'''_{min}$, or $v'''_{max} \Longleftarrow x'$ and $v''_{min} \Longrightarrow x'$. In the first case, the desired element is v''_{max}, while in the second case, the desired element is v'''_{max}.

A minimal element of the set X_2 can be found in a similar way.

It is easy to verify that the described procedure for finding a maximal element of the set X_1 or a minimal element of the set X_2 requires $O(\log m)$ time.

For $X' \subset X$, let $T_b(X')$ be an ordered balanced 2-3-tree. We present the procedure for constructing an ordered tree $T_b(X' \cup x^0)$, where $x^0 \in X \backslash X'$.

1. Find a maximal element x' of the set $X'' = \{x \in X' \mid x \Longleftarrow x^0\}$.

2. If the direct predecessor v of the vertex x' has two direct successors, then make x^0 the third successor. Correct the labels of the vertex v and of all of its predecessors in the usual way.

3. If v has three direct successors x', x'' and x''', then find a non-decreasing

permutation of the elements x^0, x', x'' and x'''. Make the vertices corresponding to the first two elements of the permutation direct successors of the vertex v (if they are not) and correct the labels of v. Introduce a new vertex v^0, and make the vertices corresponding to the last two elements of the permutation direct successors of v^0. If the direct predecessor v' of the vertex v has two direct successors, make v^0 the third one and correct the labels. If v' has three direct successors, find a permutation π of these successors and of the vertex v^0 such that $v''_{min} \Longrightarrow v'''_{max}$ for any vertices v'' and v''' with v'' being on the left of v''' in π. Then the procedure is similar to the case of the elements x', x'', x''' and x^0. It may be that $X'' = \emptyset$. In this case, take the vertex $\bar{v}_{min}$ as the vertex x' where $\bar{v}$ is the root of the tree $T_b(X')$.

The procedure for constructing an ordered tree $T_b(X' \cup x^0)$ can be implemented in at most $O(\log|X'|)$ time.

Finally, we present the procedure for constructing an ordered tree $T_b(X \backslash x^0)$ where $x^0 \in X$. If the direct predecessor of the vertex x^0 has three direct successors, then for constructing $T_b(X \backslash x^0)$ it suffices to delete from $T_b(X)$ the vertex x^0 together with the entering arc and to correct the labels. If the number of the direct successors is equal to two, we can follow the procedure for deleting an element from a set described in Section 2.6, keeping the tree ordered whenever one of the direct successors of a vertex is "transferred" to another vertex.

Sometimes, the relation $\Longrightarrow$ is defined over a set X by associating each element $x_i \in X$ with a real number α_i. Here $x_i \Longrightarrow x_j$ if and only if $\alpha_i \geq \alpha_j$ (in other situations if and only if $\alpha_i \leq \alpha_j$). In this case, the problem arises of finding a minimal element of the set $X' = \{x_i \in X | \alpha_i \geq \beta\}$ or that of finding a maximal element of the set $X'' = \{x_i \in X | \alpha_i \leq \beta\}$. Here β is a given real number and, in the general case, β need not belong to the set $\{\alpha_1, \alpha_2, \ldots, \alpha_m\}$. For solving such problems it is also convenient to use an ordered balanced 2-3-tree assigning the corresponding α_i along with the labels to intermediate vertices of the tree.

3. Polynomial Reducibility of Discrete Problems. Complexity of Algorithms

The theory of polynomial reducibility is of great importance for understanding the nature of those difficulties which arise in solving a wide range of discrete (both extremal and decision) problems. Many decision problems which have been traditionally considered as hard (e.g., the problem of determining whether a graph is Hamiltonian, the problem of the existence of a complete subgraph (a clique) with a prescribed number of vertices in a given graph, etc.) are, in fact, closely related. The existence or non-existence of an efficient algorithm for solving at least one of these so-called *NP*-complete problems implies the existence (or non-existence) of such an algorithm for all other problems. Here, an algorithm is said to be efficient if its running time is bounded by a polynomial of the input length of the problem.

A similar situation also occurs for many extremal problems (belonging to the class of so-called *NP*-hard problems). The existence of an efficient algorithm for solving at least one of the *NP*-hard problems implies the existence of such an algorithm for any *NP*-hard problem. The traveling salesman problem is an example of an *NP*-hard problem.

3.1. To introduce the concepts of an algorithm and that of its time complexity formally, we need a certain computation model. A so-called Turing machine serves as a convenient model of this type. We start with some auxiliary definitions.

An alphabet is an arbitrary finite set of characters called *letters. A word* in this alphabet is a finite non-empty sequence of the letters. *The length of a word* is the number of letters it includes (each letter is counted as many times as it appears in a word).

A deterministic Turing machine (DTM) consists of a tape, a control device, and a read-write head.

The tape is divided into *cells* and is potentially infinite from both sides. The cells are numbered ..., -2, -1, 0, 1, 2,... . Any cell can be in one of the states, each of which is in one-to-one correspondence with a letter of the alphabet $\mathfrak{C}$ (called *an external alphabet*). The total number of states is finite. The letter $c_0 \in \mathfrak{C}$ is called *blank symbol.*

At any time, *the control device* is in one of the states (number of which is finite), each denoted by a letter of *the inner alphabet* $\mathfrak{Q}$ and called *an inner machine state.* Note that $\mathfrak{C} \cap \mathfrak{Q} = \emptyset$. Two special states are distinguished: the initial state denoted by q_0, and the final state denoted by q_f.

The read-write head of the machine can move along the tape and scan exactly one of its cells at a time. The head can read a symbol in the cell and, if necessary, replace it by another.

As a rule, *an input alphabet* $\mathfrak{D}$ is defined as a proper subset of $\mathfrak{C}$. In particular, $c_0 \notin \mathfrak{D}$.

One step of a DTM consists of performing all or some of the actions listed below, depending on the control device state and the state of the tape cell being scanned:

(1) change the inner state of the machine;

(2) change the state of the cell being scanned;

(3) move the read-write head one cell to the left (L) or one cell to the right (R), or leave it at its current place (S).

In what follows, we do not distinguish between the state of a tape cell (or the state of the control device) and the corresponding letter of alphabet $\mathfrak{C}$ (or of alphabet $\mathfrak{D}$).

As a mathematical object, a DTM is determined by a string of the form ($\mathfrak{Q}$, $\mathfrak{C}$, $\mathfrak{D}$, δ, c_0, q_0, q_f). Here, δ is the mapping of some non-empty subset of the set $\mathfrak{Q}\times\mathfrak{C}$ (which does not contain pairs of the form (q_f,c_i) for $c_i \in \mathfrak{C}$) to the set $\mathfrak{Q}\times\mathfrak{C}\times\{L, R, S\}$. The mapping δ is called *the transition function.*

A state of a DTM is determined by:

(a) the sequence c_{i_1}, c_{i_2},..., c_{i_p} of the states of all tape cells, $c_{i_r} \in \mathfrak{C}$, $r = 1$, 2,..., p, (all the cells on the left of the cell having the state c_{i_1} and on the right of the cell having the state c_{i_p} are empty and are omitted from the sequence, $c_{i_1} \neq c_0$, $c_{i_p} \neq c_0$);

(b) the inner state $q \in \mathfrak{Q}$ of the control device at a given time;

(c) the state c_{i_k} of the cell being scanned;

At a time, a state of a DTM is uniquely determined by *a description* which is a word $c_{i_1}c_{i_2}...qc_{i_k}...c_{i_p}$ in alphabet $\mathfrak{C} \cup \mathfrak{Q}$. Here, the symbol $q \in \mathfrak{Q}$ precedes the symbol denoting the state of the cell being scanned at this time. The state of a DTM determined by the description of the form $c_{i_1}c_{i_2}...q_fc_{i_k}...c_{i_p}$ is called *final* (the control device is in the final state q_f). The machine stops if it reaches the final state.

Each step of a DMT can be considered as a transition of the machine from one state to another that is uniquely determined by the transition function δ. It is assumed that the machine can be driven to any prescribed state.

A DTM can be used for *processing words* written in alphabet $\mathfrak{Q}$. Let $c_1c_2...c_n$ be a word written in that alphabet. Drive the machine to the state determined by the description $q_0c_1c_2...c_n$, and let it start processing. If, after some finite number of steps, the control device reaches the state q_f, the machine stops. In this case, the DTM is said *to accept* the initial word. *The result of processing* the initial word is the word obtained from the description of the DTM in the final state, the characters c_0 and q_f being deleted.

If, after some steps, the machine reaches a state $c_1'c_2'...q'c_k'...c_\nu'$ where $q' \neq q_f$ and the pair $q'c_k'$ does not belong to the domain of δ, then the machine also stops. In this case, however, the result is not determined, and the machine *does not accept* the initial word. Situations are possible in which the machine, having started in a certain state, never stops. In this case, the result of processing is not determined either.

A non-deterministic Turing machine (NDTM) is specified by a string of the form $(\mathfrak{Q}, \mathfrak{C}, \mathfrak{D}, \Delta, c_0, q_0, q_f)$ where the symbols $\mathfrak{Q}$, $\mathfrak{C}$, $\mathfrak{D}$, c_0, q_0, q_f have the same meaning as for a DTM. The difference is in *the transition function* Δ being the mapping of some non-empty subset of the set $\mathfrak{Q}\times\mathfrak{C}$ (which does not contain pairs of the form (q_f, c_i), $c_i \in \mathfrak{C}$) to a set of subsets of the set $\mathfrak{Q}\times\mathfrak{C}\times\{L, R, S\}$.

As in the case of a DMT, *a state of a NDTM* is determined by the sequence of all states of the cells, the inner state of the control device at a given time, and the state of the cell being scanned.

The main difference between a non-deterministic Turing machine and a deterministic one is that one step of a NDTM may change the given state of the machine to any of several possible states, while, for a DTM, the number of possible new states is at most one. Therefore, having started operating in the same initial state twice, a NDTM may come to some final state at one time, and to another final state at another time, or it may never stop.

A NDTM is said *to accept* a word a if there exists a finite sequence of machine steps which drives the machine to a final state from the initial state determined by the description q_0a. If there is no such a sequence, the machine *does not accept* this word.

Let a given DTM accept a word a. The number $t(a)$ of steps of the machine required to reach the final state is called *the running time of a DTM* for processing word a.

If a NDTM accepts a word a, then, in general, there exist several sequences which drive the machine from a state q_0a to a final state.

The running time of a NDTM processing a word a is the length of the shortest sequence of machine steps which drives it from the state q_0a to a final state. The running time of a NDTM is also denoted by $t(a)$.

The function $T(n)$ is called *the time complexity* of a Turing machine (either deterministic or non-deterministic) if $T(n) = \max\{t(a) \mid a \in A_n\}$, where A_n is the set of all words of the length n this machine accepts. If the time complexity $T(n)$ of a Turing machine does not exceed some polynomial of n, then this machine is said to have a *polynomial-time complexity*.

Note that a DMT is very similar to modern computers (e.g., a transition function of a DMT can be viewed as a computer program). On the other hand, a non-deterministic machine is an absolutely abstract concept. The concept of the running time of a NDTM is also abstract. The latter concept can be given a convenient and visual interpretation by using a so-called "oracle" machine. For any feasible word, an oracle "knows" the shortest sequence of steps driving the machine to a final state. Before making a step, the machine applies to the oracle, which indicates in which of the states possible at this step the machine comes. If the NDTM does not accept the word to be processed, the oracle "lies", i.e., it indicates any of the possible states randomly. Under such an interpretation, determination of the running time of a NDTM is similar to that of a DTM, assuming that the oracle answer time is zero. We stress once again: both the NDTM and the oracle are abstract objects. The oracle can be considered as some unknown program. Being connected to the NDTM, the oracle changes it into a deterministic machine. The main difficulty is that, as a rule, we either fail to build an oracle for a NDTM or this is a very complicated program of a low speed.

3.2. *A language* in a given alphabet is a non-empty set of words of this alphabet.

A language A is called *feasible* for a given Turing machine (either deterministic or non-deterministic) if the machine accepts any word of language A. If the machine accepts those and only those words that belong to language A, the machine is said *to recognize language A*.

The class $\mathcal{P}$ is the set of all languages for each of which there is a recognizing DTM of a polynomial-time complexity. The set of all languages for each of which there exists a recognizing NDTM of a polynomial-time complexity is called *the class* $\mathcal{NP}$.

Since a deterministic machine can be viewed as a special case of a non-deterministic machine, it follows that $\mathcal{P} \subseteq \mathcal{NP}$. However, whether $\mathcal{P}$ is a proper subset of $\mathcal{NP}$ or $\mathcal{P} = \mathcal{NP}$, is still an open question. Note that the conjecture that the classes $\mathcal{P}$ and $\mathcal{NP}$ do not coincide is quite popular.

For a deterministic Turing machine M, let $M(a)$ denote the result of processing by machine M a word a written in the input alphabet of this machine. If M does not accept the word a, then $M(a)$ is not determined.

A language A^0 is called *polynomially reducible*[1] to a language A if there exists a deterministic Turing machine M which satisfies the following conditions. The machine M is of a polynomial-time complexity and processes the words written in the alphabet of language A^0 into the words written in the alphabet of language A so that $a \in A^0$ if and only if

---

[1]The presented definition of polynomial reducibility corresponds to the one given by R. M. Karp [74]; another definition given by S. A. Cook [82] is more general; however, for our purposes the presented definition is sufficient.

$M(a) \in A$.

The definition requires the existence of a DTM which recognizes some language A' such that $A^0 \subset A'$. The result of processing the words which do not belong to A' is not determined.

It is evident that the relation of polynomial reducibility defined over a set of languages is transitive.

A language A is *NP-complete* if $A \in \mathcal{NP}$ and any language in $\mathcal{NP}$ is polynomially reducible to A.

Theorem 3.1. *If a language A^0 is polynomially reducible to a language A and $A \in \mathcal{P}$, then $A^0 \in \mathcal{P}$.*

Proof. A deterministic Turing machine M that recognizes language A^0 can be constructed by implementing a series composition of the DTM M_1, which reduces A^0 to A, and the DTM M_2, which recognizes language A. A word a^0 written in the alphabet of language A^0 after processing by M_1 either becomes a word a written in the alphabet of language A or the result of this processing is not determined. In the latter case, it is clear that $a \notin A^0$. In the former case, the word a is an input for M_2. If M_2 accepts a, then the definition of polynomial reducibility implies that $a^0 \in A^0$. Otherwise, $a^0 \notin A^0$. The time complexity of the constructed machine M does not exceed the polynomial $p_1(n)+p_2(n+p_1(n))$. Here, polynomials p_1 and p_2 are the time complexity functions of the machines M_1 and M_2, respectively, and $n+p_1(n)$ is an upper bound on the maximal possible length of the result of processing a word of the length n by machine M_1. This proves the theorem.

It follows from Theorem 3.1 that *the existence of a recognizing DTM of a polynomial-time complexity for some NP-complete language implies the existence of such a machine for any language in the class $\mathcal{NP}$.*

3.3. In what follows, we concentrate on decision and extremal combinatorial problems.

A *decision problem* is a problem of recognizing properties of a certain object, to which the answer "yes" is to be given if and only if the object has these properties.

An extremal combinatorial problem can be introduced as follows. A function $F(x)$, $x \in X'$, is defined over a finite set X'. Given a subset X of set X', find an element x^0 either such that $F(x^0) = \min\{F(x) \mid x \in X\}$ (a minimization problem) or such that $F(x^0) = \max\{F(x) \mid x \in X\}$ (a maximization problem).

Decision problems of recognizing properties of objects and language recognition problems are closely related. We may encode all possible inputs of a decision problem using the words in an appropriate alphabet and consider the initial problem as a problem of recognizing the language consisting of all words corresponding to the answer "yes".

A decision problem belongs to the class $\mathcal{P}$ (or $\mathcal{NP}$) if the associated language belongs to $\mathcal{P}$ (or $\mathcal{NP}$, respectively).

Let us consider two alphabets: *binary* $\mathfrak{B}$ = {0, 1, -, [,], (,), ,} and *unary* $\mathfrak{U}$ = {1, -, [,], (,), ,}. To encode the inputs of a decision problem, we use the words specified in one of the alphabet $\mathfrak{B}$ or $\mathfrak{U}$ and determined in the following way.

In alphabet $\mathfrak{B}$:

(1) a word that is an integer k is a binary representation of k (if k is negative, then the sign "-" is used);

(2) if λ is a word that represents an integer k, then the word $[\lambda]$ is used as the name, e.g., this can be used as the number of a vertex of a graph or the number of a job;

(3) if $\lambda_1, \lambda_2, \ldots, \lambda_m$ are the words that represent objects $\Lambda_1, \Lambda_2, \ldots, \Lambda_m$, then the word $(\lambda_1, \lambda_2, \ldots, \lambda_m)$ represents the sequence $(\Lambda_1, \Lambda_2, \ldots, \Lambda_m)$.

In alphabet $\mathfrak{U}$, the words used for encoding the problem inputs are determined in a similar way as in alphabet $\mathfrak{B}$, the only difference being that now an integer k is represented not in the binary but in the *unary* form, i.e. k is represented by the word 11...1 consisting of k unit digits.

A decision problem is said to be defined in alphabet $\mathfrak{B}$ (in alphabet $\mathfrak{U}$) if the associated language is determined in alphabet $\mathfrak{B}$ (in alphabet $\mathfrak{U}$, respectively).

A decision problem B is called *NP-hard* if any problem $A \in \mathcal{NP}$ defined in alphabet $\mathfrak{B}$ is polynomially reducible to it.

A problem B is *NP-complete* if it is *NP*-hard and $B \in \mathcal{NP}$.

The usage of Turing machine as a formal model for an intuitive concept of an algorithm has a number of advantages for introducing the definitions and proving the statements. However, in what follows, we talk about algorithms (either deterministic or non-deterministic) implying any possible formalization of this concept (the Turing machines, normal Markovian algorithms or programs written in an algorithmic language).

The concept of *an elementary operation* depends on the way in which the concept of an algorithm is formalized. For a Turing machine, this is a machine step; for the algorithms designed to be run on a computer, elementary operations are such computer operations as addition, multiplication, comparison of two numbers, writing or reading a number with a known address, etc.

The time complexity of an algorithm is a function of the problem input length defined similarly to the time complexity of a Turing machine. The differences are as follows.

First, we use the concept of an elementary operation rather than that of a machine step. Second, now the final state of an algorithm is either a situation when the answer "yes" is obtained (in the case of a decision problem) or an element $x^0 \in X$ delivering an extremum to the objective function is found (in the case of an extremal problem).

An algorithm is called *polynomial-time* if its time complexity does not exceed some polynomial of the length of the problem input encoded in the alphabet $\mathfrak{B}$.

The concept of the NP-hardness, defined for decision problems, can also be used for the extremal combinatorial problems.

Associate an extremal problem B with the following decision problem B': determine whether there exists an element x' in a given set X such that $F(x') \leq y$ (or $F(x') \geq y$ in the case of a problem of maximization) for a given real number y. It is clear that, if x^0 is a solution of problem B, then an element $x' \in X$ such that $F(x') \leq y$ exists if and only if $F(x^0) \leq y$.

Hence, we may talk about *polynomial reducibility of a decision problem B' to the corresponding extremal problem B*. Similarly, due to the transitivity of the polynomial reducibility relation, we may talk about *the polynomial reducibility of an arbitrary decision problem A to a given extremal problem B* via reducibility of A to the decision problem B'.

An extremal problem B is called *NP-hard* if any decision problem $A \in \mathcal{NP}$ defined in alphabet $\mathfrak{B}$ reduces to it in polynomial time. To prove the NP-hardness of an extremal problem, it suffices to prove the NP-hardness of the corresponding decision problem.

It is obvious that *the existence of a polynomial-time algorithm for solving a NP-hard problem implies that each problem of class $\mathcal{NP}$* (including each NP-complete problem) *is solvable in polynomial time.*

The fact that a problem belongs to the class of NP-hard problems is one of its most important characteristics. Assuming that the $\mathcal{P} \neq \mathcal{NP}$ conjecture is correct, the existence of a polynomial-time algorithm for solving any NP-hard problem becomes impossible. Therefore, the NP-hardness of a problem is one of strong arguments to justify such approaches as the design of approximation or heuristic algorithms, applying enumeration schemes (such as the branch-and-bound method), as well as studying special cases of a problem.

3.4. While the problems can be divided into NP-hard and polynomially solvable (i.e., having polynomial-time algorithms for their solution), the NP-hard problems, in turn, can be subdivided into NP-hard in the strong sense problems and those having pseudopolynomial-time solution algorithms.

The concept of $\mathcal{NP}$-hardness in the strong sense is of great importance in complexity analysis of a large number of problems. First, to prove that a problem B is $\mathcal{NP}$-hard in the strong sense, it suffices to construct a so-called pseudopolynomial (rather than polynomial) reduction of an $\mathcal{NP}$-hard (in the strong sense) problem A to problem B. Second, the fact that a problem is $\mathcal{NP}$-hard in the strong sense is the evidence [56] that no fast ε-approximation algorithm exists for its solution (unless $\mathcal{P} = \mathcal{NP}$).

Let b be an input of a decision problem B. This input can be encoded either in alphabet $\mathfrak{B}$ or in alphabet $\mathfrak{U}$. If all inputs of problem B are encoded in alphabet $\mathfrak{B}$ (or alphabet $\mathfrak{U}$), the problem is said to be determined in alphabet $\mathfrak{B}$ (or alphabet $\mathfrak{U}$, respectively). It is clear that the length of input b depends on the alphabet used. Let $L_{\mathfrak{B}}(b)$ (or $L_{\mathfrak{U}}(b)$) denote the length of the input b in alphabet $\mathfrak{B}$ (or alphabet $\mathfrak{U}$).

An algorithm for solving problem B is said to be *pseudopolynomial-time* if, for an input b of the problem, its running time does not exceed some polynomial of $L_{\mathfrak{U}}(b)$. Note that *any polynomial-time algorithm is also pseudopolynomial-time.*

For a problem B and a polynomial p, let B_p denote such a subproblem of problem B that for any of its inputs b the inequality $L_{\mathfrak{U}}(b) \leq p(L_{\mathfrak{B}}(b))$ holds. Note that the only difference between a subproblem and the original problem is that, for *a subproblem*, the set of all inputs is a subset of the set of all inputs of the original problem.

It is obvious that any pseudopolynomial-time algorithm for solving problem B is a polynomial-time algorithm for solving problem B_p. Therefore, unless $\mathcal{P} = \mathcal{NP}$, neither a polynomial-time nor a pseudopolynomial-time algorithm for solving problem B exists if problem B_p is $\mathcal{NP}$-hard.

A decision problem B is called *$\mathcal{NP}$-hard in the strong sense* if there exists such a polynomial p that a problem B_p is $\mathcal{NP}$-hard. If in this case, $B \in \mathcal{NP}$, then problem B is *$\mathcal{NP}$-complete in the strong sense.*

An extremal combinatorial problem is called $\mathcal{NP}$-hard in the strong sense if the corresponding decision problem is $\mathcal{NP}$-hard in the strong sense.

To prove the $\mathcal{NP}$-hardness of a decision problem B, it suffices (due to the transitivity of the polynomial reducibility relation) to show that some $\mathcal{NP}$-hard problem A^0 determined in alphabet $\mathfrak{B}$ is polynomially reducible to it. A problem A^0 used for proving the $\mathcal{NP}$-hardness of other problems is called *standard.* A similar approach is used to prove the $\mathcal{NP}$-hardness in the strong sense.

An input of a decision problem determined either in alphabet $\mathfrak{B}$ or in alphabet $\mathfrak{U}$ can be considered as a word in the corresponding alphabet ($\mathfrak{B}$ or $\mathfrak{U}$, respectively). If Φ is a word-processing algorithm, then $\Phi(a)$ stands for the result of word a processed by this algorithm.

A problem A is said to be *pseudopolynomially reducible* to a problem B if a deterministic algorithm Φ exists for processing the inputs of problem A into the inputs of problem B such that

(1) the answer "yes" corresponds to an input a of the problem A if and only if the answer "yes" corresponds to the input $\Phi(a)$;

(2) the running time of algorithm Φ does not exceed some polynomial of $L_{\mathfrak{U}}(a)$;

(3) there exist such polynomials p' and p'' that for any input of problem A the relation $p'(L_{\mathfrak{U}}(a)) \geq L_{\mathfrak{U}}(\Phi(a))$ and $p''(L_{\mathfrak{B}}(\Phi(a))) \geq L_{\mathfrak{B}}(a)$ hold.

Theorem 3.2. *Let a problem A be NP-hard in the strong sense. If there is pseudopolynomial reduction of problem A to problem B, then the problem B is NP-hard in the strong sense.*

Proof. Since problem A is NP-hard in the strong sense, it follows that there exists a polynomial p such that the problem A_p is NP-hard. We may assume that p has only positive coefficients. Otherwise, a polynomial p_0 with positive coefficients exists such that for any non-negative x the relation $p_0(x) \geq p(x)$ holds, and problem A_p is a subproblem of problem A_{p_0}.

Let Φ be an algorithm which implements the pseudopolynomial reduction of problem A to problem B, while p' and p'' be polynomials from the definition of the pseudopolynomial reduction. As above, we may assume that the coefficients p' and p'' are positive. We show that there exist both a polynomial q and a subproblem B_q of the problem B such that: (1) for any input b of problem B_p the relation $L_{\mathfrak{U}}(b) \leq q(L_{\mathfrak{B}}(b))$ holds; and (2) problem A_p is polynomially reducible to B_q. Define $q(x) = p'(p(p''(x)))$. For each input a of problem A_p, find input $\Phi(a)$ of problem B, and let B_q denote the subproblem of problem B determined by all such inputs $\Phi(a)$. For all $\Phi(a)$, where a is an input of problem A_p, the relation $L_{\mathfrak{U}}(\Phi(a)) \leq q(L_{\mathfrak{B}}(\Phi(a)))$ holds. In fact, the definition of the polynomials p' and p'' implies that

$$L_{\mathfrak{U}}(\Phi(a)) \leq p'(L_{\mathfrak{U}}(a)) \leq p'(p(L_{\mathfrak{B}}(a))) \leq p'(p(p''(L_{\mathfrak{B}}(\Phi(a))))) = q(L_{\mathfrak{B}}(\Phi(a))).$$

It is obvious that algorithm Φ implements pseudopolynomial reduction of problem A_p to problem B_q, however, for any input a of problem A_p the relation $L_{\mathfrak{U}}(a) \leq p(L_{\mathfrak{B}}(a))$ holds. Therefore, this reduction is polynomial.

Thus, problem B_q is NP-hard and problem B is NP-hard in the strong sense which proves the theorem.

If an NP-hard decision problem B is such that for any of its inputs b and some polynomial p the relation $L_{\mathfrak{U}}(b) \leq p(L_{\mathfrak{B}}(b))$ holds, then problem B is NP-hard in the strong sense.

Corollary 3.1. *Let a problem A be NP-hard in the strong sense. If A is polynomially reducible to a problem B, and there exists a polynomial p' such that for any input a of problem A the relation $p'(L_{\mathfrak{U}}(a)) \geq L_{\mathfrak{U}}(\Phi(a))$ holds where Φ is an algorithm which implements the reduction A to B, then problem B is NP-hard in the strong sense.*

In fact, the running time of algorithm Φ is bounded by a polynomial p of $L_{\mathfrak{B}}(a)$, and, hence, by a polynomial of $L_{\mathfrak{U}}(a)$. Therefore, Φ also implements the pseudopolynomial reduction of A to B (p can be taken as the polynomial p'').

4. Bibliography and Review

The terminology from set theory and binary relations theory mainly corresponds to the monographs by Kostrikin [81] and Schreider [186], while the terminology from graph theory corresponds to the monographs by Harary [163] and Berge [15]. The properties of series-parallel graphs are studied by Valdes et al. [429, 430] and Gordon [44]. In [429, 430], an algorithm for recognizing whether a graph is series-parallel is presented. That algorithm is essentially based on the results obtained in [248]. If a graph is series-parallel, the algorithm constructs its complete decomposition tree. The running time of the algorithm is linear with respect to the number of the vertices and the arcs of a graph in question. An algorithm for constructing a complete decomposition tree of an arbitrary circuit-free graph is given in [178].

Section 2 is based on the material presented in Section 4.9–4.12 of the monograph by Aho et al. [7]. In that monograph, the reader can find additional information on balanced 2-3-trees and other data structures. These and related topics are also discussed in the monographs by Knuth [77] and Reingold et al. [133].

In presenting the topics discussed in Section 3, the authors have mainly followed the monographs by Mal'tsev [106], Garey and Johnson [56], as well as the monograph [7]. The interested reader may find relevant information on history of this question in [56] (Sections 1.4, 1.5, 5.2). Theorem 3.1 is due to Karp [74], while Theorem 3.2 is given by Garey and Johnson [275].

CHAPTER 2

POLYNOMIALLY SOLVABLE PROBLEMS

This chapter discusses sequencing and scheduling problems for which efficient algorithms are known, i.e., the algorithms whose running time is bounded by a polynomial function of the problem input length.

The first eight sections consider systems with a single machine or several identical machines. Section 9 studies systems with uniform and unrelated parallel machines.

In Section 1, sufficient conditions are established for the existence of optimal schedules with no preemption at times different from the release dates. Section 2 presents the necessary and sufficient conditions for the existence of the schedules that are feasible with respect to given deadlines, and describes the algorithms for finding these schedules. It is assumed that the set of jobs is not ordered and that preemption is allowed.

The single-machine scheduling problem of minimizing the maximum cost (the minimax criterion) is considered in Section 3, and Section 4 studies the problem of minimizing the total cost (the minisum criterion) for job processing. Note that a number of polynomially solvable special cases of the latter problem are described in Chapter 3.

Sections 5 and 6 provide results for the problem of finding a time-optimal schedule for parallel machine processing the jobs of an ordered set, assuming that either the reduction graph of precedence relation is tree-like, or that the number of machines is equal to 2. In Section 5, the processing times for all jobs are assumed to be equal and preemption is forbidden, while in Section 6 the processing times are arbitrary but preemption is allowed. Section 6 also considers the case in which no precedence relation is defined over the set of jobs. Section 7 describes algorithms for finding a multi-processor schedule that is feasible with respect to the deadlines under precedence constraints, provided that the processing times are equal. Again, it is supposed that either the reduction graph of a precedence relation is tree-like or that the number of machines is equal to two. The problem of minimizing the maximal lateness for parallel identical machines is studied in Section 8.

Section 9 is devoted to the problems of minimizing the total and the maximal cost for parallel (either uniform or unrelated) machine processing.

1. Preemption

In this section, sufficient conditions are established for the existence of optimal schedules with no preemption at times different from the release dates.

1.1. The jobs of a set $N = \{1, 2,..., n\}$ are processed on M parallel identical machines. The release date of a job $i \in N$ is $d_i \geq 0$, its processing time is $t_i > 0$. The processing of each job may be interrupted and resumed at a later time on any available machine. It is supposed that preemption does not involve time or any other expenses, and the total length of time intervals in which a job i is processed is equal to t_i.

A partial order $\rightarrow$ is defined over set N to determine the sequencing constraints for job processing. Let G denote the reduction graph of relation $\rightarrow$.

A schedule $s = s(t) = \{s_1(t), s_2(t),..., s_M(t)\}$ that is feasible with respect to the defined precedence relation must satisfy the following conditions: if $i \rightarrow j$ and $s_L(t') = i$ for some L, then $s_H(t) \neq j$ for all $t < t'$ and for all $1 \leq H \leq M$. In particular, it follows that if, for some L, H, and Q, which need not to be distinct, and some $t' < t'' < t'''$, the relations $s_L(t') = s_H(t''') = i$ and $s_Q(t'') = j$ hold, then neither $i \rightarrow j$ nor $j \rightarrow i$ is possible, i.e., $i \sim j$.

Since preemption is allowed, it follows that there may exist $1 \leq i \leq n$, $1 \leq L \neq H \leq M$, and $0 \leq t' < t'' < t''' < \infty$ such that at least one of the following conditions holds: (1) $s_L(t') = s_L(t''') = i$ but $s_L(t'') \neq i$; (2) $s_L(t') = s_H(t'') = i$. If, in this case, $s_L(t'+\delta) \neq j$ for any sufficiently small $\delta > 0$, then the processing of job i is interrupted at time t'. The preemption of the job i at time t' allows the resumption of the processing of this job at the same time on another machine.

In what follows, it is assumed that the number of preemptions in the processing of each job is finite and, hence, the number of the break-points of each of the functions $s_L(t)$, $L = 1, 2,..., M$, is finite as well.

The quality of a schedule s is characterized by the value of a real function

$F(x) = F(x_1, x_2,..., x_n)$ evaluated at $x = \bar{t}(s)$, where $\bar{t}(s) = (\bar{t}_1(s), \bar{t}_2(s),..., \bar{t}_n(s))$ is the vector of the completion times of the jobs in schedule s. It is obvious that $\bar{t}_i(s)$ is the largest value of t such that there exists a $L \in \{1, 2,..., M\}$ for which $s_L(t) = i$. A feasible (with respect to $\twoheadrightarrow$) schedule with the smallest value of $F(x)$ is called an *optimal* schedule.

1.2. In a general case, an optimal schedule is a preemptive one. We present sufficient conditions for the existence of optimal schedules for single-machine processing with no preemption at times different from d_i, $i = 1, 2,..., n$.

Theorem 1.1. *If $M = 1$ and $F(x)$ is a non-decreasing (for $x > 0$) function, then there exists an optimal schedule without preemption at times different from d_i, $i = 1, 2,..., n$.*

Proof. To prove the theorem, it suffices to show that for any feasible (with respect to $\twoheadrightarrow$) schedule s there exists a feasible schedule s^* with no preemption at times different from d_i, $i = 1, 2,..., n$, and such that $F(\bar{t}(s^*)) \le F(\bar{t}(s))$.

1. Let $d^{(1)} < d^{(2)} <...< d^{(v)}$ be the sequence of pairwise distinct values of d_i, $i = 1, 2,..., n$. Let the time intervals $(d^{(1)}, d^{(2)}]$, $(d^{(2)}, d^{(3)}]$,..., $(d^{(v)}, \infty)$ be denoted by $\beta_1, \beta_2,..., \beta_v$, respectively.

We introduce the following operations of transforming a schedule.

Operation $O_1(t', t'', \hat{t})$, $0 \le t' < t'' < \hat{t}$. Denote $\Delta = t''-t'$. Define $s'(t) = s(t+\Delta)$ in the interval $(t', \hat{t}-\Delta]$; $s'(t) = s(t-(\hat{t}-t''))$ in the interval $(\hat{t}-\Delta, \hat{t}]$, and $s'(t) = s(t)$ in the remaining intervals. If s' is a schedule, then this is said to be obtained from schedule s as a result of applying operation $O_1(t', t'', \hat{t})$.

Operation $O_2(i, j, t', t'', \hat{t})$, $0 \le t' < t'' < \hat{t}$, $i, j \in N$, $i \ne j$, is used when $s(t) = i$ in the interval $(t', t'']$ and $s(t) = j$ at some $t > \hat{t}$. Let $(t^{(1)}, t^{(2)}]$ be one of the intervals in which $s(t) = j$, $\hat{t} < t^{(1)} < t^{(2)}$. If $t^{(2)}-t^{(1)} \ge t''-t'$, then define $s'(t) = j$ in the interval $(t', t'']$, $s'(t) = i$ in the interval $(t^{(1)}, t^{(1)}+(t''-t')]$, and $s'(t) = s(t)$ in the rest of the intervals. If $t^{(2)}-t^{(1)} < t''-t'$, then transform s into $\tilde{s}$ by setting $\tilde{s}(t) = j$ in the interval $(t', t'+t^{(2)}-t^{(1)}]$, $\tilde{s}(t) = i$ in the interval $(t^{(1)}, t^{(2)}]$, and $\tilde{s}(t) = s(t)$ in the rest of the intervals. Taking $\tilde{s}$ as s and choosing the interval $(t'+t^{(2)}-t^{(1)}, t'']$ as $(t', t'']$, repeat the transformations described above until either $\tilde{s}(t) = j$ in $(t', t'']$ or $\tilde{s}(t) \ne j$ for all $t > \hat{t}$. Denote the resulting function $\tilde{s}(t)$ by $s'(t)$. If s' is a schedule, then this is said to be obtained from schedule s as a result of applying operation $O_2(i, j, t', t'', \hat{t})$.

2. Without loss of generality, we may consider only such schedules s for which the

condition $s(t') = s(t'') = k \neq 0$, where $t', t'' \in \beta_l$, $1 \leq l \leq v$, and $t' < t''$, implies that $s(t) = k$ for all $t' \leq t \leq t''$.

In fact, if $s(t) = k$ in the subintervals $(t_1', t_2']$ and $(t_1'', t_2'']$ of interval β_l, where $t_2' < t_1''$ but $s(t) \neq k$ for t, $t_2' < t \leq t_1''$, then applying operation $O_1(t_1', t_2', t_2'')$ to schedule s gives a new schedule s' with $s'(t) = k$ in the interval $(t_1'' - (t_2' - t_1'), t_2'']$. The schedule s' is, obviously, feasible with respect to $\longrightarrow$ and $\bar{t}(s') \leq \bar{t}(s)$, therefore, $F(\bar{t}(s')) \leq F(\bar{t}(s))$.

3. If in some interval β_l, $l < v$, the processing of $m \geq 2$ jobs is interrupted (with resumption in subsequent intervals), then schedule s can be transformed into schedule s', having at least the same quality and being feasible with respect to $\longrightarrow$, such that, in s', the processing of at most $m-1$ jobs is interrupted in the interval under consideration.

Let the processing of jobs i and j be interrupted in the interval β_l, $l < v$, and $s(t) = i$ in the interval $(t_1', t_2'] \subset \beta_l$, while $s(t) = j$ in the interval $(t_1'', t_2''] \subset \beta_l$. Due to Item 2 of this proof, it follows that $s(t) = i$ for some $t > d^{(l+1)}$ and $s(t) = j$ for some $t > d^{(l+1)}$. Suppose, for example, that $\bar{t}_i(s) > \bar{t}_j(s)$.

If $t_1' \geq t_2''$, we apply operation $O_2(i, j, t_1', t_2', d^{(l+1)})$ to schedule s and obtain a new schedule s' in which either the processing of job j is completed in the interval β_l (i.e., $\bar{t}_j(s') \leq d^{(l+1)}$) or job i is not processed in this interval. Schedule s' is feasible with respect to $\longrightarrow$ and $F(\bar{t}(s')) \leq F(\bar{t}(s))$.

If $t_2' \leq t_1''$, then applying operation $O_1(t_1', t_2', t_2'')$ to schedule s gives a feasible (with respect to $\longrightarrow$) schedule $\tilde{s}$ with $F(\bar{t}(\tilde{s})) \leq F(\bar{t}(s))$, satisfying the conditions of the previous case.

4. If in the interval β_l, $l < v$, there is only one job j processed with preemption (the processing of j is resumed in some subsequent interval), then schedule s can be transformed into a new feasible (with respect to $\longrightarrow$) schedule s' with $F(\bar{t}(s')) \leq F(\bar{t}(s))$, either with no preemption in the interval β_l or with a preemption at time $d^{(l+1)}$.

Suppose that $s(t) = j$ in the interval $(t_1', t_2'] \subset \beta_l$, $t_2' < d^{(l+1)}$, and $s(t) = j$ at some $t > d^{(l+1)}$.

If either $s(d^{(l+1)}) = 0$ or the processing of some job is completed at time $d^{(l+1)}$, then applying operation $O_1(t_1', t_2', d^{(l+1)})$ to schedule s gives the desired schedule s'.

Let $s(t) = i$ in the interval $(t_1'', t_2'']$, $t_1'' < d^{(l+1)} < t_2''$. If $\bar{t}_i(s) = t_2''$, then apply operation $O_1(t_1', t_2', t_2'')$ to schedule s. If $\bar{t}_i(s) > t_2''$, two cases are possible: either $\bar{t}_i(s) < \bar{t}_j(s)$ or $\bar{t}_i(s) > \bar{t}_j(s)$. In the former case, apply operation $O_1(t_1', t_2', t_2'')$ to schedule s, and, if $t_2'' - (t_2' - t_1') < d^{(l+1)}$, then apply operation $O_2(j, i, t_2'' - (t_2' - t_1'), d^{(l+1)}, t_2'')$ to the obtained schedule. In the latter case, apply operation

$O_1(t_1', t_2', t_1'')$ to schedule s, and operation $O_2(i, j, t_1'', d^{(l+1)}, t_2'')$ to the resulting schedule. In any case, we obtain the desired schedule s'.

5. Since the number of preemptions is finite, we conclude that after a finite number of the described transformation steps the original schedule s can be transformed into a schedule s^* which either is non-preemptive or in this schedule preemptions happen only at times $d^{(l)}$, $l = 2, 3, \ldots, v$. Note that the intervals β_l should be considered one after another, moving from left to right.

Each of the obtained schedules is feasible with respect to $\rightarrow$ and has at least the same quality as the original one. This proves the theorem.

The theorem gives an *exact upper bound* (equal to $v-1$ where v is the number of distinct release dates d_i, $i = 1, 2, \ldots, n$) on the smallest number of preemptions in an optimal single-machine schedule.

We give an example in which an optimal schedule has exactly $v-1$ preemptions and there is no optimal schedule having fewer preemptions.

Define $M = 1$, $n = 3$, $d_1 = 0$, $d_2 = 1$, $d_3 = 2$, $t_1 = t_2 = t_3 = 2$, $F(x) = x_1+5x_2-20x_3$. In the case under consideration, there exists the unique optimal schedule presented in Fig. 1.1.

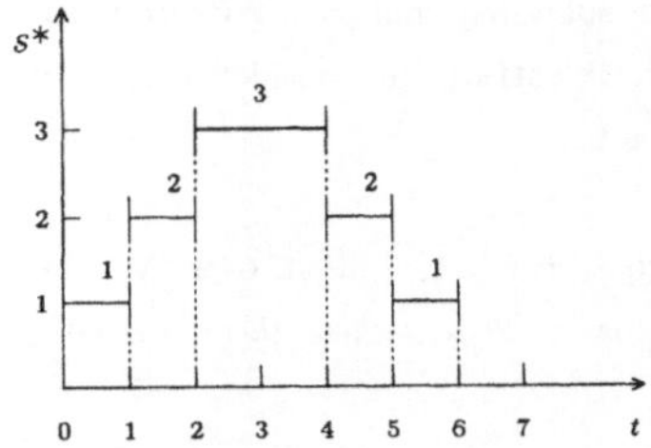

Fig. 1.1

In this schedule, the processing of job 1 is interrupted at time $t = d_2 = 1$, while the processing of job 2 is interrupted at time $t = d_3 = 2$. The processing of these jobs is resumed at times $t = 5$ and $t = 4$, respectively.

Corollary. *If $d_i = d$, $i = 1, 2, \ldots, n$, then for $M = 1$ and a non-decreasing function $F(x)$, there exists a non-preemptive optimal schedule.*

1.3. Now we consider the multi-machine case.

Let us introduce the concept of an e-quasi-concave function of n variables.

A function $F(x)$, $x = (x_1, x_2, \ldots, x_n)$ is called *concave* if for any vectors

$x^{(1)}$, $x^{(2)} \in E^n$ and a number λ, $0 \le \lambda \le 1$, the following inequality

$$F(\lambda x^{(1)} + (1-\lambda)x^{(2)}) \ge \lambda F(x^{(1)}) + (1-\lambda)F(x^{(2)}) \tag{1.1}$$

holds. Here E^n is the set of all n-dimensional vectors.

A function $F(x)$ is *quasi-concave* if for any vectors $x^{(1)}$, $x^{(2)} \in E^n$ and a number λ, $0 \le \lambda \le 1$, the inequality

$$F(\lambda x^{(1)} + (1-\lambda)x^{(2)}) \ge \min\{F(x^{(1)}),\ F(x^{(2)})\} \tag{1.2}$$

holds.

Let E_0^n be the set of all n-dimensional vectors e, whose components are the numbers 0, 1, and -1.

A function $F(x)$ is *e-quasi-concave* if for any vectors $x^{(1)} \in E^n$, $e \in E_0^n$, and any numbers α and λ, $\alpha > 0$, $0 \le \lambda \le 1$, inequality (1.2) holds where $x^{(2)} = x^{(1)} + \alpha e$.

By definition, a concave function is quasi-concave, and a quasi-concave one is e-quasi-concave as well. As can be easily seen, there exist e-quasi-concave functions which are not quasi-concave, and quasi-concave functions which are not concave.

Note that, since function $F(x)$ characterizes the quality of a schedule, it suffices to demand that it should possess some required properties on some subset of E^n, rather than on the entire set. In particular, it suffices to consider vectors $x > 0$ which do not contain more than M equal components.

Theorem 1.2. *If* $M \ge 2$, $d_i = d$, $i = 1, 2, \ldots, n$, $G = (N, \varnothing)$ *and* $F(x)$ *is a non-decreasing e-quasi-concave function (for* $x > 0$*), then there exists an optimal non-preemptive schedule.*

Proof. To prove the theorem, it suffices to show that for any schedule s there exists a non-preemptive schedule s^* such that $F(\bar{t}(s^*)) \le F(\bar{t}(s))$.

1. Without loss of generality, assume $d = 0$. Let us introduce the following operations of schedule transformation.

Operation $O_1(Q, R, t')$, $1 \le Q \ne R \le M$, $t' \ge 0$. The schedule $s'(t) = \{s_1'(t), s_2'(t), \ldots, s_M'(t)\}$ is said to be obtained from schedule s by applying operation $O_1(Q, R, t')$ if $s_L'(t) = s_L(t)$ for all $0 \le t < \infty$ and all $L \ne Q, R$; $s_Q'(t) = s_Q(t)$ and $s_R'(t) = s_R(t)$ in the interval $[0, t']$; $s_Q'(t) = s_R(t)$ and $s_R'(t) = s_Q(t)$ for all $t' < t < \infty$. This operation interchanges the machines Q and R starting at time t'. Since the machines are identical, it follows that $F(\bar{t}(s')) = F(\bar{t}(s))$.

Operation $O_2(Q, t', \pm a)$, $1 \le Q \le M$, $t' \ge 0$, $a > 0$. This operation either increases or reduces the idle time on some machine Q by a specific value a. Applying this operation to schedule s yields the family of functions $s'(t) = \{s_1'(t), s_2'(t),..., s_M'(t)\}$ where $s_L'(t) = s_L(t)$ for all $0 \le t < \infty$ and all $L \ne Q$; $s_Q'(t) = s_Q(t)$ in the interval $[0, t']$, $s_Q'(t) = 0$ in the interval $(t', t'+a]$ and $s_Q'(t) = s_Q(t-a)$ for all $t'+a < t < \infty$ if a is positive; $s_Q'(t) = s_Q(t)$ in the interval $[0, t'-a]$ and $s_Q'(t) = s_Q(t+a)$ for all $t'-a < t < \infty$ if a is negative. If s' is a schedule, then $F(\bar{t}(s')) \le F(\bar{t}(s))$ for a negative a.

Operation $O_3(Q, R, t', t'', a)$, $1 \le Q, R \le M$, $t'' \ge 0$, $t' \ge a > 0$. Consider a family of functions $s'(t) = \{s_1'(t), s_2'(t),..., s_M'(t)\}$, where $s_L'(t) = s_L(t)$ for all $0 \le t < \infty$ and all $L \ne Q, R$; $s_Q'(t) = s_Q(t)$ in the interval $[0, t'-a]$, and $s_Q'(t) = s_Q(t+a)$ for all $t'-a < t < \infty$; $s_R'(t) = s_R(t)$ in the interval $[0, t'']$; $s_R'(t) = s_Q(t+t'-t''-a)$ in the interval $(t'', t''+a]$ and $s_R'(t) = s_R(t-a)$ for all $t''+a < t < \infty$. If $Q = R$ and $t' \le t''$, define $s_Q'(t) = s_Q(t)$ in the intervals $[0, t'-a]$ and $(t''+a, \infty)$, $s_Q'(t) = s_Q(t+a)$ in the interval $(t'-a, t'']$ and $s_Q'(t) = s_Q(t+t'-t''-a)$ in the interval $(t'', t''+a]$. If s' is a schedule, this is said to be obtained from schedule s by applying operation $O_3(Q, R, t', t'', a)$.

Operation $O_4(Q, t', t'', \hat{t})$, $1 \le Q \le M$, $0 \le t' < t'' < \hat{t}$. Denote $\Delta = t''-t'$. Define $s_L'(t) = s_L(t)$ for all $1 \le L \ne Q \le M$ and for all $0 \le t < \infty$; $s_Q'(t) = s_Q(t+\Delta)$ in the interval $(t', \hat{t}-\Delta]$, $s_Q'(t) = s_Q(t-(\hat{t}-t''))$ in the interval $(t-\Delta, \hat{t}]$ and $s'(t) = s(t)$ in the remaining intervals. If s' is a schedule, this is said to be obtained from schedule s by applying operation $O_4(Q, t', t'', \hat{t})$.

2. Without loss of generality, we may consider only such schedules s for which either $s_L(t) \ne 0$ in some interval $[0, T_L]$ and $s_L(t) = 0$ for $t > T_L$, or $s_L(t) = 0$ for all $t \ge 0$, $L = 1, 2,..., M$.

Take, for example, $s_R(t') = 0$ and $s_R(t) \ne 0$ for some $t > t' \ge 0$. Since schedule s has a finite number of preemptions, it follows that both R and t' can be chosen such that t' is the largest possible. Suppose that $s_L(t') = \nu_L$ and $s_L(t'+\delta) = \mu_L$, $L = 1, 2,..., M$. The values of ν_L and μ_L need not to be different. Choose a positive δ such that $s_L(t'+\delta_1) = \mu_L$ for all $0 < \delta_1 \le \delta$, $L = 1, 2,..., M$.

If there is such a Q, $1 \le Q \le M$, that $\nu_Q \ne 0$ and $\mu_Q = 0$, apply operation $O_1(R, Q, t')$ to schedule s to obtain a new schedule s'.

If all $\nu_L = 0$, $L = 1, 2,..., M$, then choose the largest a such that $s_L(t) = 0$ in the interval $(t'-a, t']$ for all $L = 1, 2,..., M$. Apply operation $O_2(R, t', -a)$ to schedule s to obtain a new schedule s'.

If none of the mentioned situations takes place, then there exists a H, $1 \le H \le M$, such that $\mu_H \ne 0$ and $\mu_H \ne \nu_L$ for all $L = 1, 2,..., M$. In particular, it may happen that $H = R$. Apply operation $O_1(R, H, t')$ to schedule s, and operation $O_2(R, t', -(t'-a))$ to the obtained schedule where a is the largest value such that $s_L(t) \neq \mu_H$ in the interval $(t'-a, t)$ for all $L = 1, 2, \ldots, M$. Denote the resulting schedule by s'.

In any case, the idle time on machine R is reduced without increasing the idle times of the other machines. Since $\overline{t}(s') \leq \overline{t}(s)$ and $F(x)$ is a non-decreasing function, it follows that $F(\overline{t}(s')) \leq F(\overline{t}(s))$. By repeating similar arguments finitely many times, we come to the desired conclusion.

3. Let schedule s allow preemptions only at time $t = t^{(1)}$, and at this moment the processing of $v \leq M$ jobs $k_1, k_2, \ldots, k_v$ is interrupted. A job k_j is processed for t_j' time units on a machine Q_j, and then for t_j'' time units on a machine R_j. If $Q_j = R_j$, then we have $s_{Q_j}(t^{(1)}+\delta) \neq k_j$ for a sufficiently small $\delta > 0$. Let Δ_j denote the length of the time interval between time $t^{(1)}$ and the time at which the processing of job k_j is resumed. Define $\Delta_{j^*} = \min\{\Delta_j \,|\, 1 \leq j \leq v\}$.

Suppose that $Q_{j^*} = R_{j^*}$. Apply operation $O_4(Q_{j^*}, t^{(1)}-t_{j^*}', t^{(1)}, t^{(1)}+\Delta_{j^*})$ to schedule s. As a result, schedule s' with $\overline{t}(s') \leq \overline{t}(s)$ is obtained.

Suppose now that $Q_{j^*} \neq R_{j^*}$. If $\Delta_{j^*} = 0$, then by applying operation $O_1(R_{j^*}, Q_{j^*}, t^{(1)})$ to schedule s we obtain schedule s' with $\overline{t}(s') = \overline{t}(s)$.

If $\Delta_{j^*} > 0$, then apply operation $O_3(R_{j^*}, Q_{j^*}, t^{(1)}+\Delta_{j^*}+t_{j^*}'', t^{(1)}, t_{j^*}'')$ to schedule s. The resulting family of functions $s^{(1)}$ is a schedule, because there is no preemption in schedule s for $t > t^{(1)}$. If $F(\overline{t}(s^{(1)})) \leq F(\overline{t}(s))$, denote $s^{(1)}$ by s'.

Suppose that $F(\overline{t}(s^{(1)})) > F(\overline{t}(s))$. If, in schedule s, the processing of at least one of the jobs $k_1, k_2, \ldots, k_v$ is resumed on machine Q_{j^*}, then let Θ denote the length of the time interval between $t^{(1)}$ and the time at which the processing of the first of these jobs is resumed. It is clear that $\Theta \geq \Delta_{j^*}$ and $s_{Q_{j^*}}(t) \neq k_j$ for all $t^{(1)} < t \leq t^{(1)}+\Theta$ and all $1 \leq j \leq v$. If $s_{Q_{j^*}}(t) \neq k_j$ for all $t > t^{(1)}$ and all $1 \leq j \leq v$, then define $\Theta = W$, where W is a sufficiently large number. Denote $\Theta' = \min\{\Theta, t_{j^*}'\}$. Apply operation $O_3(Q_{j^*}, R_{j^*}, t^{(1)}-t_{j^*}'+\Theta', t^{(1)}+\Delta_{j^*}, \Theta')$ to schedule s. The resulting family of functions $s^{(2)}$ is a schedule. The vectors $\overline{t}(s^{(1)})$ and $\overline{t}(s^{(2)})$ are connected with the vector $\overline{t}(s)$ by the relation $\overline{t}(s^{(1)}) = \overline{t}(s)+et_{j^*}''$ and $\overline{t}(s^{(2)}) = \overline{t}(s)-e\Theta'$ for some vector $e \in E_0^n$. Since function $F(x)$ is e-quasi-concave and $F(\overline{t}(s^{(1)})) > F(\overline{t}(s))$ by assumption, it follows that $F(\overline{t}(s^{(2)})) \leq F(\overline{t}(s))$.

If $\Theta \geq t_{j^*}'$, then denote $s^{(2)}$ by s'. If $\Theta < t_{j^*}'$, apply operation $O_4(Q_{j^*}, t^{(1)}-t_{j^*}', t^{(1)}-\Theta, t^{(1)})$ to schedule $s^{(2)}$ to obtain a schedule $s^{(3)}$ with $\overline{t}(s^{(3)}) \leq \overline{t}(s^{(2)})$. Let $s_L^{(3)}(t^{(1)}) = s_{Q_{j^*}}^{(3)}(t^{(1)}+\delta)$ for any sufficiently small $\delta > 0$. Apply operation $O_1(Q_{j^*}, L, t^{(1)})$ to schedule $s^{(3)}$. Denote the resulting schedule by s'.

Thus, in any case, we are able to find a schedule s' which has at least the same quality as the original schedule s, and also allows preemptions only at time $t = t^{(1)}$. However, at this time, the processing of at most $v-1$ jobs is interrupted Therefore, there is also a non-preemptive schedule having at least the same quality as the original schedule s.

4. To complete the proof, it suffices to show that if, in schedule s, preemptions happen only at times $t^{(1)}$, $t^{(2)}$,..., $t^{(u)}$, then there exists a schedule s' which allows preemptions only at times $t^{(1)}$, $t^{(2)}$,..., $t^{(u-1)}$, and $F(\overline{t}(s')) \leq F(\overline{t}(s))$.

Schedule s satisfies the conditions of the previous item for $t > t^{(u-1)}$. Therefore, we may define $t^{(u-1)} = 0$ and use the above considerations. As a result, we obtain a schedule which has no preemption at $t > t^{(u-1)}$, coincides with the original schedule at $t \leq t^{(u-1)}$, and has at least the same quality. This proves the theorem.

The proof of Theorem 1.2 immediately implies the *existence of a schedule* $s^*(t) = \{s_1^*(t), s_2^*(t),..., s_M^*(t)\}$ *which has the mentioned properties and also the property that either* $s_L^*(t) \neq 0$ *in some interval* $(d, T_L]$ *and* $s_L^*(t) = 0$ *for* $t > T_L$, *or* $s_L^*(t) = 0$ *for all* $t \geq d$, $L = 1, 2,..., M$.

Note that if at least one of the conditions of this theorem is violated, then, in general, the search for an optimal schedule may not be restricted to non-preemptive schedules. Below, we present the corresponding examples.

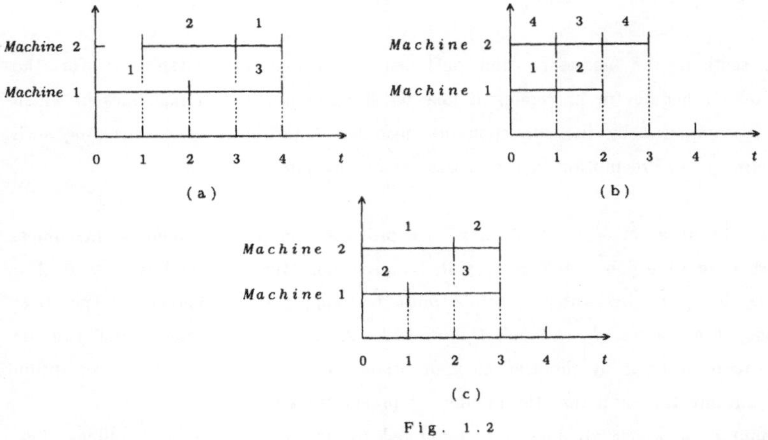

Fig. 1.2

(a) *The values of* d_i *are different*. Define $M = 2$; $n = 3$; $d_1 = 0$; $d_2 = 1$, $d_3 = 2$; $t_1 = 3$; $t_2 = t_3 = 2$; $F(x_1, x_2, x_3) = x_1+2x_2+3x_3$. In the case under consideration, for any non-preemptive schedule s we have $F(\overline{t}(s)) \geq 24$. On the other hand, for the schedule $\tilde{s}$ shown in Fig. 1.2a, we have $F(\overline{t}(\tilde{s})) = 22$. In this schedule, the processing of job 1 is interrupted at time $t = d_3 = 2$ to be resumed on the other machine at time $t = 3$.

(b) *The set of jobs is ordered*, i.e., $G \neq (N, \varnothing)$. Define $M = 2$; $n = 4$; $d_i = 0$, $i = 1, 2, 3, 4$; $t_1 = t_2 = t_3 = 1$, $t_4 = 2$, $F(x_1, x_2, x_3, x_4) = 2(x_1+x_2+x_3)+x_4$, and assume that $1 \longrightarrow 2$ and $1 \longrightarrow 3$. In this case, $F(\overline{t}(s)) = 14$ corresponds to the best non-preemptive schedule s. An optimal schedule s^* with $F(\overline{t}(s^*)) = 13$ is shown in Fig. 1.2b. In this schedule, the processing of job 4 is interrupted at time $t = 1$ to be resumed on the same machine at time $t = 2$.

(c) *Function* $F(x)$ *is not e-quasi-concave.* Let $M = 2$; $n = 3$; $d_i = 0$; $i = 1, 2, 3$; $t_1 = t_2 = t_3 = 2$; $F(x_1, x_2, x_3) = x_1^2+x_2^2+x_3^2$. In this case, $F(x)$ is non-decreasing in the positive octant but it is not an e-quasi-concave function.

In fact, for $x^{(1)} = (0, 1, 2)$, $e = (0, 1, -1)$, $\alpha = 1$ and $\lambda = 1/2$, we have $x^{(2)} = (0, 2, 1)$ and $F(\lambda x^{(1)}+(1-\lambda)x^{(2)}) = F(0, 3/2, 3/2) = 9/2 < \min\{F(0, 1, 2), F(0, 2, 1)\} = 5$. One of the optimal schedules is shown in Fig. 1.2c, where the processing of job 2 is interrupted at time $t = 1$ to be resumed on the other machine at time $t = 2$. The value of $F(x)$ corresponding to this schedule is 22, while for all non-preemptive schedules we have $F(x) \geq 24$.

2. Deadline-Feasible Schedules

In this section, the necessary and sufficient conditions are established for the existence of a schedule for processing n jobs on M parallel identical machines in which each job is completed by the corresponding deadline. Algorithms for constructing such schedules are given. Preemption in the processing of any job is allowed.

2.1. The jobs of a set $N = \{1, 2,..., n\}$ are processed on parallel identical machines. The release date of a job $i \in N$ is $d_i \geq 0$, its processing time is equal to $t_i > 0$. The deadline $D_i \geq d_i+t_i$, by which a job i must be completed, is known. In practical applications, the values d_i, t_i and D_i, $i = 1, 2,..., n$, are rational and can be considered to be integers by choosing an appropriate scale. It is assumed that preemption does not consume time and that the number of preemptions is finite.

A schedule s in which all jobs are completed by the corresponding deadlines, i.e., $\overline{t}_i(s) \leq D_i$, $i = 1, 2,..., n$, is called *feasible* (*with respect to deadlines*). Here $\overline{t}_i(s)$ is the completion time of job i in a schedule s.

We present the necessary and sufficient conditions for the existence of feasible schedules, and show how to find them (if such schedules exist).

2.2. If set N of jobs can be divided into two subsets N_1 and N_2 such that $\max\{D_i | i \in N_1\} \leq \min\{d_i | i \in N_2\}$, then a feasible schedule for processing the jobs of set N exists if and only if feasible schedules exist for processing the jobs of each subset N_1 and N_2. In what follows, it is supposed that such a situation does not arise.

Let $e_1 < e_2 < \ldots < e_{p+1}$, $p \leq 2n-1$, be a set of all pairwise distinct values of d_i and D_i, $i = 1, 2, \ldots, n$, $E_k = (e_k, e_{k+1}]$ and $\Delta_k = e_{k+1} - e_k$, $k = 1, 2, \ldots, p$. Let $n(k)$. denote the number of all jobs $i \in N$ such that $E_k \subseteq (d_i, D_i]$.

Suppose that there exists such a l, $1 \leq l \leq p$, and such job $j \in N$ that $n(l) \leq M$ and $E_l \subseteq (d_j, D_j]$. It is obvious that job j can be processed in the time interval E_l on any machine without affecting the processing of the other jobs. If $\Delta_l \geq t_j$, then delete job j from set N. If $\Delta_l < t_j$, reduce the processing time of job j by Δ_l. Perform these operations for all $j \in N$, such that $E_l \subseteq (d_j, D_j]$. As a result, we obtain a new set N' of jobs and new processing times of the jobs in this set.

For each $i \in N'$ such that $D_i \geq e_{l+1}$, we reduce the deadline D_i by Δ_l, and for each $i \in N'$ such that $d_i \geq e_{l+1}$, we also reduce its release date d_i by Δ_l.

As can be easily seen, a feasible schedule for processing the jobs of set N exists if and only if there exists a feasible schedule for processing the jobs of set N' (with d_i, t_i, D_i changed as described above). Taking set N' as N, we can repeat the above arguments until either $N' = \emptyset$ is obtained, or $n(k) > M$ for all k.

In the former case, we conclude that a feasible schedule s does exist, and the described procedure is, in fact, a procedure for finding such a schedule. In each step, we analyze an interval $E_l = (e_l, e_{l+1}]$ with $n(l) \leq M$ and the set $N_l = \{j_1, j_2, \ldots, j_{n(l)}\}$ of jobs which can be processed in this interval. Since $|N_l| = n(l) \leq M$, it follows that, for $\Delta_l \leq t_{j_L}$, we may define $s_L(t) = j_L$ in the interval E_l, while for $\Delta_l > t_{j_L}$ we define $s_L(t) = j_L$ in the interval $(e_l, e_l + t_{j_L}]$ and $s_L(t) = 0$ in the interval $(e_l + t_{j_L}, e_{l+1}]$, $L = 1, 2, \ldots, n(l)$. If $n(l) < M$, then $s_L(t) = 0$ in the interval E_l, $L = n(l)+1, n(l)+2, \ldots, M$. In this case, a feasible schedule is found in $O(n^2)$ time.

In the latter case, we come to the problem of a smaller dimension where $n(k) > M$ for all intervals E_k.

2.3. Associate the set of time intervals $\{E_1, E_2, \ldots, E_p\}$, the set of jobs $\{1, 2, \ldots, n\}$ and the sequence t_i, $i = 1, 2, \ldots, n$, with a network Γ (see Fig. 2.1) containing the source vertex x_0, the sink vertex z, and the intermediate vertices $x_1, x_2, \ldots, x_p, y_1, y_2, \ldots, y_n$. A vertex x_k corresponds to interval E_k, a vertex y_i corresponds to job i.

Connect vertices x_k and y_i by the arc of the capacity $c(x_k, y_i) = \Delta_k$ if and only if $E_k \subseteq (d_i, D_i]$; connect vertices x_0 and x_k by the arc of the capacity $c(x_0, x_k) = M\Delta_k$; and connect vertices y_i and z by the arc of the capacity $c(y_i, z) = t_i$, $k = 1, 2,..., p$, $i = 1, 2,..., n$. The arcs (x_0, x_k), $k = 1, 2,..., p$, are called the input arcs, while the arcs (y_i, z), $i = 1, 2,..., n$, are called the output arcs of the network. Note that the network Γ can be constructed in at most $O(n^2)$ time.

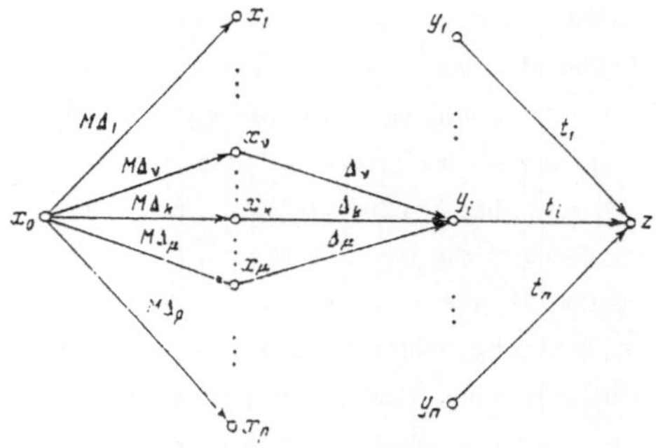

Fig. 2.1

Each deadline-feasible schedule s determines the flow f which saturates the output arcs of the network Γ. In fact, let $\tau_{ik}(s)$ be the total processing time of job i in the interval E_k in schedule s, $i = 1, 2,..., n$, $k = 1, 2,..., p$. It is obvious that $\tau_{ik}(s) \leq \Delta_k$ holds for all i and k; the inequality $\sum_{k=1}^{p} \tau_{ik}(s) = t_i$ holds for any $i \in N$, the inequality $\sum_{i=1}^{n} \tau_{ik}(s) \leq M\Delta_k$ holds for any interval E_k, and, besides, the equality $\sum_{k=1}^{p} \sum_{i=1}^{n} \tau_{ik}(s) = \sum_{i=1}^{n} t_i$ holds.

Define $f(x_0, x_k) = \sum_{i=1}^{n} \tau_{ik}(s)$ for each arc (x_0, x_k), define $f(x_k, y_i) = \tau_{ik}(s)$ for each arc (x_k, y_i), and define $f(y_i, z) = t_i$ for each arc (y_i, z). Note that the value of function f corresponding to any arc does not exceed its capacity, and, moreover, for each output arc of the network, this is equal to the capacity. Besides, for any intermediate vertex v, the sum of the values of function f over all arcs entering v is equal to the sum of its values over all arcs leaving v. The sum of the values of function f over all input arcs of the network and the sum over all output arcs are both equal to $\sum_{i=1}^{n} t_i$. Therefore, function f is a flow (with the value of $\sum_{i=1}^{n} t_i$) which saturates the output arcs of the network Γ.

On the other hand, *each flow f which saturates the output arcs of the network Γ determines a deadline-feasible schedule.* In this case, the flow along an arc (x_k, y_i) is interpreted as the total processing time of job i in interval E_k. Note that $f(x_k, y_i) \le \Delta_k$, $\sum_{k=1}^{p} f(x_k, y_i) = t_i$ and $\sum_{i=1}^{n} f(x_k, y_i) \le M\Delta_k$. Given a flow along the arcs (x_k, y_i), $i = 1, 2, \ldots, n$, for each vertex x_k, a schedule for the interval E_k can be constructed. This can be done by the following algorithm called the *packing* algorithm.

Let the jobs of a set $\tilde{N}$ have to be processed in a time interval $E = (e', e'']$. The jobs are processed on M parallel identical machines. The processing time of a job $i \in \tilde{N}$ is τ_i, the conditions $\tau_i \le \Delta$ hold for all $i \in \tilde{N}$, and, moreover, $\sum_{i \in \tilde{N}} \tau_i \le M\Delta$, where $\Delta = e'' - e'$.

Let $\pi = (i_1, i_2, \ldots, i_{|\tilde{N}|})$ denote an arbitrary permutation of the elements of set $\tilde{N}$. Define a function $\sigma(t)$ in the interval $(e', e'+M\Delta]$, assuming that $\sigma(t) = i_1$ in the interval $(e', e'+\tau_{i_1}]$, $\sigma(t) = i_k$ in the interval $\left(e'+\sum_{j=1}^{k-1}\tau_{i_j}, e'+\sum_{j=1}^{k}\tau_{i_j}\right]$, $k = 2, 3, \ldots, |\tilde{N}|$, and set $\sigma(t) = 0$ in the interval $\left(e'+\sum_{i \in \tilde{N}}\tau_i, e'+M\Delta\right]$ if $\sum_{i \in \tilde{N}}\tau_i < M\Delta$. A schedule $s(t) = \{s_1(t), s_2(t), \ldots, s_M(t)\}$ for processing the jobs of set $\tilde{N}$ in interval E is said to be constructed by the packing algorithm if in this interval $s_L(t) = \sigma(t+(L-1)\Delta)$, $L = 1, 2, \ldots, M$.

It is clear that such a schedule in interval E *can be found in at most $O(|\tilde{N}|)$ time.* In this schedule, *the number of preemptions does not exceed $M-1$.*

Having constructed the schedule for each interval E_k by the packing algorithm and having "concatenated" the schedules for the intervals $E_1, E_2, \ldots, E_p$, we obtain a deadline-feasible schedule for the jobs of set N. Finding such schedules requires no more than $O(np)$ time, i.e., at most $O(n^2)$ time. The resulting schedule has at most $n(p-1)+(M-1)p$ preemptions. In fact, while constructing a schedule for each interval E_k, we obtain at most $M-1$ preemptions, while "concatenating" the resulting schedules involves at most $n(p-1)$ preemptions.

Finding a maximal flow in a network with n vertices requires $O(n^3)$ time [2]. If the value of the resulting flow in the network Γ is $\sum_{i=1}^{n} t_i$, then there is a deadline-feasible schedule which can be found in at most $O(n^2)$ time and which has at most $n(p-1)+(M-1)p$ preemptions. Otherwise, a feasible schedule does not exist.

2.4. We now establish the necessary and sufficient conditions for the existence of deadline-feasible schedules.

For $\tilde{N} \subseteq N$, let $E(\tilde{N})$ denote the set of the numbers of all intervals E_k, each obeying the condition $E_k \subseteq (d_i, D_i]$ at least for one $i \in \tilde{N}$, and let $\tilde{n}(k)$ denote the number of all jobs $i \in \tilde{N}$ such that $E_k \subseteq (d_i, D_i]$.

Due to the saturation theorem [15], a flow, which saturates the output arcs of the network Γ, exists if and only if the inequalities

$$\sum_{i \in \tilde{N}} t_i \leq \sum_{k \in E(\tilde{N})} \Delta_k \min\{M, \tilde{n}(k)\} \tag{2.1}$$

hold for all $\tilde{N} \subseteq N$.

Thus, the following statement holds.

Theorem 2.1. *A deadline-feasible schedule exists if and only if inequalities* (2.1) *hold for all* $\tilde{N} \subseteq N$.

Since $t_i \leq D_i - d_i$ for all $i \in N$, the subsets $\tilde{N} \subseteq N$ which contain at least two elements should be considered. The total number of inequalities (2.1) is equal to $2^n - (n+1)$.

We show that the inequalities (2.1) hold for all $\tilde{N} \subseteq N$ if they hold for some specially chosen subsets $\tilde{N} \subseteq N$.

Let us choose an arbitrary subset $\overline{N} \subseteq N$. Represent $E(\overline{N})$ as $E(\overline{N}) = E^{(1)} \cup E^{(2)}$, where $E^{(1)} \cap E^{(2)} = \varnothing$, $E^{(1)} = \{\nu, \nu+1, \ldots, \mu\}$, $1 \leq \nu \leq \mu \leq p$, $E^{(2)} = \{k \in E(\overline{N}) \mid k \geq \mu+2\}$. Suppose that $E^{(2)} \neq \varnothing$. Then set $\overline{N}$ can be divided into two non-empty disjoint subsets $\overline{N}_1$ and $\overline{N}_2$ so that $E(\overline{N}_1) = E^{(1)}$ and $E(\overline{N}_2) = E^{(2)}$. In fact, if there is such an $i \in \overline{N}$ that $E(i) \cap E^{(1)} \neq \varnothing$ and $E(i) \cap E^{(2)} \neq \varnothing$ then, due to the definition of $E(i)$, we obtain $\mu+1 \subseteq E(\overline{N})$. As can be easily seen, if inequality (2.1) holds for $\tilde{N} = \overline{N}_1$ and $\tilde{N} = \overline{N}_2$, then it also holds for $\tilde{N} = \overline{N}$.

Therefore, the inequalities (2.1) hold for all $\tilde{N} \subseteq N$ if and only if they hold for all $\tilde{N} \subseteq N$ satisfying the condition: there exist such ν and μ that $1 \leq \nu \leq \mu \leq p$ and $E(\tilde{N}) = \{\nu, \nu+1, \ldots, \mu\}$.

The following procedure can be used for finding the required sets $\tilde{N}$. Choose arbitrary ν and μ, $1 \leq \nu \leq \mu \leq p$. Define $c = \{\nu, \nu+1, \ldots, \mu\}$. Find the set N' of all $i \in N$ for which $E(i) \subseteq c$. If $E(N') = c$, then define $N_{\nu\mu} = N'$. In this case, the pair ν, μ is called *essential*. Let $\overline{N}_{\nu\mu}$ denote the set of all proper subsets N'' of the set $N_{\nu\mu}$ satisfying the condition $E(N'') = c$. The set $N_{\nu\mu}$ and all subsets in $\overline{N}_{\nu\mu}$ are the desired sets $\tilde{N}$. Applying this procedure to all pairs ν, μ, $1 \leq \nu \leq \mu \leq p$, we find all sets $\tilde{N}$ for which inequalities (2.1) should be verified.

If the values of M, d_i, t_i, D_i, $i = 1, 2, \ldots, n$, are such that for all essential pairs ν, μ the inequalities (2.1) hold for any $\tilde{N} \in \overline{N}_{\nu\mu}$ if they hold for $\tilde{N} = N_{\nu\mu}$, then we say that the *regularity condition* holds. In this case, *the number of inequalities* (2.1) *to be verified does not exceed* $n(n+1)/2$. In fact, if a pair ν, μ is essential, then $e_\nu \in \{d_1, d_2,..., d_n\}$ and $e_{\mu+1} \in \{D_1, D_2,..., D_n\}$. The largest number of essential pairs is obtained if all d_i and D_i, $i = 1, 2,..., n$, are different. By numbering the jobs in increasing order of D_i, we come to the conclusion that for any essential pair ν, μ where $e_\nu = d_i$, $e_{\mu+1} = D_j$ the inequality $i \le j$ holds.

2.5. Let us consider the case $M = 1$. In this case, inequality (2.1) can be written in the form

$$\sum_{i\in\tilde{N}} t_i \le \sum_{k\in E(\tilde{N})} \Delta_k \tag{2.2}$$

for all $\tilde{N} \subseteq N$.

Let the jobs be numbered in non-decreasing order of the deadlines. Let N_k^l denote the set of all jobs $i \in N$ for which $d_i \ge d_k$ and $D_i \le D_l$.

Since, in the case under consideration, the regularity condition holds, it follows that *inequalities* (2.2) *hold for all* $\tilde{N} \subseteq N$ *if and only if*

$$\sum_{i\in N_k^l} t_i \le D_l - d_k \tag{2.3}$$

for all $1 \le k \le l \le n$.

It can easily be shown that inequalities (2.3) for all $1 \le k \le l \le n$ can be verified in *at most* $O(n^2)$ *time*.

Note that if $d_i \le d_{i+1}$, $i = 1, 2,..., n-1$, then (2.3) can be written in the form

$$\sum_{i=k}^{l} t_i \le D_l - d_k \tag{2.4}$$

for all $1 \le k \le l \le n$.

If $d_i \ge d_{i+1}$, $i = 1, 2,..., n-1$, then inequalities (2.3) hold for all $1 \le k \le l \le n$, if and only if

$$\sum_{i=1}^{k} t_i \le D_k - d_k,\ k = 1, 2,..., n. \tag{2.5}$$

If $d_i = d$, $i = 1, 2,..., n$, then (2.5) becomes

$$\sum_{i=1}^{k} t_i \le D_k - d,\ k = 1, 2,...,n. \tag{2.6}$$

We describe an $O(n\log n)$ algorithm for finding a feasible schedule s. The algorithm extends the known rule of job processing in non-decreasing order of deadlines (the *EDD* rule) to the case of different release dates, i.e., according to the algorithm, the available job which has the smallest deadline, is selected to start processing.

Let $\{d^{(1)}, d^{(2)}, ..., d^{(v)}\}$ be a set of all distinct values of d_i and $d^{(1)} < d^{(2)} < ... < d^{(v)} < d^{(v+1)} = W$, where W is a sufficiently large number.

In the first step, define $\tau = d^{(1)}$, $N_0 = \{i \mid i \in N, d_i = d^{(1)}\}$ and $s(t) = 0$ for $0 \leq t \leq d^{(1)}$.

In each step, we have a certain time τ (suppose that $d^{(u-1)} \leq \tau < d^{(u)}$, $2 \leq u \leq v+1$) and some set N_0 of jobs. Choose a job $j \in N_0$ with the smallest number (i.e., with the earliest deadline). Define $s(t) = j$ for all $\tau < t \leq \min\{d^{(u)}, \tau+t_j\}$, and, if $\tau+t_j < d^{(u)}$ and $|N_0| = 1$, define $s(t) = 0$ for all $\tau+t_j < t \leq d^{(u)}$.

If $\tau+t_j > d^{(u)}$, then add to N_0 all jobs $i \in N$ with $d_i = d^{(u)}$ and redefine t_j to become equal to $t_j-(d^{(u)}-\tau)$. If either $\tau+t_j < d^{(u)}$ and $|N_0| = 1$, or $\tau+t_j = d^{(u)}$, then delete job j from N_0 and add all jobs $i \in N$ with $d_i = d^{(u)}$. In any case, define $\tau = d^{(u)}$. If $\tau+t_j < d^{(u)}$ and $|N_0| > 1$, then delete job j from N_0 and set τ equal to $\tau+t_j$.

As a result, we obtain a new time τ, a new set N_0, and go to the next step. The schedule s is constructed when $N_0 = \emptyset$.

We show that *if there exists a feasible schedule, then the schedule s found by the described algorithm is feasible*. It suffices to show that if s is not a feasible schedule, then at least one of inequalities (2.3) is violated.

Let l be a job with the smallest number for which the deadline is violated in the schedule s, i.e., $\bar{t}_i(s) \leq D_i$, $i = 1, 2, ..., l-1$, and $\bar{t}_l(s) > D_l$. Set $t' = \bar{t}_l(s)$. Let r be the number of a step of the algorithm, in which $s(t') = l$ is obtained, and τ_r be the value of τ at which we enter step r.

Define $t'' = \max\{t \mid t < \tau_r, s(t) = 0\}$. It is easy to check that $t'' = d_i$ for some $i \in N$ and $d_i \leq d_l$.

If all jobs chosen to be processed in the first $r-1$ steps have numbers less than l, then define $\hat{t} = t''$.

Let p, $p < r$, be a step of the algorithm with the largest number, in which a job l' such that $l' > l$ is chosen for processing. If τ_p is the value of τ we enter step p, then $d_l \geq d^{(q+1)}$ (where $d^{(q)} \leq \tau_p < d^{(q+1)}$) and $s(t) \neq i$ holds in the interval $(d^{(q+1)}, t']$ if $d_i < d^{(q+1)}$. In fact, if there exists a job l'' with $d_{l''} < d^{(q+1)}$ which is processed in this interval, then $l'' > l'$ (otherwise, job l'' rather than l' would have been chosen in step p). This, however, contradicts the fact that p is the last step before step r such that a job with the number larger than l is chosen to be processed. Define $\hat{t} = \max\{t'', d^{(q+1)}\}$.

Let k be a job with the smallest number, for which $d_k = \hat{t}$. It is clear that $k \leq l$. In the interval $(d_k, t']$ only such jobs are processed, for which $d_i \geq d_k$ and $i \leq l$, i.e. $D_i \leq D_l$. Therefore,

$$d_k + \sum_{i \in N_k^l} t_i \geq t' > D_l$$

and hence, for jobs k and l, inequality (2.3) is violated.

We show that *the running time* of the described algorithm for finding a feasible schedule is $O(n\log n)$. Sort the jobs in non-decreasing order of d_i (this requires $O(n\log n)$ time, see Section 2.7 of Chapter 1). Find the set N_0 of all jobs $i \in N$ with $d_i = d^{(1)}$. Define a binary relation $\Longrightarrow$ over set N, assuming that $i \Longrightarrow j$ if and only if $i < j$. It is clear that relation $\Longrightarrow$ is a total strict order and, hence, a total pseudo-order. In the first step of the algorithm, we represent the set $N_0 \subseteq N$ ordered according to relation $\Longrightarrow$ as a balanced 2-3-tree (this takes $O(n)$ time; see Section 2.3 of Chapter 1).

The number of steps of the algorithm does not exceed $2n-1$ because, in each step, either processing of some job is completed or a new job ready for processing is added to the set N_0.

In each step, choosing job $j \in N_0$ with the smallest number (i.e., finding a maximal with respect to $\Longrightarrow$ element of set N_0) takes a constant time (in fact, one elementary operation is required; see Section 2 of Chapter 1). Either deleting a job from N_0 or adding a new job to N_0 requires $O(\log n)$ time. Changing the processing time of job j is equivalent to deleting job j from N_0 followed by inserting job j with a new processing time to N_0. This also requires at most $O(\log n)$ time.

Hence, it follows that the total running time required for finding a schedule s does not exceed $O(n\log n)$.

Remark 1. If $d_i \leq d_{i+1}$, then schedule s is non-preemptive. Therefore, *conditions* (2.4) *and* (2.6) *are necessary and sufficient for the existence of a single-machine deadline-feasible non-preemptive schedule.*

Remark 2. A feasible schedule for a *partially ordered* set of jobs (as before, $M = 1$ and preemption is allowed) can be found by an $O(n^2)$ algorithm described in Sections 3.6 and 3.7 of this chapter.

2.6. Let $M \geq 1$, $d_i = d$, $D_i = D$, $i = 1, 2,\ldots, n$. As before, it is assumed that $t_i \leq D-d$, $i = 1, 2,\ldots, n$.

Inequalities (2.1) can be written as

$$\sum_{i \in \tilde{N}} t_i \leq (D-d) \min\{M, |\tilde{N}|\}, \ \tilde{N} \subseteq N.$$

Since, in this case, the regularity condition is satisfied, it follows that *a feasible schedule exists if and only if*

$$\sum_{i=1}^{n} t_i \leq (D-d)M. \tag{2.7}$$

It is obvious that verifying this inequality takes *at most* $O(n)$ *time*. If a feasible schedule exists, it can be found by the packing algorithm applied to the set N of jobs in the time interval $(d, D]$ (see Section 2.3). This also takes $O(n)$ *time*. In the resulting schedule, *the number of preemptions does not exceed* $M-1$.

2.7. We now consider the case $M \geq 1$, assuming that either $D_i = D$ or $d_i = d$, $i = 1, 2, \dots, n$. These situations are equivalent, since *a feasible schedule for processing jobs with parameters* d_i *and* $D_i = D$ *exists if and only if there is a feasible schedule for processing jobs with parameters* $d_i' = d$ *and* $D_i' = D+d-d_i$.

In what follows, without loss of generality, we consider the case $M \geq 1$, $d_i = 0$, $i = 1, 2, \dots, n$. It is again assumed that $t_i \leq D_i$, $i = 1, 2, \dots, n$.

Let the jobs be numbered in non-decreasing order of D_i. For $\tilde{N} \subseteq N$, assume that $\tilde{N} = \{i_1, i_2, \dots, i_l\}$, where $i_j < i_k$ if $j < k$. Inequality (2.1) can be written in the form

$$\sum_{j=1}^{l} t_{i_j} \leq D_{i_1}\min\{l, M\} + (D_{i_2}-D_{i_1}) \min\{l-1, M\} + \dots + \\ + (D_{i_{l-1}}-D_{i_{l-2}}) \min\{2, M\} + (D_{i_l}-D_{i_{l-1}}). \tag{2.8}$$

Since $t_i \leq D_i$, $i = 1, 2, \dots, n$, it follows that inequality (2.8) holds for any set $\tilde{N}$ with $|\tilde{N}| = l \leq M$. If $|\tilde{N}| = l > M$, then inequality (2.8) can be written in the form

$$\sum_{j=1}^{l} t_{i_j} \leq \sum_{j=l-M+1}^{l} D_{i_j}. \tag{2.9}$$

This inequality holds if and only if

$$\sum_{k=1}^{i_{l-M+1}} t_k + \sum_{j=l-M+2}^{l} t_{i_j} \leq \sum_{j=l-M+1}^{l} D_{i_j}.$$

Thus, a feasible schedule exists if and only if

$$\sum_{k=1}^{i_1} t_k + \sum_{j=2}^{M} t_{i_j} \leq \sum_{j=1}^{M} D_{i_j} \tag{2.10}$$

holds for all $\tilde{N} = \{i_1, i_2, \dots, i_M\} \subseteq N$. The total number of these inequalities is $\binom{M}{n}$.

In this case, the regularity condition may be, in general, violated. In fact, consider the intervals $E_k = (D_{k-1}, D_k]$ of the length $\Delta_k = D_k - D_{k-1}$, $k = 1, 2, \dots, n$, where $D_0 = 0$, and the case of $\Delta_k = 0$ is included. For each essential pair ν, μ, we have $\nu = 1$ and

$\mu \in \{1, 2,..., n\}$. Therefore, the set $N_{\nu\mu}$ is of the form $N_{1\mu} = \{1, 2,..., \mu\}$, $\mu = 1, 2,..., n$, and for each $N'' \in \bar{N}_{\nu\mu}$, we have $\mu \in N''$. Let $M = 2$, $t_1 = t_3 = 1$, $t_2 = 2$, $t_4 = 5$, $D_1 = 1$, $D_2 = 2$, $D_3 = 4$, $D_4 = 5$. A direct verification shows that inequality (2.1) holds for $N_{11} = \{1\}$, $N_{12} = \{1, 2\}$, $N_{13} = \{1, 2, 3\}$, $N_{14} = \{1, 2, 3, 4\}$, but this does not hold for $N'' = \{1, 2, 4\} \subseteq N_{14}$.

We show that the regularity condition holds if $t_i = t$, $i = 1, 2,..., n$. If $\tilde{N} = N_{1\mu} = \{1, 2,..., \mu\}$, $\mu > M$, then inequality (2.9) can be written as

$$\mu t \leq \sum_{k=\mu-M+1}^{\mu} D_k. \tag{2.11}$$

Suppose that this inequality holds for all $\mu > M$. Choose an arbitrary μ and an arbitrary set $N'' = \{i_1, i_2,..., i_l\} \in \bar{N}_{1\mu}$, $l > M$, $i_1 < i_2 < ... < i_l = \mu$. Let a job i_{l-M+1} have the number p. Define $\mu' = p+M-1$. Since $i_{l-M+1} \geq l-M+1$, we have $\mu' \geq 1$ and $lt \leq \mu' t \leq \sum_{k=\mu'-M+1}^{\mu'} D_k = \sum_{k=p}^{\mu} D_k \leq \sum_{j=l-M+1}^{l} D_{i_j}$. Hence, inequality (2.9) also holds for $\tilde{N} = N''$.

Thus, *a feasible schedule for* $M \geq 1$, $d_i = 0$, $t_i = t < D_1$, $i = 1, 2,..., n$, *exists if and only if inequalities* (2.11) *hold for all* $\mu > M$. The number of these inequalities is $n-M$.

2.8. Let, as before, $M \geq 1$, $d_i = 0$, $t_i \leq D_i$, $i = 1, 2,..., n$. We describe an $O(n\log n)$ *algorithm* for finding a feasible schedule. This algorithm is a natural generalization of the packing algorithm.

Let $D^{(1)} < D^{(2)} < ... < D^{(v)}$ be all pairwise distinct values of D_i. Let N_u denote the set of all jobs $i \in N$ with $D_i = D^{(u)}$, $u = 1, 2,..., v$. Define $T_L^{(1)} = 0$, $L = 1, 2,..., M$.

The algorithm consists of v steps. In each step u, $u = 1, 2,..., v$, we are given $D^{(u)}$, $T_L^{(u)}$, $L = 1, 2,..., M$, and a job set N_u. A step of the algorithm involves $|N_u|$ iterations, at each of which one job $i \in N_u$ is assigned for processing.

At the first iteration of step u, define $D = D^{(u)}$, $T_L = T_L^{(u)}$, $\delta_L = D-T_L$, $L = 1, 2,..., M$, $\tilde{N} = N_u$. At each iteration of this step, take an arbitrary job $i \in \tilde{N}$.

(a) If $t_i > \delta_L$, $L = 1, 2,..., M$, then, as shown below, there is no feasible schedule.

(b) If $t_i \leq \delta_L$ for all machines L for which $\delta_L \neq 0$, then define $s_P(t) = i$ in the interval $(T_P, T_P+t_i]$. Here P is a machine with the smallest $\delta_P \neq 0$ (if there are several of them, take any). Modify T_P and δ_P, assuming them to be equal to T_P+t_i and δ_P-t_i, respectively.

(c) Suppose that the conditions in (a) and (b) do not hold. Let P be a machine with the largest δ_P such that $t_i \geq \delta_P$. If $t_i > \delta_P$, assume that Q is a machine with the smallest δ_Q such that $t_i < \delta_Q$. If there are several machines which satisfy the above conditions, take any of them as P or Q. Define $s_P(t) = i$ in the interval $(T_P, D]$, and, if $t_i > \delta_P$, then define $s_Q(t) = i$ in the interval $(T_Q, T_Q+t_i-\delta_P]$. Modify T_P and δ_P, setting them to be equal to D and 0, respectively. Modify T_Q and δ_Q, setting them to be equal to $T_Q+t_i-\delta_P$ and $\delta_Q+\delta_P-t_i$, respectively.

Delete job i from $\tilde{N}$, proceed to the next iteration, and so on, until $\tilde{N} = \emptyset$ is obtained. In that case, go to the next step $u+1$, assuming $T_L^{(u+1)} = T_L$, $L = 1, 2,..., M$. Having performed step v, define $s_L(t) = 0$ for $t > T_L^{(v+1)}$, $L = 1, 2,..., M$.

Note that if conditions (b) and (c) hold for each iteration of this algorithm, we obtain a feasible schedule s. Otherwise, the algorithm stops as soon as, at some iteration, conditions (a) hold.

We show that the *running time* of this algorithm is $O(n\log n)$.

Each iteration is associated with a set R of all pairwise distinct values of T_L considered at that iteration. Since, for any T' and T'' in R, either $T' < T''$ or $T'' < T'$ holds, it follows that set R is ordered by the relation $<$ and can be represented as a balanced 2-3-tree (see Section 2 of Chapter 1). Each $T \in R$ is associated with a terminal vertex (a leaf) of the balanced 2-3-tree and with a list of numbers L of the machines for which $T_L = T$. At the first iteration of the first step, we have $R = \{T\}$ where $T = 0$. This value of T corresponds to the tree consisting of a single vertex, and to the list $\{1, 2,..., M\}$ of machines.

At each iteration, the search for the cases, in which either $t_i > \delta_L = D-T_L$ holds for all $L = 1, 2,..., M$, or $t_i \leq D-T_L$ holds for all L such that $D-T_L \neq 0$, reduces to finding either the smallest element T' the set R or the largest element T'' such that $T'' < D$. (We may take any machine in the list corresponding to the value of T'' as machine P). If none of these cases takes place, it is required to find machines P and Q. To do that, it suffices to find the smallest element $\bar{T} \in R$ such that $t_i \geq D-\bar{T}$, and, if $t_i > D-\bar{T}$, the largest element $\hat{T} \in R$ such that $t_i < D-\hat{T}$. All these operations can be implemented in $O(\log M)$ time (see Section 2.8 of Chapter 1).

The modification of the value T_P reduces to deleting the number P from the list of machines corresponding to $\bar{T}$, and, if the obtained list is empty, to deleting the element $\bar{T}$ from R.

Let a modified value of T_P be equal to $\bar{T}'$. If $\bar{T}' \in R$, then the number P should be added to the list of machines corresponding to $\bar{T}'$. If set R does not contain $\bar{T}'$, then $\bar{T}'$ should be inserted into R, and the list $\{P\}$ of machines corresponding to $\bar{T}'$ should be formed. The value of T_Q is modified in the same way. These operations also require $O(\log M)$ time.

Since the total number of iterations is n, finding a schedule s takes $O(n\log M)$ time (if the jobs are pre-sorted in non-decreasing order of their deadlines). Taking into account the running time required to sort the jobs in non-decreasing order of D_i, we conclude that the time complexity of the algorithm is $O(n\log n + n\log M)$ or, equivalently, $O(n\log n)$, due to $M < n$.

2.9. We show that *if a feasible schedule does exist, then the algorithm described in Section* 2.8 *finds such a schedule s*. In other words, if the algorithm does not find a schedule (i.e., at some iteration $t_i > \delta_L$, $L = 1, 2,\ldots, M$, holds for the chosen job i), then there is no feasible schedule.

Let $s^{(1)}$ be some feasible schedule for processing the jobs of set N, and let $N^{(1)}$ be a set of jobs processed according to this schedule in the interval $(0, D^{(1)}]$. It is clear that $N_1 \subseteq N^{(1)}$.

Note that $t_i \le D^{(1)}$, $i \in N_1$, and the first step of the algorithm under consideration is, essentially, the packing algorithm applied to the set N_1 of jobs in the interval $(0, D^{(1)}]$ (see Section 2.3). If $N^{(1)} \ne N_1$, then choose all jobs $i \in N^{(1)}$ such that, in the schedule s, they are processed within the interval $(0, D^{(1)}]$ for $t_i' < t_i$ time units (it is obvious that each $i \notin N_1$). Define the processing times of these jobs to be equal to t_i', and apply the packing algorithm to the set $N^{(1)}$ in the interval $(0, D^{(1)}]$. Here, a permutation which starts with all jobs of set N_1 can be chosen as permutation π of the elements of set $N^{(1)}$. Denote the resulting schedule for the jobs of set $N^{(1)}$ in the interval $(0, D^{(1)}]$ by $\bar{s}^{(1)}$. By defining $s^{(2)}(t) = \bar{s}^{(1)}(t)$ in the interval $(0, D^{(1)}]$ and $s^{(2)}(t) = s^{(1)}(t)$ beyond this interval, we obtain a feasible schedule $s^{(2)}$. It is clear that $s^{(2)}$ is such a feasible schedule that $s_L^{(2)}(t) = s_L(t)$ for $0 \le t \le T_L^{(2)}$, $L = 1, 2,\ldots, M$, where $T_L^{(2)}$ are the values of T_L, $L = 1, 2,\ldots, M$, obtained after the first step of the algorithm.

Let $s^{(u)}$ denote a feasible schedule such that $s_L^{(u)}(t) = s_L(t)$ for $0 \le t \le T_L^{(u)}$, $L = 1, 2,\ldots, M$, where $T_L^{(u)}$ are the values of T_L, $L = 1, 2,\ldots, M$, obtained after the $(u-1)$th step of the algorithm. We show that, in this case, we may pass from the schedule $s^{(u)}$ to a feasible schedule $s^{(u+1)}$ such that: (1) $s_L^{(u+1)}(t) = s_L(t)$ for $0 \le t \le T_L^{(u+1)}$; (2) $s_L^{(u+1)}(t) = s_L^{(u)}(t)$ for $t > D^{(u)}$, $L = 1, 2,\ldots, M$. It is evident that $s_L^{(u+1)}(t) = s_L^{(u)}(t)$ for $0 \le t \le T_L^{(u)}$, $L = 1, 2,\ldots, M$.

As shown below, in order to prove this, it suffices to prove the following *Statement A*: If there is a schedule $\hat{s}$ for processing the jobs of some set $\tilde{N}$ such that the conditions $\hat{s}_L(t) = 0$ are satisfied for all $t \le T_L \le D$ and all $t > D$, $L = 1, 2,\ldots, M$, then the schedule $\tilde{s}$ for processing the jobs of this set, constructed by the procedure to be performed in each step of the algorithm, also satisfies the above conditions.

If Statement A holds then we can pass from the schedule $s^{(u)}$ to the schedule $s^{(u+1)}$ in the following way. Let $\tilde{N}$ denote the set of all those $i \in N$ for which there exist such L and t, $1 \le L \le M$, $T_L^{(u)} < t \le D^{(u)}$, that $s_L^{(u)}(t) = i$. It is obvious that $N_u \subseteq \tilde{N}$. If $\tilde{N} \ne N_u$ and a job $i \in \tilde{N} \backslash N_u$ is processed in the interval $(0, D^{(u)}]$ for $t_i' \le t_i$ time units, then choose t_i' and $D^{(u)}$ as the processing times and the deadlines, respectively, for all $i \in \tilde{N} \backslash N_u$. Apply the procedure performed at each step of the algorithm to the set $\tilde{N}$, choosing the jobs of set N_u first.

Now, we proceed to a direct proof of Statement A. The proof is by induction with respect to the number $\tilde{n}$ of jobs in $\tilde{N}$. The statement holds for $\tilde{n} = 1$. Suppose that it is valid for all $\tilde{n}$, $1 \le \tilde{n} < l$, and show that the statement also holds for $\tilde{n} = l$.

Without loss of generality, assume that $T_L \ge T_{L+1}$, $L = 1, 2,..., M-1$. Represent the interval $(T_M, D]$ as a family of subintervals of length Δ such that schedule $\hat{s}$ is non-preemptive within these subintervals and each time T_L, $\delta_L \ne 0$, is the beginning of some subinterval (this can be done because $\hat{s}$ has a finite number of preemptions). Let the obtained time intervals of length Δ be numbered by the integers $1, 2,..., q$, starting with the interval $(T_M, T_M+\Delta]$. An interval with the number α is of the form $(T_M+(\alpha-1)\Delta, T_M+\alpha\Delta]$.

Let i be the job chosen at the first iteration of the algorithm for finding schedule $\tilde{s}$. The existence of the schedule $\hat{s}$ implies that there is a machine L for which $t_i \le \delta_L$. If, in schedule $\tilde{s}$, job i is processed on two machines (P and Q), then $Q = P+1$. Let t' and t'' be the completion times of job i on machines P and $P+1$, respectively. If i is not processed on machine $P+1$, then define $t'' = T_{P+1}$.

Let us transform schedule $\hat{s}$ (see Fig. 2.2a) to a new schedule s' (see Fig. 2.2b) in the following way. If in a time interval α (of length Δ) with $T_M+\alpha\Delta > T_P$ we have that $\hat{s}_K(t) = i$, $K \ne P$, then define $s_P'(t) = \hat{s}_K(t)$ and $s_K'(t) = \hat{s}_P(t)$ in the interval α. If $\hat{s}_K(t) = i$ in the interval α with $T_M+\alpha\Delta \le T_P$, and V is a machine with the smallest number for which $T_V < T_M+\alpha\Delta$, then define $s_V'(t) = \hat{s}_K(t)$ and $s_K'(t) = \hat{s}_V(t)$ in the interval α. In other cases, define $s_L'(t) = \hat{s}_L(t)$, $L = 1, 2,..., M$. It is easy to verify that s' is a schedule.

The time T_L is called the ready time of machine L. This machine is said to be ready in the interval α if $T_L < T_M+\alpha\Delta$.

Let us introduce two operations for transforming schedule s' into a new schedule s''. *Operation* $O_1(\alpha, \beta)$ is applied when the same number of machines are ready in the intervals with numbers α and β ($\alpha < \beta$) (for example, intervals 7 and 8, or 15 and 16 in Fig. 2.2b). This operation interchanges these intervals: $s_L''(t) = s_L'(t+(\beta-\alpha)\Delta)$ and $s_L''(t+(\beta-\alpha)\Delta) = s_L'(t)$ for $T_M+(\alpha-1)\Delta < t \le T_M+\alpha\Delta$ and $s_L''(t) = s_L'(t)$ for other values of t, $L = 1, 2,..., M$.

It is easy to verify that s'' is a schedule.

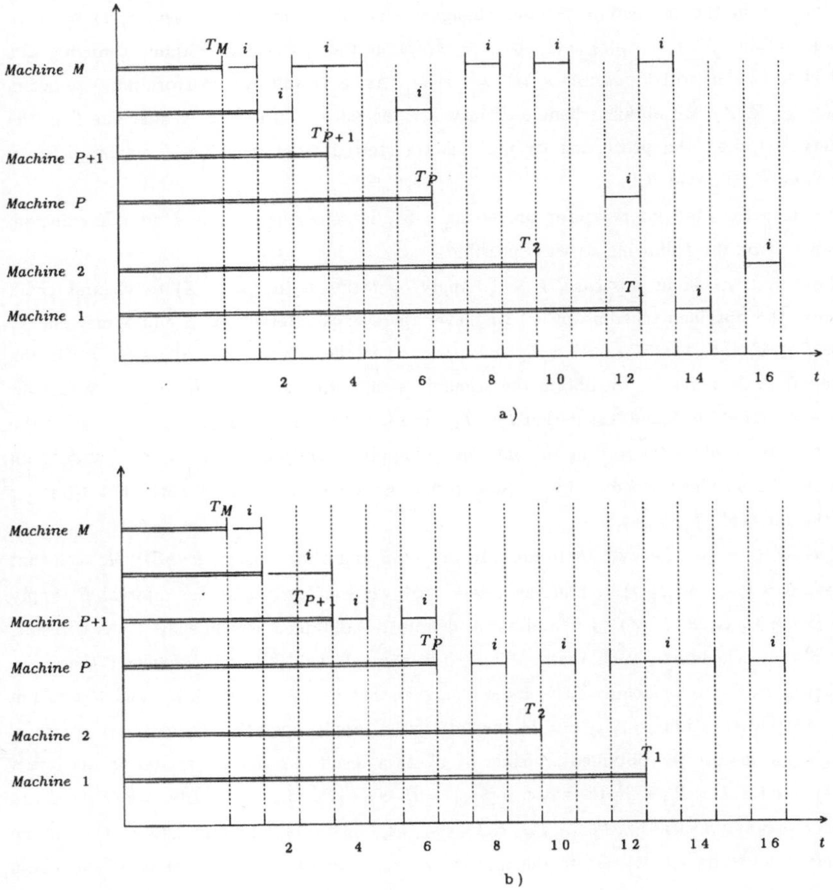

Fig. 2.2

Operation $O_2(\alpha, \beta, R, Z)$ (where α, β are the numbers of intervals, $\alpha < \beta$, while R, Z are the numbers of machines, such that $T_M + \alpha\Delta \leq T_R < T_M + \beta\Delta$) is applied when a different number of machines are ready in the intervals α and β, and the relation $s_Z'(t) = i$ holds in the interval α, while $s_L'(t) \neq i$ holds in the interval β, $L = 1, 2, \ldots, M$; for example, in Fig. 2.2b, one may choose $\alpha = 1$, $\beta = 5$, $Z = M$, $R = P+1$ or $\alpha = 4$, $\beta = 9$, $Z = P+1$, $R = P$. In this case, there exists a machine V such that $s_V'(t) = k$ holds in the interval β where either $k = 0$ or $k \in N$ and $s'_L(t) \neq k$ for $L = 1, 2, \ldots, M$ in the interval α. Operation $O_2(\alpha, \beta, R, Z)$ is performed in two stages. Define $\bar{s}'_Z(t) = k$ in the interval α and $\bar{s}'_V(t) = i$ in the interval β without changing schedule s' in other cases: $\bar{s}'_L(t) = s'_L(t)$. Then define $s''_V(t) = \bar{s}'_R(t)$ and $s''_R(t) = \bar{s}'_V(t)$ in the interval β without changing the schedule $\bar{s}'$ in other cases: $s''_L(t) = \bar{s}'_L(t)$. As a result of performing operation $O_2(\alpha, \beta, R, Z)$, we obtain schedule s'' in which the job i is processed on machine R in the interval β, i.e., the processing of job i is transferred from machine Z (interval α) to machine R (interval β).

Suppose that the intervals for processing job i in schedules s' and $\tilde{s}$ do not coincide. Then one of the following cases is possible.

Case 1. $t_i < \delta_P$. In this case, $P = 1$. Apply operation $O_2(\alpha, \beta, 1, Z)$ to s' (and again denote the obtained schedule by s') whenever there exist intervals α, β and a machine Z, $1 < Z \leq M$, such that $T_M+\beta\Delta > T_1$ and $s'_Z(t) = i$ in the interval α, while $s'_1(t) \neq i$ in the interval β. As a result, we obtain the schedule s' such that $s'_L(t) \neq i$ for $L \neq 1$. Whenever there are intervals α and β such that $T_1 < T_M+\alpha\Delta \leq T_1+t_i < T_M+\beta\Delta \leq D$, and $s'_1(t) \neq i$ in the interval α, while $s'_1(t) = i$ in the interval β, apply operation $O_1(\alpha, \beta)$ to s' and again denote the obtained schedule by s'. As a result, we obtain schedule s' such that $s'_1(t) = i$ in the interval $(T_1, T_1+t_i]$.

Case 2. $t_i \geq \delta_P$. Whenever there are intervals α, β and a machine Z, $P < Z \leq M$, such that $T_M+\beta\Delta > T_P$, and $s'_Z(t) = i$ in the interval α, while $s'_P(t) \neq i$ in the interval β, apply operation $O_2(\alpha, \beta, P, Z)$ to s' and again denote the obtained schedule by s'. As a result, we obtain schedule s' such that $s'_P(t) = i$ in the interval $(T_P, D]$. Whenever there are intervals α, β and a machine Z, $P+1 < Z \leq M$, such that $T_{P+1} < T_M+\beta\Delta \leq T_P$, and $s'_Z(t) = i$ in the interval α, while $s'_{P+1}(t) \neq i$ in the interval β, apply operation $O_2(\alpha, \beta, P, Z)$ to s' and again denote the obtained schedule by s'. As a result, we obtain schedule s' such that $s'_L(t) \neq i$ for $L > P+1$. Note that $t_i < \delta_{P+1} = D-T_P+(T_P-T_{P+1})$. Finally, when there are intervals α and β such that $T_{P+1} < T_M+\alpha\Delta \leq T_{P+1}+t_i-\delta_P < T_M+\beta\Delta \leq T_P$, and $s'_{P+1}(t) \neq i$ in the interval α, while $s'_{P+1}(t) = i$ in the interval β, apply operation $O_1(\alpha, \beta)$ to s' and again denote the obtained schedule by s'. As a result, we obtain a schedule s' such that $s'_{P+1}(t) = i$ in the interval $(T_{P+1}, T_{P+1}+t_i-\delta_P]$ and $s'_P(t) = i$ in the interval $(T_P, D]$. Note that it follows from $t_i < \delta_{P+1} = D-T_{P+1}$ that $T_{P+1}+t_i-\delta_P = T_{P+1}+t_i-D+T_P < T_P$.

In both cases, we obtain a schedule s' such that the intervals for processing job i in this schedule coincide with the intervals for processing this job in schedule $\tilde{s}$.

By defining $T_P = t'$, $T_{P+1} = t''$ and temporarily disregarding job i, we come to the case of $l-1$ jobs (with the new values of T_L). Taking into account the inductive assumption, we conclude that schedule $\tilde{s}$ for processing the jobs of set $\tilde{N}$, which is constructed according to the procedure to be performed in each step of the algorithm, is in fact the desired one. This completes the proof of Statement *A*.

Remark. The maximal number of preemptions in processing the jobs in the schedule found by the algorithm described in Section 2.8 is $n-1$.

It can be easily seen that the first job in the schedule is processed with no preemption, while the processing of each subsequent job can be interrupted at most once. The maximum number of preemptions can be reduced to $n-2$ by constructing a schedule for the first $n-1$ jobs by the above algorithm, and by assigning the last job to be processed on the machine with the smallest value of T_L.

3. Single Machine. Maximal Cost

In this section, the problem of minimizing the maximal cost for scheduling n jobs on a single machine is considered. Various assumptions are made with regard to the release dates, the due dates, cost functions, and other parameters.

3.1. The jobs of a set $N = \{1, 2,..., n\}$ are processed on a single machine. Preemption in processing any job is allowed. A job $i \in N$ is available not earlier than the release time $d_i \geq 0$, its processing time is $t_i > 0$, and the due date is $D_i \geq 0$. A precedence relation $\rightarrow$ is defined over set N which describes a feasible order of job processing. The reduction graph of that relation is denoted by $G = (N, U)$. Each job $i \in N$ is associated with a non-decreasing real function $\varphi_i(t)$ which represents the cost for having job i completed at time t.

It is required to find a feasible (with respect to $\rightarrow$) schedule s^* which minimizes the function

$$F_{max}(s) = \max\{\varphi_i(\bar{t}_i(s)) \mid i \in N\} \tag{3.1}$$

over all schedules s feasible with respect to $\rightarrow$ where $\bar{t}_i(s)$ is the completion time of job i in schedule s.

3.2. Suppose that $d_i = 0$, $i = 1, 2,..., n$. In this case, the search for an optimal schedule s^* can be restricted to considering the class of schedules according to which each job is processed without preemption (see Section 1 of this chapter). Each of these schedules is specified by a permutation $\pi = (i_1, i_2, \ldots, i_n)$ of the elements of N (feasible with respect to $\rightarrow$, i.e., the relation $i_\nu \rightarrow i_\mu$ implies that $\nu < \mu$). Let the set of all feasible permutations be denoted by $\mathcal{P}_n(G)$.

It is required to find a permutation π^* in the set $\mathcal{P}_n(G)$ with the smallest value of the function

$$F_{max}(\pi) = \max\{\varphi_i(\bar{t}_i(\pi)) \mid i \in N\}, \tag{3.2}$$

where $\bar{t}_i(\pi)$ is the completion time of job i if the jobs are processed according to the sequence π, i.e., $\bar{t}_{i_k}(\pi) = \sum_{j=1}^{k} t_{i_j}$. Such a permutation π^* is called *optimal.*

Let Q^- denote the set of all minimal (with respect to the order relation $\rightarrow$ defined over N) elements of a set $Q \subseteq N$.

Theorem 3.1. *In the case* $d_i = 0$, $i = 1, 2, \ldots, n$, *a permutation* $\pi = (i_1, i_2, \ldots, i_n)$ *such that* $i_k \in J_k^-$ *for* $k = 1, 2, \ldots, n$, *where* $J_k = \{i_1, i_2, \ldots, i_k\}$, *and*

$$\varphi_{i_k}\left[\sum_{j=1}^{k} t_{i_j}\right] = \min\left\{\varphi_l\left[\sum_{j=1}^{k} t_{i_j}\right] \middle| \, l \in J_k^-\right\}, \tag{3.3}$$

is optimal.

Proof. Permutation π is feasible because, if otherwise, there exist indices k and j, $k > j$, such that $i_k \rightarrow i_j$ and, therefore, $i_k \notin J_k^-$.

Let $\pi^* = (i_1^*, i_2^*, \ldots, i_n^*)$ be an optimal permutation. We show that π^* can be transformed into π without increasing the value of function (3.2). Suppose that for some k, $1 \le k \le n$, the relations $i_k^* \neq i_k$ and $i_j^* = i_j$ hold for all $j > k$. It suffices to show that $F_{max}(\pi') = F_{max}(\pi^*)$, where $\pi' = (\sigma, i_k, i_{k+1}, \ldots, i_n)$ and σ is the sequence $(i_1^*, i_2^*, \ldots, i_k^*)$ without the element i_k. It is obvious that $\pi' \in \mathcal{P}_n(G)$. Since $\{i_1, i_2, \ldots, i_k\} = \{i_1^*, i_2^*, \ldots, i_k^*\}$, it follows from (3.3) that

$$\varphi_{i_k}(\bar{t}_{i_k}(\pi')) = \varphi_{i_k}\left[\sum_{j=1}^{k} t_{i_j^*}\right] \le \varphi_{i_k^*}\left[\sum_{j=1}^{k} t_{i_j^*}\right] = \varphi_{i_k^*}(\bar{t}_{i_k^*}(\pi^*)) \le F_{max}(\pi^*).$$

Since $\bar{t}_{i_j^*}(\pi') \le \bar{t}_{i_j^*}(\pi^*)$ for $i_j^* \in \{\sigma\}$, $\bar{t}_{i_j^*}(\pi') = \bar{t}_{i_j^*}(\pi^*)$ for $j > k$ and all functions $\varphi_i(t)$ are non-decreasing, we conclude that $\varphi_i(\bar{t}_i(\pi')) \le \varphi_i(\bar{t}_i(\pi^*)) \le F_{max}(\pi^*)$ for all $i \in N$, $i \neq i_k$. Therefore, $F_{max}(\pi') \le F_{max}(\pi^*)$, but since π^* is an optimal permutation, we have $F_{max}(\pi') = F_{max}(\pi^*)$. This proves the theorem.

Theorem 3.1 immediately implies *an algorithm* for finding an optimal permutation in n steps.

Define $J_n = N$. Find such an $i_n \in J_n^-$ that $\varphi_{i_n}(T_n) = \min\{\varphi_l(T_n) \mid l \in J_n^-\}$, where $T_n =$

$\sum_{i \in J_n} t_i$.

Similarly, define $J_{n-1} = J_n \backslash \{i_n\}$. Find such an $i_{n-1} \in J^-_{n-1}$ that $\varphi_{i_{n-1}}(T_{n-1}) = \min\{\varphi_l(T_{n-1}) \mid l \in J^-_{n-1}\}$, where $T_{n-1} = \sum_{i \in J_{n-1}} t_i$, and so on.

Repeating this process, we eventually find a required optimal sequence $\pi^* = (i_1, i_2, \dots, i_n)$.

The running time of the algorithm is $O(n^2)$. In each step, finding a minimal (with respect to $\longrightarrow$) element of a set and deleting one of them from that set requires $O(n)$ time (see Section 1.4 of Chapter 1). Thus, the total time for these operations in all n steps does not exceed $O(n^2)$. In each step $r = n-k+1$, $r = 1, 2, \dots, n$, of the algorithm, at most k values of the cost functions have to be computed and at most $k-1$ comparisons of these values have to be performed. Therefore, the total number of cost function evaluations does not exceed $n(n+1)/2$, while the total number of their comparisons does not exceed $n(n-1)/2$. Hence, the algorithm requires at most $O(n^2)$ time (provided that computing a cost function value takes a constant time).

3.3. We consider some *special cases* of the problem of minimizing the maximal cost, assuming, as before, that $d_i = 0$, $i = 1, 2, \dots, n$.

Let the cost functions $\varphi_i(t)$ be such that for any ν, $\mu \in N$, either $\varphi_\nu(t) \le \varphi_\mu(t)$ hold for all $t \in (0, T]$ or $\varphi_\nu(t) \ge \varphi_\mu(t)$ hold for all $t \in (0, T]$. Here $T = \sum_{i \in N} t_i$. Let the jobs be numbered in such a way that $\varphi_1(t) \ge \varphi_2(t) \ge \dots \ge \varphi_n(t)$ for all $t \in (0, T]$.

In the case under consideration, in a step $r = n-k+1$, $r = 1, 2, \dots, n$, of the algorithm for finding an optimal permutation, it suffices to take an element of the set J^-_k with the largest number as the element i_k. In this case, the running time of the algorithm is still $O(n^2)$, but computation of the cost function values is not required. If $G = (N, \varnothing)$, then the permutation $\pi^* = (1, 2, \dots, n)$ is optimal. In this case, an optimal permutation is found by numbering the jobs in *at most* $O(n\log n)$ *time*.

These are some examples of the cost functions that have the described property: (a) $\varphi_i(t) = \varphi(t) + \alpha_i$, $i = 1, 2, \dots, n$; (b) $\varphi_i(t) = \alpha_i \varphi(t)$, $\alpha_i > 0$, $i = 1, 2, \dots, n$, $\varphi(t) \ge 0$, $t \in (0, T]$; (c) $\varphi_i(t) = \varphi(t+\alpha_i)$, $i = 1, 2, \dots, n$. Here φ is a non-decreasing function defined over the interval $(0, T]$. In each of these cases, the jobs should be numbered in non-increasing order of α_i.

If the due dates D_i, $i = 1, 2, \dots, n$, are given, then non-decreasing functions of the lateness $L_i = \bar{t}_i - D_i$ are normally used as the cost functions. If, in this case $\varphi_i(t) = \varphi(t - D_i)$, $i = 1, 2, \dots, n$, and φ is a non-decreasing function, then the cost functions belong to the type (c), and the jobs should be numbered in non-decreasing order of the due dates.

Therefore, if $G = (N, \emptyset)$ and $D_1 \le D_2 \le \ldots \le D_n$, then the permutation $\pi^* = (1, 2, \ldots, n)$ is *optimal* for the problem of:

- minimizing *the maximal lateness* (the case of $\varphi_i(t) = t - D_i$);
- minimizing *the maximal tardiness* (the case of $\varphi_i(t) = \max\{t - D_i, 0\}$);
- finding a schedule *without* late jobs (the case of $\varphi_i(t) = \mathrm{sgn}(\max\{t - D_i, 0\})$).

3.4. Let $d_i \ge 0$, $\varphi_i(t) = \varphi(t - D_i)$, $i = 1, 2, \ldots, n$, where φ is a non-decreasing function. Preemption in processing each job is forbidden.

It is clear that a permutation $\pi^* \in \mathcal{P}_n(G)$ which minimizes *the maximal lateness* $L_{max}(\pi) = \max\{\bar{t}_i(\pi) - D_i \mid i \in N\}$ also minimizes *the maximal cost* $F_{max}(\pi) = \max\{\varphi(\bar{t}_i(\pi) - D_i) \mid i \in N\}$. Consider two situations. In the first, the jobs have the parameters d_i', t_i, D_i' and are processed according to the sequence $\pi' = (i_1, i_2, \ldots, i_n)$, while in the second situation, the jobs have the parameters d_i'', t_i, D_i'' and are processed according to the sequence $\pi'' = (i_n, i_{n-1}, \ldots, i_1)$. Let us find sufficient conditions for the maximal latenesses to be equal in both cases.

It can be easily shown (for example, by induction with respect to l) that, in the first situation, the completion time of a job i_l is $\bar{t}_{i_l}(\pi') = \max\{\bar{t}_{i_{l-1}}(\pi'), d_{i_l}'\} + t_{i_l} = \max\left\{d_{i_k}' + \sum_{j=k}^{l} t_{i_j} \mid k = 1, 2, \ldots, l\right\}$, where $\bar{t}_{i_0}(\pi') = 0$. Hence, in the first situation, the maximal lateness is $L_{max}' = \max\left\{d_{i_k}' + \sum_{j=k}^{l} t_{i_j} - D_{i_l}' \mid 1 \le k \le l \le n\right\}$. Similarly, in the second situation, the maximal lateness is $L_{max}'' = \max\left\{d_{i_l}'' + \sum_{j=k}^{l} t_{i_j} - D_{i_k}'' \mid 1 \le k \le l \le n\right\}$.

If the equality $d_\nu' - D_\mu' = d_\mu'' - D_\nu''$ holds for any ν and μ, $1 \le \nu, \mu \le n$, then $L_{max}' = L_{max}''$.

Thus, if $d_i' = C - D_i''$ and $D_i' = C - d_i''$, $i = 1, 2, \ldots, n$, then in both cases, the maximal latenesses are the same. Here C is an arbitrary constant.

In a number of cases, this observation allows the solution procedure for the problem with $d_i = 0$, $i = 1, 2, \ldots, n$, to be extended to problems with $d_i \ge 0$, $i = 1, 2, \ldots, n$.

In fact, consider the following *Problem A*. Let $d_i = d_i'' \ge 0$, $D_i = D_i'' = D$, $\varphi_i(t) = \varphi(t - D_i)$, $i = 1, 2, \ldots, n$, where $\varphi(x)$ is a non-decreasing function. Preemption in job processing is forbidden. A precedence relation $\longrightarrow$ with the reduction graph G is defined over the set $N = \{1, 2, \ldots, n\}$ of jobs.

It is required to find a permutation π in the set $\mathcal{P}_n(G)$ of permutations (feasible with respect to $\longrightarrow$) which minimizes the function

$$F_{max}(\pi) = \max\{\varphi(\bar{t}_i - D_i) \mid i \in N\}. \tag{3.4}$$

Note that a special case of Problem A ($D_i = 0$, $i = 1, 2,..., n$, $\varphi(t) = t$) is the problem of finding a time-optimal schedule.

Let us consider *Problem B* of minimizing function (3.4) over the set $\mathcal{P}_n(G')$, provided that $d_i = d_i' = 0$, $D_i = D_i' = D - d_i''$, $i = 1, 2,..., n$, and G' is the reduction graph of the precedence relation $\Longrightarrow$ defined over the set N which is inverse to the order $\longrightarrow$ (i.e. $\nu \Longrightarrow \mu$ if and only if $\mu \longrightarrow \nu$).

Since for $C = D$ the relations $d_i' = C - D_i''$ and $D_i' = C - d_i''$, $i = 1, 2,..., n$, hold, we conclude that, if $\pi^{(B)} = (i_1, i_2,..., i_n)$ is a solution of Problem B, then $\pi^{(A)} = (i_n, i_{n-1},..., i_1)$ is a solution of Problem A. In particular, if $G = (N, \varnothing)$, then it follows from the previous item of this section, that processing the jobs in non-decreasing order of their due dates solves Problem B. Therefore, to solve Problem A, it suffices to process the jobs *in non-decreasing order of their release dates.*

3.5. We now consider the problem of minimizing the maximal cost for processing n jobs on a single machine assuming that the release dates are different, the values of t_i and d_i, $i = 1, 2,..., n$, are rational and can be regarded as integers by choosing an appropriate scale. Preemption is allowed. It is assumed that the cost functions $\varphi_i(t)$ are arbitrary non-decreasing functions. The precedence relation $\longrightarrow$ is defined over set $N = \{1, 2,..., n\}$ of jobs. We look for an optimal schedule in the class of schedules that are feasible with respect to $\longrightarrow$.

Let the unit length time intervals starting at $t = 0$ be numbered by the integers 1, 2,... . Due to Theorem 1.1 (see Section 1 of this chapter), there exists an optimal schedule which is either non-preemptive or preemptions happen only at the release dates. Thus, it suffices to consider the schedules $s(t)$ such that $s(t) = const$ in each unit time interval. In other words, in order to specify a schedule, it suffices to assign (obeying certain conditions) one of the numbers 0, 1, 2,..., n to each unit interval.

Let $B^0(i)$ denote the set of all direct predecessors of i in N (i.e., all those $k \in N$ for which $k \longrightarrow i$ and there is no j such that $k \longrightarrow j$ and $j \longrightarrow i$). Define $\bar{d}_i = d_i$ if $B^0(i) = \varnothing$, and, otherwise, define $\bar{d}_i = \max\{d_i, \max\{\bar{d}_k + t_k \mid k \in B^0(i)\}\}$. It is clear that the processing of job i cannot start before $\bar{d}_i$.

Let the jobs be numbered so that $\bar{d}_1 \le \bar{d}_2 \le ... \le \bar{d}_n$. Define $N_i = \{k \mid k \in N, k \ge i\}$. Let $T(s)$ denotes the makespan (i.e., the maximum completion time) for schedule s. The maximal cost $F_{max}(s)$ for schedule s is calculated by formula (3.1).

Lemma 3.1. *There exists a schedule s^* which minimizes both $T(s)$ and $F_{max}(s)$ such that*

$$T(s^*) = \max\left\{\bar{d}_i + \sum_{k \in N_i} t_k \,\middle|\, i \in N\right\}.$$

Proof. Note that $\max\left\{\bar{d}_i + \sum_{k \in N_i} t_k \,\middle|\, i \in N\right\}$ is the time before which the processing of all jobs of set N cannot be completed. Therefore, if, for some schedule $\tilde{s}$, the equality $T(\tilde{s}) = \bar{d}_i + \sum_{k \in N_i} t_k$ is obtained for a certain $i \in N$, then $\tilde{s}$ is a schedule with the smallest $T(s)$.

Let s' be a schedule with the smallest value of $F_{max}(s)$. If $s'(t) \neq 0$ holds for all unit intervals with the numbers $\bar{d}_1+1, \bar{d}_1+2, \ldots, T(s')$, then s' is the desired schedule s^*.

Suppose that for some of the above intervals $s'(t) = 0$ holds. Among these intervals choose the one with the largest number τ. If there is such a job $i \in N$ that $\bar{d}_i = \tau$ and the processing of all jobs $1, 2, \ldots, i-1$ is completed before the time τ, then $T(s') = \bar{d}_i + \sum_{k \in N_i} t_k$ and s' is the desired schedule s^*. Otherwise, schedule s' can be transformed into the schedule s'' such that $s'(t) = s''(t)$ in the unit intervals with the numbers $1, 2, \ldots, \tau-1$ and $s''(t) \neq 0$ in the interval τ. Moreover, $T(s'') \leq T(s')$ and $F_{max}(s'') = F_{max}(s')$.

In fact, if there is no such $i \in N$ that $\bar{d}_i = \tau$, then we may define $s''(t) = s'(t)$ in all intervals except τ and $\tau+1$, while defining $s''(t) = s'(t+1)$ in the interval τ and $s''(t) = 0$ in the interval $\tau+1$. Suppose that $\bar{d}_i = \tau$ for some $i \in N$ and there is such a job j that $\bar{d}_j < \tau$ and the processing of job j is completed in the interval with the number $\tau' > \tau$. If there are several such jobs, then the one with the smallest number may be chosen as job j. Define $s''(t) = s'(t)$ in all intervals besides τ and τ', while defining $s''(t) = j$ in the interval τ and $s''(t) = 0$ in the interval τ'.

It is clear that, in any case, schedule s'' is feasible, and, besides, $T(s'') \leq T(s')$ and $F_{max}(s'') = F_{max}(s')$.

Repeating these considerations finitely many times, we either conclude that s' is a desired schedule s^* or obtain a schedule s^0 such that $F_{max}(s^0) = F_{max}(s')$, $s^0(t) \neq 0$ in the intervals with the numbers $\tau, \tau+1, \ldots, T(s')-1$ and $s^0(t) = 0$ in the interval with the number $T(s')$ and in the subsequent intervals, i.e. $T(s^0) < T(s')$. This proves the lemma.

It follows from Lemma 3.1 that the search for a schedule which minimizes function $F_{max}(s)$ can be restricted to considering the class of *time-optimal* schedules. A schedule s^* which minimizes both $T(s)$ and $F_{max}(s)$ is called *optimal*.

For an optimal schedule, the makespan is equal to

$$T = \max\left\{\bar{d}_i + \sum_{k \in N_i} t_k \,\middle|\, i \in N\right\}. \tag{3.5}$$

Let l denote the largest $i \in N$ for which the maximum is achieved in (3.5). Suppose that $u \in N_l^-$ and

$$\varphi_u(T) = \min\{\varphi_i(T) \,|\, i \in N_l^-\}. \tag{3.6}$$

For a time-optimal schedule, the makespan for the jobs of set $N\backslash u$ is given by

$$T' = \max\left\{\bar{d}_i + \sum_{k \in N_i \backslash u} t_k \,\middle|\, i \in N\backslash u\right\}. \tag{3.7}$$

Along with the initial problem, consider the following *reduced* scheduling problem. If $t_u \le T-T'$, delete job u from set N. If $t_u > T-T'$, then job u is given a new processing time equal to $t_u' = t_u-(T-T')$ and a new cost function $\varphi_u'(t) = -W$, where W is a sufficiently large number. Leave the parameters of other jobs unchanged. Let s' be an optimal (i.e. minimizing both $T(s)$ and $F_{max}(s)$) schedule for the reduced problem. We show that $T(s') = T'$.

In fact, if $t_u \le T-T'$, then $T(s') = T'$ by definition. Let $t_u > T-T'$. We have $T' = T-(T-T') = \bar{d}_l + \sum_{k \in N_l \backslash u} t_k + t_u - (T-T') = \bar{d}_l + \sum_{k \in N_l \backslash u} t_k + t_u'$. It is obvious that $T(s') \ge T'$. If $T(s') = \bar{d}_j + \sum_{k \in N_j \backslash u} t_k + t_u' > T'$ for some $j \le u$, then by adding $T-T'$ to both sides of this inequality, we obtain $\bar{d}_j + \sum_{k \in N_j} t_k > T$, which contradicts (3.5). If $T(s') = \bar{d}_j + \sum_{k \in N_j} t_k > T'$ for some $j > u$, then $\bar{d}_j + \sum_{k \in N_j \backslash u} t_k > T'$, which contradicts (3.7). Therefore, $T(s') = T'$.

Theorem 3.2. *Let s' be an optimal schedule for the reduced problem and $T'' = \max\{d_u, T'\}$. Then a schedule s such that $s(t) = u$ in the interval $(T'', T]$ and $s(t) = s'(t)$ in other intervals is an optimal one for the initial problem.*

Proof. We show that among optimal schedules for the initial problem there exists a schedule $\tilde{s}$ such that $\tilde{s}(t) = u$ in the interval $(T'', T]$.

Suppose that $T' \le d_u$, i.e., for any job $i \in N\backslash u$ the inequality $\bar{d}_i + \sum_{k \in N_i \backslash u} t_k \le d_u$ holds. Hence, $\bar{d}_u = d_u$ and $\bar{d}_i + \sum_{k \in N_i} t_k \le d_u + t_u$ for any $i \in N$, i.e., $T = d_u + t_u$. Therefore, for any schedule s such that $T(s) = T$, we have $s(t) = u$ in the interval $(d_u, T]$.

Suppose that $T' > d_u$. Consider a schedule $\bar{s}$ optimal for the initial problem and such that $\bar{s}(t) \ne u$ in the unit interval with the number q, $q > T'$, and $\bar{s}(t) = u$ in the intervals with the numbers $q+1, q+2, \dots, T$. The case $q = T$ is also possible.

Note that $t_u \ge T-T'$, since otherwise $T'+t_u < T$, which contradicts (3.5) due to the inequality $d_u < T'$. Therefore, there exists an interval with the number p, $p < q$, in which $\bar{s}(t) = u$. Here, p can be chosen in such a way that $\bar{s}(t) \ne u$ in the intervals $p+1, p+2, \dots, q-1$.

Since the maximum in (3.5) is attained at $l \le u$ we have $\overline{s}(t) \ne 0$ in the interval $(\overline{d}_l, T]$ and, hence, in the interval $(p+1, q]$ as well. Among the jobs processed in the time interval $(p+1, q]$ choose the job with the smallest number v. Note that $\overline{d}_v < p$, otherwise $\overline{d}_v + \sum_{k \in N_v \setminus u} t_k > T'$. Let job v be completed at time r. It is clear that $p < r \le q$.

We construct a schedule $\overline{\overline{s}}$ by defining $\overline{\overline{s}}(t) = v$ in the interval p, $\overline{\overline{s}}(t) = u$ in the interval with the number r and $\overline{\overline{s}}(t) = \overline{s}(t)$ in other intervals. This schedule is feasible, and, besides, $T(\overline{\overline{s}}) = T(\overline{s})$ and $F_{max}(\overline{\overline{s}}) \le F_{max}(\overline{s})$ because $\varphi_u(T) \le \varphi_v(T)$.

Repeating these considerations finitely many times, we obtain an optimal schedule $\hat{s}$ such that $\hat{s}(t) = u$ in the intervals $q, q+1, \dots, T$. Thus, this is a desired schedule $\tilde{s}$.

The schedule $\tilde{s}$ determines a schedule $\tilde{s}'$ for the reduced problem (by defining $\tilde{s}'(t) = 0$ in the interval $(T', T]$ and $\tilde{s}'(t) = \tilde{s}(t)$ in other intervals). If s' is an optimal schedule for the reduced problem, then for schedule s for the initial problem such that $s(t) = u$ in the interval $(T'', T]$ and $s(t) = s'(t)$ in other intervals, the following holds: $F_{max}(s) = \max\{F_{max}(s'), \varphi_u(T)\} \le \max\{F_{max}(\tilde{s}'), \varphi_u(T)\} = F_{max}(\tilde{s})$. This proves the theorem.

3.6. *An algorithm* for finding an optimal schedule s^* follows directly from Theorem 3.2. In each step, calculate T by formula (3.5), find job l with the largest number among the jobs for which the maximum is attained in (3.5), and job $u \in N_l^-$ for which (3.6) holds. Find T' by formula (3.7) and compute $T'' = \max\{d_u, T'\}$. Define $s^*(t) = u$ in the interval $(T'', T]$ and formulate the following reduced problem: if $t_u \le T - T'$, then delete u from N, if $t_u > T - T'$, then define the cost function $\varphi_u(t) = -W$ in the interval $(0, T']$ and set the processing time of job u to be equal to $t_u - (T - T')$. Go to the next step, and so on, until $N = \varnothing$ is obtained. Define $s^*(t) = 0$ in all intervals for which $s^*(t)$ is not yet determined.

It is clear that schedule s^* is found in a finite number of steps. Moreover, as shown below, the number of steps in the algorithm is at most $2n-1$.

We now show that *the running time* of the algorithm for finding an optimal schedule s^* is $O(n^2)$.

For each $i \in N$, finding the set $B^0(i)$ and computing $\overline{d}_i$ requires at most $O(n)$ time. Thus, for all $i \in N$, this takes $O(n^2)$ time.

It is clear that, in each step of the algorithm, computing T, T', T'' and finding job l takes $O(n)$ time. Finding set N_l^-, as well as finding and deleting job $u \in N_l^-$ can be done in $O(n)$ time (see Section 1.4 of Chapter 1), provided that the computation of the cost function value requires a constant time.

We now show that *the number of steps in the algorithm* is at most $2n-1$. First, we show that, if, in schedule s^*, preemption happens at time t', then $t' = d_i$ for some $i \in N$.

Consider an arbitrary step r of the algorithm. Let N be a set of jobs to be considered in this step, and let l_r and u_r be the jobs l and u found in this step. Assume that T and T' are calculated by formulas (3.5) and (3.7), respectively. It is obvious that in the interval $(\overline{d}_{l_r}, T]$ only the jobs of the set N_{l_r} are processed. If $t_{u_r} \le T - T'$, then job u_r is processed in the time interval $(T - t_{u_r}, T]$ with no preemption. Let $t_{u_r} > T - T'$ and j be the job with the smallest number for which $\overline{d}_j + \sum_{k \in N_j \setminus u} t_k = T'$ holds. Then $\overline{d}_j = d_j$, since, otherwise, $\overline{d}_k + t_k = \overline{d}_j$ for some $k \in B^0(j)$, $k < j$, and $\overline{d}_k + \sum_{i \in N_k} t_i = T'$. Note that $\overline{d}_{u_r} < d_j$ (otherwise, it follows from $T' + t_{u_r} > T$ that $\overline{d}_j + \sum_{k \in N_j} t_k > T$, which contradicts (3.5)) and, hence, $u_r \notin N_j$. It is evident that in the interval $(d_j, T']$ only the jobs of set N_j are processed. Therefore, the processing of job u_r is interrupted.

We show that this interruption takes place at time d_j. In fact, in step r, define $\varphi_{u_r}(t) = -W$ in the interval $(0, T']$ and, therefore, the inequality

$$\varphi_{u_r}(d_j) < \varphi_i(d_j) \tag{3.8}$$

holds for all $i \in N_{l_r} \setminus N_j$. Having performed a certain number of steps of the algorithm (i.e., having found a schedule for processing the jobs of set N_j in the interval $(d_j, T']$), we obtain the reduced problem for which the set $\tilde{N}$ of the jobs still to be processed coincides with $N \setminus N_j$. In the next step, we obtain $T = d_j$. Taking into account (3.8) and the fact that in the interval $(\overline{d}_l, d_j]$ only the jobs of set $N_{l_r} \setminus N_j$ can be processed, we conclude that, in this step, job u_r is chosen as job u, i.e. $s^*(t) = u_r$ in the interval with the number d_j. Thus, the processing of job u is interrupted at time d_j.

Any job processed with preemption in schedule s^* can be given similar consideration.

Thus, if, in schedule s^*, there is a preemption at time t', then $t' = d_i$ for some $i \in N$ and, hence, *the total number of preemptions does not exceed* $n-1$. Since job n is processed with no preemption, it follows that the number of steps in the algorithm is at most $2(n-1)+1 = 2n-1$.

Thus, the running time of the algorithm for finding schedule s^* is $O(n^2)$.

3.7. We now consider some *special cases* of the problem of minimizing the maximal cost. Let the cost functions $\varphi_i(t)$ be such that for any ν, $\mu \in N$, either $\varphi_\nu(t) \le \varphi_\mu(t)$ holds for all $t \in (0, T]$ or $\varphi_\nu(t) \ge \varphi_\mu(t)$ holds for all $t \in (0, T]$, where T is computed by formula (3.5). Examples of the cost functions having this property are given in Section 3.3.

In this case, the running time of the algorithm is still $O(n^2)$ but computation the cost functions in each step of the algorithm is not required. To see this, associate each job

$i \in N$ with an integer $J(i)$ (called the job index) so that the relation $\varphi_\nu(t) \leq \varphi_\mu(t)$, $t \in (0, T]$ implies $J(\nu) \leq J(\mu)$. In each step of the algorithm, finding a job $u \in N_l^-$ for which (3.6) holds reduces to finding the job with the smallest index in the set N_l^-. This can be done in at most $O(n)$ time.

One of the functions that has the required property is $\varphi_i(t) = \text{sign}(\max\{t-D_i, 0\})$ (see Section 3.3). Therefore, the proposed algorithm can be used for finding *a schedule for a partially ordered set of jobs that is feasible with respect to deadlines.*

3.8. To conclude this section, note that if, in each step of the algorithm, the relation $t_u \leq T-T'$ holds for job u, then the resulting schedule s^* is non-preemptive. In this case, the algorithm can also be used for solving the problem assuming that *preemption is forbidden* .

Various *sufficient conditions* can be formulated under which this situation takes place. Below, we present some of them.

(a) Let d denote a common divisor of the numbers $d_1, d_2,\ldots, d_n$. If $t_i = d$, $i = 1, 2,\ldots, n$, then s^* is non-preemptive. In fact, in this case T and T' are multiples of d and, hence, $T-T' \geq d$. In particular, s^* is non-preemptive if $t_i = 1$, $i = 1, 2,\ldots, n$.

(b) Schedule s^* is non-preemptive if $\bar{d}_i+t_i \leq \bar{d}_{i+1}$, $i = 1, 2,\ldots, n-1$. In fact, in this case, in step r of the algorithm, we have $T = \bar{d}_{n-r+1}+t_{n-r+1}$, $T' = \bar{d}_{n-r}+t_{n-r}$ and, consequently, $u = n-r+1$ and $T-T' \geq t_u$.

(c) Schedule s^* is non-preemptive if $\varphi_i(t) = \varphi(t-\bar{d}_i)$, $i = 1, 2,\ldots, n$, and φ is a non-decreasing function. In fact, in this case, in each step, the job with the highest number is chosen as job u. Suppose that, in some step, $T' = \bar{d}_j + \sum_{k \in N_j \setminus u} t_k$ holds for some $j \in N$. Then $T'+t_u \leq T$ (otherwise, $T'+t_u = \bar{d}_j + \sum_{k \in N_j} t_k > T$ which contradicts (3.5)).

Note that due to Lemma 3.1, the smallest value of $T(s)$ corresponds to schedule s^*, therefore *a time-optimal* non-preemptive schedule can be obtained, for example, by setting $\varphi_i(t) = t-\bar{d}_i$, $i = 1, 2,\ldots, n$.

If a precedence relation $\rightarrow$ is not specified over set N, then $\bar{d}_i = d_i$, $i = 1, 2,\ldots, n$, and $\varphi_i(t) = \varphi(t-d_i)$ is a function of the flow time of job i. Thus, the smallest value of the *maximal job flow time* and, therefore, of any *non-decreasing function* of this time is achieved when the jobs are processed with no preemption according to the sequence $(1, 2,\ldots, n)$. Recall that, in this case, we have $d_1 \leq d_2 \leq \ldots \leq d_n$.

4. Single Machine. Total Cost

This section considers a number of polynomially solvable single–machine scheduling problems to minimize the total cost.

4.1. The jobs of set $N = \{1, 2,..., n\}$ are processed on a single machine. The release date of a job $i \in N$ is $d_i \geq 0$, its processing time is $t_i > 0$, and its due date is $D_i \geq 0$. Preemption in the processing of any job is allowed. Each job $i \in N$ is associated with a cost function $\varphi_i(t)$ that is non–decreasing (in the planning interval

It is required to find a schedule s^* for processing the jobs of set N which minimizes the function

$$F_{\Sigma}(s) = \sum_{i=1}^{n} \varphi_i(\bar{t}_i(s)) \tag{4.1}$$

Here $\bar{t}_i(s)$ is the completion time of a job i in schedule s. The schedule s^* is called *optimal*.

In the following, we consider the problems of finding optimal schedules if

(a) $\varphi_i(t) = \alpha_i u_i(t)$, where $u_i(t) = 0$ if $t \leq D_i$, $u_i(t) = 1$ if $t > D_i$; $\alpha_i > 0$, $i = 1, 2,..., n$; it is assumed that d_i and D_i are related in the following way: if $d_\nu < d_\mu$, then $D_\nu \leq D_\mu$, $1 \leq \nu, \mu \leq n$; this problem is usually called *the problem of minimizing the weighted number of late jobs*;

(b) $\varphi_i(t)$ are arbitrary (non-decreasing) functions, d_i and D_i are integers, and $t_i = 1$, $i = 1, 2,..., n$;

(c) $\varphi_i(t) = \varphi(t) + \beta_i$, $i = 1, 2,..., n$, where $\varphi(t)$ is a non-decreasing function.

Note that the situation in which the jobs are simultaneously available (i.e., $d_i = 0$, $i = 1, 2,..., n$) and (non-decreasing) cost functions belong to exactly one of the following classes: (1) $\varphi_i(t) = \alpha_i t + \beta_i$, (2) $\varphi_i(t) = \alpha_i \exp(\gamma t) + \beta_i$, and (3) $\varphi_i(t) = \varphi(t) + \beta_i$ is considered in Chapter 3.

4.2. Consider the first of the problems mentioned above. Let d_i and D_i be such that for all $1 \leq \nu, \mu \leq n$, the condition $d_\nu < d_\mu$ implies $D_\nu \leq D_\mu$. Let the jobs be numbered in such a way that the inequalities $d_1 \leq d_2 \leq ... \leq d_n$ and $D_1 \leq D_2 \leq ... \leq D_n$ hold.

A schedule s with no late jobs exists if and only if the inequalities

$$\sum_{i=k}^{l} t_i \leq D_l - d_k \tag{4.2}$$

for any $1 \leq k \leq l \leq n$ (see Section 2.5 of this chapter).

For a non-preemptive schedule determined by a permutation $\pi = (i_1, i_2,..., i_n)$, the completion time of a job i_j, $j = 1, 2,..., n$, is given by

$$\bar{t}_{i_j}(\pi) = \max\{d_{i_j}, \bar{t}_{i_{j-1}}(\pi)\}+t_{i_j}, \ \bar{t}_{i_0}(\pi) = 0. \tag{4.3}$$

For $N' \subseteq N$, let $\overrightarrow{\pi_{N'}}$ denote a permutation of the elements of N' in which the jobs are sorted in numerical order. If the jobs are processed according to the sequence $\overrightarrow{\pi_N} = (1, 2,..., n)$, then it follows from (4.3) that

$$\bar{t}_l(\overrightarrow{\pi_N}) = \max\left\{d_k + \sum_{i=k}^{l} t_i \,\middle|\, 1 \le k \le l\right\} \tag{4.4}$$

for all $l = 1, 2,..., n$.

Comparing (4.2) and (4.4), we come to the following conclusion.

A schedule s with no late jobs exists if and only if for the sequence $\overrightarrow{\pi_N}$ *the inequalities* $\bar{t}_l(\overrightarrow{\pi_N}) \le D_l$ *hold for all* $l = 1, 2,..., n$.

Therefore, if in the sequence $\overrightarrow{\pi_N}$ at least one of the due dates is violated, then *there is no schedule* (either preemptive or non-preemptive) without late jobs.

Let s be some schedule for processing the jobs of set N, and R be a set of late jobs, i.e., jobs which are completed after their due dates in schedule s. Let R' denote a set of late jobs assuming that the jobs are processed according to the sequence $(\overrightarrow{\pi_{N\setminus R}}, \pi_R)$ where π_R is an arbitrary permutation of the elements of R.

We show that $R' \subseteq R$. Let us find a schedule $\bar{s}$ for processing the jobs of the set $N\setminus R$ assuming $\bar{s}(t) = 0$, if $s(t) \in R$ and, otherwise, setting $\bar{s}(t) = s(t)$. It is obvious that in schedule $\bar{s}$, all jobs of set $N\setminus R$ are completed by their due dates. Therefore, there are no late jobs if the jobs are processed according to the sequence $\overrightarrow{\pi_{N\setminus R}}$, i.e. $R' \subseteq R$.

Let $\varphi_i(t) = \alpha_i u_i(t)$, where $u_i(t) = 0$ if $t \le D_i$, $u_i(t) = 1$ if $t > D_i$; $\alpha_i > 0$, $i = 1, 2,..., n$. If R is the set of late jobs schedule s, then $F_\Sigma(s) = \sum_{i\in R} \alpha_i$.

If s^* is an optimal schedule and R^* is the set of late jobs in this schedule, then a schedule $\tilde{s}$ determined by the permutation $(\overrightarrow{\pi_{N\setminus R^*}}, \pi_{R^*})$ is also optimal for any sequence π_{R^*} of the elements of set R^*.

In fact, if $\tilde{R}$ is a set of late jobs in schedule $\tilde{s}$, then $\tilde{R} \subseteq R^*$ and $F_\Sigma(\tilde{s}) = \sum_{i\in\tilde{R}} \alpha_i \le \sum_{i\in R^*} \alpha_i = F_\Sigma(s^*)$.

Thus, in the case under consideration, to find an optimal schedule, it suffices to find a set $R^* \subseteq N$ with the smallest value of $f(R) = \sum_{i\in R} \alpha_i$ such that the processing of the jobs of set $N\setminus R^*$ in numerical order does not imply violation of the due dates. Such a set R^* is called *optimal*. In general, there may be several such sets: $R_1^*, R_2^*,..., R_v^*$. Denote

$H^* = \{R_1^*, R_2^*,..., R_v^*\}$.

In the following, along with the original problem of finding a set R^*, we consider the *reduced* problems derived from the original one by removing a certain subset of jobs.

Let $R \subset R_i^* \in H^*$ and $\overline{R}^*$ be an optimal set for the reduced problem obtained from the original one by removing the set R of jobs. Then the set $R' = R \cup \overline{R}^*$ is optimal for the original problem, i.e. $R' \in H^*$. In fact, none of the jobs in set $N\backslash R'$ is late if they are processed according to the sequence $\overrightarrow{\pi}_{N\backslash R'}$, and

$$f(R') = f(R)+f(\overline{R}^*) \leq f(R)+f(R_i^*\backslash R) = f(R_i^*).$$

Let $\pi = (i_1, i_2,..., i_k)$, and $T(\pi)$ denote the completion time of job i_k, assuming that the jobs of the set $\{i_1, i_2,..., i_k\}$ are processed according to the sequence π. Let $\pi\backslash i_\nu$, $1 \leq \nu \leq k$, denote the permutation obtained from π by removing an element i_ν, i.e., Let $\pi\backslash i_\nu$, $1 \leq \nu \leq k$, denote the permutation obtained from π by deleting an element i_ν, i.e., $\pi\backslash i_\nu = (i_1, i_2,..., i_{\nu-1}, i_{\nu+1},..., i_k)$. Similarly, $\pi\backslash N'$ is the permutation obtained from π by deleting the elements of a set $N' \subseteq \{\pi\} = \{i_1, i_2,..., i_k\}$.

Theorem 4.1. *Let* $\overrightarrow{\pi} = (1, 2,..., n)$, $\overline{t}_j(\overrightarrow{\pi}) \leq D_j$, $j = 1, 2,..., k-1$, $\overline{t}_k(\overrightarrow{\pi}) > D_k$ *and* $\pi = (1, 2,..., k)$. *If there exists such a* μ, $1 \leq \mu \leq k$, *that* $T(\pi\backslash\mu) \leq T(\pi\backslash\nu)$ *and* $\alpha_\mu \leq \alpha_\nu$ *for all* $1 \leq \nu \leq k$, *then there exists such an optimal set* $\tilde{R}^*$ *that* $\mu \in \tilde{R}^*$.

Proof. Note that in the case under consideration, any optimal set $R^* \in H^*$ contains at least one job ν, $1 \leq \nu \leq k$.

We prove the theorem by induction with respect to the number of jobs n. It is obvious that for $n = 1$ the theorem holds. Suppose that the theorem holds $n \leq n_0$. We show that this also holds for $n = n_0+1$.

1. Let $H^* = \{R_1^*, R_2^*,..., R_v^*\}$. In each of the sets R_l^*, choose the job with the highest number θ_l. Define $\theta = \max\{\theta_l | 1 \leq l \leq v\}$.

(a) Suppose that $k < \theta$. Delete the job θ from set N. If the conditions of the theorem hold for the original problem, then they still hold for the obtained reduced problem. Since, in the reduced problem, we have $n = n_0$, it follows that there exists an optimal set which contains job μ. Adding job θ to this set, we obtain the desired optimal set $\tilde{R}^*$.

(b) Let $k = n$. In this case, job n is the only late job in the sequence $\pi = (1, 2,..., n)$. We have $T(\pi\backslash\mu) \leq T(\pi\backslash n) = \overline{t}_{n-1}(\pi) \leq D_{n-1} \leq D_n$. Since $\alpha_\mu \leq \alpha_\nu$ for all $1 \leq \nu \leq n$, we may define $\tilde{R}^* = \{\mu\}$.

2. In what follows, we assume that $\theta \leq k < n$.

Denote $z_j(\pi_{N'}) = \max\{0, \overline{t}_j(\pi_{N'})-D_j\}$, where $j \in N' \subseteq N$. Define $z(\overrightarrow{\pi}) = \max\{z_j(\overrightarrow{\pi})|$

$k \le j \le n\}$.

Since, in the case under consideration, any optimal set R^* is contained in $\{\pi\}$, we have $T(\pi)-T(\pi\backslash R^*) \ge z(\vec{\pi})$. On the other hand, if $T(\pi)-T(\pi\backslash R) \ge z(\vec{\pi})$ for some set $R \subseteq \{\pi\}$ and $\sum_{i \in R} \alpha_i = \sum_{i \in R^*} \alpha_i$, then $R \in H^*$.

(a) Let $k \le n-2$. Formulate a new problem obtained by deleting the jobs $k+1$, $k+2, \dots, n$ from set N, followed by adding a new job $k+1$ with $d_{k+1} = d_k$, $D_{k+1} = T(\pi)$, $t_{k+1} = z(\vec{\pi})$ and $\alpha_{k+1} = W$, where W is a sufficiently large number. Let $\bar{N}$ denote the obtained set of jobs.

It is obvious that, for any $R^* \in H^*$, no job of the set $\bar{N}\backslash R^*$ is late if these jobs are processed according to the sequence $\vec{\pi}_{\bar{N}\backslash R^*}$. If $\bar{R}^*$ is an optimal set for the new problem, then $\sum_{i \in \bar{R}^*} \alpha_i = \sum_{i \in R^*} \alpha_i$. Since $\alpha_{k+1} = W$, it follows that $\bar{R}^* \subseteq \{\pi\}$. We have that

$$\bar{t}_{k+1}(\vec{\pi}_{\bar{N}\backslash \bar{R}^*}) = T(\pi\backslash \bar{R}^*)+t_{k+1} = T(\pi\backslash \bar{R}^*)+z(\vec{\pi}) \le D_{k+1} = T(\pi)$$

holds. Therefore, $T(\pi)-T(\pi\backslash \bar{R}^*) > z(\vec{\pi})$. Thus, $\bar{R}^*$ is an optimal set for the original problem.

For the new problem, we have $|N'| = k+1 \le n-1 = n_0$. If the theorem conditions hold for the original problem, then they still hold for the new problem. By induction, there exists a set $\bar{R}^*$ which contains μ.

(b) Let $k = n-1$ and $z_{n-1}(\vec{\pi}) \ge z_n(\vec{\pi})$. We have $T(\pi\backslash\mu) \le T(\pi\backslash(n-1)) \le D_{n-2} \le D_{n-1}$ and $T(\pi) = D_{n-1}+z_{n-1}(\pi)$. Hence, $T(\pi)-T(\pi\backslash\mu) \ge z_{n-1}(\pi) = z(\vec{\pi})$. Therefore, we may set $\tilde{R}^* = \{\mu\}$.

(c) Suppose that $k = n-1$ and $z_{n-1}(\vec{\pi}) < z_n(\vec{\pi})$. Let R^* be some optimal set which does not contain μ, and let γ be a job in R^* with the lowest number. Denote $\bar{R} = R^*\backslash\gamma$. To prove the theorem, it suffices to show that the set $R = \bar{R} \cup \mu$ is optimal.

Since $\alpha_\mu \le \alpha_\gamma$, we have $\sum_{i \in R} \alpha_i \le \sum_{i \in R^*} \alpha_i$. Therefore the set R is optimal if $T(\pi\backslash R) \le D_{n-1}$ and $T(\vec{\pi}\backslash R) \le D_n$. Since $T(\pi\backslash\mu) \le T(\pi\backslash(n-1)) \le D_{n-2} \le D_{n-1}$ and $\mu \in R$, we have $T(\pi\backslash R) \le D_{n-1}$. It is obvious that $T(\vec{\pi}\backslash R^*) \le D_n$. Therefore, we have only to show that

$$T(\vec{\pi}\backslash R) \le T(\vec{\pi}\backslash R^*). \tag{4.5}$$

We show that, in the case under consideration, the inequality $T(\pi\backslash\mu) \le T(\pi\backslash\gamma)$ implies

$$T(\vec{\pi}\backslash\mu) \le T(\vec{\pi}\backslash\gamma) \tag{4.6}$$

which, in turn, implies

$$T(\vec{\pi}\backslash\{\mu, r\}) \le T(\vec{\pi}\backslash\{\gamma, r\}) \tag{4.7}$$

for any $r \in \bar{R}$.

Inequality (4.6) follows from the inequality $T(\pi\backslash\mu) \le T(\pi\backslash\gamma)$ and from the obvious relation $T(\overrightarrow{\pi}\backslash\nu) = \max\{T(\pi\backslash\nu),\ d_n\}+t_n$ valid for any $\nu \in \{\pi\}$.

If $\overline{R} = \emptyset$, then (4.6) gives (4.5), and the theorem is proved. Therefore, in the following, we assume that $\overline{R} \neq \emptyset$ and $\gamma < k$.

Let us prove that inequality (4.7) holds. We introduce the following notation: $\Delta(\overrightarrow{\pi}\backslash i) = T(\overrightarrow{\pi})-T(\overrightarrow{\pi}\backslash i)$ and $\Delta(\overrightarrow{\pi}\backslash i\backslash j) = T(\overrightarrow{\pi}\backslash i)-T(\overrightarrow{\pi}\backslash\{i,\ j\})$, $i,\ j \in \{\pi\}$. Define $\delta_i = \overline{t}_{i-1}(\overrightarrow{\pi})-d_i$ for $i = 2, 3,\ldots, n$, $\delta_i^r = \overline{t}_{i-1}(\overrightarrow{\pi}\backslash r)-d_i$ for $i \geq r+2$, and $\delta_{r+1}^r = \overline{t}_{r-1}(\overrightarrow{\pi}\backslash r)-d_{r+1}$.

Note that $\delta_i > 0$ for all $i \geq \gamma+1$ (otherwise, deleting γ from π does not affect the completion times of jobs $n-1$ and n and, hence, $\gamma \notin R^*$). The inequality $T(\pi\backslash\mu) \leq T(\pi\backslash(n-1))$ implies that $\delta_i \geq \delta_{n-1}$ and, therefore, $\delta_i > 0$ for all $i \geq \mu+1$.

It is clear that

$$\Delta(\overrightarrow{\pi}\backslash\mu) = \min\{t_\mu,\ \min\{\delta_i\,|\,\mu+1 \leq i \leq n\}\},$$
$$\Delta(\overrightarrow{\pi}\backslash\gamma) = \min\{t_\gamma,\ \min\{\delta_i\,|\,\gamma+1 \leq i \leq n\}\}.$$

It follows from (4.6) that

$$\Delta(\overrightarrow{\pi}\backslash\mu) \geq \Delta(\overrightarrow{\pi}\backslash\gamma). \tag{4.8}$$

Note that $\delta_i^r > 0$ (otherwise, $\gamma \notin R^*$) and $\delta_i > \delta_i^r$ for $i \geq r+1$. Also, observe that $\delta_r \geq \delta_{r+1}^r$, since

$$\delta_{r+1}^r = \overline{t}_{r-1}(\overrightarrow{\pi}\backslash r)-d_{r+1} = \overline{t}_{r-1}(\overrightarrow{\pi})-d_{r+1} \leq \overline{t}_{r-1}(\overrightarrow{\pi})-d_r.$$

Suppose that $r > \max\{\gamma,\ \mu\}$. Then

$$\Delta(\overrightarrow{\pi}\backslash r\backslash\mu) = \min\{t_\mu,\ \min\{\delta_i\,|\,\mu+1 \leq i \leq r-1\},\ \min\{\delta_i^r\,|\,r+1 \leq i \leq n\}\},$$

$$\Delta(\overrightarrow{\pi}\backslash r\backslash\gamma) = \min\{t_\gamma,\ \min\{\delta_i\,|\,\gamma+1 \leq i \leq r-1\},\ \min\{\delta_i^r\,|\,r+1 \leq i \leq n\}\}.$$

Denote $a = \min\{t_\mu,\ \min\{\delta_i\,|\,\mu+1 \leq i \leq r-1\}\}$, $b = \min\{t_\gamma,\ \min\{\delta_i\,|\,\gamma+1 \leq i \leq r-1\}\}$, $c = \min\{\delta_i\,|\,r \leq i \leq n\}$, $d = \min\{\delta_i^r\,|\,r+1 \leq i \leq n\}$. Then $\Delta(\overrightarrow{\pi}\backslash\mu) = \min\{a,\ c\}$, $\Delta(\overrightarrow{\pi}\backslash\gamma) = \min\{b,\ c\}$, $\Delta(\overrightarrow{\pi}\backslash r\backslash\mu) = \min\{a,\ d\}$, $\Delta(\overrightarrow{\pi}\backslash r\backslash\gamma) = \min\{b,\ d\}$.

Inequality (4.8) can be written as

$$\min\{a,\ c\} \geq \min\{b,\ c\}. \tag{4.9}$$

The inequalities $\delta_i > \delta_i^r$, $i \geq r+1$, and $\delta_r \geq \delta_{r+1}^r$ imply $c \geq d$. We show that, in this case, the inequality

$$\min\{a,\ d\} \geq \min\{b,\ d\}. \tag{4.10}$$

holds. If $c \leq a$, then $\min\{a,\ d\} = d$. If $c > a$, then it follows from (4.9) that $a \geq b$, hence (4.10) holds.

Inequality (4.10) implies $\Delta(\overrightarrow{\pi}\backslash r\backslash\mu) \geq \Delta(\overrightarrow{\pi}\backslash r\backslash\gamma)$ and, therefore, relation (4.7) holds.

Suppose that $\gamma < r < \mu$. In this case

$$\Delta(\overrightarrow{\pi}\backslash r\backslash \mu) = \min\{t_\mu, \min\{\delta_i^r \mid \mu+1 \le i \le n\}\},$$
$$\Delta(\overrightarrow{\pi}\backslash r\backslash \gamma) = \min\{t_\gamma, \min\{\delta_i \mid \gamma+1 \le i \le r-1\}, \min\{\delta_i^r \mid r+1 \le i \le n\}\}.$$

Denote $A = t_\mu$, $B = \min\{t_\gamma, \min\{\delta_i \mid \gamma+1 \le i \le r-1\}\}$, $C = \min\{\delta_i \mid \mu+1 \le i \le n\}$, $D = \min\{\delta_i \mid r+1 \le i \le \mu\}$, $E = \delta_r$, $C' = \min\{\delta_i^r \mid \mu+1 \le i \le n\}$, $D' = \min\{\delta_i^r \mid r+1 \le i \le \mu\}$. Then

$$\Delta(\overrightarrow{\pi}\backslash \mu) = \min\{A, C\},\ \Delta(\overrightarrow{\pi}\backslash \gamma) = \min\{B, E, D, C\},$$
$$\Delta(\overrightarrow{\pi}\backslash r\backslash \mu) = \min\{A, C'\},\ \Delta(\overrightarrow{\pi}\backslash r\backslash \gamma) = \min\{B, D', C'\}.$$

Inequality (4.8) can be written as

$$\min\{A, C\} \ge \min\{B, E, D, C\}. \tag{4.11}$$

It follows from $\delta_i > \delta_i^r$, $i \ge r+1$, that $D > D'$ and $C > C'$, while $\delta_r \ge \delta_{r+1}^r$ implies $E \ge D'$.

We show that it follows from (4.11) that

$$\min\{A, C'\} \ge \min\{B, D', C'\}. \tag{4.12}$$

In fact, if $C' \le A$, then $C' \ge \min\{B, D', C'\}$. Otherwise, $\min\{A, C'\} = A = \min\{A, C\} \ge \min\{B, E, D, C\} \ge \min\{B, D', C'\}$.

Inequality (4.12) implies $\Delta(\overrightarrow{\pi}\backslash r\backslash \mu) \ge \Delta(\overrightarrow{\pi}\backslash r\backslash \gamma)$ and, therefore, relation (4.7) holds.

We now pass to the direct proof of relation (4.5). Delete job r from the set N. It is obvious that set $R^*\backslash r$ is optimal for the obtained reduced problem. Since $|R^*| \ge 2$, we have $z_n(\overrightarrow{\pi}_{N\backslash r}) > 0$. If job n is the only late job, then Item (1b) of this proof implies $\tilde{R}^* = \{\mu, r\}$. Suppose that $z_{n-1}(\overrightarrow{\pi}_{N\backslash r}) > 0$. Inequality (4.7) implies $T(\overrightarrow{\pi}_{N\backslash r}\backslash \mu) \le T(\overrightarrow{\pi}_{N\backslash r}\backslash \gamma)$. It follows from the latter inequality (see (4.6) and (4.7)) that $T(\overrightarrow{\pi}_{N\backslash r}\backslash\{\mu, r_1\}) \le T(\overrightarrow{\pi}_{N\backslash r}\backslash\{\gamma, r_1\})$, where $r_1 \in \overline{R}\backslash r$. The last inequality can be written as

$$T(\overrightarrow{\pi}\backslash\{\mu, r, r_1\}) \le T(\overrightarrow{\pi}\backslash\{\gamma, r, r_1\}).$$

Repeating similar considerations finitely many times, we conclude that relation (4.5) holds. The theorem is proved.

Corollary 4.1. *Let $R \subset R^* \in H^*$, $\overrightarrow{\pi}_{N\backslash R} = (i_1, i_2, \ldots, i_k, \ldots, i_r)$, $\bar{t}_{i_j}(\overrightarrow{\pi}_{N\backslash R}) \le D_{i_j}$, $j = 1, 2, \ldots, k-1$, $\bar{t}_{i_k}(\overrightarrow{\pi}_{N\backslash R}) > D_{i_k}$ and $\pi = (i_1, i_2, \ldots, i_k)$. If there exists a μ, $1 \le \mu \le k$, such that $T(\pi\backslash i_\mu) \le T(\pi\backslash i_\nu)$ and $\alpha_{i_\mu} \le \alpha_{i_\nu}$ for all $1 \le \nu \le k$, then there exists a set $\tilde{R}^* \in H^*$ such that $R \cup i_\mu \subseteq \tilde{R}^*$.*

This statement directly follows from Theorem 4.1 and the above remark on the relation between optimal solutions of the original and reduced problems.

4.3. In a general case, the search for an optimal set R^* involves a large number of variants. In enumerative solution methods, applying Corollary 4.1 can frequently reduce this search. In this section, several *special cases* of the problem are considered for which the set R^* can be found as a result of systematic application of Corollary 4.1.

In these cases, *the algorithm* for finding the set R^* is as follows. It is assumed, as before, that for all $1 \leq \nu, \mu \leq n$ the condition $d_\nu < d_\mu$ implies $D_\nu \leq D_\mu$. Let the jobs be numbered in such a way that the inequalities $d_1 \leq d_2 \leq \ldots \leq d_n$ and $D_1 \leq D_2 \leq \ldots \leq D_n$ hold. Define $R = \emptyset$. In each step, find the first late job i_k, provided that the jobs are processed according to the sequence $\overrightarrow{\pi}_{N \setminus R} = (i_1, i_2, \ldots, i_r)$. If there are no late jobs, then $R^* = R$, otherwise, define $\pi = (i_1, i_2, \ldots, i_k)$. In the situations considered below, in each step of the algorithm a job i_μ can be found such that $T(\pi \setminus i_\mu) \leq T(\pi \setminus i_\nu)$ and $\alpha_{i_\mu} \leq \alpha_{i_\nu}$ for all $1 \leq \nu \leq k$. Find this job, redefine R to be equal to $R \cup i_\mu$ and go to the next step. It is obvious that the number of steps in the algorithm is at most n. We show that *the running time* of the algorithm is at most $O(n^2)$.

Numbering the jobs in such a way that the inequalities $d_1 \leq d_2 \leq \ldots \leq d_n$ and $D_1 \leq D_2 \leq \ldots \leq D_n$ hold (or verifying that this numbering is impossible) takes $O(n\log n)$ time (see Section 2.1 of Chapter 1).

In each step, the procedure for finding the job i_k can be implemented as follows. Let u and v be numbers of the jobs i_k and i_μ, respectively, and σ be a subsequence of π found in the previous step. Let $\overrightarrow{\pi}_{N \setminus R} = (i_1, i_2, \ldots, i_r)$. Compute $\overline{t}_{i_j}(\overrightarrow{\pi}_{N \setminus R})$ for $i_j \geq u+1$ by formula (4.3), assuming $\overline{t}_{i_{j-1}}(\overrightarrow{\pi}_{N \setminus R}) = T(\sigma \setminus v)$ for $i_j = u+1$ (in the first step, assume $u = 0$ and $t_{i_0}(\overrightarrow{\pi}_{N \setminus R}) = 0$). Comparing $\overline{t}_{i_j}(\overrightarrow{\pi}_{N \setminus R})$ and D_{i_j}, $i_j \geq u+1$, choose the first job, for which the inequality $\overline{t}_{i_j}(\overrightarrow{\pi}_{N \setminus R}) > D_{i_j}$ holds, as the job i_k. It is obvious that finding jobs i_k (in all steps of the algorithm) requires at most $O(n)$ time.

Consider the procedure of choosing the job i_μ such that $T(\pi \setminus i_\mu) \leq T(\pi \setminus i_\nu)$ holds for all $1 \leq \nu \leq k$. If $k = 1$, then $i_\mu = i_k$. If $k > 1$, the job i_μ can be found in k iterations. At the first iteration, define $p = i_1$, form two dummy sequences σ_1 and σ_1', and define $T(\sigma_1) = T(\sigma_1') = 0$. At iteration l, $l = 2, 3, \ldots, k$, define $\sigma_l = (\sigma_{l-1}, i_l)$ and $\sigma_l' = (\sigma_{l-1}', i_{l-1})$. Compute $T(\sigma_l) = \max\{T(\sigma_{l-1}), d_{i_l}\} + t_{i_l}$,and $T(\sigma_l') = \max\{T(\sigma_{l-1}'), d_{i_{l-1}}\} + t_{i_{l-1}}$. If $T(\sigma_l) > T(\sigma_l')$, set p equal to i_l, the sequence σ_l equal to σ_l', and the value of $T(\sigma_l)$ equal to $T(\sigma_l')$. It is clear that job p found after the kth iteration satisfies the condition $T(\pi \setminus p) \leq T(\pi \setminus i_\nu)$ for all $1 \leq \nu \leq k$. Define $i_\mu = p$. It is easy to verify that finding job i_μ in each step takes at most $O(k)$ time, or at most $O(n)$ time.

Thus, in the case under consideration, finding an optimal set R^* requires at most $O(n^2)$ time.

(a) Consider the problem on minimizing *the number of late jobs with the same release dates*. In this case, $d_1 = d_2 = ... = d_n = 0$, $\alpha_1 = \alpha_2 = ... = \alpha_n = 1$.

Let the jobs be numbered in non-decreasing order of their due dates: $D_1 \le D_2 \le ... \le D_n$. Since $T(\pi \backslash i_\nu) = T(\pi) - t_{i_\nu}$, $1 \le \nu \le k$, it follows that the job $i \in \{\pi\}$ with the longest processing time t_i has to be chosen as i_ν.

Example. Let $N = \{1, 2, 3, 4, 5, 6\}$, $d_i = 0$, $\alpha_i = 1$, $i = 1, 2, ..., 6$; the values of t_i and D_i are given in Table 4.1.

Table 4.1

i	1	2	3	4	5	6
t_i	4	1	3	2	3	1
D_i	4	5	6	7	7	8

The set R, the sequence $\vec{\pi}_{N\backslash R} = (i_1, i_2, ..., i_r)$ and the values of $\bar{t}_{i_j}(\vec{\pi}_{N\backslash R})$ and D_{i_j}, $j = 1, 2, ..., r$, for each step of the algorithm are shown in Table 4.2. This table also contains the values of i_k, $\pi = (i_1, i_2, ..., i_k)$ and i_μ obtained in each step. Note that, in the second step, either job 3 or job 5 can be chosen as i_μ. Here, we have chosen $i_\mu = 3$.

We have $R^* = \{1, 3\}$. The schedules defined by the permutations $\pi_1^* = (2, 4, 5, 6, 1, 3)$ and $\pi_2^* = (2, 4, 6, 3, 1)$ are optimal. There are two late jobs.

Table 4.2

Step	1						2					3			
R	∅						1					1, 3			
$\vec{\pi}_{N\backslash R} = (i_1, i_2, ..., i_r)$	1	2	3	4	5	6	2	3	4	5	6	2	4	5	6
$\bar{t}_{i_j}(\vec{\pi}_{N\backslash R})$	4	5	8	10	13	14	1	4	6	9	10	1	3	6	7
D_{i_j}	4	5	6	7	7	8	5	6	7	7	8	5	7	7	8
i_k	3						5					-			
$\pi = (i_1, i_2, ..., i_k)$	(1, 2, 3)						(2, 3, 4, 5)					-			
i_μ	1						3								

In the case under consideration, finding an optimal set R^* can be implemented in *at most* $O(n\log n)$ *time* by using balanced 2-3-trees for data representation (see Section 2 of Chapter 1). Define the total pseudo-order $\Longrightarrow$ over set N in the following way: $i \Longrightarrow j$ if and only if $t_i \geq t_j$. In each step of the algorithm an ordered set $\{\pi\}$ is represented as a balanced 2-3-tree. Then finding job i_μ (which is a maximal element with respect to $\Longrightarrow$) takes constant time; in fact, one elementary operation is required. Deleting job i_μ from $\{\pi\}$ takes $O(\log n)$ time. Therefore, finding job i_μ in all steps of the algorithm requires at most $O(n\log n)$ time.

We show that constructing balanced 2-3-trees in all steps of the algorithm can be done in at most $O(n\log n)$ time. In the first step, constructing the tree takes $O(n)$ time. Let u be the number of job i_k and N' be a set $\{\pi\}\backslash i_\mu$ found in some step of the algorithm. In the next step, finding the set $\{\pi\}$ involves including jobs $u+1$, $u+2,\ldots$, v into set N', where v is the number of the job i_k in this step. Obtaining the balanced 2-3-tree corresponding to the set $\{\pi\}$ from the tree for the set N' can be done in $O(n_i\log n)$ time, where $n_i = v-u$. Thus, representing the sets $\{\pi\}$ by the balanced 2-3-trees in all steps of the algorithm requires at most $O(n\log n)$ time.

(b) Suppose that $d_i = 0$, $i = 1, 2,\ldots, n$, and, for any $1 \leq i, j \leq n$, the condition $t_i < t_j$ implies $\alpha_i \geq \alpha_j$. Let the jobs be numbered in non-decreasing order of their due dates, i.e., $D_1 \leq D_2 \leq \ldots \leq D_n$.

Since $T(\pi\backslash i_\nu) = T(\pi)-t_{i_\nu}$, $1 \leq \nu \leq k$, it follows that the job $i \in \{\pi\}$ with the longest processing time t_i and the smallest weight α_i has to be chosen as the job i_μ. In this case, *finding an optimal set requires at most $O(n\log n)$ time.*

(c) Consider the problem on minimizing *the number of late jobs with different release dates.* As before, assume that the jobs are numbered in such a way that $d_1 \leq d_2 \leq \ldots \leq d_n$ and $D_1 \leq D_2 \leq \ldots \leq D_n$.

Since, in this case $\alpha_1 = \alpha_2 = \ldots = \alpha_n = 1$, it follows that for finding the job i_μ it suffices to compute $T(\pi\backslash i_\nu)$ for all $1 \leq \nu \leq k$ and to choose the job with the smallest of these values.

Example. Let $N = \{1, 2, 3, 4, 5, 6\}$, $\alpha_i = 1$, $i = 1, 2,\ldots, 6$. The values of t_i, d_i and D_i are given in Table 4.3.

Table 4.3

i	1	2	3	4	5	6
t_i	1	2	2	3	2	1
d_i	1	2	3	3	4	7
D_i	3	4	5	7	8	9

(1) Define $R = \varnothing$. The sequence $\overrightarrow{\pi_{N\backslash R}} = (i_1, i_2, \dots, i_r) = (1, 2, 3, 4, 5, 6)$. To find the job i_k, compute $\overline{t}_{i_j}(\overrightarrow{\pi_{N\backslash R}})$ by formula (4.3) and compare it with D_{i_j}: $\overline{t}_1(\overrightarrow{\pi_{N\backslash R}}) = 2 < D_1$, $\overline{t}_2(\overrightarrow{\pi_{N\backslash R}}) = 4 = D_2$, $\overline{t}_3(\overrightarrow{\pi_{N\backslash R}}) = 6 > D_3$. Thus, $i_k = 3$ and $\pi = (1, 2, 3)$.

For finding the job i_μ, we use the procedure described above when analyzing the running time for constructing an optimal set R^*. At the first iteration, we have $p = 1$, $\sigma_1 = \sigma_1' = (\varnothing)$, $T(\sigma_1) = T(\sigma_1') = 0$. At the second iteration, $\sigma_2 = (2)$, $\sigma_2' = (1)$, $T(\sigma_2) = 4$ and $T(\sigma_2') = 2$. Since $T(\sigma_2) > T(\sigma_2')$, define $p = 2$, $\sigma_2 = \sigma_2' = (1)$ and $T(\sigma_2) = T(\sigma_2') = 2$. At the third iteration $\sigma_3 = (1, 3)$, $\sigma_3' = (1, 2)$; $T(\sigma_3) = 5$ and $T(\sigma_3') = 4$. Since $T(\sigma_3) > T(\sigma_3')$, define $p = 3$, $\sigma_3 = \sigma_3' = (1, 2)$, $T(\sigma_3) = T(\sigma_3') = 4$. Define $i_\mu = p = 3$ and $R = \{3\}$.

(2) We have $R = \{3\}$, $\overrightarrow{\pi_{N\backslash R}} = (1, 2, 4, 5, 6)$, $i_k = 5$, $\pi = (1, 2, 4, 5)$ and $i_\mu = 4$. Define $R = \{3, 4\}$.

(3) If the jobs are processed according to the sequence $\overrightarrow{\pi_{N\backslash R}} = (1, 2, 5, 6)$, there are no late jobs. Therefore, $R^* = \{3, 4\}$ and the schedules specified by the permutations $\pi_1^* = (1, 2, 5, 6, 3, 4)$ and $\pi_2^* = (1, 2, 5, 6, 4, 3)$ are optimal. The number of late jobs is 2.

(d) Suppose that the jobs can be numbered so that $d_1 \le d_2 \le \dots \le d_n$, $D_1 \le D_2 \le \dots \le D_n$, $t_1 \le t_2 \le \dots \le t_n$, $\alpha_1 \ge \alpha_2 \ge \dots \ge \alpha_n$.

In this case, while finding an optimal set R^*, the job i_k can be chosen as i_μ. In fact, $\alpha_{i_k} \le \alpha_{i_\nu}$ and $T(\pi\backslash i_k) \le T(\pi) - t_{i_k} \le T(\pi) - t_{i_\nu} \le T(\pi\backslash i_\nu)$, $1 \le \nu \le k-1$.

As shown above, finding the jobs i_k in all steps of the algorithm can be done in $O(n)$ time. Therefore, in the case under consideration, *finding an optimal set requires at most* $O(n\log n)$ *time.*

(e) Suppose that the jobs can be numbered in such a way that $d_1 \le d_2 \le \dots \le d_n$, $D_1 \le D_2 \le \dots \le D_n$, $\alpha_1 \ge \alpha_2 \ge \dots \ge \alpha_n$ and $t_i \le d_{i+1} - d_i$, $i = 1, 2, \dots, n-1$.

In this case, $T(\pi\backslash i_k) \le T(\pi) - t_{i_k}$, $T(\pi\backslash i_\nu) = T(\pi)$ and $\alpha_{i_k} \le \alpha_{i_\nu}$, $1 \le \nu \le k-1$. Therefore, job i_k can be taken as i_μ. *Finding an optimal set requires* $O(n\log n)$ *time.*

(f) Suppose that the jobs can be numbered so that $D_1 \le D_2 \le \dots \le D_n$, $\alpha_1 \le \alpha_2 \le \dots \le \alpha_n$, $d_i = d_{i-1} + t$, $2(n-i)t \le t_i \le 2(n-i)t + t$, $i = 1, 2, \dots, n$, $d_0 = 0$, $t > 0$.

In this case, we have

$$T(\pi\backslash i_1) = T(\pi) - (t_{i_1} - t) \le T(\pi) - (2(n-i_1)t - t),$$

$$T(\pi\backslash i_\nu) = T(\pi) - t_i \ge T(\pi) - (2(n-i_\nu)t + t),\ 2 \le \nu \le k.$$

Since $i_1 \le i_\nu - 1$, we have $T(\pi \backslash i_1) \le T(\pi \backslash i_\nu)$ for all $2 \le \nu \le k$. Since $\alpha_{i_1} \le \alpha_{i_\nu}$, $2 \le \nu \le k$, it follows that the job i_1 can be chosen as i_μ. *Finding an optimal set takes* $O(n\log n)$ *time.*

4.4. Consider the problem of minimizing the total cost that differs from the problem considered in Section 4.2 in the following: (1) $d_i = 0$, $i = 1, 2,..., n$, and (2) the jobs of a given set $Q \subseteq N$ must be completed before their corresponding due dates.

The cost functions are of the form $\varphi_i(t) = \alpha_i u_i(t)$, where $u_i(t) = 0$, if $t \le D_i$ and $u_i(t) = 1$, if $t > D_i$; $\alpha_i > 0$, $i \notin Q$. It is required to find a schedule s^* with the lowest total cost, provided that jobs of set Q do not violate their due dates. Such a schedule is called *optimal*.

Let the jobs be numbered in non-decreasing order of their due dates. If R^* is a set of late jobs in schedule s^*, then, similarly to Section 4.2, a schedule $\tilde{s}$ determined by a permutation $(\overrightarrow{\pi}_{N \backslash R^*}, \pi_{R^*})$ is optimal for any sequence π_{R^*} of the jobs of R^*. Thus, the problem reduces to finding such a set R^* of jobs such that (a) $R^* \subseteq N \backslash Q$; (b) jobs of the set $N \backslash R^*$ processed in numerical order do not violate their due dates, and (c) for any set R satisfying the conditions (a) and (b), the lowest value of $f(R) = \sum_{i \in R} \alpha_i$ corresponds to the set R^*. The set R^* is called *optimal* for the problem under consideration.

Let $Q = \{q_1, q_2,..., q_p\}$, where $q_1 < q_2 < ... < q_p$. If $\sum_{l=1}^{j} t_{q_l} > D_{q_j}$ for some $1 \le j \le p$, then the problem has no solution. Then, we assume that $\sum_{l=1}^{j} t_{q_l} \le D_{q_j}$ for $j = 1, 2,..., p$. Let us modify the job due dates in the following way. Define $\bar{D}_i = D_i$ for each job i, $q_p \le i \le n$. Define $\bar{D}_i = \min\{D_i, \bar{D}_{q_j} - t_{q_j}\}$ for each job i, $q_{j-1} \le i \le q_j - 1$, where $j = p$, $p-1,..., 1$ and $q_0 = 1$.

Let us consider *the reduced problem* obtained from the original one by removing the jobs of set Q, followed by making corresponding changes to the due dates for the remaining jobs. Assign the due date D'_i to a job $i \in N' = N \backslash Q$ in the following way. Define $D'_i = \bar{D}_i$ for each job i, $i < q_1$. Define $D'_i = \bar{D}_i - \sum_{l=1}^{j} t_{q_l}$ for each job i, $q_j < i < q_{j+1}$, where $1 \le j \le p$ and $q_{p+1} = n+1$.

It can be easily shown that the condition $\sum_{l=1}^{j} t_{q_l} \le D_{q_j}$, $j = 1,...,p$, implies that $D'_i \ge 0$ for all $i \in N'$. We show that, for the reduced problem, the relation $D'_\nu \le D'_\mu$ holds for $\nu < \mu$. To do this, it suffices to show that $D'_\nu \le D'_\mu$, provided that job ν directly precedes job μ in the sequence $\overrightarrow{\pi}_{N'}$.

Suppose that $q_j < \nu < \mu < q_{j+1}$, $1 \le j \le p$. Then it follows from $D_\nu \le D_\mu$ that $\bar{D}_\nu \le \bar{D}_\mu$, and this implies $D'_\nu \leq D'_\mu$. It is easy to verify that for $1 \leq \nu < \mu < q_1$, $D'_\nu \leq D'_\mu$ as well.

Suppose that $\nu < q_j < \mu < q_{j+1}$, $1 \leq j \leq p$. If $j = 1$, then $D'_\nu = \bar{D}_\nu = \min\{D_\nu, \bar{D}_{q_1} - t_{q_1}\}$. If $p > 1$, then $D'_\nu = \min\{D_\nu, D_{q_1} - t_{q_1}, \bar{D}_{q_2} - t_{q_1} - t_{q_2}\}$. If $p = 1$, then $D'_\nu = \min\{D_\nu, D_{q_1} - t_{q_1}\}$. On the other hand, $D'_\mu = \bar{D}_\mu - t_{q_1}$. If $p > 1$, then $D'_\mu = \min\{D_\mu - t_{q_1}, \bar{D}_{q_2} - t_{q_1} - t_{q_2}\}$. If $p = 1$, then $D'_\mu = D_\mu - t_{q_1}$. Since $D_{q_1} \leq D_\mu$, we have $D'_\nu \leq D'_\mu$.

If $j > 1$, then

$$D'_\nu = \bar{D}_\nu - \sum_{l=1}^{j-1} t_{q_l} = \min\left\{D_\nu - \sum_{l=1}^{j-1} t_{q_l},\ \bar{D}_{q_j} - \sum_{l=1}^{j} t_{q_l}\right\}.$$

If $p > j$, then

$$D'_\nu = \min\left\{D_\nu - \sum_{l=1}^{j-1} t_{q_l},\ D_{q_j} - \sum_{l=1}^{j} t_{q_l},\ \bar{D}_{q_{j+1}} - \sum_{l=1}^{j+1} t_{q_l}\right\}.$$

If $p = j$, then

$$D'_\nu = \min\left\{D_\nu - \sum_{l=1}^{j-1} t_{q_l},\ D_{q_j} - \sum_{l=1}^{j} t_{q_l}\right\}.$$

On the other hand,

$$D'_\mu = \bar{D}_\mu - \sum_{l=1}^{j} t_{q_l}.$$

If $p > j$, then

$$D'_\mu = \min\left\{D_\mu - \sum_{l=1}^{j} t_{q_l},\ \bar{D}_{q_{j+1}} - \sum_{l=1}^{j+1} t_{q_l}\right\}.$$

If $p = j$, then

$$D'_\mu = D_\mu - \sum_{l=1}^{j} t_{q_l}.$$

Since $D_{q_j} \leq D_\mu$, we have $D'_\nu \leq D'_\mu$.

Similarly, we can show that $D'_\nu \leq D'_\mu$ for $\nu < q_j$, $q_k < \mu < q_{k+1}$, $1 \leq j < k \leq p$.

Since for the reduced problem, the inequality $0 \leq D'_\nu \leq D'_\mu$ holds for any jobs $\nu, \mu \in N'$, $\nu < \mu$, it follows that this problem belongs to the class of problems considered in Sections 4.2 and 4.3. In particular, if for the reduced problem the condition $t_\nu < t_\mu$ implies $\alpha_\nu \geq \alpha_\mu$ for any $\nu, \mu \in N'$, then an optimal set can be found in $O(n'\log n')$ time, where $n' = |N'|$ (see Item (b) of Section 4.3).

Theorem. 4.2. *A set R' optimal for a reduced problem is optimal for the original problem.*

Proof. Let R^* be an optimal set for the original problem. It is obvious that $R^* \subseteq N'$.

To prove the theorem, it suffices to show that (a) $\overline{t}_i(\overrightarrow{\pi}_{N\setminus R^*}) \le D'_i$ for all $i \in N\setminus R^*$, and (b) $\overline{t}_i(\overrightarrow{\pi}_{N\setminus R'}) \le D_i$ for all $i \in N\setminus R'$. In fact, relation (a) implies $\sum_{i\in R'} \alpha_i \le \sum_{i\in R^*} \alpha_i$, while the latter inequality and (b) imply optimality of the R' for the original problem.

(1) First, we prove that relation (a) holds. For any $i \in N\setminus R^*$ the inequality $\overline{t}_i(\overrightarrow{\pi}_{N\setminus R^*}) \le D_i$ holds, while for i, $q_{j-1} \le i < q_j - 1$, $j = 1, 2,\dots, p$, $\overline{t}_i(\overrightarrow{\pi}_{N\setminus R^*}) \le D_{q_k} - \sum_{l=1}^{k} t_{q_l}$, $j \le k \le p$, is valid. Hence, $\overline{t}_i(\overrightarrow{\pi}_{N\setminus R^*}) \le \overline{D}_i$, $i \in N\setminus R^*$.

For a job i, $i \le q_1 - 1$, we have $\overline{t}_i(\overrightarrow{\pi}_{N\setminus R^*}) = \overline{t}_i(\overrightarrow{\pi}_{N\setminus R^*}) \le \overline{D}_i = D'_i$. For a job i, $q_j < i < q_{j+1}$, we have that

$$\overline{t}_i(\overrightarrow{\pi}_{N\setminus R^*}) = \overline{t}_i(\overrightarrow{\pi}_{N\setminus R^*}) - \sum_{l=1}^{j} t_{q_l} \le \overline{D}_i - \sum_{l=1}^{j} t_{q_l} = D'_i$$

holds. Therefore, $\overline{t}_i(\overrightarrow{\pi}_{N\setminus R^*}) \le D'_i$ for all $i \in N\setminus R^*$.

(2) Now we prove that relation (b) also holds. The inequality $\overline{t}_i(\overrightarrow{\pi}_{N\setminus R'}) \le D'_i$ holds for any $i \in N\setminus R'$. Therefore, for $i \in N'$, $i < q_1$, we have $\overline{t}_i(\overrightarrow{\pi}_{N\setminus R'}) = \overline{t}_i(\overrightarrow{\pi}_{N\setminus R'}) \le D'_i = \overline{D}_i \le D_i$, while for $i \in N'$, $q_j > i > q_{j+1}$, $j = 1, 2,\dots, p$, we have that

$$\overline{t}_i(\overrightarrow{\pi}_{N\setminus R'}) = \overline{t}_i(\overrightarrow{\pi}_{N\setminus R'}) + \sum_{l=1}^{j} t_{q_l} \le D'_i + \sum_{l=1}^{j} t_{q_l} = \overline{D}_i \le D_i$$

holds.

We show that $\overline{t}_{q_j}(\overrightarrow{\pi}_{N\setminus R'}) \le D_{q_j}$, $j = 1, 2,\dots, p$. If $q_j < i$ for all $i \in N'$, then

$$\overline{t}_{q_j}(\overrightarrow{\pi}_{N\setminus R'}) = \sum_{l=1}^{j} t_{q_l} \le D_{q_j}.$$

Otherwise, let ξ be a job in N' with the largest number for which $\xi < q_j$ holds. Then

$$\overline{t}_{q_j}(\overrightarrow{\pi}_{N\setminus R'}) = \overline{t}_\xi(\overrightarrow{\pi}_{N\setminus R'}) + \sum_{l=1}^{j} t_{q_l} \le D'_\xi + \sum_{l=1}^{j} t_{q_l}.$$

If $\xi < q_k$ for all $k = 1, 2,\dots, p$, then $D'_\xi = \overline{D}_\xi$ and

$$\overline{t}_{q_j}(\overrightarrow{\pi}_{N\setminus R'}) \le \overline{D}_\xi + \sum_{l=1}^{j} t_{q_l} \le D_{q_j} - \sum_{l=1}^{j} t_{q_l} + \sum_{l=1}^{j} t_{q_l} = D_{q_j}.$$

If $q_{i-1} < \xi < q_i \le q_j$, then $D'_\xi = \overline{D}_\xi - \sum_{l=1}^{i-1} t_{q_l}$ and

$$\overline{t}_{q_j}(\overrightarrow{\pi}_{N\setminus R'}) \le \overline{D}_\xi - \sum_{l=1}^{i-1} t_{q_l} + \sum_{l=1}^{j} t_{q_l} \le D_{q_j} - \sum_{l=1}^{j} t_{q_l} - \sum_{l=1}^{i-1} t_{q_l} + \sum_{l=1}^{j} t_{q_l} = D_{q_j}.$$

Thus, we obtain $\overline{t}_i(\overrightarrow{\pi}_{N\setminus R'}) \le D_i$ for all $i \in N\setminus R'$. The theorem is proved.

4.5. We now consider the problem of finding a schedule s^* which minimizes (4.1), provided that d_i are integers, $t_i = 1$ and $\varphi_i(t), i = 1, 2,\dots, n$, are non-decreasing functions. Such a schedule is called *optimal.*

Since all $t_i = 1$ and there exists an optimal schedule with no preemption at times different from d_i (see Section 1 of Chapter 2), there exists an optimal *non-preemptive* schedule. A non-preemptive schedule s is specified the sequence $\pi = (i_1, i_2, \ldots, i_n)$ of jobs. In this case, the completion time for job i_j is $\overline{t}_{i_j}(\pi) = \max\{d_{i_j}, \overline{t}_{i_{j-1}}(\pi)\}+1$, $j = 1, 2, \ldots, n$, where $\overline{t}_{i_0}(\pi) = 0$.

Let the jobs be numbered in non-decreasing order of d_i. Without loss of generality, assume $d_1 = 0$. Let s be a schedule specified by the permutation $\overrightarrow{\pi_N} = (1, 2, \ldots, n)$, such that $s(t) \neq 0$ in the time intervals $(a_1, b_1]$, $(a_2, b_2], \ldots, (a_k, b_k]$, $0 = a_1 < b_1 < a_2 < b_2 < \ldots < a_k < b_k$, and $s(t) = 0$ outside these intervals. Let N_ν denote the set of jobs processed in an interval $(a_\nu, b_\nu]$, $\nu = 1, 2, \ldots, k$.

We show that there exists an optimal non-preemptive schedule $\tilde{s}$ such that $\tilde{s}(t) \in N_\nu$ in the interval $(a_\nu, b_\nu]$, $\nu = 1, 2, \ldots, k$, and $\tilde{s}(t) = 0$ outside these intervals.

Let s^* be an optimal non-preemptive schedule such that $s^*(t) \neq 0$ in the time intervals $(a_1', b_1']$, $(a_2', b_2'], \ldots, (a_l', b_l']$, $a_1' < b_1' < a_2' < b_2' < \ldots < a_l' < b_l'$, and $s^*(t) = 0$ outside these intervals. Let N_μ' denote the set of jobs processed in an interval $(a_\mu', b_\mu']$, $\mu = 1, 2, \ldots, l$.

Let the unit length time intervals starting from $d = 0$ be numbered by the integers 1, 2, ... An interval θ is of the form $(\theta-1, \theta]$.

If $a_1 < a_1'$ and $s^*(t) = 1$ in an interval θ, then construct a new schedule by defining $s'(t) = 1$ in the interval 1, $s'(t) = 0$ in the interval θ and $s'(t) = s^*(t)$ in other intervals. It is obvious that $F_\Sigma(s') = F_\Sigma(s^*)$.

If $a_1 = a_1'$, then $b_1' \leq b_1$. If $b_1' = b_1$, then $N_1' = N_1$. If $b_1' < b_1$, then $N_1' \subset N_1$ and s^* can be transformed into a schedule s'' by defining $s''(t) = j \in N_1 \backslash N_1'$ in the interval $(b_1', b_1'+1]$, $s''(t) = 0$ in the interval θ' and $s''(t) = s^*(t)$ in other intervals. Here θ' is an interval such that $s^*(t) = j$. It is obvious that $F_\Sigma(s'') = F_\Sigma(s^*)$.

Repeating these considerations, we conclude that in a finite number of steps schedule s^* can be transformed into the desired schedule $\tilde{s}$.

Thus, in the case under consideration, the problem of finding an optimal schedule is decomposed into k subproblems of finding optimal schedules for the sets $N_1, N_2, \ldots, N_k$ of jobs. For each subproblem ν, all jobs of the set N_ν are started and completed in the time interval $(a_\nu, b_\nu]$, where $a_\nu = \min\{d_i | i \in N_\nu\}$, $b_\nu = a_\nu + n_\nu$, $n_\nu = |N_\nu|$, provided that these jobs are processed according to the sequence $\overrightarrow{\pi_{N_\nu}}$ (i.e., in non-decreasing order of d_i).

We show that finding an optimal schedule for a set N_ν of jobs reduces to solving a $n_\nu \times n_\nu$ assignment problem. Without loss of generality, we assume that $n_\nu = n$, $a_\nu = d_1 = 0$,

$b_\nu = n$. Let $c_{i\theta} = \varphi_i(\theta)$ for $\theta = d_i+1, d_i+2,..., n$, and $c_{i\theta} = W$ for $\theta = 1, 2,..., d_i$, where W is a sufficiently large number. Introduce a variable $x_{i\theta}$ equal to 1 if job i is processed in interval θ; otherwise, its value is 0. We have

$$\text{minimize} \sum_{i=1}^{n} \sum_{\theta=1}^{n} c_{i\theta} x_{i\theta} \tag{4.13}$$

subject to

$$\sum_{\theta=1}^{n} x_{i\theta} = 1, \; i = 1, 2,..., n, \tag{4.14}$$

$$\sum_{i=1}^{n} x_{i\theta} = 1, \; \theta = 1, 2,..., n. \tag{4.15}$$

Conditions (4.14) imply that each job i is processed in some unit interval θ, $1 \le \theta \le n$. Conditions (4.15) imply that one of the jobs is processed in each unit length interval.

Assume that $\varphi_i(\theta)$ can be computed in a constant time (in fact, computing $\varphi_i(\theta)$ can be viewed as an elementary operation). Then preparing the input for the assignment problem (4.13)-(4.15) takes at most $O(n^2)$ time. Since an assignment problem can be solved in at most $O(n^3)$ time (see, e.g., [58]), the original problem of finding an optimal schedule can be solved in $O(n^3)$ time.

Remark 1. If the considered problem is supplemented with the condition that the processing of each job i must be completed by the *deadline* D_i, then in constructing the assignment problem it suffices to define $c_{i\theta} = \varphi_i(\theta)$ for $\theta = d_i+1, d_i+2,..., D_i$ and $c_{i\theta} = W$ for $\theta = 1, 2,..., d_i$ and $\theta = D_i+1, D_i+2,..., n$, where W is a sufficiently large number.

Remark 2. Consider the following problem of minimizing *the cumulative processing cost*. The jobs of the set N are processed on a single machine. For a job $i \in N$, the release date is d_i, its processing time is t_i; d_i and t_i being integers. Each job is associated with a non-decreasing function $\psi_i(\theta)$, where θ is the number of a unit length time interval. Preemption is allowed at integer times. If a job i is processed in unit length time intervals $\theta_1, \theta_2,..., \theta_{t_i}$, then its processing cost is $\sum_{k=1}^{t_i} \psi_i(\theta_k)$. It is required to find a schedule for which the cumulative cost for processing all jobs is minimal. This problem reduces to the the one of minimizing the total cost considered above. In fact, each job i can be considered as t_i jobs of unit length, $i^{(1)}, i^{(2)},..., i^{(t_i)}$. Define $d_{i^{(j)}} = d_i$, let the cost function $\varphi_{i^{(j)}}(t)$ be equal to $\psi_i(\theta)$ for $t \in (\theta-1, \theta]$, $i \in N$, $j = 1, 2,..., t_i$.

4.6. To conclude this section, we consider the problem of minimizing the total cost for a single machine, provided that the cost functions are $\varphi_i(t) = \varphi(t)+b_i$, $i = 1, 2,..., n$, where $\varphi(t)$ is a non-decreasing function. It is again assumed that the release date of a job i is $d_i \geq 0$, its processing time is $t_i > 0$, $i = 1, 2,..., n$, and preemption is allowed.

In the situation under consideration, there exists an optimal schedule with no preemptions at times different from d_i, $i = 1, 2,..., n$ (see Section 1of Chapter 2). We show that it can be found by the so-called SPT ("shortest processing time") rule extended to the case of different release dates.

The algorithm for constructing an optimal schedule can be described as follows. The decision to start (or to resume) the processing of a job is taken either when a new job is released or when the previous job is completed.

Let $\{d^{(1)}, d^{(2)},..., d^{(v)}\}$ be a set of all pairwise-distinct values of d_i, and $d^{(1)} < d^{(2)} < ... < d^{(v)} < d^{(v+1)} = W$, where W is a sufficiently large number.

In the first step, define $\tau = d^{(1)}$, $N_0 = \{i \mid i \in N, d_i = d^{(1)}\}$ and $s(t) = 0$ for $0 \leq t \leq d^{(1)}$. In each step, there is a certain time τ (e.g., assume $d^{(u-1)} \leq \tau < d^{(u)}$, $2 \leq u \leq v+1$) and some set N_0 of jobs. In set N_0, find a job l with the shortest processing time, i.e., $t_l = \min\{t_i \mid i \in N_0\}$. Define $s(t) = l$ for all $\tau < t \leq \min\{d^{(u)}, \tau+t_l\}$, and, if $\tau+t_l < d^{(u)}$ and $|N_0| = 1$, define $s(t) = 0$ for all $\tau+t_l < t \leq d^{(u)}$.

If $\tau+t_l > d^{(u)}$, then add all jobs $i \in N$ with $d_i = d^{(u)}$ to N_0, and let t_l be equal to $t_l-(d^{(u)}-\tau)$. If either (a) $\tau+t_l < d^{(u)}$ and $|N_0| = 1$ or (b) $\tau+t_l = d^{(u)}$, then delete job l from N_0 and add all jobs $i \in N$ with $d_i = d^{(u)}$. In any case, define $\tau = d^{(u)}$. If $\tau+t_l < d^{(u)}$ and $|N_0| > 1$, then delete job l from N_0 and set τ to be equal to $\tau+t_l$. As a result, we obtain a new time τ and a new set N_0. Go to the next step. The scheduling is completed when $N_0 = \emptyset$.

We show that *the resulting schedule is the desired optimal schedule*. The proof is by induction with respect to the number v of different release dates.

Let $v = 1$, i.e., the release dates for all jobs are the same (without loss of generality, assume that $d_i = d = 0$, $i = 1, 2,..., n$). In this case, there exists an optimal non-preemptive schedule (see Section 1 of Chapter 2) that is specified by the sequence $\pi^* = (i_1, i_2,..., i_n)$ of jobs. The completion time of a job i_k is $\bar{t}_{i_k}(\pi^*) = \sum_{p=1}^{k} t_{i_p}$. In the schedule constructed by the described algorithm, let the jobs be processed according to the sequence $\pi = (j_1, j_2,..., j_n)$. It is clear that $t_{j_k} \leq t_{j_{k+1}}$, $k = 1, 2,..., n-1$. If $t_{i_\nu} > t_{i_{\nu+1}}$ and $\tilde{\pi} = (i_1, i_2,..., i_{\nu-1}, i_{\nu+1}, i_\nu, i_{\nu+2},..., i_n)$, then $\bar{t}_{i_k}(\tilde{\pi}) = \bar{t}_{i_k}(\pi^*)$ for $k = 1, 2,..., \nu-1$ and $k = \nu+2, \nu+3,..., n$, $\bar{t}_{i_\nu}(\tilde{\pi}) = \bar{t}_{i_{\nu+1}}(\pi^*)$ and $\bar{t}_{i_{\nu+1}}(\tilde{\pi}) < \bar{t}_{i_\nu}(\pi^*)$. Since $\varphi(t)$ is a non-decreasing function, it follows that the schedule determined by the sequence $\tilde{\pi}$ is also optimal. Repeating these considerations finitely many times, we conclude that the sequence π determines an optimal schedule.

Suppose that this statement holds for $v = V$. We show that this also holds for $v = V+1$. Let s^* be an optimal schedule. Without loss of generality, we may assume that it belongs to the class of schedules with no preemption at times different from $d^{(i)}$, $i = 1, 2, \ldots, v$. Let s be a schedule constructed by the algorithm described above.

If the total processing time of jobs with release dates equal to $d^{(1)}$ do not exceed $d^{(2)} - d^{(1)}$ or $s(t) = s^*(t)$ in the interval $(d^{(1)}, d^{(2)}]$, then by the induction assumption we have $F_\Sigma(s) = F_\Sigma(s^*)$. In the following, it is assumed that $s^*(t) \neq 0$ in the interval $(d^{(1)}, d^{(2)}]$.

Suppose that $s(t) = s^*(t)$ for all $d^{(1)} \leq t \leq \tau$, $s(t) = i$ for all $\tau < t \leq \min\{d^{(2)}, \tau+t_i\}$, $s^*(t) = j$ for all $\tau < t \leq \min\{d^{(2)}, \tau+t_j\}$ and $t_j > t_i$.

(1) Let $\tau+t_i \geq d^{(2)}$. Find all time intervals in which either $s^*(t) = i$ or $s^*(t) = j$. It is clear that the total length of these intervals is t_i+t_j. Find a schedule s^0 by setting $s^0(t) = i$ or $s^0(t) = j$ in these intervals in such a way that the condition $s^0(t_1) = i$ and $s^0(t_2) = j$ implies $t_1 < t_2$. In other intervals, define $s^0(t) = s^*(t)$. Since $t_i < t_j$, we obtain $F_\Sigma(s^0) = F_\Sigma(s^*)$. In the interval $(d^{(1)}, d^{(2)}]$, we have $s^0(t) = s(t)$.

(2) Let $\tau+t_i < d^{(2)}$. Find a new schedule $\tilde{s}$ by setting $\tilde{s}(t) = i$ in the interval $(\tau, \tau+t_i]$, $\tilde{s}(t) = j$ in all intervals for which $s^*(t) = i$, and $\tilde{s}(t) = s^*(t)$ in other intervals. Since $t_j > t_i$, we obtain $F_\Sigma(\tilde{s}) = F_\Sigma(s^*)$.

Suppose that $\tau+t_j < d^{(2)}$. Let by $\tilde{t}$ the largest value of $t \in (d^{(1)}, d^{(2)}]$, for which $\tilde{s}(t) = j$.

If $\tilde{t} > \tau+t_i+t_j$, then find a schedule $\bar{s}$ by setting $\bar{s}(t) = \tilde{s}(t+t_j-t_i)$ in the interval $(\tau+t_i, \tilde{t}-(t_j-t_i)]$, $\bar{s}(t) = j$ in the interval $(\tilde{t}-(t_j-t_i), \tilde{t}]$ and $\bar{s}(t) = \tilde{s}(t)$ in other intervals. Again, we have $F_\Sigma(\bar{s}) = F_\Sigma(\tilde{s})$.

If $\tilde{t} = \tau+t_j$, then find a schedule $\bar{\bar{s}}$ by setting $\bar{\bar{s}}(t) = \tilde{s}(t+t_j-t_i)$ in the interval $(\tau+t_i, d^{(2)}-(t_j-t_i)]$, $\bar{\bar{s}}(t) = j$ in the interval $(d^{(2)}-(t_j-t_i), d^{(2)}]$ and $\bar{\bar{s}}(t) = \tilde{s}(t)$ in other intervals. It is clear that $F_\Sigma(\bar{\bar{s}}) = F_\Sigma(\tilde{s})$.

If $\tilde{s}(d^{(2)}) = k \neq j$ and $\tilde{s}(t) = k$ for some $t > d^{(2)}$, then select all time intervals in which either $\bar{\bar{s}}(t) = j$ or $\bar{\bar{s}}(t) = k$. The total length of these intervals is equal to t_j+t_k. Find a schedule s' by setting either $s'(t) = j$ or $s'(t) = k$ in the selected intervals in such a way that the conditions $s'(t_1) = j$ and $s'(t_2) = k$ imply either (a) $t_1 < t_2$ if $t_j \leq t_k$ or (b) $t_1 > t_2$ if $t_j > t_k$. For other intervals, define $s'(t) = \bar{\bar{s}}(t)$. It can be easily shown $F_\Sigma(s') = F_\Sigma(\bar{\bar{s}})$.

In any case, we obtain a new optimal schedule which coincides with the schedule s in the interval $(d^{(1)}, \tau+t_i]$ and has no preemption before the time $d^{(2)}$.

Repeating similar considerations finitely many times, we come to an optimal schedule $\hat{s}$ such that $\hat{s}(t) = s(t)$ in the interval $(d^{(1)}, d^{(2)}]$. Due to the inductive assumption, we conclude that the schedule constructed by the described algorithm is optimal.

To implement some procedures of the described algorithm, we can represent the data using the balanced 2-3-trees. In this case, *an optimal* schedule can *be found in at most* $O(n\log n)$ *time.*

Define a total pseudo-order $\Longrightarrow$ over the set N in the following way: $i \Longrightarrow j$ if and only if $t_i \leq t_j$. In the first step of the algorithm, represent the ordered set $N_0 \subseteq N$ as a balanced 2-3-tree. This can be done in $O(n)$ time (see Section 2.3 of Chapter 1).

The number of steps of the algorithm is at most $2n-1$ since, in each step, at least one of the following situations is occurs: (a) some job is completed; (b) a new job is added to the set N_0. In each step, finding a job $l \in N_0$ with the shortest processing time (i.e., finding an element of the set N_0 that is maximal with respect to $\Longrightarrow$) requires one elementary operation. Deleting job l from N_0 or adding a new job to N_0 takes at most $O(\log n)$ time (see Section 2 of Chapter 1). Changing the processing time of the job l is equivalent to deleting l from N_0, followed by adding l with a new processing time to N_0 (here we consider that the relation $\Longrightarrow$ is defined for a new element and any $i \in N$, $i \neq l$). This also takes at most $O(\log n)$ time. Hence, *an optimal schedule can be found in at most* $O(n\log n)$ *time.*

Example. Consider the problem of *minimizing the total flow time* for single-machine processing. This problem is a special case of the problem of minimizing the total cost (for $\varphi(t) = t$ and $b_i = 0$, $i = 1, 2,..., n$) discussed in this section. Let $n = 7$, and the processing times t_i and the release dates d_i are given in Table 4.4.

Table 4.4

i	1	2	3	4	5	6	7
t_i	4	2	2	1	3	1	1
d_i	1	1	4	7	7	11	15

We have $d^{(1)} = 1$, $d^{(2)} = 4$, $d^{(3)} = 7$, $d^{(4)} = 11$, $d^{(5)} = 15$. Define $d^{(6)} = W$, where W is a sufficiently large number. The value of τ and the set N_0 for each step of the algorithm are given in Table 4.5. This table also presents job $l \in N_0$ with the shortest processing time, the values of $s(t)$, and new value of t_l obtained in this step (if the processing time of job l is changed in this step). The resulting schedule is presented in Fig. 4.1.

Table 4.5

Step	τ	N_0	l	$s(t)$	New t_l
1	1	1, 2	2	$s(t)=0,\ 0\le t\le 1$; s $s(t)=2,\ 1<t\le 3$	-
2	3	1	1	$s(t)=1,\ 3<t\le 4$	$t_1=3$
3	4	1, 3	3	$s(t)=3,\ 4<t\le 6$	-
4	6	1	1	$s(t)=1,\ 6<t\le 7$	$t_1=2$
5	7	1, 4, 5	4	$s(t)=4,\ 7<t\le 8$	-
6	8	1, 5	1	$s(t)=1,\ 8<t\le 10$	-
7	10	5	5	$s(t)=5,\ 10<t\le 11$	$t_5=2$
8	11	5, 6	6	$s(t)=6,\ 11<t\le 12$	-
9	12	5	5	$s(t)=5,\ 12<t\le 14$; $s(t)=0,\ 14<t\le 15$	-
10	15	7	7	$s(t)=5,\ 15<t\le 16$; $s(t)=16,\ 14<t\le W$	-
11	W	$\emptyset$	-	-	-

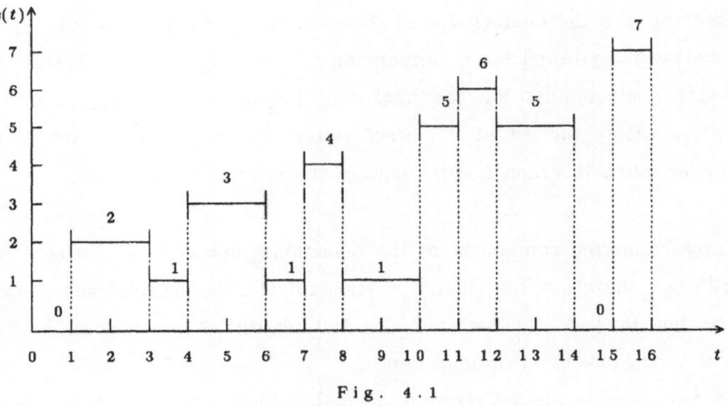

Fig. 4.1

5. Identical Machines. Maximal Completion Time. Equal Processing Times

In this section we consider the problem of finding a time-optimal schedule for identical parallel machines and a partially ordered set of jobs with equal processing times, assuming that either the reduction graph of a precedence relation is tree-like or that there are two machines.

5.1. The jobs of a set $N = \{1, 2, \ldots, n\}$ are processed on M parallel identical machines. All jobs have the same release dates (i.e., $d_i = 0$, $i = 1, 2, \ldots, n$) and equal processing times. Without loss of generality, we assume $t_i = 1$, $i = 1, 2, \ldots, n$, where t_i is the processing time of a job i. Preemption is forbidden. A precedence relation $\rightarrow$ is defined over set N to determine a feasible job processing sequence. Let the reduction graph of this relation be denoted by G. Let $\bar{t}_i(s)$ be the completion time of a job i in schedule s. It is required to find a feasible (with respect to $\rightarrow$) schedule s^* for processing the jobs of set N which minimizes the makespan (i.e., the maximal completion time of all jobs):

$$T(s) = \max\{\bar{t}_i(s) \mid i \in N\}. \tag{5.1}$$

The value $T(s)$ is called *the length* of schedule s, and schedule s^* is called a (*time-*) *optimal* schedule.

Let the unit length time intervals starting at $t = 0$ be numbered by the integers 1, 2,... . An interval with the number θ is of the form $(\theta-1, \theta]$. In what follows, we do not distinguish between a job and the corresponding vertex of graph G. As before, N^- and N^+ denote the sets of all minimal and maximal (with respect to $\rightarrow$) elements of set N. For a job i, let $A^0(i)$ denote the set of its direct successors, and let $B^0(i)$ denote the set of its direct predecessors. In graph G, $h(i)$ denotes the height of a vertex i.

5.2. Let each connected component of the reduction graph G be an intree.

We describe an algorithm for finding a schedule that is feasible with respect to $\rightarrow$, called the h-algorithm. A schedule found by this algorithm is called an h-schedule. We show that an h-schedule is an optimal one.

The number of steps in the h-algorithm is equal to the length of an h-schedule. A step θ consists of at most $M+1$ iterations. At each of these iterations (except the last one), a job is assigned to be processed in the unit time interval θ. At the last iteration, the transition to the next step is performed. The total number of iterations is $T(s)+n-1$, where $T(s)$ is the length of an h-schedule.

First, define $s_L(t) = 0$ for $L = 1, 2,..., M$, $t \geq 0$ and consider all jobs of set N to be unmarked. Set $\theta = 1$.

In each step θ, the following iterations are to be performed. Find a machine H such that $s_H(\theta) = 0$. Choose a job j with the largest height $h(j)$ among unmarked jobs of set N^+. Define $s_H(t) = j$ in the interval θ, mark job j, and go to the next iteration. If either we fail to find machine H or all jobs of N^+ have been marked, then delete the marked jobs from N. If $N \neq \emptyset$, increase θ by 1 and go to the next step. If $N = \emptyset$, then the h-schedule $s(t) = \{s_1(t), s_2(t),..., s_M(t)\}$ is constructed.

Let the vertices of graph G (i.e., the jobs of set N) be numbered as described in the case of an intree in Section 1.4 of Chapter 1. Then, at each iteration of the h-algorithm, the job with the highest number among the unmarked jobs of the set N^+ can be chosen as the job j.

Let λ be the list of the jobs of set N sorted in decreasing order of their numbering. At each iteration of the h-algorithm, choose the first unmarked element of list λ belonging to set N^+ as the job j. The resulting schedule is called a *λ-schedule* (schedules of this type are also called list schedules). It is clear that a λ-schedule found according to the list $\lambda = (n, n-1,..., 2, 1)$ is, at the same time, an h-schedule.

Example. Let $M = 3$, $N = \{1, 2,..., 12\}$, $t_i = 1$, $d_i = 0$, $i = 1, 2,..., 12$, and the reduction graph of the precedence relation defined over N is shown in Fig. 5.1a.

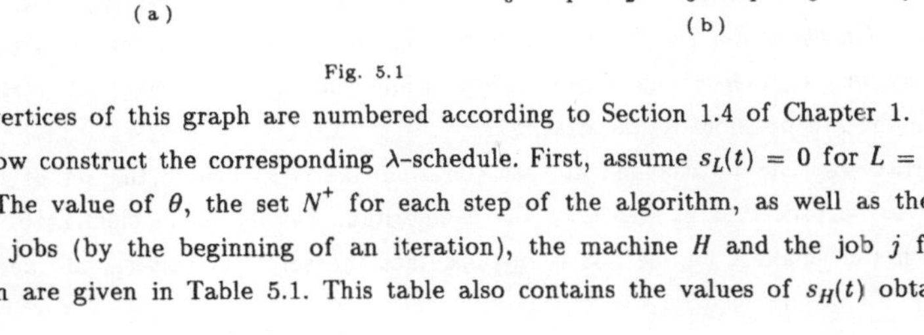

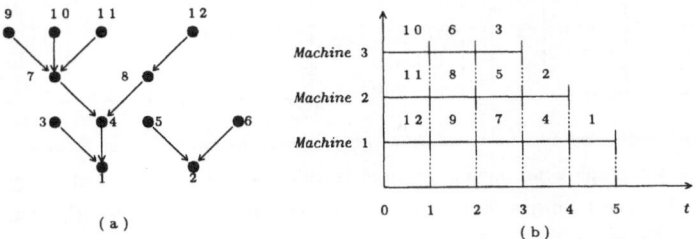

Fig. 5.1

The vertices of this graph are numbered according to Section 1.4 of Chapter 1.

We now construct the corresponding λ-schedule. First, assume $s_L(t) = 0$ for $L = 1, 2, 3$, $t \geq 0$. The value of θ, the set N^+ for each step of the algorithm, as well as the set of marked jobs (by the beginning of an iteration), the machine H and the job j for each iteration are given in Table 5.1. This table also contains the values of $s_H(t)$ obtained at each iteration. The resulting λ-schedule is presented in Fig. 5.1b. Here $T(s) = 5$.

Table 5.1

Step θ	1				2			
N^+	3, 5, 6, 9, 10, 11, 12				3, 5, 6, 8, 9			
Iteration	1	2	3	4	1	2	3	4
Marked jobs	-	12	11, 12	10, 11, 12	-	9	8, 9	6, 8, 9
H	1	2	3	-	1	2	3	-
j	12	11	10		9	8	6	
The value of $s_H(t)$ for $\theta-1<t\le\theta$	$s_1(t)=$ $=12$	$s_2(t)=$ $=11$	$s_3(t)=$ $=10$		$s_1(t)=$ $=9$	$s_2(t)=$ $=8$	$s_3(t)=$ $=6$	

Step θ	3				4			5	
N^+	3, 5, 7				2, 4			1	
Iteration	1	2	3	4	1	2	3	1	2
Marked jobs	-	7	5, 7	3, 5, 7	-	4	2, 4	-	2
H	1	2	3	-	1	2	3	1	2
j	7	5	3		4	2	-	1	-
The value of $s_H(t)$ for $\theta-1<t\le\theta$	$s_1(t)=$ $=7$	$s_2(t)=$ $=5$	$s_3(t)=$ $=3$		$s_1(t)=$ $=4$	$s_2(t)=$ $=2$		$s_1(t)=$ $=1$	

We show that *finding a λ-schedule takes at most $O(n)$ time.* It is assumed that the vertices of the graph G are numbered as in Section 1.4 of Chapter 1, and the graph G is represented in the following way. There are two one-dimensional arrays Q_B and S'_A, each consisting of n elements. The number b_k written in the kth position of the array Q_B shows how many direct predecessors vertex k has, while the direct successor of vertex k is placed in the kth position of the array S'_A.

The array Q_B is to be changed at each iteration. Let $N(\theta)$ denote the set of the jobs assigned to be processed in step θ of the h-algorithm. Let n_θ be an element of the set $N(\theta)$ with the smallest number. It is obvious that the set $N(1)$ consists of the first M

elements in the list λ, which correspond to the zero elements of the array Q_B. If there are less than M such elements, they all compose the set $N(1)$.

At each iteration of step θ, choose the element of the set $N(\theta)$ with the largest number as the job j and delete it from $N(\theta)$ (which corresponds to marking the job j). Find the direct successor i of the job j in the position j of the array S_A' and decrease the ith element of the array Q_B by 1 (which corresponds to deleting j from G). If the ith element of the array Q_B happens to be zero and $i > n_\theta$, insert the job i into the set $N(\theta+1)$. If, when entering the next step $\theta+1$, we have $|N(\theta+1)| < M$, then scan the list λ starting with the element with the number $n_\theta - 1$, and insert the elements of that list, which correspond to the zero elements of the array Q_B, into the set $N(\theta+1)$ (trying to make that set consist of M jobs, if possible). It is clear that if the data are represented in the way described, then finding the schedule requires at most $O(n)$ time.

5.3. As mentioned above, a λ-schedule is, at the same time, an h-schedule.

Theorem 5.1. *If the reduction graph G of the precedence relation $\rightarrow$ defined over set N is an intree, then an h-schedule is a time-optimal schedule for processing the jobs of set N.*

Proof. Suppose that the theorem does not hold. Then for the given number of machines M there exists the smallest (with respect to the cardinality) set N such that an h-schedule s is not time-optimal. Let $|N| = n$, $T(s) = T$ and T^* be the length of the optimal schedule for processing the jobs of set N. It follows that $T > T^*$.

Let r be the terminal vertex of the graph G. For any schedule $\tilde{s}$ of the length τ, only job r is processed in the interval τ. Hence, it follows that if $\tilde{s}_L(t) \neq 0$, $L = 1, 2,\ldots, M$, in the time interval $(0, \tau-2]$, and $\tilde{s}_H(t) \neq 0$ for some H, $1 \leq H \leq M$, in the interval $\tau-1$, then schedule $\tilde{s}$ is optimal (in this case, the jobs of set $N \backslash r$ cannot be processed within less than $\tau-1$ time units).

Let N_θ denote set N^+ obtained by the step θ, $\theta = 1, 2,\ldots, T$, of the h-algorithm. Since s is not optimal, there exists a machine H such that $s_H(t) = 0$ in some interval θ', where $\theta' \leq T-2$. Therefore, at the last iteration of step θ' all jobs of the set $N_{\theta'}$ are marked, and $|N_{\theta'}| < M$. Since G is an intree, it follows that $|N_{\theta'+1}| \leq |N_{\theta'}| < M$. Hence, in the interval $\theta'+1$ (and, therefore, in the interval $T-2$) at least one machine is idle.

Note that, for schedule s, job r is processed in the interval T, while all jobs processed in the interval $T-1$ belong to the set $B^0(r)$. Among the jobs processed in the interval $T-2$ there is a job which does not belong to the set $B^0(r)$ (otherwise, there would not be an idle machine in the interval $T-2$). Therefore, there is a job r'' processed in the interval $T-2$, such that $r'' \in B^0(r')$ holds for some $r' \in B^0(r)$.

Let $R = r \cup B^0(r)$. It is clear that by defining $s_L'(t) = 0$ if $s_L(t) = i$, $i \in R$, and $s_L'(t) = s_L(t)$ in other cases, we obtain an h-schedule s' for processing the jobs of set $N' = N \backslash R$. In this case, $T(s') = T-2$, and the height of each job is two units less than that for the initial problem.

The reduction graph G' of the relation $\longrightarrow$ which corresponds to the set N' can be transformed into a tree by adding a new vertex $\bar{r}$ and connecting the terminal vertices of the graph G' with $\bar{r}$ by the arcs leaving these vertices.

We can obtain an h-schedule s'' for processing the jobs of set $N'' = N' \cup \bar{r}$ from the schedule s' by setting $s_1''(t) = \bar{r}$ and $s_L''(t) = s_L'(t)$ for $L = 2, 3, \ldots, M$ in the interval $T-1$ and $s_L''(t) = s_L'(t)$, $L = 1, 2, \ldots, M$, in other intervals. It is clear that $T(s'') = T-1$ and $|N''| \leq n-1$.

An optimal schedule s^* for the jobs of set N has the length T^*. Defining $\bar{s}_L(t) = 0$ if $s_L^*(t) = i$, $i \in R$ (except the case $L = 1$ and $T^*-2 < t \leq T^*-1$), $\bar{s}_1(t) = \bar{r}$ in the interval T^*-1 and $\bar{s}_L(t) = s_L^*(t)$ in other cases, $L = 1, 2, \ldots, M$, we obtain a schedule $\bar{s}$ for processing the jobs of set N'' having the length $T^*-1 < T-1$. Therefore, the h-schedule s'' for processing the jobs of set N'' where $|N''| \leq n-1$ is not optimal. We have come to a contradiction. This proves the theorem.

Corollary 5.1. *If each of the connected components of the graph G is an intree, then an h-schedule is a time-optimal schedule for processing the jobs of set N.*

Proof. Let us add a new job $\bar{r}$ to the set N and assume that $i \longrightarrow \bar{r}$ for all $i \in N$. The reduction graph of the relation $\longrightarrow$ specified on the set $N \cup \bar{r}$ is an intree. Construct an h-schedule s' for processing the set $N \cup \bar{r}$. Due to theorem 5.1 this schedule is optimal. In schedule s', the job $\bar{r}$ is processed last, say, in the time interval τ. The jobs of set N are processed in the intervals $1, 2, \ldots, \tau-1$. Defining $s_L(t) = 0$ in the interval τ and $s_L(t) = s_L'(t)$, $L = 1, 2, \ldots, M$, in other intervals, we obtain an h-schedule s for processing the jobs of set N. This schedule is optimal, because otherwise schedule s' would not be optimal. This proves the corollary.

Suppose that each connected component of the graph G is an outtree. Reverse the orientation of each arc of this graph. As a result, we obtain the graph G' such that each of its components is an intree. It is clear that the graph G' is the reduction graph of the precedence relation which is the inverse of the initial one. Using the graph G', find an h-schedule s'. Having found schedule s' (and, hence, having found its length $T(s')$), find a schedule s, by setting $s_L(t) = s'_L(t)$ for $t > T(s')$ and $s_L(t) = s'_L(t+T(s')-2\theta+1)$, $L = 1, 2,\ldots, M$, for $\theta-1 < t \leq \theta$, $\theta = 1, 2,\ldots, T(s')$. The schedule s is called an $\tilde{h}$*-schedule*. The lengths of schedules s and s' are equal, and the feasibility of s' with respect to the graph G' implies the feasibility of s with respect to G (and vice versa). Thus, the following statement follows.

Corollary 5.2. *If each connected component of the graph G is an outtree, then an $\tilde{h}$-schedule is a time-optimal schedule for processing the jobs of set N.*

5.4. We now consider the case when the reduction graph G of the precedence relation $\rightarrow$ defined over N is an arbitrary circuit-free directed graph, but the number of machines $M = 2$.

Let $\nu = (\nu_1, \nu_2,\ldots, \nu_k)$ and $\mu = (\mu_1, \mu_2,\ldots, \mu_l)$ be sequences of integers, and k, $l \geq 0$. If $k = 0$, the sequence ν is empty. Sequence ν is said to be *lexicographically smaller* than sequence μ if: (1) there is such i, $1 \leq i \leq k$, that for all j, $1 \leq j < i$, $\nu_j = \mu_j$ and $\nu_i < \mu_i$ hold or (2) $\nu_j = \mu_j$, $j = 1, 2,\ldots, k$, and $k \leq l$.

Let the vertices of the graph G be numbered in the following way. Assign number 1 to one of the terminal vertices. Let numbers $1, 2,\ldots, j-1$ be assigned and Q be a set of such non-numbered vertices which have no non-numbered successors. For each vertex $i \in Q$, construct a sequence $a(i)$ of all its direct successors (i.e., the jobs of a set $A^0(i)$), taking the elements in decreasing numerical order. Assign the number j to one of the jobs $i \in Q$ with the lexicographically smallest sequence $a(i)$.

Renumbering the vertices of the graph G in the described way requires at most $O(n^2)$ time. In fact, suppose that the vertices of the graph G are numbered arbitrarily and G is given by its adjacency matrix. Find all terminal vertices of the graph, and make the list Q_A of n elements such that the number $|A^0(i)|$ is placed in the ith position. It is obvious that this requires at most $O(n^2)$ time. We show how to change the current vertex numbering into the one described above.

Let $\tilde{Q}$ be a queue of vertices ready to be assigned new numbers (these vertices are either terminal or the new numbers have been assigned to all of their successors). At the beginning, $\tilde{Q}$ consists of all terminal vertices of a graph. Form the list L consisting of vertices which have not been given new numbers but which have direct successors with new numbers. Each vertex appears in the list L at most once. Initially, the list L is empty.

The algorithm for renumbering the vertices consists of n steps, each corresponding to the assignment of a new number to some vertex. In each step, assign the next number to the first element in the queue $\tilde{Q}$. Suppose that this element is q. Delete q from $\tilde{Q}$. Adjust the list L in the following way. Using the adjacency matrix, for each element i in the list L verify whether i belongs to the set $B^0(q)$. If $i \in B^0(q)$, mark this element in the list L and in the adjacency matrix. Form a sequence L' consisting of two parts. In the first part, arrange arbitrarily the elements of the set $B^0(q)$ which are not included in L (to do this, scan the column q of the adjacency matrix and remove the marks from this matrix). In the second part, arrange the marked elements of the list L (in the same order as in the list L). Change the list L by deleting the marked elements and adding the sequence L' to the rear of L. It is easy to check that, in each step, constructing the list L takes $O(n)$ time. In the list L, the elements i are arranged in lexicographically increasing order of the sequences $\bar{a}(i)$. Here $\bar{a}(i)$ denotes the decreasing sequence of the numbers of those direct successors of a job i which have been given new numbers (up to the step under consideration); in particular, $\bar{a}(i) = a(i)$ if all direct successors of a job i have been given new numbers. The described arrangement of the list L does not require the sequences $a(i)$ to be obtained as such.

For each job $j \in B^0(q)$, reduce the number in the jth position of the list Q_A by one. Scanning the list L from the front to the rear, choose such elements i that 0 is placed in the ith position of the list Q_A, delete them from L and add them to $\tilde{Q}$. Having performed this procedure for all jobs of the list L, we obtain a new list L, a new queue $\tilde{Q}$ and go to the next step. It is obvious that the running time of each step in the algorithm is at most $O(n)$ and, hence, numbering all vertices of the graph takes *at most* $O(n^2)$ *time*.

Example. Let the reduction graph of a precedence relation defined over set N be shown in Fig. 5.2, and the initial numbering of its vertices is given by the letters A, B,..., J.

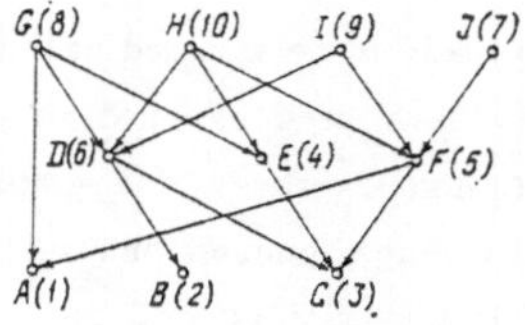

Fig. 5.2

Table 5.2

Step	New number of vertex	$B^0(q)$	L'	L after being corrected	L at the end of a step	Changes in list Q_A	$\tilde{Q}$
1	$A=1$	$F\underline{G}$	$F\underline{G}$	$F\underline{G}$	$F\underline{G}$	$[F]=1$, $[G]=2$	BC
2	$B=2$	D	D	$F\underline{G}D$	$F\underline{G}D$	$[D]=1$	C
3	$C=3$	DEF	$E\underline{F}D$	$\underline{G}EFD$	$\underline{G}$	$[D]=[E]=$ $=[F]=0$	EFD
4	$E=4$	$\underline{G}H$	$\underline{HG}$	HG	HG	$[G]=1$, $[H]=2$	FD
5	$F=5$	HIJ	$IJ\underline{H}$	$GIJH$	GIH	$[H]=[I]=$ $=1$, $[J]=0$	DJ
6	$D=6$	$\underline{G}HI$	$\underline{GIH}$	$\underline{G}IH$	–	$[G]=[H]=$ $=[I]=0$	$JGIH$
7	$J=7$	–	–	–	–	–	GIH
8	$G=8$	–	–	–	–	–	IH
9	$I=9$	–	–	–	–	–	H
10	$H=10$	–	–	–	–	–	–

The set of terminal vertices is $\{A, B, C\}$, and the list Q_A is of the form (0, 0, 0, 2, 1, 2, 3, 3, 2, 1). Initially $\tilde{Q} = (A, B, C)$ and $L = (\emptyset)$. For each step of the algorithm of obtaining the new numbering of the vertices, Table 5.2 gives the new number of a vertex q, set $B^0(q)$, sequence L' where the marked elements of the list L are underlined, list L (after being corrected as well as at the end of a step). This table also contains the new values of the elements of the list Q_A obtained at the end of a step and the queue $\tilde{Q}$. The number in position p in the list Q_A is denoted by $[p]$, $p = A, B, \ldots, J$. The new numbers of vertices are shown in Fig. 5.2 (in parentheses).

5.5. We describe an *algorithm* for finding a schedule that is feasible with respect to $\longrightarrow$ which is called (by analogy with Section 5.2) *a λ-schedule*. Then we show that a λ-schedule is a time-optimal schedule in the case of two machines.

Suppose that the vertices of the graph G are numbered by the integers $1, 2, \ldots, n$ as described in Section 5.4, and that $\lambda = (n, n-1, \ldots, 1)$.

The number of steps in the algorithm for constructing the λ-schedule s is equal to the length $T(s)$ of the schedule. In a step θ, $\theta = 1, 2, \ldots, T(s)$, a schedule in the interval θ is to be constructed, and the jobs assigned to be processed in this interval are deleted from set N. For step θ, let i and j be the jobs with the largest numbers in the sets N^+ and $N^+ \backslash i$, respectively. In the interval θ, define $s_1(t) = i$, and, if $|N^+| > 1$, define $s_2(t) = j$, otherwise, define $s_2(t) = 0$. Delete i from set N. If $|N^+| > 1$, then delete j as well. If $N \neq \varnothing$, increase θ by 1 and go to the next step. If $N = \varnothing$, define $s_1(t) = s_2(t) = 0$ for $t > \theta$. As a result, we obtain the λ-schedule $s(t) = \{s_1(t), s_2(t)\}$.

Finding the set N^+ and deleting elements i and j from N requires at most $O(n)$ time in each step of the algorithm (see Section 1.4 of Chapter 1). Therefore, *the running time* of *the algorithm is at most* $O(n^2)$.

Example. Let $M = 2$, $N = \{1, 2, \ldots, 17\}$, $t_i = 1$, $d_i = 0$, $i = 1, 2, \ldots, 17$, and the reduction graph G of precedence relation defined over N is given in Fig. 5.3a. The jobs are numbered by the algorithm described in Section 5.4. Note that the subgraph of the graph G induced by the set of vertices $\{1, 2, \ldots, 10\}$ coincides with the graph considered in Section 5.4.

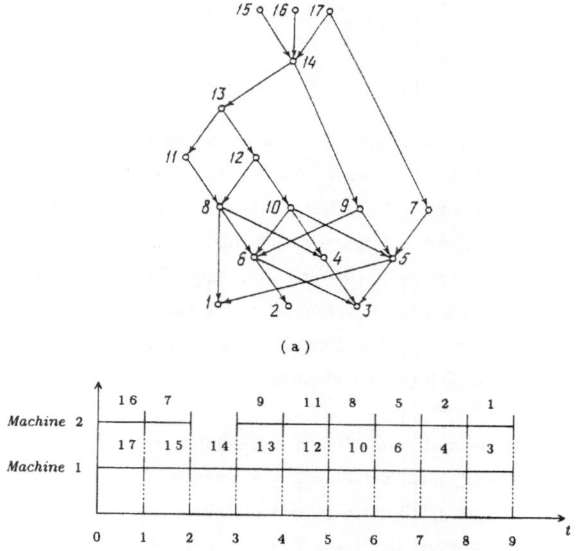

(b)

Fig. 5.3

For each step θ of the algorithm, Table 5.3 gives the set N^+, the numbers of jobs i and j, as well as the obtained values of $s_1(t)$ and $s_2(t)$. The constructed λ-schedule is shown in Fig. 5.3b.

Table 5.3

Step θ	1	2	3	4	5	6	7	8	9
N^+	15, 16, 17	7, 15	14	9, 13	11, 12	8, 10	4, 5, 6	1, 2, 4	1, 3
i	17	15	14	13	12	10	6	4	3
j	16	7	—	9	11	8	5	2	1
$s_1(t)$ for $\theta-1<t\le\theta$	17	15	14	13	12	10	6	4	3
$s_2(t)$ for $\theta-1<t\le\theta$	16	7	0	9	11	8	5	2	1

5.6. We now prove the following statement:

Theorem 5.2. *A λ-schedule is a time-optimal schedule for the two-machine processing of the jobs of set N.*

Proof. Let s denote the λ-schedule found by the algorithm described in Section 5.5.

Suppose that, in schedule s, a job $k \in N$ is processed in the time interval δ_k. Note that if the job i is processed on the first machine and $\delta_i \le \delta_k$, then $i > k$. In fact, let $\delta_i = \theta$ and let N_θ denote the set N^+ obtained in the step θ of the algorithm. If $k \in N_\theta$, then $i > k$ according to the procedure of constructing the λ-schedule. If $k \notin N_\theta$, then there exists a job $l \in N_\theta$ such that $l \longrightarrow k$. According to the procedure of λ-scheduling, we have $i > l$, and the numbering of the vertices implies that for any l and k such that $l \longrightarrow k$, $l > k$ holds. Hence, $i > k$.

If $s_L(t) = 0$ in the interval θ, machine L is said to processed a dummy job 0 in this interval. Let $\bar{N}$ be the set N of jobs with the dummy job 0 included.

Define the jobs $p_0, p_1, \dots, p_u, p_v \in N$, $v = 0, 1, \dots, u$, $r_0, r_1, \dots, r_u, r_v \in \bar{N}$, $v = 0, 1, \dots, u$, and the sets of jobs $P_0, P_1, \dots, P_u, P_v \subset N$, $v = 0, 1, \dots, u$, in the following way.

Let p_0 and r_0 be jobs processed in schedule s in the time interval $T(s)$ on the first and the second machine, respectively. Note that $p_0 > r_0$. Suppose that $p_{v-1}, p_{v-2}, \dots, p_0$ and $r_{v-1}, r_{v-2}, \dots, r_0$ have been defined. Let r_v denote such a job of the set $\bar{N}$ that $r_v < p_{v-1}$, $\delta_{r_v} < \delta_{p_{v-1}}$ and there is no such job $k \in \bar{N}$ that $\delta_{r_v} \le \delta_k < \delta_{p_{v-1}}$ and $k < p_{v-1}$. It is clear that the job r_v is processed on the second machine (if the job i is processed on the first machine and $\delta_i < \delta_{p_{v-1}}$, then $i > p_{v-1}$). Let p_v denote the job processed on the first machine in the interval δ_{r_v} (thus, $\delta_{p_v} = \delta_{r_v}$). Suppose that the jobs p_u, $p_{u-1},\ldots, p_0$ and $r_u, r_{u-1},\ldots, r_0$ have been defined, and there is no job r_{u+1} (i.e. either $\delta_{p_u} = 1$ or $k > p_u$ for all k such that $\delta_k < \delta_{p_u}$). For all v, $0 \le v \le u$, define $P_v = \{k \in N \mid \delta_{p_{v+1}} < \delta_k < \delta_{p_v}\} \cup p_v$. Define also $P_u = \{k \in N \mid \delta_k < \delta_{p_u}\} \cup p_u$.

The values p_v, r_v and P_v, $v = 0, 1,\ldots, u$, for the schedule in Fig. 5.3b are shown in Fig. 5.4.

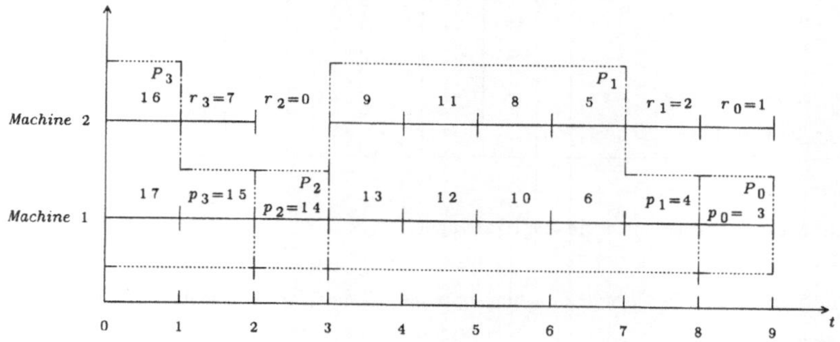

Fig. 5.4

We show that $k \longrightarrow k'$ for all $k \in P_v$ and $k' \in P_{v-1}$, $v = 1, 2,\ldots, u$.

First, we show that $p_v \longrightarrow k'$ for all $k' \in P_{v-1}$. By definition of the job r_v, for that job the inequality $r_v < p_{v-1}$ holds, and $k' \ge p_{v-1}$ holds for all $k' \in P_{v-1}$. Consequently, $r_v < k'$. Let $\tilde{N}$ denote the set N^+ obtained by the step δ_{r_v}. The definition of a λ-schedule implies that r_v is the job with the largest number in the set $\tilde{N}\backslash p_v$. Hence, for any $k' \in P_{v-1}$ it follows that $k' \notin \tilde{N}$. Thus, $p_r \longrightarrow k'$.

Let $k \ne p_v$. First, assume that $k \in P_v^-$. The definition of the set P_v implies that $k > p_v$. Let $a(k)$ and $a(p_v)$ be the sequences of all direct successors of the jobs k and p_v, respectively, sorted in decreasing numerical order. The inequality $k > p_v$ implies that $a(p_v)$ is lexicographically smaller than the sequence $a(k)$.

We show that the first $|P_{v-1}^+|$ elements of the sequence $a(p_v)$ are jobs of the set P_{v-1}^+. In fact, for any $k' \in P_{v-1}$ the inequality $k' > p_{v-1}$ holds, and for any j such that $\delta_j \ge \delta_{p_{v-1}}$ the inequality $p_{v-1} > j$ holds. Hence, the elements of set P_{v-1} have the largest numbers among all the jobs processed after the interval δ_{p_v}. Since $p_v \longrightarrow k'$ for all $k' \in P_{v-1}$, we have $p_v \subseteq B^0(l)$ for all $l \in P_{v-1}^+$, and the first $|P_{v-1}^+|$ elements of the sequence $a(p_v)$ are the elements of set P_{v-1}^+.

If the sequence $a(p_v)$ is lexicographically smaller than $a(k)$, the condition $k \in B^0(l)$ is satisfied for any $l \in P^+_{v-1}$, i.e., $k \longrightarrow k'$ for any $k' \in P_{v-1}$.

Finally, if $k \neq p_v$ and $k \notin P^-_v$, then $k \longrightarrow j$ for some $j \in P^-_v$, which implies $k \longrightarrow k'$ for any $k' \in P_{v-1}$.

Now, it can be easily shown that the λ-schedule is optimal. For any $k \in P_v$ and $k' \in P_{v'}$, where $u \geq v > v' \geq 0$, $k \longrightarrow k'$ holds, i.e., all jobs of the set P_v must be completed before the jobs of the set $P_{v'}$ start. Each of the sets P_v, $v = 0, 1, \ldots, u$, contains an odd number of jobs. Let P_v include $2n_v - 1$ jobs. Evidently, it takes at least n_v time units to process all jobs of set P_v, and for any schedule $\tilde{s}$ (feasible with respect to $\longrightarrow$) for processing the jobs of set N, $T(\tilde{s}) \geq \sum_{v=0}^{u} n_v$ holds. Since $T(s) = \sum_{v=0}^{u} n_v$ holds for schedule s, this schedule is optimal. This proves the theorem.

Remark. As can be seen from the example below, *in general, a λ-schedule need not be optimal if $M \neq 2$.*

Let $N = \{1, 2, \ldots, 11\}$, $M = 3$, $t_i = 1$, $d_i = 0$, $i = 1, 2, \ldots, 11$, and the reduction graph G of the relation $\longrightarrow$ defined over set N be shown in Fig. 5.5. The jobs of set N are numbered as described in Section 5.4.

Figure 5.6a presents the λ-schedule s constructed by the algorithm described in Section 5.5, while Fig. 5.6b shows an optimal schedule s^*. We have $T(s^*) = 4$, $T(s) = 5$.

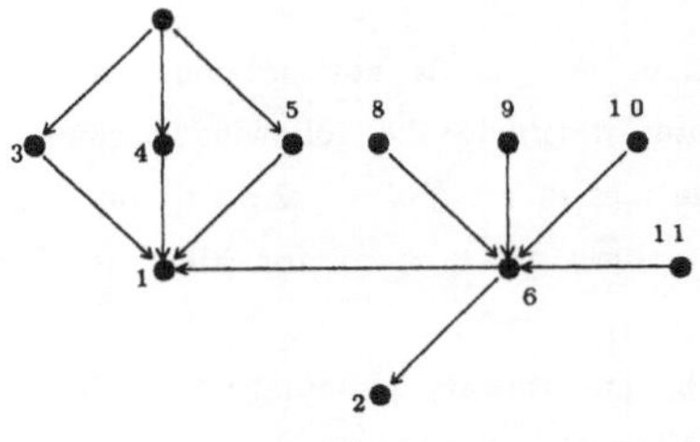

Fig. 5.5.

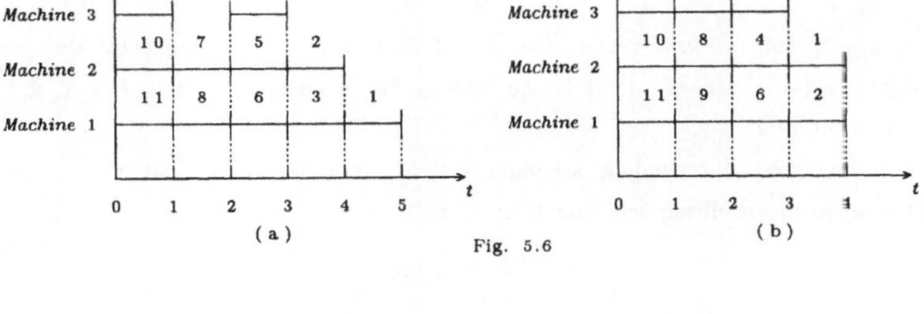

(a) Fig. 5.6 (b)

6. Identical Machines. Maximal Completion Time. Preemption

In this section we consider the problem of finding a time-optimal preemptive schedule for processing n jobs on M parallel identical machines. Polynomial algorithms are given for the cases: (a) the set of jobs is not ordered; (b) the reduction graph of the precedence relation is tree-like; (c) $M = 2$ and the reduction graph is arbitrary. In the last two cases the job processing times are assumed to be commensurable.

6.1. The jobs of a set $N = \{1, 2,..., n\}$ are processed on M parallel identical machines. The processing time of a job $i \in N$ is $t_i > 0$. All jobs have the same release dates. Without loss of generality, we assume that the release date is $d = 0$. Preemption is allowed. It is assumed that preemptions do not consume time, and that their number is finite.

The precedence relation $\rightarrow$ is defined over set N to specify a possible order of job processing. The reduction graph of this relation is denoted by $G = (N, U)$. If $\bar{t}_i(s)$ is the completion time of the job i in a schedule s, then $T(s) = \max\{\bar{t}_i(s) \mid i \in N\}$ is, evidently, the maximal completion time for schedule s (the length of schedule s). It is required to find *a time-optimal schedule* s^*, i.e., a schedule which is the shortest among all feasible (with respect to $\rightarrow$) schedules.

6.2. Suppose that the set of jobs is not ordered, i.e., $G = (N, \varnothing)$. Recall that Section 2.3 of this chapter described the following packing algorithm for finding a schedule for processing the jobs of set $N = \{1, 2,..., n\}$ on M parallel identical machines in the interval $(e', e'']$ subject to $t_i \le \Delta$ for all $i \in N$ and $\sum_{i \in N} t_i \le M\Delta$ (here $\Delta = e'' - e'$).

Let $\pi = (i_1, i_2,..., i_n)$ be an arbitrary permutation of the elements of set N. In the interval $(e', e'+M\Delta]$, define the function $\sigma(t)$ assuming $\sigma(t) = i_1$ in the interval $(e', e'+t_{i_1}]$, $\sigma(t) = i_k$ in the interval $\left(e'+\sum_{j=1}^{k-1} t_{i_j}, e'+\sum_{j=1}^{k} t_{i_j}\right]$, $k = 2, 3,..., n$, and, if $\sum_{i \in N} t_i < M\Delta$, then $\sigma(t) = 0$ in the interval $\left(e'+\sum_{i \in N} t_i, e'+M\Delta\right]$. A schedule $s(t) = \{s_1(t), s_2(t),..., s_M(t)\}$ for processing the jobs of set N is said to be found by the packing algorithm if $s_L(t) = \sigma(t+(L-1)\Delta)$ in the interval $(e', e'']$ and $s_L(t) = 0$, $L = 1, 2,..., M$, outside this interval.

The running time for finding schedule $s(t)$ is at most $O(n)$, and *the number of preemptions* in the resulting schedule is at most $M-1$.

If the length T^* of an optimal schedule is known, then the schedule s^* can be found by applying the packing algorithm in the interval $(0, T^*]$. It is clear that the value of T^* cannot be less than $T^0 = \max\left\{\max\{t_i \mid i \in N\}, \sum_{i \in N} t_i/M\right\}$. On the other hand, the packing algorithm applied to the interval $(0, T^0]$ finds a schedule for processing the jobs of set N with the length T^0. Thus, $T^* = T^0$.

We show that $M-1$ (the maximal number of preemptions for the schedule obtained by the packing algorithm) is a tight lower bound on the number of preemptions in an optimal schedule. In other words, there exists an instance of the problem under consideration such that any optimal schedule contains at least $M-1$ preemptions.

Let $N = \{1, 2, \dots, M+1\}$ and $t_i = M$ for all $i \in N$. The packing algorithm finds an optimal schedule of length $M+1$ without idle machines in the interval $(0, M+1]$. It is obvious that any optimal schedule does not allow idle time in this interval. Suppose that there exists an optimal schedule $\tilde{s}$ with a number of preemptions less than $M-1$. Then at least two machines (say, machines K and L) process the jobs without preemption. Furthermore, these machines process some jobs k and l in the interval $(0, M]$ without preemption. Therefore, there are times t and t' such that $M \le t < t' \le M+1$, and in the interval $(t, t']$, machine K processes some job i, while machine L processes job j. In the interval $(t, t']$, the other $M-2$ machines can process only the jobs of set $N\backslash\{i, j, k, l\}$, i.e., at most $M-3$ jobs. Thus, in the interval $(t, t']$, at least one machine is idle, and schedule $\tilde{s}$ cannot be optimal.

6.3. Let the precedence relation $\rightarrow$ be defined over set N of jobs, and $G = (N, U)$ be the reduction graph of this relation. Each vertex i of graph G is given the weight t_i (i.e., the processing time of job i).

In the following, we do not distinguish between a job $i \in N$ and the corresponding vertex of graph G. Since no misunderstanding arises, the concepts of the processing time t_i of job i and the weight t_i of vertex i are considered to be equivalent. We also use, for example, the expression "a schedule for the graph G" (instead of "a schedule that is feasible with respect to $\rightarrow$ for processing the jobs of set N").

Throughout this section, it is assumed that all t_i are *commensurable*, i.e., there is a real number w such that $t_i = l_i w$, where l_i are natural numbers, $i = 1, 2, \dots, n$.

Let us consider the graph $G_w = (N_w, U_w)$ obtained from G by replacing each vertex $i \in N$ by the chain of l_i vertices $i_1, i_2, \dots, i_{l_i}$, $(i_{j-1}, i_j) \in U_w$, $j = 2, 3, \dots, l_i$. In this case, we replace all arcs entering a vertex i of graph G by the arcs entering the vertex i_1 in graph G_w, and the arcs leaving a vertex i, by the arcs leaving i_{l_i} in G_w. Notice that all jobs of set N_w have equal processing times ω.

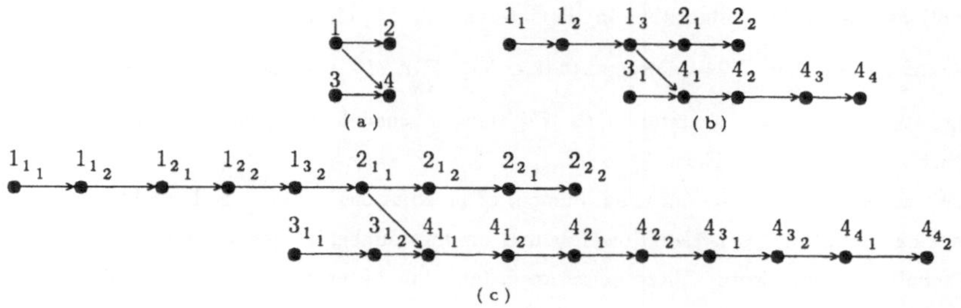

Fig. 6.1

In turn, each vertex of graph G_w can be represented as a chain of p vertices of equal weight w/p. Let $G_{w/p}$ denote the resulting graph. The graphs G_w and $G_{w/p}$ corresponding to the graph G in Fig. 6.1a are shown in Figs. 6.1b and 6.1c. Here $t_1 = 7.5$, $t_2 = 5$, $t_3 = 2.5$, $t_4 = 10$, $w = 2.5$, and $p = 2$.

It is easy to see that non-preemptive schedules for each of the graphs G_w or $G_{w/p}$ are (in general, preemptive) schedules for the graph G.

For a graph $\hat{G}$, let $T^*(\hat{G})$ and $\tilde{T}^*(\hat{G})$ be the lengths of non-preemptive and preemptive optimal schedules, respectively.

Theorem 6.1. *For* $p = 1, 2, \ldots$ *the relation*

$$\tilde{T}^*(G) \le T^*(G_{w/p}) \le \tilde{T}^*(G) + c/p$$

holds where the value of c *depends only on* n *and* w.

Proof. Let s be a preemptive optimal schedule for graph G, and $\tau_1 < \tau_2 < \ldots < \tau_m$ denote the sequence of time moments at which at least one job is completed in this schedule. Assume $\tau_0 = 0$. For a k, $1 \le k \le m$, consider the time interval $I_k = (\tau_{k-1}, \tau_k]$. It is obvious that all jobs processed in this interval are incomparable (with respect to $\longrightarrow$). Suppose that the jobs processed in the interval I_k are $j_1, j_2, \ldots, j_{n_k}$. Let δ_l denote the total processing time of job j_l in this interval. Regarding j_l as a job with the processing time δ_l, $l = 1, 2, \ldots, n_k$, find a schedule for processing the jobs $j_1, j_2, \ldots, j_{n_k}$ in the interval I_k by the packing algorithm. Let s' be the schedule obtained by the packing algorithm applied to all intervals I_k, $k = 1, 2, \ldots, m$. It is clear that $T(s') = T(s)$.

For schedule s', let us call a time interval $(t', t'']$ the *assignment interval* if in this interval $s_L'(t) = const$, $L = 1, 2, \ldots, M$, and there exist both H and Q, $1 \le H \le Q \le M$, such that $s_H'(t') \ne s_H'(t'+\delta)$ and $s_Q'(t'') \ne s_Q'(t''+\delta)$ for a sufficiently small $\delta > 0$ (i.e., at times t' and t'' another job is assigned to be processed). Since in schedule s' the processing of a job is interrupted at most twice in an interval I_k (or at most once if the completion time of a job is τ_k), there are at most $2n$ assignment intervals in an interval I_k.

Since there are no restrictions on the times of possible preemptions, the length of subintervals of the processing of each job in the schedule s' need not be a multiple of w/p. Let us increase (if necessarily) the length of each assignment interval so that it becomes a multiple of w/p. The length of each interval I_k increases by at most $2nw/p$, while the length of the whole "schedule" increases by at most $2n^2w/p$. Here, we use the quotation marks to point out that the processing of each job i may take longer than is actually necessary (more than the processing time t_i). We call this new "schedule" an *extended* schedule.

If $t_i = l_i w$, then job i is processed in the extended schedule at least within $l_i p$ subintervals of the length w/p. Let $i_1, i_2, \ldots, i_{l_i p}$ be the vertices of the graph $G_{w/p}$ which correspond to job i. Let us find a (non-preemptive) schedule $\tilde{s}$ for the graph $G_{w/p}$ such that the jobs $i_1, i_2, \ldots, i_{l_i p}$ are processed in the first $l_i p$ intervals in which job i is processed in the extended schedule. In the remaining intervals of processing job i the relation $\tilde{s}_L(t) = 0$ holds for all appropriate L. We have $T(\tilde{s}) \le T(s') + 2n^2w/p$.

It is obvious that $T^*(G_{w/p}) \le T(\tilde{s})$ and, hence, $T^*(G_{w/p}) \le \tilde{T}^*(G) + 2n^2w/p$.

An optimal non-preemptive schedule for the graph $G_{w/p}$ is some schedule, presumably a preemptive one, for the graph G. Hence, $\tilde{T}^*(G) \le T^*(G_{w/p})$. This proves the theorem.

This theorem allows us *to approximate with any desired accuracy* (by choosing an appropriate p) an optimal preemptive schedule for G using an optimal non-preemptive schedule for $G_{w/p}$ with equal weights of vertices.

6.4. We now introduce the concept of a schedule for the *machine-sharing* processing of jobs.

Consider a system of M parallel identical machines as a processing system which uses total power M. Assume that at any time some power $\alpha(i)$, $0 \le \alpha(i) \le 1$, can be used in the processing of a job i. In this case, the total power to be used at each time cannot exceed M.

In the situation under consideration, a machine can process more than one job at a time and uses some portion of its power for each job (machine sharing). A job i is processed in a time interval $(t', t'']$ if and only if at each time $t \in (t', t'']$ non-zero power is to be used in the for processing of this job. It is assumed that the processing of each job i

can be defined by specifying a finite number of time intervals such that the power to be used in each interval for processing job i is constant.

Let δ_1, $\delta_2, \ldots$, δ_l be the lengths of all time intervals in which a job $i \in N$ is processed, and $\alpha_1(i)$, $\alpha_2(i), \ldots$, $\alpha_l(i)$ be portions of power to be used in the processing job i in these intervals. Then the relation $\sum_{k=1}^{l} \alpha_k(i)\delta_k = t_i$ holds.

Without going into formalities, *a machine-sharing schedule* s_α is a sequence of time intervals such that for each of them a set $N' \subseteq N$ of jobs together with portions of power to be used in this interval for processing each job of the set N' are indicated. In particular, the case $N' = \emptyset$ is possible.

It is assumed that the number of mentioned intervals is finite, the total power to be spent in each interval does not exceed M, and assigning the new portions of power happens at the left end of an interval. In the schedule s_α, preemption is allowed in processing each job, and precedence constraints must be satisfied (if $i \longrightarrow j$, then in schedule s_α the processing of job j starts only after job i is completed).

As usually, for a schedule s_α, *the length* $T(s_\alpha)$ denotes the time taken to process all jobs. Since the jobs are processed since the time $t = 0$, $T(s_\alpha)$ is in fact the completion time of the last job. A schedule s_α^* of the shortest length is called *optimal* (or time-optimal) schedule.

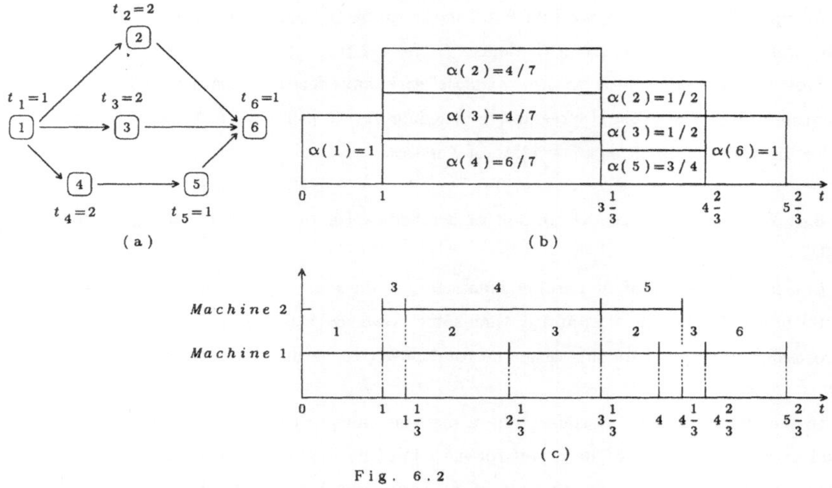

Fig. 6.2

For the graph G shown Fig. 6.2a, one of the schedules s_α is given in Fig. 6.2b. Here

$M = 2$. The sequence of intervals $(0, 1]$, $(1, 3\frac{1}{3}]$, $(3\frac{1}{3}, 4\frac{2}{3}]$, $(4\frac{2}{3}, 5\frac{2}{3}]$, and $(5\frac{2}{3}, \infty)$ corresponds to this schedule. In the interval $(0, 1]$ job 1 is processed, and the power allocated for its processing in this interval is $\alpha(1) = 1$. In the interval $(1, 3\frac{1}{3}]$, jobs 2, 3 and 4 are processed, and here $\alpha(2) = \alpha(3) = \frac{4}{7}$, $\alpha(4) = \frac{6}{7}$. In each interval, the total power to be used does not exceed $M = 2$. Job 2 is processed in the intervals $(1, 3\frac{1}{3}]$ and $(3\frac{1}{3}, 4\frac{2}{3}]$. For this job, we have $\frac{4}{7} \times 2\frac{1}{3} + \frac{1}{2} \times 1\frac{1}{3} = 2$. Similar relations also hold for the other jobs. The precedence constraints in job processing defined by the graph G are satisfied. Processing is non-preemptive. The value of $T(s_\alpha) = 5\frac{2}{3}$.

Let S_α denote the set of all machine-sharing schedules s_α, while S denote the set of all schedules s (in the usual sense) for processing the jobs of set N. In both cases, preemption is allowed. Since each schedule $s \in S$ is at the same time a machine-sharing schedule (the case $\alpha(i) = 1$ for all $i \in N$), we have $S \subset S_\alpha$.

We show that *any schedule* $s_\alpha \in S_\alpha$ *may be transformed into a schedule* $s \in S$ *such that* $T(s) \leq T(s_\alpha)$, and this takes at most $O(n^2)$ time.

Let $\tau_0 = 0$ and $\tau_1 < \tau_2 <\ldots< \tau_m$ be the times moments at which at least one job is completed in schedule s_α. For a k, $1 \leq k \leq m$, let us consider the interval $I_k = (\tau_{k-1}, \tau_k]$ and the set $\tilde{N}_k$ of jobs processed in s_α in this interval. It is obvious that the jobs of set $\tilde{N}_k$ are incomparable (with respect to $\longrightarrow$).

Let $\gamma_1, \gamma_2,\ldots, \gamma_l$ be the lengths of the subintervals for processing a job $j \in \tilde{N}_k$ in the interval I_k, and the amounts of power to be used in the processing of job j in these subintervals be $\alpha_1(j), \alpha_2(j),\ldots, \alpha_l(j)$, respectively. The value $\Delta_j = \sum_{i=1}^{l} \alpha_i(j)\gamma_i$ may be considered as an ordinary processing time of job j in the interval I_k. Since $\alpha_i(j) \leq 1$, we have $\Delta_j \leq \tau_{k+1} - \tau_k$. At any time, the total power does not exceed M, therefore, $\sum_{j \in \tilde{N}} \Delta_j \leq M(\tau_{k+1} - \tau_k)$. Hence, we can construct a schedule $s \in S$ for processing the jobs of set $\tilde{N}_k$ in the interval I_k by the packing algorithm. By "concatenating" the schedules for the intervals I_k, $k = 1, 2,\ldots, m$, we obtain the schedule s such that $T(s) \leq T(s_\alpha)$. Since the running time of the packing algorithm is $O(n)$, it takes at most $O(n^2)$ time to transform a given schedule s_α into a schedule s.

The schedule s found by the packing algorithm from the schedule s_α, presented in Fig. 6.2b, is given in Fig. 6.2c.

6.5. As follows from Section 6.4, the problem of finding a (time-) optimal schedule $s^* \in S$ reduces to the problem of finding an optimal machine-sharing schedule $s_\alpha^* \in S_\alpha$. In this case $T(s^*) = T(s_\alpha^*)$, and finding s^* from a known s_α^* requires at most $O(n^2)$ time.

Let us describe *an* $O(n^2)$ *algorithm* for finding a schedule $\hat{s}_\alpha \in S_\alpha$ and show that the schedule $\hat{s}_\alpha$ is optimal at least when (1) the graph G is tree-like; (2) the graph G is arbitrary but $M = 2$.

Let the weight of a vertex i of graph G be equal to the processing time t_i of job i. *The weighted length of a path* in graph G is the sum of weights of the vertices in this path. Find the weighted lengths of all paths from a vertex i to the terminal vertices of graph G. The largest of the found values is called *the weighted height* $H(i)$ of vertex i. If $t_i = t$, $i \in N$, then $H(i) = th(i)$ where $h(i)$ is the height of vertex i in graph G.

As before, N^+ denotes the set of all maximal (with respect to $\rightarrow$) elements of set N. Let us divide the set N^+ into subsets $N_1^+, N_2^+, \dots, N_u^+$ of vertices with equal weighted heights, and order these subsets in decreasing values of $H(N_k^+)$, where $H(N_k^+)$ is the weighted height of the vertices in set N_k^+, $k = 1, 2, \dots, u$.

Let m be the largest integer for which the relation

$$\sum_{k=1}^{m} |N_k^+| \le M \tag{6.1}$$

holds. If $|N_1^+| > M$, assume $m = 0$.

We now describe the algorithm for finding a schedule $\hat{s}_\alpha$.

In the first step, set $\underline{t} = 0$. In each step, find the time $\bar{t} > \underline{t}$ and assign the jobs of the set N^+ for processing in the time interval $(\underline{t}, \bar{t}]$ according to the following rule. Each job i of the set $\bigcup_{k=1}^{m} N_k^+$ is given power $\alpha(i) = 1$. If $\sum_{k=1}^{m} |N_k^+| < M$, then the remaining power $a = M - \sum_{k=1}^{m} |N_k^+|$ is equally distributed between the jobs in the set N_{m+1}^+, i.e., for each $i \in N_{m+1}^+$ the equality $\alpha(i) = a/b$ holds where $b = |N_{m+1}^+|$. For the other jobs define $\alpha(i) = 0$. It is clear that a and b are integers and $a < b$.

Note that if $m = u$, then $\alpha(i) = 1$ for all $i \in N^+$. If $m = 0$, i.e., $|N_1^+| > M$, then $a = M$ and $\alpha(i) = M/|N_1^+|$ for all jobs $i \in N_1^+$.

Let us introduce a parameter τ, $0 < \tau < \infty$, and define $t_i^\tau = t_i - \alpha(i)\tau$ provided that a (constant) power is to be used for processing job i in the interval $(\underline{t}, \underline{t}+\tau]$. It is natural to interpret t_i^τ as the total processing time (for $\alpha(i) = 0$) or the remaining processing time (for $\alpha(i) > 0$) of job i in the interval $(\underline{t}+\tau, \infty)$. If in the interval $(\underline{t}, \underline{t}+\tau]$ we have $\alpha(i) = \alpha$ for all $i \in N_k^+$, then define $H^\tau(N_k^+) = H(N_k^+) - \alpha\tau$.

Let $\bar{t}$ denote the smallest value of $\underline{t}+\tau$ for which at least one of the following events occurs: (1) $t_i^\tau = 0$ for some $i \in N^+$; (2) either $H^\tau(N_m^+) = H^\tau(N_{m+1}^+)$ or $H^\tau(N_{m+1}^+) = H^\tau(N_{m+2}^+)$. Finding $\bar{t}$ does not involve essential difficulties.

In fact, if $m = u$, i.e., if $N_{m+1}^+ = \emptyset$, then $\bar{t} = \underline{t}+A$ where

$$A = \min\left\{t_i \middle| i \in \bigcup_{k=1}^{m} N_k\right\}$$

and at time $\bar{t}$ event 1 occurs.

If $m = 0$, then $\bar{t} = \underline{t} + \min\{B, C\}$, where

$$B = \min\{bt_i/a \mid i \in N_{m+1}^+\},\ C = b[H(N_{m+1}^+) - N(N_{m+2}^+)]/a.$$

In this case, at time $\bar{t}$ either event 1 happens (if $B < C$), or event 2 (if $B > C$) occurs, or both events take place simultaneously (if $B = C$).

If $0 < m < u$ and $a = 0$, then $\bar{t} = \underline{t} + \min\{A, D\}$ where

$$D = b\{H(N_m^+) - H(N_{m+1}^+)\}/(b-a).$$

In this case, at time $\bar{t}$ either event 1 happens (if $A < D$) or event 2 happens (if $A > D$), or both events take place simultaneously (if $A = D$).

Finally, if $0 < m < u$, $a > 0$, then $\bar{t} = \underline{t} + \min\{A, B, C, D\}$. Here, event 1 takes place if $\min\{A, B, C, D\}$ is equal either to A or to B, and event 2 occurs if $\min\{A, B, C, D\}$ is equal either to C or to D.

Using the found value of $\bar{t}$, we obtain the new processing times t_i equal to t_i^τ for $\tau = \bar{t} - \underline{t}$ and remove from N (i.e., from graph G) all jobs (vertices) with zero processing times.

Again, denote the obtained set of jobs and the graph by N and G, respectively, let $\underline{t}$ be equal to the found value of $\bar{t}$, and go to the next step of the algorithm. Finding a schedule $\hat{s}_\alpha$ is completed when the current set N is empty.

Example. Let $M = 3$ and the graph G be a tree (see Fig. 6.3a).

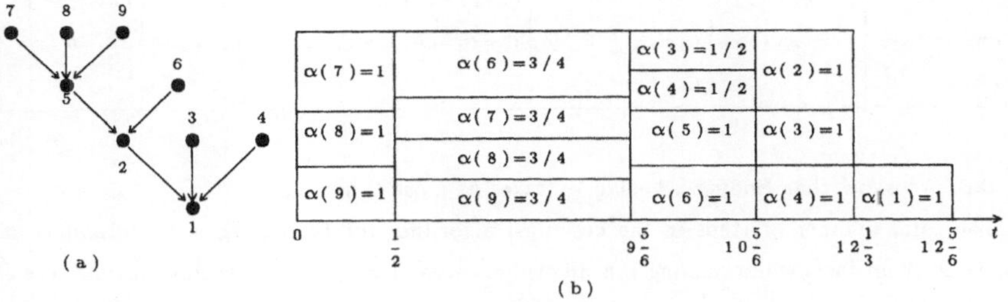

Fig. 6.3

The values of t_i, $i = 1, 2, \ldots, 9$, are given in Table 6.1. Here $w = 1/2$.

For each step of the algorithm, Table 6.2 gives the set $N^+ = N_1^+ \cup N_2^+ \cup \ldots \cup N_u^+$, the value of m, the heights $H(N_m^+)$, $H(N_{m+1}^+)$, and $H(N_{m+2}^+)$, the amount of power $\alpha(i)$ to be spent, as well as time $\bar{t}$ and the current values of t_i. At time $\bar{t} = 1/2$, event 2 occurs (here $\tau = 1/2$,

$N_m^+ = N_1^+ = \{7, 8, 9\}$, $N_{m+1}^+ = N_2^+ = \{6\}$, $H^\tau(N_m^+) = 10\frac{1}{2} - \frac{1}{2} = 10$, $H^\tau(N_{m+1}^+) = 10$ and, hence, $H^\tau(N_m^+) = H^\tau(N_{m+1}^+)$). At time moments $9\frac{5}{6}$, $12\frac{1}{5}$, and $12\frac{5}{6}$ event 1 occurs. At time $10\frac{5}{6}$, events 1 and 2 take place. The resulting schedule $\hat{s}_\alpha$ is shown in Fig. 6.3b. We have $T(\hat{s}_\alpha) = 12\frac{5}{6}$.

Table 6.1

i	1	2	3	4	5	6	7	8	9
t_i	$\frac{1}{2}$	$1\frac{1}{2}$	2	2	1	8	$7\frac{1}{2}$	$7\frac{1}{2}$	$7\frac{1}{2}$

Table 6.2

Step	1	2	3	4	5
$N^+ = N_1^+ \cup \cup N_2^+ \cup \ldots N_u^+$	$\{7, 8, 9\} \cup \cup \{6\} \cup \{3, 4\}$	$\{6, 7, 8, 9\} \cup \cup \{3, 4\}$	$\{5, 6\} \cup \{3, 4\}$	$\{2, 3, 4\}$	$\{1\}$
m	1	0	1	1	1
$H(N_m^+)$	$10\frac{1}{2}$	-	3	2	$\frac{1}{2}$
$H(N_{m+1}^+)$	10	10	$2\frac{1}{2}$	-	-
$H(N_{m+2}^+)$	$2\frac{1}{2}$	$2\frac{1}{2}$	-	-	-
$\alpha(i)$	$\alpha(7) = \alpha(8) = = \alpha(9) = 1$	$\alpha(6) = \alpha(7) = = \alpha(8) = \alpha(9) = = 3/4$	$\alpha(5) = \alpha(6) = 1$, $\alpha(3) = \alpha(4) = 1/2$	$\alpha(2) = \alpha(3) = = \alpha(4) = 1$	$\alpha(1) = 1$
$\bar{t}$	$\frac{1}{2}$ (event 2)	$9\frac{5}{6}$ (event 1)	$10\frac{5}{6}$ (events 1 and 2)	$12\frac{1}{3}$ (event 1)	$12\frac{5}{6}$ (event 1)
New values of t_i	$t_7 = t_8 = t_9 = 7$	$t_6 = 1$, $t_7 = t_8 = t_9 = 0$	$t_3 = t_4 = 1\frac{1}{2}$, $t_5 = t_6 = 0$	$t_2 = t_3 = t_4 = 0$	$t_1 = 0$

6.6. We show that finding schedule $\hat{s}_\alpha$ takes at most $O(n^2)$ time.

The total number of steps in the described algorithm for finding $\hat{s}_\alpha$ is finite and is at most $O(n)$. In fact, while running this algorithm, event 1 may happen at most n times (as a result of this event, at least one element is deleted from set N). Event 2 may also take place at most n times (as a result of which the weighted heights of some vertices of graph G become equal, these being reduced by the same value in subsequent steps).

The height of each vertex can be found by numbering the vertices as described in Section

5.4 of this chapter (this takes $O(n^2)$ time), followed by computing $H(i)$ for $i = 1, 2,...$ by the formula $H(i) = t_i+\max\{H(j) \mid j \in A^0(i)\}$, where $A^0(i)$ is the set of the direct successors of vertex i. It is clear that finding the heights of vertices requires at most $O(n^2)$ time.

Finding the set N^+ can be implemented in at most $O(n)$ time (see Section 1.4 of Chapter 1). Define the total pseudo-order $\Longrightarrow$ over set N, assuming $i \Longrightarrow j$ if and only if $H(i) \geq H(j)$. Using a balanced 2-3-tree (see Section 2 of Chapter 1) to represent the set N^+, sort the jobs in this set in non-increasing order of $H(i)$ (this takes at most $O(n_r \log n_r)$ time, where n_r is the number of elements added to set N^+ in step r). In each step, finding the value of m by formula (6.1) and the sets $\bigcup_{k=1}^{m} N_k^+$, N_{m+1}^+, N_{m+2}^+ takes at most $O(n)$ time. It is clear that, in each step, computation of the values of $\bar{t}$, the new values of t_i and $H(i)$ also takes at most $O(n)$ time. Deleting an element from N and finding new elements of set N^+ can be done in at most $O(n)$ time (see Section 1.4 of Chapter 1).

Thus, each step r of the algorithm takes at most $O(n)+O(n_r \log n_r)$ time. Since the number of steps does not exceed $O(n)$, and $O(n_1 \log n_1)+O(n_2 \log n_2) + ... + O(n_r \log n_r) + ...$ does not exceed $O(n \log n)$, schedule $\hat{s}_\alpha$ can be found in at most $O(n^2)$ time.

6.7. We show that in the case when graph G is *an intree*, the schedule $\hat{s}_\alpha$ obtained by the algorithm described in Section 6.5 is *a time-optimal* schedule.

For the case when G is *an arbitrary circuit-free graph* and $M = 2$, the proof is similar.

The scheme of the proof is as follows. Given an initial intree $\mathcal{T}^-$ with the weights of vertices $t_i = l_i w$, where l_i are natural numbers, $i = 1, 2,..., n$, and w is a real number, for any natural p one can construct a tree $\mathcal{T}^-_{w/p}$ such that each vertex i of $\mathcal{T}^-$ is replaced by a chain consisting of pl_i vertices of equal weight w/p. For $\mathcal{T}^-_{w/p}$, an optimal non-preemptive schedule can be found by the h-algorithm described in Section 5.2 of this chapter. Denote the resulting h-schedule by $s^*_{w/p}$, and its length by $T^*(\mathcal{T}^-_{w/p})$. Furthermore, we show that there exists a natural number z such that $T^*(\mathcal{T}^-_{w/pz}) = T(\hat{s}_\alpha)$ for any natural p. Hence, Theorem 6.1 implies that $\tilde{T}^*(\mathcal{T}^-) = T(\hat{s}_\alpha)$, i.e., $\hat{s}_\alpha$ is an optimal schedule.

We give the proof in five steps.

1. Let $t^{(r)}$ denote time $\bar{t}$ and $G^{(r)} = (N^{(r)}, U^{(r)})$ denote graph G obtained after performing r steps of the algorithm for finding schedule $\hat{s}_\alpha$. Let $t_i^{(r)}$ be the weights of the vertices of $G^{(r)}$.

We show that (a) $t^{(r)} = \rho^{(r)} w$, where $\rho^{(r)}$ is a rational number; (b) $t_i^{(r)} = l_i^{(r)} w^{(r)}$ for all $i \in N^{(r)}$, where $l_i^{(r)}$ are natural numbers, and $w^{(r)} = \gamma^{(r)} w$, where $\gamma^{(r)}$ is a rational number.

We restrict our consideration to the case $r = 1$. The statement (a) directly follows from the description of the algorithm for finding schedule $\hat{s}_\alpha$. Let us prove statement (b). As a result of performing the first step of the algorithm, the value of t_i is decreased either by $\rho^{(1)}w$, if $i \in \bigcup_{k=1}^{m} N_k^+$, or by $(a/b)\rho^{(1)}w$ if $i \in N_{m+1}^+$, while for the remaining $i \in N$ the values of t_i do not change. Hence, for any $i \in N^{(1)}$ the relation $t_i^{(1)} = \rho_i w$ holds, where ρ_i is a rational number, i.e. $t_i^{(1)} = (q_i/v_i)w$ where q_i and v_i are natural numbers. Define $v^{(1)} = v_1 v_2 \cdots v_{n^{(1)}}$ where $n^{(1)}$ is the number of elements of set $N^{(1)}$. Then

$$t_i^{(1)} = \frac{q_i v_1 \cdots v_{i-1} v_{i+1} \cdots v_{n^{(1)}}}{v^{(1)}} w = \frac{q_i^{(1)}}{v^{(1)}} w,$$

where $q_i^{(1)}$ and $v^{(1)}$ are natural numbers. Let us represent $q_i^{(1)}$ as $q_i^{(1)} = l_i^{(1)} q^{(1)}$, where $q^{(1)}$ is the greatest common divisor of the numbers $q_i^{(1)}$, $i = 1, 2,\ldots, n^{(1)}$. Denote $\gamma^{(1)} = q^{(1)}/v^{(1)}$. Then $t_i^{(1)} = l_i^{(1)} w^{(1)}$, where $w^{(1)} = \gamma^{(1)} w$, $l_i^{(1)}$ is natural and $\gamma^{(1)}$ is rational.

2. We show that there exists a natural number y such that $G_{w/py}^{(1)}$ exists for all natural numbers p.

The graph $G_{w/py}^{(1)}$ can be defined for each natural number p if $w/py = w^{(1)}/p^{(1)}$ for some natural $p^{(1)}$. Since $t^{(1)} = \rho^{(1)}w$, we have $t^{(1)} = (c^{(1)}/d^{(1)})w$, where $c^{(1)}$ and $d^{(1)}$ are natural numbers. Let $\alpha(i)$ be an amount of power assigned to job $i \in N$ in the first step of the algorithm for finding schedule $\hat{s}_\alpha$. If, in this case, $N_{m+1}^+ \neq \emptyset$, then for $i \in N_{m+1}^+$ we have $\alpha(i) = a/b$.

Define $y = bd^{(1)}v^{(1)}$. Then $w/py = w/bd^{(1)}pv^{(1)} = w^{(1)}/bd^{(1)}pq^{(1)}$ and $p^{(1)} = bd^{(1)}pq^{(1)}$.

If $N_{m+1}^+ = \emptyset$, then define $y = d^{(1)}v^{(1)}$. We have $w/py = w/d^{(1)}pv^{(1)} = w^{(1)}/d^{(1)}pq^{(1)}$ and $p^{(1)} = d^{(1)}/pq^{(1)}$. Note that here for the proof it would be sufficient to define $y = v^{(1)}$ in both cases (then $p^{(1)} = pq^{(1)}$). However, below we use the values of y presented above.

3. Let $T^{(1)}$ be a tree obtained from $\mathcal{T}^-$ as a result of the first step of the algorithm for finding schedule $\hat{s}_\alpha$, $t_i^{(1)} = l_i^{(1)} w^{(1)}$ be weights of the vertices of the tree $T^{(1)}$, and $t^{(1)} = (c^{(1)}/d^{(1)})w$, where $w^{(1)} = \gamma^{(1)}w$, $\gamma^{(1)} = q^{(1)}/v^{(1)}$, $l_i^{(1)}$, $c^{(1)}$, $d^{(1)}$, $q^{(1)}$, $v^{(1)}$ are natural numbers.

We have $t^{(1)} = (c^{(1)}/d^{(1)})w = (c^{(1)}py/d^{(1)})(w/py)$, where y is a natural number defined in the previous item of the proof. Let us construct an optimal non-preemptive schedule $s_{w/py}^*$ for the tree $\mathcal{T}_{w/py}^-$ using the h-algorithm described in Section 5.2 of this chapter. While finding schedule $s_{w/py}^*$, we regard $\mathcal{T}_{w/py}^-$ to be a tree with unit weights of all vertices assuming that w/py is taken as a time unit. Let T' be a tree obtained from $\mathcal{T}_{w/py}^-$ as a result of performing $c^{(1)}py/d^{(1)}$ steps of the h-algorithm (i.e. the one obtained at time $t^{(1)}$).

We show that T' is isomorphic to the tree $T^{(1)}_{w/py}$. To prove this we assume that in the first step of the algorithm for finding schedule $\hat{s}_\alpha$, the set N^+_{m+1} is not empty (if $N^+_{m+1} = \emptyset$ the proof is similar).

It is obvious that T' is a subtree of the tree $\mathcal{T}^-_{w/py}$, and $T^{(1)}_{w/py}$ is isomorphic to a subtree of the tree $\mathcal{T}^-_{w/py}$.

Let $\xi_i = \{i_1, i_2, ..., i_l\}$ be the set of the vertices of the tree $\mathcal{T}^-_{w/py}$ derived from vertex i of the tree $\mathcal{T}^-$. If, in the graph $T^{(1)}_{w/py}$, there is a vertex which corresponds to the vertex i_j, denote it by i_j'. Let ξ_i' denote the set of vertices i_j'. Define $\xi_i'' = \{i_1, i_2, ..., i_l\} \cap N'$, where N' is the set of vertices of the graph T'.

To prove isomorphism of T' and $T^{(1)}_{w/py}$ it suffices to show that $|\xi_i'| = |\xi_i''|$ for all $i \in N$.

In schedule $\hat{s}_\alpha$, the amount of power to be used in the processing of a job $i \in \bigcup_{k=1}^{m} N^+_k$ in the interval $(0, t^{(1)}]$ is $\alpha(i) = 1$. Hence, by time $t^{(1)}$ the processing time of such a job decreases by $t^{(1)} = (c^{(1)}/d^{(1)})w = (c^{(1)}py/d^{(1)})(w/py)$. Therefore, ξ_i' contains $c^{(1)}py/d^{(1)}$ elements less than ξ_i. Defining $y = bd^{(1)}v^{(1)}$, we obtain $|\xi_i'| = |\xi_i| - bc^{(1)}pv^{(1)}$ for $i \in \bigcup_{k=1}^{m} N^+_k$.

The amount of power assigned for the processing of each job $i \in N^+_{m+1}$ in the interval $(0, t^{(1)}]$ is $\alpha(i) = a/b$, and the processing time of such a job by the time $t^{(1)}$ decreases by $\alpha(i)t^{(1)} = (ac^{(1)}/(bd^{(1)}))w = (ac^{(1)}py/(bd^{(1)}))w/py$. Therefore, $|\xi_i'| = |\xi_i| - ac^{(1)}pv^{(1)}$ for $i \in N^+_{m+1}$.

The rest of the jobs $i \in N$ are not processed in the interval $(0, t^{(1)}]$ in schedule $\hat{s}_\alpha$. Therefore, for each of them we have $|\xi_i'| = |\xi_i|$.

We show that similar relations hold for ξ_i''. Let $\overline{N}_1, \overline{N}_2, ..., \overline{N}_u$ be such subsets of the vertices of the tree $\mathcal{T}^-_{w/py}$ that $\xi_i \subseteq \overline{N}_k$ if and only if $i \in N^+_k$, $k = 1, 2, ..., u$. While performing the h-algorithm, the vertices are removed from $\overline{N}_k$ as the corresponding jobs are processed. Since all vertices of the tree $\mathcal{T}^-_{w/py}$ have the weight w/py, the jobs are completed at discrete times: w/py, $2w/py$, $3w/py$, etc. The interval $(0, t^{(1)}]$ includes $t^{(1)}/(w/py) = bc^{(1)}pv^{(1)}$ intervals of the length w/py. It is easy to check that according to the h-algorithm exactly one job of each set $\xi_i \subseteq \bigcup_{k=1}^{m} \overline{N}_k$ and one job of each of a sets $\xi_i \subseteq \overline{N}_{m+1}$ are processed in each of these intervals (and in each interval, a vertices with the largest heights are chosen from b sets $\xi_i \subseteq \overline{N}_{m+1}$). The jobs of the other sets ξ_i, $i \in N$, are not processed in the interval $(0, t^{(1)}]$. Note that $abc^{(1)}pv^{(1)}/b = ac^{(1)}pv^{(1)}$ jobs of each set $\xi_i \subseteq \overline{N}_{m+1}$ are processed.

Consequently, if $i \in \bigcup_{k=1}^{m} N^+_k$ (i.e., $\xi_i \subseteq \bigcup_{k=1}^{m} \overline{N}_k$), then at time $t^{(1)}$, ξ_i'' contains $bc^{(1)}pv^{(1)}$

vertices fewer than ξ_i, i.e., $|\xi_i''| = |\xi_i| - bc^{(1)}pv^{(1)}$. If $i \in N_{m+1}^+$ (i.e. $\xi_i \subseteq \bar{N}_{m+1}$), then $|\xi_i''| = |\xi_i| - ac^{(1)}pv^{(1)}$. For the remaining jobs $i \in N$, $|\xi_i''| = |\xi_i|$ holds. Comparing $|\xi_i''|$ and $|\xi_i'|$ yields $|\xi_i''| = |\xi_i'|$ for all $i \in N$.

4. We show that there exists a natural number z such that $T^*(\mathcal{T}^-_{w/pz}) = T(\hat{s}_\alpha)$ for any natural p. Here $T^*(\mathcal{T}^-_{w/pz})$ is the length of an optimal schedule $s^*_{w/pz}$ found by the h-algorithm applied to the graph $\mathcal{T}^-_{w/pz}$, and $T(\hat{s}_\alpha)$ is the length of schedule $\hat{s}_\alpha$.

We prove this statement by induction with respect to the number of events 1 and 2 that take place while finding schedule $\hat{s}_\alpha$. Note that the number of events 1 and 2 is finite, and the weights of the vertices of the graph obtained after either event 1 and 2 happens remain commensurable (according to step 1 of this proof). It is obvious that if only one event 1 or 2 takes place, then this is event 1, which happens at time $T(\hat{s}_\alpha)$. In this case, the tree $\mathcal{T}^-$ consists of a single vertex, and any natural number can act as z.

Let the statement be valid if at most $\mu - 1$ events 1 or 2 take place, and suppose that μ events 1 or 2 take place while finding schedule $\hat{s}_\alpha$. Assume, as before, that the first of these events occur at time $t^{(1)}$, and let the weights of the vertices i of the tree $T^{(1)}$ (obtained from $\mathcal{T}^-$ at time $t^{(1)}$ while finding schedule $\hat{s}_\alpha$) be equal to $t_i^{(1)} = l_i^{(1)}w^{(1)}$, where $w^{(1)} = (q^{(1)}/v^{(1)})w$ and $l_i^{(1)}$, $q^{(1)}$ and $v^{(1)}$ are natural numbers. Denote by s' the schedule for $T^{(1)}$ found by the algorithm described in Section 6.5. Then we have

$$T(\hat{s}_\alpha) = t^{(1)} + T(s'). \tag{6.2}$$

It is clear that, while finding schedule s', at most $\mu - 1$ events 1 or 2 take place, and using the inductive assumption, a natural number z' can be found such that $T^*(T^{(1)}_{w^{(1)}/p'z'}) = T(s')$ for $p' = 1, 2, \ldots$, where $T^*(T^{(1)}_{w^{(1)}/p'z'})$ is the length of a non-preemptive optimal schedule found by the h-algorithm applied to graph $T^{(1)}_{w^{(1)}/p'z'}$.

If this holds for any natural p', then it also holds for $p' = q^{(1)}, 2q^{(1)}, \ldots$. Defining $p' = pq^{(1)}$ and taking into account that $w^{(1)} = (q^{(1)}/v^{(1)})w$, we obtain

$$T^*(T^{(1)}_{w/pv^{(1)}z'}) = T(s'),\ p = 1, 2, \ldots \tag{6.3}$$

Let y be defined as in step 2 of this proof, and let z be the smallest common multiple for y and $v^{(1)}z'$. Then step 3 of this proof implies that $T^{(1)}_{w/pz}$ is the tree obtained from $T_{w/pz}$ at time $t^{(1)}$ while finding $s^*_{w/pz}$ by the h-algorithm. Hence, it follows that

$$T^*(\mathcal{T}^-_{w/pz}) = t^{(1)} + T^*(T^{(1)}_{w/pz}),\ p = 1, 2, \ldots \tag{6.4}$$

From (6.3) and (6.4), we obtain

$$T^*(\mathcal{T}^-_{w/pz}) = t^{(1)} + T(s'),\ p = 1, 2, \ldots \tag{6.5}$$

Finally, it follows from (6.2) and (6.5) that

$T^*(\mathcal{T}^-_{w/pz}) = T(\hat{s}_\alpha),\ p = 1, 2,...$

5. Thus, there exists a natural number z such that $T^*(\mathcal{T}^-_{w/pz}) = T(\hat{s}_\alpha)$ for all p. Theorem 6.1 implies that $\tilde{T}^*(\mathcal{T}^-) = T(\hat{s}_\alpha)$, i.e., $\hat{s}_\alpha$ is an optimal schedule.

A machine-sharing schedule $\hat{s}_\alpha$ can be transformed into a preemptive schedule s^* without sharing the machines such that $T(s^*) \le T(\hat{s}_\alpha)$. Therefore, $T(s^*) = \tilde{T}^*(\mathcal{T}^-)$. As shown above, at most $O(n^2)$ time is required for finding $\hat{s}_\alpha$ and transforming it into s^*.

Remark 1. Let graph $G = (N, U)$ be a *forest* such that each connected component is an *intree*. Add a new vertex j with $t_j = w$ to set N, and include the arcs, leaving the roots of all trees and entering the vertex j, into the set U. The resulting graph G is, evidently, an intree. If s^* is an time-optimal schedule for G', then the schedule $\bar{s}$ such that $\bar{s}_L(t) = s^*_L(t)$, if $s^*_L(t) \ne j$, and $\bar{s}_L(t) = 0$, otherwise, $L = 1, 2,..., M$, is optimal for G. The relation $T(\bar{s}) = T(s^*) - w$ holds.

Remark 2. Let graph $G = (N, U)$ be a *forest* such that each connected component is an *outtree*. Denote by G'' the graph obtained from G by inverting the orientation of all its arcs. Let s^* be an optimal schedule for G'' and $T(s^*) = T$. Denote s' the set of M piecewise-constant functions $\{s'_1(t), s'_2(t),..., s'_M(t)\}$ such that $s'_L(t) = s^*_L(t)$, $L = 1, 2,..., M$, if t is not a point of discontinuity of a function $s^*_L(t)$. At the points t of discontinuity of a function $s^*_L(t)$, define $s'_L(t) = s'_L(t+\delta)$ for a sufficiently small $\delta > 0$ (i.e., unlike $s^*_L(t)$, the functions $s'_L(t)$ are right-semicontinuous rather than left-semicontinuous). Then, defining $\bar{s}_L(t) = s'_L(T-t)$ for $t \in (0, T]$ and $\bar{s}_L(t) = 0$ for $t > T$, $L = 1, 2,..., M$, we obtain schedule $\bar{s}$ which is optimal for the original graph G.

7. Identical Machines. Due Dates. Equal Processing Times

This section studies the problems of finding a deadline-feasible schedules for parallel identical machines and partially ordered sets of jobs with equal processing times. It is assumed that either the reduction graph of a precedence relation is an intree or the number of machines is two. The algorithms presented in this section can also be used for finding time-optimal schedules (along with the algorithms described in Section 5 of this chapter).

7.1. The jobs of a set $N = \{1, 2,..., n\}$ are processed on M parallel identical machines.

Their release dates are the same $d_i = d = 0$, $i = 1, 2,..., n$. The processing times t_i are equal. Without loss of generality, we assume that $t_i = 1$, $i = 1, 2,..., n$. For each job i, a deadline D_i is specified by which this job must be completed.

Preemption in the processing of each job is forbidden. A precedence relation $\rightarrow$ is defined over set N to describe a possible sequence of job processing. Let G denote the reduction graph of this relation $\rightarrow$. In what follows, we do not distinguish between a job $i \in N$ and the corresponding vertex of graph G.

As before, N^- and N^+ denote the sets of all minimal and maximal (with respect to $\rightarrow$) elements of set N, $B(i)$ and $A(i)$ are the sets of all jobs j such that $j \rightarrow i$ and $i \rightarrow j$ hold, respectively; $i \rightarrowtail k$ is used to denote that job k is a direct successor of job i.

It is required to find a schedule s that is feasible with respect to $\rightarrow$ for processing the jobs of set N such that $\bar{t}_i(s) \le D_i$, $i = 1, 2,..., n$. Here $\bar{t}_i(s)$ is the completion time of job i in schedule s. Such a schedule is called *deadline-feasible*.

Similar to Section 5 of this chapter, we introduce the useful concept of a λ-schedule. Assign numbers 1, 2, ... to unit length time intervals, starting at $t = 0$. The interval θ is of the form $(\theta-1, \theta]$. Let us introduce the list $\lambda = (i_1, i_2,..., i_n)$ of jobs (i.e., a permutation of jobs) and determine the schedule specified by the list λ in the following way.

To start with, assume $\theta = 1$, $s_L(t) = 0$ for $L = 1, 2,..., M$, $t \ge 0$ and assume that all elements of the list λ are unmarked. In each step, the following operations are to be made.

Find a machine H with the smallest number such that $s_H(\theta) = 0$. Find the first unmarked job j in the list λ which belongs to the set N^+. Mark job j, define $s_H(t) = j$ in the interval θ and go to the next step. If we fail to find either machine H or job j, remove the marked jobs from the list λ and set N; go to the next step, having increased θ by 1. A desired schedule is found when the list λ (and, hence, set N) becomes empty.

A schedule constructed by this algorithm is called a *λ-schedule*.

One elementary operation is required to verify whether an element of the list λ belongs to the set N^+ (assuming that the graph is given by a list of predecessors, see Section 1.4 of Chapter 1). Deleting each maximal (with respect to $\rightarrow$) element from set N requires at most $O(n)$ time (see Section 1.4 of Chapter 1). Hence, finding a λ-schedule takes *at most $O(n^2)$ time.*

Example A. Let $N = \{1, 2,..., 11\}$, $M = 2$, $t_i = 1$, $d_i = 0$, $i = 1, 2,..., 11$; and let the graph G be given in Fig. 7.1a, $\lambda = (1, 2,..., 11)$. The corresponding λ-schedule is given in Fig. 7.1b.

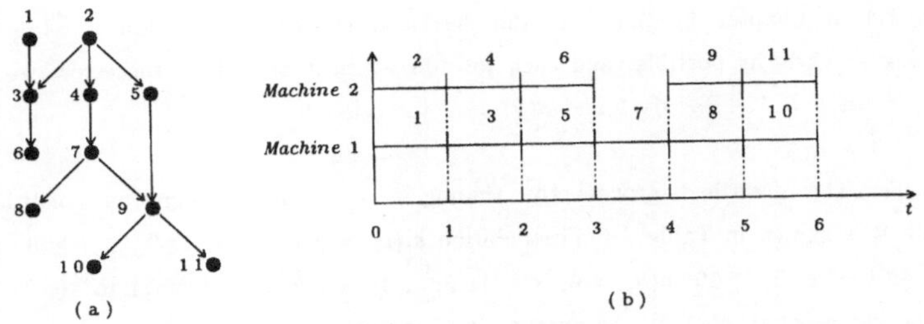

Fig. 7.1

Example B. Let $N = \{1, 2,..., 10\}$, $M = 3$, $t_i = 1$, $d_i = 0$, $i = 1, 2,..., 10$, and the graph G be as given in Fig. 7.2a, $\lambda = (1, 2,..., 10)$. The corresponding λ-schedule is given in Fig. 7.2b.

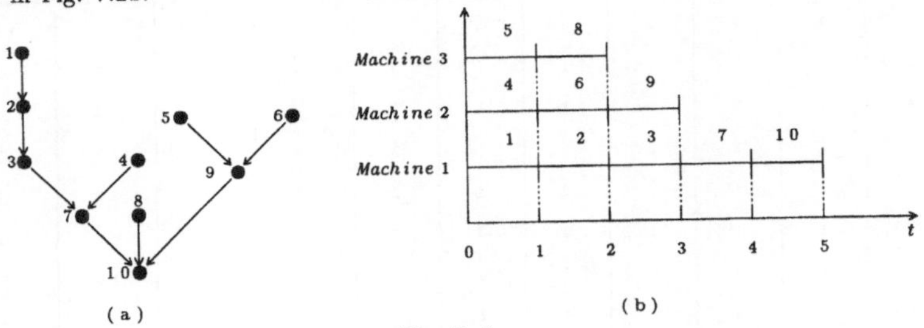

Fig. 7.2

If the list $\lambda = (i_1, i_2,..., i_n)$ has the property that for any job $k \in N$ all jobs of the set $B(k)$ are on the left of k in the list λ, and the graph G is an intree, then a λ-schedule can be found in *at most* $O(n)$ *time* in the following way.

First, define $s_L(t) = 0$ for $L = 1, 2,..., M$, $t \geq 0$. Introduce the variables $\delta(k)$, $k = 1, 2,..., n$, $n(\theta)$, $\theta = 1, 2,..., n$, and η. The value of variable $\delta(k)$ is one unit greater than the largest number of a unit time interval to which a job of the set $B(k)$ is assigned for processing. A variable $n(\theta)$ says how many jobs are assigned for processing in the interval θ. The variable η is equal to the smallest value of θ such that $n(\theta) < M$. Start with $\delta(k) = \eta = 1$, $k = 1, 2,..., n$, and $n(\theta) = 0$, $\theta = 1, 2,..., n$.

For each $k = i_j$, from $j = 1$ to $j = n$, perform the following. Define $\theta = \max\{\delta(k), \eta\}$. Let H be a machine with the smallest number, for which $s_H(\theta) = 0$. Define $s_H(t) = k$ in the interval θ. Increase $n(\theta)$ by 1, and if $n(\theta) = M$ is obtained, then assume $\eta = \theta+1$. Let $k \rightarrowtail l$. Define the value of $\delta(l)$ to be equal to $\max\{\delta(l), \theta+1\}$.

It is easy to verify that the schedule obtained this way is a λ-schedule and this can be found in most $O(n)$ time. Note that for any $k \in N$, finding the job l such that $k \rightarrowtail l$ requires just one elementary operation if the graph G is given by the array S_A (see Section 1.4 of Chapter 1). Since, in the described procedure, any job is assigned for processing as early as possible, and each job has at most one direct successor, we obtain $n(\theta) = M$ for $\theta = 1, 2, \ldots, \eta - 1$ and $n(\theta) < M$ for other θ.

Example. The described process for finding a λ-schedule under the conditions of Example B is shown in Table 7.1. First, define $s_L(t) = 0$ for $L = 1, 2, 3$, $t \geq 0$; $\eta = 1$, $\delta(k) = 1$, $k = 1, 2, \ldots, 10$, $n(\theta) = 0$, $\theta = 1, 2, \ldots, 10$. For each k from 1 to 10, Table 7.1 gives the values of η, $\delta(k)$, θ, the number H of a machine assigned for processing job k in the interval θ. Besides, the value of $n(\theta)$ obtained after assigning job k for processing, the number of the job l such that $k \rightarrowtail l$, and a new value of $\delta(l)$ are shown.

Table 7.1

k	1	2	3	4	5	6	7	8	9	10
η	1	1	1	1	1	2	2	2	3	3
$\delta(k)$	1	2	3	1	1	1	4	1	2	5
θ	1	2	3	1	1	2	4	2	3	5
H	1	1	1	2	3	2	1	3	2	1
$n(\theta)$	1	1	1	2	3	2	1	3	2	1
l	2	3	7	7	9	9	10	10	10	-
$\delta(l)$	2	3	4	4	2	2	5	5	5	-

7.2. Let us consider the problem of finding a deadline-feasible schedule, assuming that the graph G is an intree. Recall that $d_i = 0$ and $t_i = 1$, $i = 1, 2, \ldots, n$.

If $i \longrightarrow j$, then the processing of i and j without violating the deadlines would require job i to be completed not later that by time $D_j - 1$. Therefore, the deadline for the job i can be set equal to $\min\{D_i, D_j - 1\}$. Using this fact, the following algorithm may be proposed for modifying the deadlines.

The algorithm consists of n steps. In the first step, define $D'_r = D_r$ for the root r of the tree. In each subsequent step, choose a job i that the value of D'_i has not yet determined, but D'_j has been determined for its direct successor j. Define $D'_i = \min\{D_i, D'_j - 1\}$.

Finding the new deadlines D'_i requires *at most* $O(n)$ *time* (if, for example, the graph G is given by the list of predecessors using the arrays Q_B and S_B; see Section 1.4 of Chapter

1).

It is easy to verify that *the schedule is feasible with respect to modified deadlines* D_i' *if and only if it is feasible with respect to the initial deadlines* D_i.

In fact, since $D_i' \le D_i$, $i = 1, 2, \dots, n$, the schedule that is feasible with respect to D_i' is also feasible with respect to D_i. Assume that the schedule s is feasible with respect to D_i but this is not feasible with respect to D_i'. Without loss of generality, we may assume that in schedule s each job is processed in a single unit length time interval of the form $(\theta-1, \theta]$. Among the jobs for which the modified deadlines are violated, choose a job k processed in a unit time interval with the largest number θ. It follows that $D_k' < \theta \le D_k$. If $k \rightarrowtail j$, then $D_k' = D_j' - 1$. Since schedule s is feasible with respect to $\longrightarrow$, job j is processed in the interval $\delta \ge \theta+1$. Job k has been chosen so that $\delta \le D_j'$. Thus, we obtain a contradiction in that both $\theta+1 \le \delta \le D_j'$ and $\theta > D_k' = D_j'-1$ hold.

Thus, in speaking about a schedule that is feasible with respect to deadlines, we need not specify whether these deadlines are the original or modified ones.

Theorem 7.1. *A deadline-feasible schedule exists if and only if a* λ*-schedule corresponding to the list* $\lambda = (i_1, i_2, \dots, i_n)$, *where* $D_{i_j}' \le D_{i_{j+1}}'$, $j = 1, 2, \dots, n-1$, *is deadline-feasible.*

Proof. Let $\lambda = (i_1, i_2, \dots, i_n)$, $D_{i_j}' \le D_{i_{j+1}}'$, $j = 1, 2, \dots, n-1$, and assume that a λ-schedule s is not deadline-feasible. Among the jobs for which the modified deadlines are violated, choose a job i processed in a unit time interval with the smallest number θ. We have $\theta > D_i' \ge \lfloor D_i' \rfloor$. Here $\lfloor x \rfloor$ is the largest integer which does not exceed x. For schedule s, let δ, $\delta \le \lfloor D_i' \rfloor$, be an interval with the largest number where fewer than M jobs with the deadlines $D_j' \le D_i'$ are processed.

If the interval δ does not exist, then the theorem is proved because there are at least $M\lfloor D_i' \rfloor + 1$ jobs which must be completed by time D_i', and, hence, there is no feasible schedule.

Suppose that the interval δ exists. We show that this assumption results in a contradiction. The way in which λ-schedule s has been found implies that the following conditions hold for this schedule: (a) there is a job k, $k \longrightarrow i$, which is processed in the interval δ (otherwise, i should have been processed in the interval δ); (b) if $D_j' \le D_i'$ and job j is processed in the interval with the number larger than δ, then there exists such a job l that is processed in the interval δ and $l \longrightarrow j$ (otherwise, j should have been processed in the interval δ). Consider two cases: $\delta = \lfloor D_i' \rfloor$ and $\delta < \lfloor D_i' \rfloor$.

If $\delta = \lfloor D_i' \rfloor$, then job k is completed at time $\lfloor D_i' \rfloor < \theta$. Since $k \longrightarrow i$, the procedure for finding the modified deadlines implies that $D'_k \le D'_i - 1 < \lfloor D'_i \rfloor$. Hence, for job k, a modified deadline is violated, and we obtain the contradiction to the choice of job i.

Let $\delta < \lfloor D'_i \rfloor$. Then, in the interval $\delta+1$, M jobs j with deadlines $D'_j \le D'_i$ are processed. For each of these jobs there exists a job l, $l \rightarrow j$, which is processed in the interval δ. It follows from the procedure for modifying the deadlines that $D'_l \le D'_j - 1 \le D'_i - 1 < \lfloor D'_i \rfloor$. Since at most $M-1$ such jobs l can be processed in the interval δ, at least two jobs have the same predecessor. This is impossible for an intree. The obtained contradiction proves the theorem.

Thus, if G is an intree, $d_i = 0$ and $t_i = 1$, $i = 1, 2,..., n$, then *for finding a deadline-feasible schedule* (if such a schedule exists) it suffices: (a) to compute the modified deadlines D'_i (this takes at most $O(n)$ time); (b) to obtain the list λ of jobs sorted in non-decreasing order of D'_i (this can be done in $O(n\log n)$ time, see Section 2.7 of Chapter 1); (c) to find a λ-schedule (this requires at most $O(n)$ time since for all $k \in N$ the jobs of the set $B(k)$ are on the left of a job k in the list λ). Thus, *in at most $O(n\log n)$ time* either a deadline-feasible schedule is found or we conclude that no such schedule exists.

Example. Under the conditions of Example B, let the values of deadlines D_i for the jobs be as given in Table 7.2. The obtained modified deadlines D'_i are also given in the table.

Table 7.2

i	1	2	3	4	5	6	7	8	9	10
D_i	6	5	6	4	9	8	7	3	2	8
D'_i	4	5	6	4	1	1	7	3	2	8

The list λ of jobs sorted in non-decreasing order of D'_i is of the form: $\lambda = (5, 6, 9, 8, 1, 4, 2, 3, 7, 10)$. The corresponding λ-schedule is given in Fig. 7.3. This schedule is feasible with respect to the deadlines.

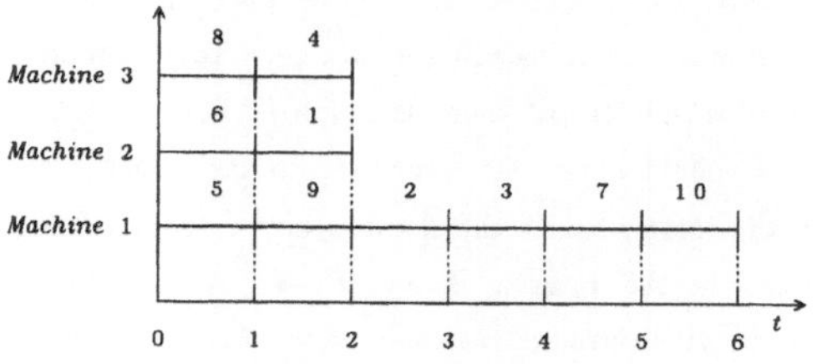

Fig. 7.3

In conclusion, note that if all deadlines are *equal*, i.e., $D_i = D$, $i = 1, 2, \ldots, n$, then the procedure of modifying the deadlines yields $D_i' = D-(h_i-1)$, where h_i is the height of a vertex i in graph G. In this case, the list λ of the jobs sorted in non-decreasing order of the modified deadlines coincides with the list of jobs ordered in non-increasing order of the heights, and is independent of D. If D is taken as the shortest deadline for which there exists a feasible schedule s, then this schedule is, evidently, a time-optimal schedule. Therefore, in the case under consideration (G is *an intree*, $d_i = 0$, $t_i = 1$, $i = 1, 2, \ldots, n$), a time-optimal schedule is the λ-schedule corresponding to the list λ with the jobs sorted *in non-increasing order* of the heights. If the vertices of the intree are numbered as in Section 1.4 of Chapter 1 (i.e., the root has number 1; then all vertices with a height of two are numbered; after them the vertices with a height of three are numbered, and so on), then the list λ is of the form: $\lambda = (n, n-1, \ldots, 2, 1)$. This is consistent with the result obtained in Section 5.2 of this chapter.

Example. Under the conditions of Example B, the list λ of jobs sorted in non-increasing order of the heights is $\lambda = (1, 2, 3, 4, 5, 6, 7, 8, 9, 10)$. The corresponding λ-schedule given in Fig. 7.2b is a time-optimal schedule.

7.3. Suppose that the reduction graph G of a precedence relation defined over set N is an arbitrary directed circuit-free graph, but the number of machines is $M = 2$. As before, it is assumed that $d_i = 0$ and $t_i = 1$, $i = 1, 2, \ldots, n$.

Note that if, for some job i, there are k jobs of the set $A(i)$ whose deadlines do not exceed D, then the processing of job i in any deadline-feasible schedule must be completed by no later than $D - \lceil k/2 \rceil$, where $\lceil x \rceil$ is the smallest integer greater than or equal to x. Therefore, we may define the deadline for job i equal to $\min\{D_i, D - \lceil k/2 \rceil\}$. Using this fact, the following *algorithm* for modifying the deadlines can be offered.

Start with the modified deadline $D_j' = D_j$ for all $j \in N^-$. In each step, choose a job $i \in N$ for which the modified deadline has not yet been determined but for all jobs of the set $A(i)$ modified deadlines have been computed. Let $D^{(1)}, D^{(2)}, \ldots, D^{(l)}$ be the sequence of all distinct modified deadlines corresponding to the jobs of $A(i)$, and $g(i, D^{(k)})$ denote the number of elements of the set $A(i)$ whose modified deadlines do not exceed $D^{(k)}$, $k = 1, 2, \ldots, l$. Define $D_i' = \min\left\{D_i, \min\{D^{(k)} - \lceil \tfrac{1}{2} g(i, D^{(k)}) \rceil \mid 1 \le k \le l\right\}$.

Example. Under the conditions of Example A, let the values of the original deadlines be given in Table 7.3.

Table 7.3

i	1	2	3	4	5	6	7	8	9	10	11
D_i	2	3	8	9	8	7	5	6	10	7	7

The results of the computation of the modified deadlines are shown in Table 7.4. For each step in the algorithm, we give: job i for which the modified deadline is being calculated; the set $A(i)$; the set $D^{(1)}$, $D^{(2)}$,..., $D^{(l)}$ of all different values of D_j' for $j \in A(i)$; the number $g(i, D^{(k)})$ of the elements of the set $A(i)$ whose modified deadlines do not exceed $D^{(k)}$, $k = 1, 2,..., l$. The initial deadline D_i and the obtained modified deadline D_i' are also presented. For example, for job $i = 2$, the modified deadline is equal to 2, since this job corresponds to 9 elements of the set $A(i)$ whose modified deadlines do not exceed 7.

Table 7.4

Step	1	2	3	4	5	6	7	8	9	10	11
i	6	8	11	11	3	9	1	5	7	4	2
$A(i)$	∅	∅	∅	∅	{6}	{10, 11}	{3, 6}	{9, 10, 11}	{8, 9, 10, 11}	{7, 8, 9, 10, 11}	{3, 4, ..., 11}
$D^{(1)}; D^{(2)}; ...; D^{(l)}$	-	-	-	-	7	7	6; 7	6; 7	6; 7	5; 6; 7	4; 5; 6; 7
$g(i, D^{(1)})$; ...; $g(i, D^{(l)})$	-	-	-	-	1	2	1; 2	1; 3	2; 4	1; 3; 5	1; 3; 6; 9
D_i	7	6	7	7	8	10	2	8	5	9	3
D_i'	7	6	7	7	6	6	2	5	5	4	2

We show that modifying the deadlines can be done in *at most* $O(n^2)$ *time* if the relation $\longrightarrow$ is in a transitively closed form. The algorithm for constructing the transitive closure G' of a graph G is described in [260] (the running time is $O(n^{\log 7})$) and in [5] (the running time is $O(n^3/\log n)$).

Suppose that relation $\longrightarrow$ is given by the graph G', defined by the adjacency matrix R. Recall that graph G' contains an arc (i, j) if and only if $i \longrightarrow j$. The sum of the elements of the ith row of matrix R is equal to $c_i = |A(i)|$, $i = 1, 2,..., n$.

For each row i of matrix R, form a list $L(i)$ to contain the modified deadlines of the jobs in set $A(i)$ keeping them sorted in non-increasing order of their values. Initially, all lists $L(i)$ are empty.

If $c_i = 0$, then $i \in N^-$. Note that finding the set N^- requires at most $O(n^2)$ time. Define $D_i' = D_i$ for all $i \in N^-$.

Define a total pseudo-order $\Longrightarrow$ over the set N^-, assuming $i \Longrightarrow j$ if $D_i' \geq D_j'$. Let us represent the set N^- as a balanced 2-3-tree T (this takes at most $O(n)$ time, see Section 2.3 of Chapter 1).

In T, choose a maximal (with respect to $\Longrightarrow$) element j (this element has the largest value of the modified deadline). Delete j from T (this takes at most $O(\log n)$ time, see Section 2.6 of Chapter 1). Insert the value of D_j' at the end of the lists $L(i)$ for those rows i which contain 1 in column j. Reduce by 1 the values of c_i corresponding to these rows. Note that, for each maximal element j in T all these transformations require at most $O(n)$ time.

If, for some i, we obtain $c_i = 0$, then this implies that for all jobs in the set $A(i)$ the modified deadlines have been computed. In this case, the list $L(i)$ contains all modified deadlines for the jobs in the set $A(i)$ in non-increasing order of their values. It is easy to verify that, given a list $L(i)$, the modified deadline D_i' for job i can be computed in at most $O(n)$ time. Having computed D_i', insert element i into the current 2-3-tree T (this takes at most $O(\log n)$ time).

Choose the next maximal with respect to $\Longrightarrow$ element in T, and so on, until all modified deadlines are found.

It is easy to check that the total running time required for computing the modified deadlines for all jobs is at most $O(n^2)$.

We show now that *a schedule is feasible with respect to the modified deadlines if and only if it is feasible with respect to the original deadlines.*

Since $D_i' \leq D_i$, $i = 1, 2, \ldots, n$, a schedule that is feasible with respect to D_i' is also feasible with respect to D_i. Suppose that a schedule s is feasible with respect to D_i but is not feasible with respect to D_i'. Without loss of generality, we may assume that in schedule s each job is processed in a single unit time interval of the form $(\theta-1, \theta]$. Among the jobs for which the modified deadlines are violated, choose a job i to be processed in the unit time interval with the largest number θ. It follows that $D_i' < \theta \leq D_i$.

The procedure for finding the modified deadlines implies that there exists number $g(i, D')$ of jobs in the set $A(i)$ whose modified deadlines do not exceed D' such that $D_i' = D' - \lceil \frac{1}{2} g(i, D') \rceil$ holds. Since all jobs in the set $A(i)$ are processed without violating the deadlines and each of them is processed in a unit time interval with numbers larger than θ, $g(i, D')$ jobs should be processed in the interval $(\theta+1, D']$. However, $D_i' < \theta$, therefore, $D'-\theta < \lceil \frac{1}{2} g(i, D') \rceil$ and, hence, in the interval $(\theta+1, D']$ in schedule s more than two jobs should be processed in some unit time interval. We have come to a contradiction.

Thus, in speaking about a schedule that is feasible with respect to deadlines, we need not specify whether these deadlines are the original or modified ones.

Theorem 7.2. *A deadline-feasible schedule exists if and only if a λ-schedule corresponding to the list $\lambda = (i_1, i_2, \ldots, i_n)$ where $D'_{i_j} \le D'_{i_{j+1}}$, $j = 1, 2, \ldots, n-1$, is feasible.*

Proof. Let $\lambda = (i_1, i_2, \ldots, i_n)$, $D'_{i_j} \le D'_{i_{j+1}}$, $j = 1, 2, \ldots, n-1$, and suppose that the λ-schedule s is not feasible with respect to the deadlines. Among the jobs for which the modified deadlines are violated, choose a job i processed in a unit time interval with the smallest number θ. We have $\theta > D'_i \ge \lfloor D'_i \rfloor$.

Note that if $j \longrightarrow l$, then $D'_j < D'_l$. The procedure for finding a λ-schedule implies that in each unit length time interval with a number less than θ, at least one job, whose deadline does not exceed D'_i, is processed. Let us consider two cases.

Case 1. Suppose there exist unit time intervals with numbers less than θ such that only one of the jobs processed in each of these intervals has a modified deadline less than or equal to D'_i. Let δ be an interval with the largest number among these intervals, and let k be the job such that $D'_k \le D'_i$ processed in the interval δ. Then in each of the unit intervals $\delta+1, \delta+2, \ldots, \theta-1$ two jobs are processed whose deadlines do not exceed D'_i. The number of these jobs including the job i is $2(\theta-\delta)-1$. For each of these jobs, job k is a predecessor (otherwise, in a λ-schedule one of these $2(\theta-\delta)-1$ jobs should have been processed in the interval δ). Thus, for job k there exist at least $2(\theta-\delta)-1$ jobs in the set $A(k)$ whose modified deadlines do not exceed D'_i. Therefore, for the modified deadline of job k the relation $D'_k \le D'_i - \lceil (2(\theta-\delta)-1)/2 \rceil = D'_i - \theta + \delta$ should hold. Since $\theta > D'_i$, we have $D'_k < \delta$. Thus, for job k, the modified deadline is violated, which contradicts the way in which job i has been chosen.

Case 2. Suppose that, in schedule s, in each of the intervals $1, 2,, \ldots, \theta-1$ two jobs are processed whose modified deadlines do not exceed D'_i. Then there are at least $2\theta-1$ jobs whose deadlines do not exceed D'_i. Therefore, if a schedule, that is feasible with respect to the modified deadlines, exists then $2\lfloor D'_i \rfloor \ge 2\theta-1$. Since $\theta > \lfloor D'_i \rfloor$, we have $\theta \ge \lfloor D'_i \rfloor + 1$ and $2\theta-2 \ge 2\lfloor D'_i \rfloor$, which implies $2\theta-1 > 2\lfloor D'_i \rfloor$. Therefore, a schedule that is feasible with respect to the deadlines does not exist. This proves the theorem.

Thus, if $M = 2$, $d_i = 0$, $t_i = 1$, $i = 1, 2,..., n$, and the precedence relation $\longrightarrow$ is given in the transitively closed form, then *for finding a deadline-feasible schedule* (if this exists) it suffices: (a) to compute the modified deadlines D_i' (this takes at most $O(n^2)$ time); (b) to form the list λ of the jobs sorted in non-decreasing order of D_i' (this requires at most $O(n\log n)$ time, see Section 2.7 of Chapter 1); (c) to find a λ-schedule (this can be done in at most $O(n^2)$ time). Thus, *in at most* $O(n^2)$ *time* either a deadline-feasible schedule is found or we conclude that no such schedule exists.

Example. For the set of jobs of the previous example, the list λ is $\lambda = (1, 2, 4, 5, 7, 3, 8, 9, 6, 10, 11)$, and a deadline-feasible schedule is shown in Fig. 7.4.

It is easy to verify that if all original deadlines are *the same*, i.e., $D_i = D$, $i = 1, 2,..., n$, then the list λ of jobs (sorted in non-decreasing order of the modified deadlines) does not depend on D and is only determined by the form of graph G.

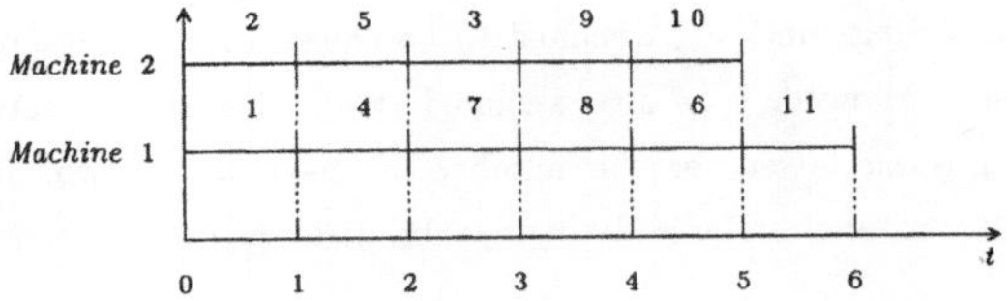

Fig. 7.4

Consequently, *a* time-*optimal* schedule for a partially ordered set of jobs with equal (unit) processing times to be processed on two parallel identical machines can be constructed (in at most $O(n^2)$ time) by defining all original deadlines be the same (for example, $D_i = n$, $i = 1, 2,..., n$), followed by using the described algorithm for a deadline-feasible schedule.

Remark 1. The algorithm for finding a feasible schedule described in Section 7.3 can also be applied when *the release dates* d_i *are different, and all deadlines are the same* ($D_i = D$, $i = 1, 2,..., n$). Change the orientation of each arc of the graph G' of the precedence relation $\longrightarrow$. As a result, we obtain a graph $\tilde{G}'$ which is the reduction graph of the precedence relation that the inverse of the original one. Define $\tilde{d}_i = 0$, $\tilde{D}_i = D - d_i$, $i = 1, 2,..., n$. Let $\tilde{s}$ be a schedule that is feasible with respect to the deadlines $\tilde{D}_i$, and $T(\tilde{s}) = \max\{\bar{t}_i(\tilde{s}) \mid i \in N\}$ be the length of this schedule. Then the desired schedule s can be obtained by defining $s_L(t) = \tilde{s}_L(t)$ for $t > T(\tilde{s})$ and $s_L(t) = \tilde{s}_L(t + T(\tilde{s}) - 29 + 1)$, $L = 1, 2,..., M$, for $\theta - 1 < t \leq \theta$, $\theta = 1, 2,..., T(\tilde{s})$.

A similar approach can be used for an arbitrary number of machines, provided that graph G is an *outtree*, the release dates are different, and all deadlines are the same.

Remark 2. The algorithm given in Sections 7.2 and 7.3 can be used to find a deadline-feasible schedule having the length not greater than a given number D. To do this, it suffices to define the original deadlines that exceed D to be equal to D. A *time-optimal schedule* (among the deadline-feasible ones) can be obtained by choosing an appropriate (integer) value of D in the interval $[n/M, n]$ by the binary search method [7]. In this case, the above algorithms are applied at most $O(\log n)$ times.

8. Identical Machines. Maximum Lateness

This section considers the problem of finding a job processing schedule for parallel identical machines to minimize the maximum lateness. In the case of a partially ordered set of jobs, the processing times are assumed to be equal, the release dates are the same, and no preemption is allowed. It is also assumed that either the reduction graph of the precedence relation is an intree, or the number of machines is two. In the case of a non-ordered set of jobs, their release dates may be different, and preemption is allowed.

8.1. The jobs of a set $N = \{1, 2,..., n\}$ are processed on M parallel identical machines. The release date of the job $i \in N$ is $d_i \geq 0$, its processing time is $t_i > 0$, and its due date is $D_i \geq 0$. A precedence relation $\longrightarrow$ is defined over set N to determine a possible sequence for job processing. Let $G = (N, U)$ denote the reduction graph of this relation. The schedule s is feasible with respect to $\longrightarrow$ if for any $i, j \in N$ such that $i \longrightarrow j$ the relation $s_H(t') = i$, $1 \leq H \leq M$, implies $s_L(t) \neq j$, $L = 1, 2,..., M$, for all $t \leq t'$.

A schedule s^* that is feasible with respect to $\longrightarrow$ is called *optimal* if it minimizes the function

$$L_{max}(s) = \max\{L_i(s) \mid i \in N\}, \tag{8.1}$$

where $L_i(s) = \bar{t}_i(s) - D_i$ is *the lateness* of job i, $\bar{t}_i(s)$ is the completion time of job i in a schedule s.

The value $L^* = L_{max}(s^*)$ is called *the optimal value* of the maximum lateness.

The following general observation can be made before proceeding to a description of the algorithms for optimal scheduling.

For any schedule s that is feasible with respect to $\longrightarrow$, the inequalities $\bar{t}_i(s) \leq$

$D_i+L_{max}(s)$ and $L^* \le L_{max}(s)$ hold. Hence, there is no schedule s that is feasible with respect to $\longrightarrow$ such that $\bar{t}_i(s) \le D_i+\tau$ for $\tau < L^*$, $i = 1, 2,..., n$. Thus, the problem of finding an optimal schedule reduces to one of finding the smallest value of τ for which there exists a schedule that is feasible both with respect to $\longrightarrow$ and with respect to the modified deadlines $D_i' = D_i+\tau$. This schedule is a desired optimal schedule s^*, and L^* is equal to the obtained value of τ.

In this section, the following cases of optimal scheduling are considered:

(a) $d_i = 0$, $t_i = 1$, $i = 1, 2,..., n$, no preemption is allowed, and the graph G is an intree;

(b) $d_i = 0$, $t_i = 1$, $i = 1, 2,..., n$, no preemption is allowed, and $M = 2$;

(c) d_i, t_i and D_i are integers, $i = 1, 2,..., n$, $G = (N, \emptyset)$, and preemption is allowed.

8.2. Let us consider cases (a) and (b). If the value of L^* is known, then for finding an optimal schedule we may use the algorithm for finding schedules that are feasible both with respect to $\longrightarrow$ and with respect to the deadlines equal to D_i+L^* (see Sections 7.2 and 7.3 of this chapter). Each of these algorithms has the property such that for any $\delta \ge 0$ and the deadlines equal to $D_i+L^*+\delta$, the same schedule is found. In fact, changing each deadline by the same value does not change the list λ and, hence, the corresponding λ-schedule.

Thus, for finding an optimal schedule in cases (a) and (b) it suffices to choose the values of D_i+W as the deadlines, where W is a sufficiently large number, and to use the corresponding algorithms from Section 7. Recall that the running time of the first algorithm (to be applied in case (a)) is $O(n\log n)$, while that of the second one (to be applied in case (b)) is $O(n^2)$.

8.3. Let us consider case (c). A solution to the problem under consideration can be found by choosing trial values of τ and verifying whether there exists a schedule s which is feasible with respect to the deadlines equal to $D_i+\tau$. If such a schedule does exist, then the current value of τ is called *feasible*. Starting with some infeasible value of τ, increase τ until the smallest feasible value of τ is obtained. Due to the observation made in Section 8.1, this value of τ is L^*.

In the case under consideration, scheduling without violating the deadlines can be done by using a network flow model (see Section 2.3 of this chapter). For each trial value of τ, a flow network model is to be constructed in the following way. Let $\{e_1, e_2,..., e_{2n}\}$ (where $e_1 \le e_2 \le ... \le e_{2n}$) be a set of values of d_i and $D_i+\tau$, $i = 1, 2,.., n$, and let

$E_k = (e_k, e_{k+1}]$, $k = 1, 2,..., 2n-1$. If $d_i = D_j+\tau$ for some i and j, then e with a smaller subscript corresponds to d_i. The network Γ contains the source x_0, connected with the vertices x_k, $k = 1, 2,..., 2n-1$ (corresponding to the intervals E_k), by the arcs of a capacity $M(e_{k+1}-e_k)$, and the sink z, where the arcs of a capacity t_i enter from the vertices y_i, $i = 1, 2,..., n$ (corresponding to the jobs i). An arc (x_k, y_i) of the capacity $e_{k+1}-e_k$ is present if and only if $d_i \le e_k$ and $e_{k+1} \le D_i+\tau$. A trial value of τ is feasible if and only if the value of a maximal flow in the network is $\Theta = \sum_{i=1}^{n} t_i$ (i.e., the arcs entering the sink are saturated). Recall that finding a maximal flow requires at most $O(n^3)$ time, while the subsequent construction of a schedule that is feasible with respect to the deadlines $D_i+\tau$ (if it does exist) takes at most $O(n^2)$ time (see Section 2.3 of this chapter).

It is clear that the structure of the network Γ depends on τ. The value of τ is called *critical* if such i and j exist that $D_i+\tau = d_j$. The structure of the network remains the same for all τ between two successive critical values.

The value of $\tau = L^*$ is to be found in two stages.

At the first stage, find τ_0, the largest infeasible critical value of τ. Since there are at most n^2 critical values, finding τ_0 by the binary search method [7] involves verification of at most $\log n^2$ (i.e., at most $O(\log n)$) values of τ, and this can be done in at most $O(n^3 \log n)$ time.

At the second stage, perform the following procedure for $\tau = \tau_\nu$ beginning with $\nu = 0$. Find the maximal flow value and the total capacity of a minimum cut of the network corresponding to τ_ν. Let R_ν be a cut with the total minimum capacity Θ_ν. If $\Theta_\nu < \Theta$, then increase the value of τ_ν so that the capacity of the cut R_ν becomes equal to Θ. Denote the obtained value of τ by $\tau_{\nu+1}$ and repeat the procedure. As a result, the increasing sequence of τ_ν is obtained. The process terminates at a step r for which $\Theta_r = \Theta$ and, hence, τ_r is the smallest feasible value of τ, i.e., $\tau_r = L^*$.

We show that, at the second stage, the number of steps does not exceed $O(\min\{n^2, \log n + \log t_{max}\})$, where $t_{max} = \max\{t_i | i \in N\}$.

Verify how the capacity Θ_ν of the cut R_ν of the network corresponding to an infeasible value of τ_ν changes when τ_ν increases by $\eta > 0$. An interval $E_k = (e_k, e_{k+1}]$ is said to belong to class 1 if $e_k = d_i$ and $e_{k+1} = D_j+\tau_\nu$; to class 2 if $e_k = D_i+\tau_\nu$ and $e_{k+1} = d_j$; to class 3 in all other cases. When τ_ν increases by η, the capacity of the arc leaving a vertex x_k increases by η if E_k belongs to class 1, decreases by η if E_k belongs to class 2, and remains unchanged if E_k belongs to class 3. The situation is similar for the arcs (x_0, x_k), but here the capacity is changed by $M\eta$. It is obvious that the total capacity of the cut R_ν changes by $\gamma_\nu\eta$, where γ_ν is an integer. Since τ_0 is the largest infeasible critical value of τ, the value of the total capacity of the cut R_ν should grow with τ_ν, i.e. $\gamma_\nu \geq 1$.

Making the transition from τ_ν to $\tau_{\nu+1}$, define $\eta = \lceil(\Theta-\Theta_\nu)/\gamma_\nu\rceil$ and $\tau_{\nu+1} = \tau_\nu+\eta$. Here $\lceil x\rceil$ is the smallest integer greater than or equal to x.

Consider how the capacities of the network cuts change when making the transition from τ_ν to $\tau_{\nu+1}$. The pair (Θ', γ) corresponds to each cut where Θ' is the total capacity of the cut, and γ is an integer such that the total capacity of the cut changes by $\gamma\eta$ when τ_ν increases by η. When making the transition from τ_ν to $\tau_{\nu+1}$, the values of γ for the cuts remain unchanged but the values of Θ' may change. If $\gamma \geq \gamma_\nu$, then the total cut capacity grows up to a value which is not less than Θ. Hence, it follows that $\gamma_{\nu+1} < \gamma_\nu$, $\nu = 1, 2,\dots, r-1$. It is easy to check that γ_0 does not exceed $O(n^2)$. Since γ_ν are integers and $\gamma_\nu \geq 1$, it follows that the number of steps r is at most $O(n^2)$.

Now we give a more accurate estimation of the number of steps. For all cuts corresponding to the pairs (Θ', γ) such that $\Theta' < \Theta$, the inequality $\gamma > 0$ holds. In fact, since τ_0 is the largest infeasible critical value of τ and $\tau_\nu > \tau_0$, it follows that if τ_ν increases by some value η', then the capacities of all cuts become not less than Θ. Since the values of γ do not change while making the transition from τ_ν to $\tau_{\nu+1}$, we conclude that $\Theta' < \Theta$ implies $\gamma > 0$.

Let c_ν be the number of all possible (Θ', γ) pairs such that $\Theta' \leq \Theta$ for $\tau = \tau_\nu$. Since the number of all possible pairs such that $\Theta_\nu < \Theta' \leq \Theta$ and $0 \leq \gamma \leq \gamma_\nu$ is $(\Theta-\Theta_\nu)(\gamma_\nu+1)$, it follows that $c_\nu \geq (\Theta-\Theta_\nu)(\gamma_\nu+1)$.

While making the transition from τ_ν to $\tau_{\nu+1}$, the capacity Θ' such that $\Theta-(\Theta-\Theta_\nu)\gamma/\gamma_\nu \leq \Theta' < \Theta$ of each cut with $\gamma < \gamma_\nu$ increases to the value not less than Θ. Therefore, having made the transition from τ_ν to $\tau_{\nu+1}$ we obtain that the number of possible pairs (Θ', γ) such that $\Theta' \leq \Theta$ becomes equal to $\Theta-(\Theta-\Theta_\nu)\gamma/\gamma_\nu-\Theta_\nu$ for each γ, $0 \leq \gamma < \gamma_\nu$. Moreover, for all γ we obtain

$$c_{\nu+1} = (\Theta-\Theta_\nu)\gamma_\nu - \frac{\Theta-\Theta_\nu}{\gamma_\nu}\sum_{\gamma=0}^{\gamma_\nu-1}\gamma = (\Theta-\Theta_\nu)(\gamma_\nu - \frac{\gamma_\nu-1}{2}) = \frac{(\Theta-\Theta_\nu)(\gamma_\nu+1)}{2}.$$

Hence, it follows that

$$\frac{c_{\nu+1}}{c_\nu} \leq \frac{(\Theta-\Theta_\nu)(\gamma_\nu+1)}{2(\Theta-\Theta_\nu)(\gamma_\nu+1)} = \frac{1}{2}.$$

Thus, in each step, the number of possible pairs (γ, Θ') such that $\Theta' \leq \Theta$ is reduced at least twice. Taking into account that γ does not exceed $O(n^2)$, we conclude that the number of steps is at most $O(\log(n^2\Theta))$, i.e., this does not exceed $O(\log n + \log t_{max})$, where

$t_{max} = \max\{t_i \mid i \in N\}$. Thus, the number of steps at the second stage is at most $O(\min\{n^2, \log n + \log t_{max}\})$.

Therefore, in case (c), the total running time required for optimal scheduling is at most $O(n^3 \min\{n^2, \log n + \log t_{max}\})$.

8.4. We show that the problem of minimizing *the makespan with arbitrary* d_i reduces to that of minimizing the maximum lateness for $d_i = 0$, $i = 1, 2,..., n$.

Consider two scheduling problems of minimizing the value of $L_{max}(s)$. In the first, the jobs have the parameters $d_i' = 0$, t_i', D_i', $i = 1, 2,..., n$, and the precedence relation $\longrightarrow$ is defined over set N. In the second problem, the jobs have the parameters d_i'', t_i'', D_i'', where $d_i'' = D - D_i'$, $t_i'' = t_i'$, $D_i'' = D$, $D = \max\{D_i' \mid i \in N\}$, and the precedence relation $\Longrightarrow$ is defined over set N being reverse to $\longrightarrow$ (i.e., $i \Longrightarrow j$ if and only if $j \longrightarrow i$). These problems are called *conjugate*.

Let s' be an optimal schedule for the first problem and $L_{max}(s') = L'$. Denote by $\bar{s}$ the set of M piecewise-constant functions $\{\bar{s}_1(t), \bar{s}_2(t),..., \bar{s}_M(t)\}$ such that $\bar{s}_L(t) = s_L'(t)$, $L = 1, 2,..., M$, if t is not a discontinuity point of the function $s_L'(t)$. At a discontinuity point t of the function $s_L'(t)$, define $\bar{s}_L(t) = \bar{s}_L(t+\delta)$ for a sufficiently small $\delta > 0$ (so that unlike $s_L'(t)$, the functions $\bar{s}_L(t)$ are right-semicontinuous rather than left-semicontinuous). Defining $s_L''(t) = \bar{s}_L(D+L'-t)$ for $t \in (0, D+L']$ and $s_L''(t) = 0$ for $t > D+L'$, $L = 1, 2,..., M$, we obtain a feasible schedule s'' for the second problem. It is easy to verify that $L_{max}(s'') = \max\{\bar{t}_i(s'') - D_i'' \mid i \in N\} = L'$, and s'' is an optimal schedule for the second problem. Since $D_i'' = D$, $i = 1, 2,..., n$, it follows that schedule s'' is also time-optimal schedule.

Therefore, to solve the problem of finding a time-optimal schedule with arbitrary d_i it suffices to solve its conjugate problem of finding a schedule minimizing $L_{max}(s)$ for $d_i' = 0$, $D_i' = D - d_i$, $D = \max\{d_i \mid i \in N\}$, $i = 1, 2,..., n$.

Notice that *the smallest maximal tardiness* corresponds to the schedule to which the smallest $L_{max}(s)$ corresponds. In fact, the maximal tardiness $z_{max}(s) = \max\{0, \max\{L_i(s) \mid i \in N\}\}$ coincides with $L_{max}(s)$ for $L_{max}(s) \geq 0$. It is clear that this remark also applies to the function $F(s) = \max\{\varphi(L_i(s)) \mid i \in N\}$ where φ is any non-decreasing function. Thus, the algorithms described here can be used to find a schedule with the smallest $z_{max}(s)$ or $\max\{\varphi(L_i(s)) \mid i \in N\}$.

9. Uniform and Unrelated Parallel Machines. Total and Maximal Costs

This section studies a number of polynomially solvable problems of minimizing either the total or the maximal cost of processing jobs on uniform and unrelated parallel machines. In particular, the problems to minimize the makespan are considered.

9.1. The jobs of a set $N = \{1, 2,..., n\}$ are processed on M parallel machines. All jobs have the same release dates $d_i = 0$, $i = 1, 2,..., n$. The processing time of a job i on a machine L is $t_{iL} > 0$. Each job i is associated with a non-decreasing cost function $\varphi_i(t)$.

To minimize the total cost, it is required to find a schedule s^* for processing the jobs set N such that the function

$$F_{\Sigma}(s) = \sum_{i=1}^{n} \varphi_i(\bar{t}_i(s)) \tag{9.1}$$

accepts the smallest value.

To minimize the maximal cost, it is required to find a schedule s^* for processing the jobs of set N such that the function

$$F_{max}(s) = \max \{\varphi_i(\bar{t}_i(s)) \mid i \in N\} \tag{9.2}$$

accepts the smallest value.

Here $\bar{t}_i(s)$ denotes the completion time of a job i in a schedule s. For each of the problems, schedule s^* is called *optimal*.

If preemption in processing each job is forbidden, the schedule is specified by partitioning set N into subsets $N_1, N_2,..., N_M$ (some of them may be empty) and by determining the sequence for processing the jobs of each set N_L on the corresponding machine L, $L = 1, 2,..., M$.

In what follows, we consider the following problems of minimizing the total cost, assuming that preemption is forbidden:

(a) $\varphi_i(t) = t$, $i = 1, 2,..., n$;

(b) $\varphi_i(t) = t$, $t_{iL} = a_L t_i$, $t_i > 0$, $a_L > 0$, $i = 1, 2,..., n$; $L = 1, 2,..., M$;

(c) $\varphi_i(t)$ is a non-decreasing function; $t_{iL} = a_L$, $a_L > 0$, $i = 1, 2,..., n$, $L = 1, 2,..., M$;

(d) $\varphi_i(t) = \alpha_i u_i(t)$, $t_{iL} = 1$, $i = 1, 2,..., n$, $L = 1, 2,..., M$ (here $u_i(t) = 0$ if $t \le D_i$, $u_i(t) = 1$ if $t > D_i$; $\alpha_i > 0$, $D_i \ge 0$ is the due date of a job i; D_i, $i = 1, 2,..., n$, are integers).

We also consider the following problems of minimizing the maximal cost:

(a) $\varphi_i(t)$ is a non-decreasing function, $t_{iL} = a_L$, $a_L > 0$, $i = 1, 2,..., n$, $L = 1, 2,..., M$, and preemption is forbidden;

(b) $\varphi_i(t) = t$, $i = 1, 2,..., n$;

(c) $\varphi_i(t) = t - D_i$, $i = 1, 2,..., n$.

To conclude this section, we consider a natural generalization of the total cost minimization problems when machine ready times and machine-usage cost functions are involved.

9.2. Let us consider the first of the mentioned problems. It is required to find a schedule s^* which minimizes the function $\sum_{i \in N} \overline{t}_i(s^*)$.

Let s be some schedule such that a job i is processed on a machine L and this machine processes $k-1$ jobs, $1 \le k \le n$, after job i. Then the processing time t_{iL} contributes (as a summand) to the value of $\overline{t}_i$, and to the values of $\overline{t}_j$ for all $k-1$ jobs j processed on machine L after job i. The sum $\sum_{i \in N} \overline{t}_i(s)$ may be represented as kt_{iL} plus the terms independent of t_{iL}. If job i is the last to be processed on machine L, then the factor at t_{iL} is equal to 1; if this job is the one before last, then the factor is 2, and so on.

Let us introduce a variable x_{iLk} equal to 1 if job i is processed on machine L which processes $k-1$ jobs after i. Otherwise, $x_{iLk} = 0$. Then the problem under consideration may be formulated as the following transportation problem:

$$\sum_{i=1}^{n} \sum_{L=1}^{M} \sum_{k=1}^{n} kt_{iL}x_{iLk} \longrightarrow \min \tag{9.3}$$

subject to

$$\sum_{L=1}^{M} \sum_{k=1}^{n} x_{iLk} = 1, \; i = 1, 2,..., n, \tag{9.4}$$

$$\sum_{i=1}^{n} x_{iLk} \le 1, \; L = 1, 2,..., M, \; k = 1, 2,..., n, \tag{9.5}$$

$$x_{iLk} \ge 0, \; i = 1, 2,..., n, \; L = 1, 2,..., M, \; k = 1, 2,..., n. \tag{9.6}$$

Condition (9.4) implies that each job i should be processed by one of the machines and occupies a certain position in the sequence of jobs corresponding to this machine. Condition (9.5) implies that in the sequence of jobs corresponding to any machine, each position is occupied by at most one job.

The formulated problem can be reduced to that discussed in [57], and can be solved in *$O(n^3)$ time.*

9.3. Let the processing time of a job i on a machine L be $t_{iL} = a_L t_i$ where $a_L > 0$, $t_i > 0$, $i = 1, 2,..., n$, $L = 1, 2,..., M$, i.e., a machine L has the processing speed $1/a_L$. As before, it is required to find a schedule that minimizes the total flow time $\sum_{i \in N} \bar{t}_i(s)$.

Consider the following problem. Let $\pi = (i_1, i_2,..., i_n)$ be some permutation of the elements of set $\{1, 2,..., n\}$. Given two n-dimensional vectors $\alpha = (\alpha_1, \alpha_2, .., \alpha_n)$ and $\beta = (\beta_1, \beta_2,..., \beta_n)$ with real components, define

$$f(\pi) = \sum_{k=1}^{n} \alpha_k \beta_{i_k}$$

It is required to find a permutation π^* of the elements of set $\{1, 2,.. , n\}$ which minimizes $f(\pi)$. Without loss of generality, assume that the components of vectors α and β are numbered in such a way that $\alpha_1 \geq \alpha_2 \geq ... \geq \alpha_n$.

Consider a permutation π' which differs from π in that the elements i_k and i_{k+1} are interchanged. We have

$$f(\pi') - f(\pi) = (\alpha_k - \alpha_{k+1})(\beta_{i_{k+1}} - \beta_{i_k})$$

If $\beta_{i_k} \leq \beta_{i_{k+1}}$, then this difference is non-negative. Thus, by sorting the values β_l in non-decreasing order we obtain the desired permutation $\pi^* = (i_1^*, i_2^*,..., i_n^*)$. Here $\beta_{i_k^*} \leq \beta_{i_{k+1}^*}$ for all $k = 1, 2,..., n-1$.

Thereby, the function $f(\pi)$ reaches its smallest value if smaller β_{i_k}'s correspond to larger α_k's.

Let us return to the scheduling problem under consideration. Without loss of generality, assume that the jobs are numbered so that $t_1 \geq t_2 \geq ... \geq t_n$. Let us construct a $(M \times n)$-matrix $A = \|\alpha_{Lk}\|$, assuming $\alpha_{Lk} = ka_L$, i.e.

$$A = \begin{Vmatrix} a_1 & 2a_1 & & na_1 \\ a_2 & 2a_2 & \dots & na_2 \\ \cdot & \cdot & \dots & \ddots \\ a_M & 2a_M & & na_M \end{Vmatrix}$$

Sort the elements of matrix A in non-decreasing order and denote by β_j the jth element of the obtained sequence, so that $\beta_1 \leq \beta_2 \leq ... \leq \beta_{Mn}$.

Associate an element β_i of matrix A with job $i \in N$. If $\beta_i = \alpha_{Lk}$, then job i is processed on machine L and occupies the kth place from the end in the sequence of jobs on that machine. Note that if an element α_{Lk}, $k > 1$, corresponds to job i, then for any $\alpha_{Lk'}$, $k' < k$, there exists a corresponding job $j \in N$ (because $\alpha_{Lk'} < \alpha_{Lk}$). As a result, the sequences of jobs to be processed on each of the machines are obtained. Thus, some non-preemptive schedule s is found. This schedule is optimal due to the considerations used above (in minimizing the function $f(\pi)$).

Numbering the jobs so that $t_1 \geq t_2 \geq ... \geq t_n$ takes $O(n\log n)$ time (see Section 2.7 of Chapter 1).

Let us estimate the running time required for finding the elements $\beta_1, \beta_2, ..., \beta_n$ of matrix A. Sort the elements $a_1, a_2, ..., a_M$ in non-decreasing order: $a_{i_1} \leq a_{i_2} \leq ... \leq a_{i_M}$. This takes at most $O(M\log M)$ time. Note that the set $B = \{\beta_1, \beta_2, ..., \beta_n\}$ can include at most n (first) elements of row i_1 of matrix A, at most $n/2$ (first) elements of row i_2, and so on, at most n/M (first) elements of row i_M. Thus, the search for the elements of set B is restricted to at most $n+n/2+n/3+...+n/M = nS_M$ elements of matrix A where $S_M = \sum_{k=1}^{M} \frac{1}{k}$ is a partial sum of the harmonic series. Let C denote the set of these elements. It is known that partial sums of the harmonic series grow as $\ln M$ or, equivalently, as $\log M$. Represent set C as a balanced 2-3-tree. This can be done in at most $O(n\log M)$ time (see Section 2.3 of Chapter 1). Finding the elements $\beta_1, \beta_2, ..., \beta_n$ using the constructed 2-3-tree, takes at most $O(n\log(n\log M))$ time.

Assuming that $M < n$, the running time required for finding an optimal schedule is at most $O(n\log n)$.

If the machines are *identical* (which is the case when all a_L are equal), the schedule s^* found by the described method is essentially a schedule constructed by the well-known SPT rule (shortest processing time): at a moment when a machine becomes idle, the job with the shortest processing time among available jobs is chosen and is assigned to be processed on that machine. Scheduling by the SPT rule also takes $O(n\log n)$ time.

9.4. Let $t_{iL} = a_L$, $i = 1, 2, ..., n$, $L = 1, 2, ..., M$, and the cost functions be non-decreasing. Then the problem of finding a schedule with the smallest value of function (9.1) reduces to the following transportation problem.

Let us introduce a variable x_{iLk} which equals 1 if job i is processed on machine L and occupies the kth place in the sequence of jobs processed on that machine. Otherwise, $x_{iLk} = 0$. Denote $c_{iLk} = \varphi_i(ka_L)$, $i = 1, 2, ..., n$, $L = 1, 2, ..., M$, $k = 1, 2, ..., n$.

It is required to minimize

$$\sum_{i=1}^{n} \sum_{L=1}^{M} \sum_{k=1}^{n} c_{iLk}x_{iLk} \tag{9.7}$$

subject to

$$\sum_{L=1}^{M} \sum_{k=1}^{n} x_{iLk} = 1,\ i = 1, 2,\dots, n, \tag{9.8}$$

$$\sum_{i=1}^{n} x_{iLk} \leq 1,\ L = 1, 2,\dots, M,\ k = 1, 2,\dots, n, \tag{9.9}$$

$$x_{iLk} \geq 0,\ i = 1, 2,\dots, n,\ L = 1, 2,\dots, M,\ k = 1, 2,\dots, n. \tag{9.10}$$

Condition (9.8) implies that each job is processed on one of the machines and occupies a certain position in the sequence of jobs processed on that machine. Condition (9.9) implies that for any machine, in the sequence of jobs processed on that machine, each position is occupied by at most one job.

Problem (9.7)-(9.10) is similar to problem (9.3)-(9.6) and its solution can be obtained in *at most* $O(n^3)$ *time*. Moreover, the problem of minimizing the maximal cost can be solved in a similar way if $t_{iL} = a_L$, $i = 1, 2,\dots, n$, $L = 1, 2,\dots, M$, but with function (9.7) changed to $\max\{c_{iLk}x_{iLk} \mid i = 1, 2,\dots, n,\ L = 1, 2,\dots, M,\ k = 1, 2,\dots, n\}$.

9.5. Let $t_{iL} = 1$, $i = 1, 2,\dots, n$, $L = 1, 2,\dots, M$, and the cost functions be of the form $\varphi_i(t) = \alpha_i u_i(t)$, where $u_i(t) = 0$ if $t \leq D_i$ and $u_i(t) = 1$ if $t > D_i$; $\alpha_i > 0$, $i = 1, 2,\dots, n$. Here the due date D_i of a job i is an integer. It is required to find a schedule which minimizes function (9.1) (*Problem I*).

Besides, consider *Problem II* which differs from Problem I in that all processing times are equal to $1/M$ and the jobs are processed on a single machine.

A schedule s for processing the jobs on M parallel machines is said to have no unjustified machine idle time if either $s_L(t) = 0$ in the interval $(0, \infty)$, or $s_L(t) \neq 0$ in the interval $(0, t']$ and $s_L(t) = 0$ for $t > t'$, $L = 1, 2,\dots, M$. Similarly, in the case of a single machine, a schedule s' does not involve unjustified machine idle time if $s'(t) \neq 0$ in the interval $[0, \sum_{i \in N} t_i]$ and $s'(t) = 0$ outside this interval.

Let S and S' be sets of all non-preemptive schedules with no unjustified idle time for Problems I and II, respectively. It is clear that there exist optimal schedules for Problems I and II in sets S and S', respectively.

Schedules s and s' for problems I and II are called conjugate if for each unit interval $(\theta-1, \theta]$, $\theta = 1, 2,\dots,$ there holds $s'(t) = s_L(\theta)$ for all $t \in (\theta-1+(L-1)/M,\ \theta-1+L/M]$ (and, vice versa, $s_L(t) = s'(\theta-1+L/M)$ for all $t \in (\theta-1, \theta]$, $L = 1, 2,\dots, M$).

It is easy to check that (a) if schedule s' belongs to S', then the conjugate schedule s belongs to S (generally speaking, the opposite is not true); (b) the values of function (9.1) coincide for conjugate schedules for Problems I and II .

Therefore, *to find an optimal schedule for Problem I, it suffices to find an optimal schedule $s^{*\prime} \in S'$ for Problem II and to construct the conjugate schedule.* The $O(n\log n)$ algorithm for finding an optimal schedule $s^{*\prime} \in S'$ for Problem II is given in Section 4.3(b) of this chapter.

Example. Consider the following Problem I: $M = 3$, $n = 10$, the values of D_i and α_i, $i = 1, 2, \ldots, 10$, are given in Table 9.1.

Table 9.1

i	1	2	3	4	5	6	7	8	9	10
D_i	1	1	1	2	2	2	2	3	3	4
α_i	5	4	6	2	3	4	5	1	3	2

An optimal schedule $s^{*\prime}$ for the corresponding Problem II is shown in Fig.9.1a. The conjugate schedule s^*, which is optimal for Problem I, is shown in Fig.9.1b. Here $F_\Sigma(s^*) = F_\Sigma(s^{*\prime}) = 2$.

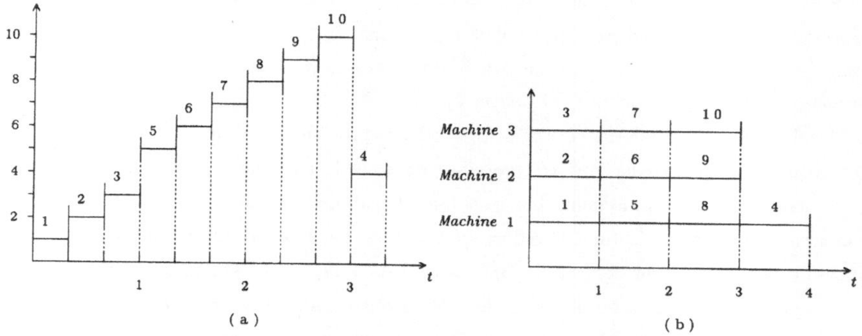

Fig. 9.1

9.6. Now we consider the problem of minimizing the maximal cost. Suppose that $\varphi_i(t) = t$, $i = 1, 2, \ldots, n$, and for each job preemption is allowed.

For a schedule s, let τ_{iL} denote the total length of time intervals in which job i is processed on machine L. The relation $\sum_{L=1}^{M} (\tau_{iL}/t_{iL}) = 1$, $i = 1, 2, \ldots, n$, must hold. Denote $T = \max\{\bar{t}_i(s) \mid i \in N\}$.

It is easy to verify that the values of T and τ_{iL} form a feasible solution for the following linear programming problem:

$$T \rightarrow \min \tag{9.11}$$

subject to

$$\sum_{L=1}^{M} \frac{\tau_{iL}}{t_{iL}} = 1, \ i = 1, 2, \dots, n, \tag{9.12}$$

$$\sum_{L=1}^{M} \tau_{iL} \le T, \ i = 1, 2, \dots, n, \tag{9.13}$$

$$\sum_{i=1}^{n} \tau_{iL} \le T, \ L = 1, 2, \dots, M, \tag{9.14}$$

$$\tau_{iL} \ge 0, \ i = 1, 2, \dots, n, \ L = 1, 2, \dots, M. \tag{9.15}$$

On the other hand, if T and τ_{iL}, $i = 1, 2, \dots, n$, $L = 1, 2, \dots, M$, is a solution of problem (9.11)-(9.15) and there is a schedule s where the total length of intervals for processing job i on machine L is τ_{iL}, and $\max\{\bar{t}_i(s) \mid i \in N\} = T$, then this schedule is a desired optimal schedule. We now show that the schedule s does exist and outline a method for finding s. Also, we show that finding this schedule takes polynomial time. Then the existence of a polynomial-time algorithm for solving a linear programming problem [166] implies the existence of such an algorithm for the problem under consideration.

Let T and $\tau = \|\tau_{iL}\|$ be a solution of problem (9.11)-(9.15). Define $\tilde{T} = T$. It follows that

$$\tilde{T} = \max\left\{\max\left\{\sum_{L=1}^{M} \tau_{iL} \,\middle|\, i = 1, 2, \dots, n\right\}, \max\left\{\sum_{i=1}^{n} \tau_{iL} \,\middle|\, L = 1, 2, \dots, M\right\}\right\}. \tag{9.16}$$

Call row i (column L) of matrix τ *dense* if $\sum_{L=1}^{M} \tau_{iL} = \tilde{T}$ ($\sum_{i=1}^{n} \tau_{iL} = \tilde{T}$, respectively). Let $V(\tau)$ be a subset of positive elements of matrix τ that contains one element of each dense row and each dense column, and at most one element of each remaining rows and columns.

Let δ be the largest number satisfying the following conditions:

(a) $\delta \le \tau_{iL}$ if $\tau_{iL} \in V(\tau)$ and τ_{iL} is an element of a dense row or dense column;

(b) $\delta \le \tau_{iL} + \tilde{T} - \sum_{L=1}^{M} \tau_{iL}$ if $\tau_{iL} \in V(\tau)$ but row i is not dense;

(c) $\delta \le \tau_{iL} + \tilde{T} - \sum_{i=1}^{n} \tau_{iL}$ if $\tau_{iL} \in V(\tau)$ but column L is not dense;

(d) $\delta \le \tilde{T} - \sum_{L=1}^{M} \tau_{iL}$ if $V(\tau)$ does not contain elements of row i;

(e) $\delta \le \tilde{T} - \sum_{i=1}^{n} \tau_{iL}$ if $V(\tau)$ does not contain elements of the column L.

For example, let $n = 4$, $M = 3$, $T = 11$ and matrix τ is of the form:

$$\tau = \begin{Vmatrix} 3 & \underline{4} & 4 \\ \underline{4} & 0 & 0 \\ 0 & 6 & 0 \\ 4 & 0 & \underline{6} \end{Vmatrix} \begin{matrix} 11 \\ 4 \\ 6 \\ 10 \end{matrix}$$

$$\begin{matrix} 11 & 10 & 10 \end{matrix}$$

Here, the sum of elements in a row (in a column) is written next to this row (column). The first row and the first column are dense. the elements of the chosen set $V(\tau)$ are underlined. We have

$$\delta \le \tau_{12} = 4, \ \delta \le \tau_{21} = 4;$$

$$\delta \le \tau_{21} + \tilde{T} - \sum_{L=1}^{3} \tau_{2L} = 4 + 11 - 4 = 11;$$

$$\delta \le \tau_{34} + \tilde{T} - \sum_{L=1}^{3} \tau_{3L} = 6 + 11 - 10 = 7;$$

$$\delta \le \tau_{12} + \tilde{T} - \sum_{i=1}^{4} \tau_{i2} = 4 + 11 - 10 = 5;$$

$$\delta \le \tau_{34} + \tilde{T} - \sum_{i=1}^{4} \tau_{i4} = 6 + 11 - 10 = 7;$$

$$\delta \le \tilde{T} - \sum_{L=1}^{3} \tau_{3L} = 11 - 6 = 5;$$

Thus, in the case under consideration, $\delta = 4$.

A desired schedule s can be constructed as follows. Start with the interval $(\eta, \ \eta+\delta]$, where $\eta = 0$, and define $s_L(t) = \tau_{iL}$ for each element $\tau_{iL} \in V(\tau)$ in the interval $(\eta, \ \eta+\min\{\tau_{iL}, \ \delta\}]$ and, if $\tau_{iL} < \delta$, define $s_L(t) = 0$ in the interval $(\eta+\tau_{iL}, \ \eta+\delta]$. If there are no elements of the column L in the set $V(\tau)$, then set $s_L(t) = 0$ in the interval $(\eta, \ \eta+\delta]$.

Define $\tau'_{iL} = \max\{0, \ \tau_{iL}-\delta\}$ if $\tau_{iL} \in V(\tau)$ and $\tau_{iL}' = \tau_{iL}$, otherwise. Let $T' = \tilde{T} - \delta$. As a result, we obtain the matrix $\tau' = \|\tau'_{iL}\|$ and the value of T' for which the relation

$$T' = \max\left\{\max\left\{\sum_{L=1}^{M} \tau'_{iL} \,\middle|\, i = 1, 2, \ldots, n\right\}, \ \max\left\{\sum_{i=1}^{n} \tau'_{iL} \,\middle|\, L = 1, 2, \ldots, M\right\}\right\}$$

holds.

Note that matrix τ' has at least one positive element less than matrix τ, or one dense (with respect to T') row (or one dense column) more than τ.

Denote τ' by τ, and T' by $\tilde{T}$. Define η be equal to $\eta+\delta$. Find new $V(\tau)$ and δ. Find schedule s in the interval $(\eta, \ \eta+\delta]$ as described above, and so on. The total number of steps is at most $r+M+n$, where r is the number of positive elements in the initial matrix τ.

For the example under consideration, we obtain successively

$$\tau = \begin{array}{c} \left\| \begin{matrix} \underline{3} & 0 & 4 \\ 0 & 0 & 0 \\ 0 & \underline{6} & 0 \\ 4 & 0 & \underline{2} \end{matrix} \right\| \\ \begin{matrix} 7 & 6 & 6 \end{matrix} \end{array} \begin{matrix} 7 \\ 0 \\ 6 \\ 6 \\ \end{matrix}, \; \tilde{T} = 7, \; \delta = 3,$$

$$\tau = \begin{array}{c} \left\| \begin{matrix} 0 & 0 & \underline{4} \\ 0 & 0 & 0 \\ 0 & \underline{3} & 0 \\ \underline{4} & 0 & 0 \end{matrix} \right\| \\ \begin{matrix} 4 & 3 & 4 \end{matrix} \end{array} \begin{matrix} 4 \\ 0 \\ 3 \\ 4 \\ \end{matrix}, \; \tilde{T} = 4, \; \delta = 4.$$

The resulting schedule s is shown in Fig. 9.2.

In each step of the described scheduling procedure, set $V(\tau)$ is to be found. We show that *for any matrix* τ (*with non-negative elements*) *and for the number* $\tilde{T}$ *satisfying condition* (9.16) *there exists a set* $V(\tau)$.

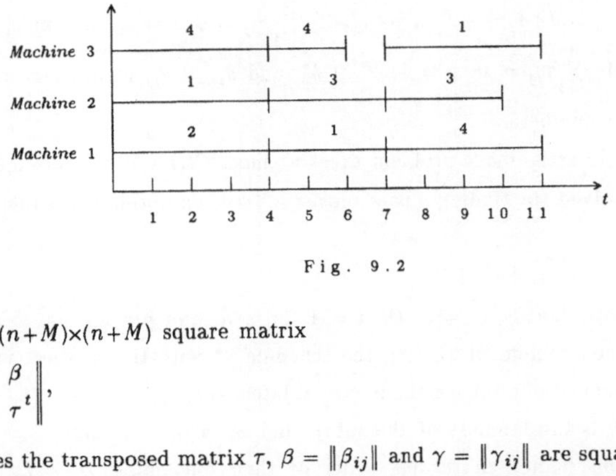

Fig. 9.2

Consider the $(n+M)\times(n+M)$ square matrix

$$u = \left\| \begin{matrix} \tau & \beta \\ \gamma & \tau^t \end{matrix} \right\|,$$

where τ^t denotes the transposed matrix τ, $\beta = \|\beta_{ij}\|$ and $\gamma = \|\gamma_{ij}\|$ are square matrices with non-negative elements of the order $n\times n$ and $M\times M$, respectively. The matrices β and γ are chosen in such a way that the sum of the elements in each row and each column of the resulting matrix u is $\tilde{T}$. The matrix $\alpha = u/\tilde{T}$ is doubly stochastic, i.e., its elements α_{ij} satisfy the following conditions: $\alpha_{ij} \geq 0$; $\sum_{j=1}^{n+M} \alpha_{ij} = 1$, $i = 1, 2, \ldots, n+M$, $\sum_{i=1}^{n+M} \alpha_{ij} = 1$, $j = 1, 2, \ldots, n+M$.

Therefore, the matrix α may be represented as

$$\alpha = \sum_{l=1}^{(n+M)!} \omega_l P_l, \ \omega_l \geq 0, \ \sum_{l=1}^{(n+M)!} \omega_l = 1, \tag{9.17}$$

where $\{P_l\}$ is the set of all $(n+M)\times(n+M)$ permutation matrices (i.e., (0, 1)-matrices with a single non-zero element in each row and each column) [14].

It is easy to check that any matrix P_l for which $\omega_l > 0$ in expression (9.17) determines some set $V(\tau)$ and, hence, there exists at least one set $V(\tau)$.

Finding a set $V(\tau)$ reduces to obtaining the matrix $u = \|u_{ij}\|$ of the above form, followed by solving the assignment problem:

$$\sum_{i=1}^{n+M} \sum_{j=1}^{n+M} a_{ij} x_{ij} \rightarrow \min$$

subject to

$$\sum_{j=1}^{n+M} x_{ij} = 1, \ i = 1, 2,..., n+M,$$

$$\sum_{i=1}^{n+M} x_{ij} = 1, \ j = 1, 2,..., n+M.$$

Here $a_{ij} = W$, $j = M+1, M+2,..., n+M$, if $\sum_{j=1}^{n} \beta_{ij} = 0$, $i = 1, 2,..., n$; $a_{ij} = W$, $i = n+1, n+2,..., n+M$, if $\sum_{i=1}^{M} \gamma_{ij} = 0$, $j = 1, 2,..., M$, and $a_{ij} = u_{ij}$ in other cases, where W is a sufficiently large number.

Since solving the assignment problem takes at most $O((n+M)^3)$ time and the number of problems to be solved for finding s is at most $r+M+n$, an optimal schedule can be found in *polynomial time*.

9.7. Suppose now that $\varphi_i(t) = t - D_i$, $i = 1, 2,..., n$, and preemption is allowed.

In this case, the problem of finding the schedule s^* with the smallest value of function (9.2) is the problem of minimizing the maximal lateness $L_{max}(s) = \max \{L_i(s) \mid i \in N\}$, where $L_i(s) = \bar{t}_i(s) - D_i$ is the lateness of the job i, and D_i is the due date, $i = 1, 2,..., n$.

Number the jobs in non-decreasing order of their due dates: $D_1 \leq D_2 \leq ... \leq D_n$. For a schedule s, let $\tau_{iL}^{(k)}$ denote the total length of time intervals in which machine L processes job i in the interval $(D_{k-1}+L_{max}(s), D_k+L_{max}(s)]$. Here $k = 1, 2,..., n$, $D_0 = -L_{max}(s)$.

It is easy to verify that the values of $L_{max}(s)$ and $\tau_{iL}^{(k)}$, $k = 1, 2,..., n$, $i = 1, 2,..., n$, $L = 1, 2,..., M$, form a feasible solution of the following linear programming problem:

$$L_{max}(s) \longrightarrow \min \tag{9.18}$$

subject to

$$\sum_{L=1}^{M} \sum_{k=1}^{i} \frac{\tau_{iL}^{(k)}}{t_{iL}} = 1,\ i = 1, 2,..., n, \tag{9.19}$$

$$\sum_{L=1}^{M} \tau_{iL}^{(1)} \le D_1 + L_{max}(s),\ i = 1, 2,..., n, \tag{9.20}$$

$$\sum_{L=1}^{M} \tau_{iL}^{(k)} \le D_k - D_{k-1},\ i = k, k+1,..., n,\ k = 2, 3,..., n, \tag{9.21}$$

$$\sum_{i=1}^{n} \tau_{iL}^{(1)} \le D_1 + L_{max}(s),\ L = 1, 2,..., M, \tag{9.22}$$

$$\sum_{i=k}^{n} \tau_{iL}^{(k)} \le D_k - D_{k-1},\ L = 1, 2,..., M,\ k = 2, 3,..., n, \tag{9.23}$$

$$\tau_{iL}^{(k)} \ge 0,\ i = 1, 2,..., n,\ k = 1, 2,..., n,\ L = 1, 2,..., M. \tag{9.24}$$

On the other hand, given a solution of problem (9.18)-(9.24), we can find a schedule with the smallest value of maximal lateness. To do this, the procedures described in Section 9.6 can be used.

9.8. Let us consider a natural extension of the problem of minimizing the total cost for processing n jobs on M parallel machines.

As before, assume that all jobs are available at time $d = 0$. The processing time of job i on machine L is t_{iL}, $i = 1, 2,..., n$, $L = 1, 2,..., M$. No preemption is allowed.

It is assumed that machine L cannot start processing before time $\tau_L \ge 0$. A function $\varphi_{iL}(t)$ is associated with job i and machine L specifying the cost to be "paid" if job i is processed on machine L and this processing is completed at time t. If machine L completes processing of all assigned jobs at time t, then the cost equal to $\tilde{\varphi}_L(t)$ has to be "paid". All cost functions are non-decreasing.

It is required to find a schedule with the lowest total cost.

Below, this problem is studied under the following additional condition: each machine L processes the assigned jobs according to the sequence of reverse numerical order (starting at time τ_L). In this case, the schedule is evidently specified by partitioning set N into subsets N_1, N_2,..., N_M (some of them may be empty).

Let us introduce the following *auxiliary* problem of optimal partitioning a finite set into subsets.

Let $F_m(N_1, N_2,..., N_M)$ be a real-valued function defined over the set of all partitions of a set $\overline{N} = \{1, 2,..., m\}$ into subsets $N_1, N_2,..., N_M$. Here $\bigcup_{L=1}^{M} N_L = \overline{N}$, and some N_L may be empty. It is assumed that functions F_m are defined for all $m \geq 0$ and have the following property:

$$F_{m+1}(N_1, N_2,..., N_{L-1}, N_L \cup \{m+1\}, N_{L+1},N_{L+2},..., N_M)$$
$$= \Phi_L^{m+1}(F_m(N_1, N_2,..., N_L,..., N_M), |N_L|),$$

where Φ_L^{m+1} a function non-decreasing with respect to its first argument, and $|N_L|$ is the cardinality of the set N_L, $L = 1, 2,..., M$, $F_0 = const.$

Given $m = n$, it is required to find a partition of set $N = \{1, 2,..., n\}$ into subsets $N_1^*, N_2^*,..., N_M^*$ with the smallest value of the function $F_n(N_1, N_2,..., N_M)$.

Let $n_1, n_2,..., n_M$ be non-negative integers such that $\sum_{L=1}^{M} n_L = m$.

Consider the functions $f_L^{m+1}(n_1, n_2,..., n_M) = \min\{F_{m+1}(N_1, N_2,..., N_{L-1}, N_L \cup \{m+1\}, N_{L+1}, N_{L+2},..., N_M) \mid |N_H| = n_H, H = 1, 2,..., M\}$.

It can be shown that for any $m \geq 0$ and $1 \leq L \leq M$ the recurrent relation

$$f_L^{m+1}(n_1, n_2,..., n_H,..., n_M) = \Phi_L^{m+1}(\min\{f_H^m(n_1, n_2,..., n_{H-1}, n_H-1,$$
$$n_{H+1},..., n_M) \mid n_H > 0, H = 1, 2,..., M\}, n_L)$$

holds. This relation allows us to arrange the successive computation of all values of f_L^m, $m = 1, 2,..., n$, $L = 1, 2,..., M$, $\sum_{L=1}^{M} n_L = m$.

If $f_{L_1}^n(n_1^*, n_2^*,..., n_M^*)$ is the smallest of the computed values of f_L^n, then $|N_H^*| = n_H^*$, $H = 1, 2,..., M$, and $n \in N_{L_1}^*$. If a minimum of $f_L^{n-1}(n_1^*, n_2^*,..., n_{L_1-1}^*, n_{L_1}^*-1, n_{L_1+1}^*,..., n_M^*)$ is attained at $L = L_2$, then $n-1 \in N_{L_2}^*$, etc. As a result, the desired partition N_1^*, $N_2^*,..., N_M^*$ is obtained. This requires at most $\binom{M+n}{M}$, i.e., no more than $O(n^M)$ computations of the values of f_L^m for a fixed M.

We now consider some *special cases* of the original problem.

(a) Let $M = 2$, $\tau_1 = \tau_2 = 0$, $t_{i1} = t_{i2} = t_i$, $\varphi_{i1}(t) = \varphi_{i2}(t) = t$, $i = 1, 2,..., n$, $\tilde{\varphi}_1(t) = \tilde{\varphi}_2(t) = 0$.

Let $F_m(N_1, N_2)$ denote the total cost of processing the jobs of set $\overline{N} = \{1, 2,..., m\}$, $m \leq n$, provided that machine 1 processes the jobs of set N_1, and machine 2 processes the jobs of set N_2, $N_1 \cup N_2 = \overline{N}$. Since each machine processes the jobs in decreasing numerical order starting at $\tau = 0$, we may add the job $m+1$ to the set N_L, $L \in \{1, 2\}$, thereby increasing the total cost by the value of $(|N_L|+1)t_{m+1}$. In other words, for all m, $0 \leq m < n$, the relation

$$F_{m+1}(N_1 \cup \{m+1\},\ N_2) = F_m(N_1,\ N_2)+(|N_1|+1)t_{m+1}$$

holds.

A similar relation is also valid for $F_{m+1}(N_1,\ N_2 \cup \{m+1\})$. Therefore, we may define

$$\Phi_L^{m+1}(F_m(N_1,\ N_2), |N_L|) = F_m(N_1,\ N_2)+(|N_L|+1)t_{m+1},$$

where $L = 1,\ 2$ and $F_0 = 0$.

Having solved the auxiliary problem with the functions Φ_L^{m+1} of the form shown above, we obtain a desired partition of set $N = \{1,\ 2,\dots,\ n\}$ into subsets.

(b) Let $M = 2$, $\varphi_{iL}(t) = t+c_{iL}$, $\tilde{\varphi}_L = \alpha_L t$, $\alpha_L > 0$, $i = 1,\ 2,\dots,\ n$, $L = 1,\ 2$.

As in the previous case, let $F_m(N_1,\ N_2)$ denote the total cost of processing the jobs of set N_1 on machine 1 and the jobs of set N_2 on machine 2; $N_1 \cup N_2 = \overline{N}$, $\overline{N} = \{1,\ 2,\dots,\ m\}$, $m \le n$.

Let us add the job $m+1$ to the set N_L, $L = 1,\ 2$. If $N_L \neq \varnothing$, then the total cost changes by $(\tau_L+t_{m+1,L}+c_{m+1,L})+|N_L|t_{m+1,L}+\alpha_L t_{m+1,L}$. If $N_L = \varnothing$, then the total cost changes by $(\tau_L+t_{m+1,L}+c_{m+1,L})+\alpha_L(\tau_L+t_{m+1,L})$.

Therefore, we may define

$$\Phi_L^{m+1}(F_m(N_1,N_2), |N_L|) = F_m(N_1,N_2)+(\alpha_L\beta_L+1)\tau_L\ +(|N_L|+\alpha_L+1)t_{m+1,L}+c_{m+1,L},$$

where $\beta_L = 1$ if $|N_L| = 0$, $\beta_L = 0$ if $|N_L| > 0$; $L = 1,\ 2$ and $F_0 = 0$.

(c) Let $M \ge 2$, $\varphi_{iL}(t) = b_L t+c_{iL}$, $\tilde{\varphi}_L = \alpha_L t$, $\alpha_L > 0$, $b_L > 0$, $i = 1,\ 2,\dots,\ n$, $L = 1,\ 2,\dots,\ M$.

It is clear that in this case

$$\Phi_L^{m+1}(F_m(N_1,\ N_2,\dots,\ N_M), |N_L|)= F_m(N_1,\ N_2,\dots,\ N_M)+(\alpha_L\beta_L+b_L)\tau_L$$
$$+(b_L|N_L|+\alpha_L+b_L)t_{m+1,L}+c_{m+1,L},$$

where $\beta_L = 1$ if $|N_L| = 0$, $\beta_L = 0$ if $|N_L| > 0$; $L = 1,\ 2,\dots,\ M$, $0 \le m < n$, $F_0 = 0$.

10. Bibliography and Review

10.1. The problems of the existence of a non-preemptive optimal schedule are discussed by McNaughton (a single machine, the total cost, non-decreasing cost functions; or parallel machines, the total cost, linear cost functions) [356] and by Rothkopf (parallel machines, the total cost, exponential cost functions) [386]. Theorem 1.1 is proved by

Tanaev and Gordon [41, 155]. The proof of Theorem 1.2 is given in [155]. An extension of Theorem 1.2 to the case of different release dates can be found in [46].

Gordon, Tanaev [47] and Horn [295] give an algorithm for finding a deadline-feasible schedule for parallel identical machines provided that preemption is allowed by reducing that scheduling problem to a maximal flow problem (Section 2.3). The case of $M = 1$ (Section 2.5) is considered by Jackson [304], Gordon [41], Vizing [28], Kopylov [79]. The problem for $M \geq 1$, $d_i = d$, $D_i = D$, $i = 1, 2,..., n$ (Section 2.6) is solved by McNaughton [356]. An $O(n\log n)$ algorithm for $M \geq 1$, $d_i = d$, $i = 1, 2,..., n$ (described in Section 2.8) is due to Sahni [395]. For $M \geq 1$, algorithms are offered for $D_i = D$, $i = 1, 2,..., n$, (the running time is $O(nM)$) [284] and for $t_i = t$, $i = 1, 2,..., n$ (the running time is $O(n^3\log n)$) [413]. The problems of finding all feasible schedules for $M \geq 1$ are discussed in [124], those for $M = 1$, in [400] (preemption is allowed) and in [258] (preemption is forbidden). Finding a non-preemptive feasible schedule is an NP-hard problem even if $M = 1$, or $M = 2$ and $d_i = 0$ (see Sections 2 and 1 of Chapter 4, respectively). For these situations, finding feasible schedules is studied in [1, 28, 83]. Several $O(n^2)$ algorithms for finding feasible schedules for preemptive processing of a partially ordered set of jobs are developed by Gordon and Tanaev [49] ($M = 1$, see Remark 2 in Section 2.5; Sections 3.6 and 3.7) and by Lawler [338] ($M > 1$, provided that either G is an intree and $d_i = 0$ or G is an outtree and $D_i = D$, $i = 1, 2,..., n$).

Vizing [28] establishes the necessary and sufficient conditions for the existence of a deadline-feasible schedule provided that preemption is allowed and a job can be processed on more than one machine at a time; sufficient conditions are also derived for non-preemptive processing without violating the deadlines. For $M = 1$, verifying the necessary and sufficient conditions for the existence of a non-preemptive feasible schedule requires at most $O(n^2)$ time either if $t_i = t$, $i = 1, 2,..., n$ [31], or if the release dates and deadlines are agreeable, as presented in Remark 1 in Section 2.5 [49]. For the first of these cases, an $O(n\log n)$ algorithm is described in [276] which finds a feasible schedule (and a time-optimal feasible schedule).

If the machines are uniform ($t_{iL} = a_L t_i$, $i = 1, 2,..., n$, $L = 1, 2,..., M$), then the algorithms for feasible preemptive scheduling are proposed in the cases: (a) $d_i = d$, $D_i = D$, the running time is $O(n+M\log M)$ [286]; (b) either $d_i = d$ and D_i are arbitrary or $D_i = D$ and d_i are arbitrary, the running time is $O(n\log n+Mn)$ [397]; (c) d_i and D_i are arbitrary, the running time is $O(M^2n^4+n^5)$ [353]. The case of $M = 2$ is considered in [338] (the running time is $O(n^2)$ if $d_i = 0$ and $O(n^3)$ if $d_i \geq 0$). In a general case of unrelated machines, the problem can be solved by linear programming reduction [339].

The problems of the existence of feasible schedules are also considered by Revchuk [130] (given "access windows" of machines) and by Lapko [85] (unlimited deterministic flows of jobs).

Algorithms for minimizing the maximal cost for $M = 1$ and $d_i = 0$, $i = 1, 2,..., n$, (Section 3.2) are proposed by Livshits [98] for a non-ordered set of jobs and by Lawler [332] for a partially ordered set of jobs. Solutions to the problems of minimizing the maximal lateness (L_{max}) and the maximal tardiness for $d_i = 0$, $i = 1, 2,..., n$, and $G = (N, \emptyset)$ (Section 3.3) is obtained by Jackson [304], and by Lawler and Moore [332, 342] for the case of a partially ordered set of jobs. Lageweg et al. [326] reduce the non-preemptive case of the problem of minimizing L_{max} for $D_i = D$, $d_i \geq 0$, $i = 1, 2,..., n$, to the case of the same release dates and different due dates (a special case of the problem discussed in Section 3.4).

Gordon [42, 43] develops an algorithm for solving the problem of minimizing the maximal cost for $M = 1$, $d_i \geq 0$, $i = 1, 2,..., n$, provided that preemption is allowed. An $O(n^2)$ algorithm for solving this problem given a partially ordered set of jobs (see Section 3.6) is designed by Gordon and Tanaev [49] and independently by Baker et al. [194]; see also [88]. A special case of this problem (that of minimizing L_{max}) is discussed in [87, 295, 326]. Lageweg et al. [326] give an algorithm for solving the problem of minimizing L_{max} for $t_i = 1$, $d_i \geq 0$, $i = 1, 2,..., n$, provided that preemption is forbidden (a special case of the problem considered in Section 3.8(a)). In [412], the latter problem with equal processing times is considered.

In [363], the problem of minimizing the maximal cost is considered, provided that the processing times may be negative, while in [327, 409], it is assumed that the cost is associated with both violating the due dates and starting processing before the time d_i.

Theorem 4.1 is proved by Tanaev and Gordon [157]. The proof of a similar statement for a special case ($\alpha_i = 1$, $i = 1, 2,..., n$) given by Kise et al. [316] contains an error (Lemma 2 [316] is not valid). The solution procedure for the problem of minimizing the number of late jobs (Section 4.3(a) and (b)) is offered by Moore (for $d_i = 0$, $i = 1, 2, ..., n$) and Kise et al. Mine [316] (for agreeable d_i and D_i); see also [354]. The solution to the problem studied in Section 4.3(b) is obtained by Gordon and Tanaev [45], as well as by Lawler [334]. Sidney [407] proposes an algorithm for the problem considered in Section 4.4, assuming $\alpha_i = 1$, $i = 1, 2,..., n$.

Lawler [331] suggests reducing the problem of minimizing the total cost for $t_i = 1$, $d_i = 0$, $i = 1, 2,..., n$, to the assignment problem (see Section 4.5). In [424], this reduction is extended to the situation where a single time variable resource system is considered instead of a single-machine system. The problems of minimizing the total cost for single-machine processing when the cost is associated with the processing of a job in each unit interval is discussed in [140, 179, 180, 181]. The case of $\varphi_i(t) = z_i$, $t_i = 1$ is considered in [384].

The solution to the problem studied in Section 4.6 with $\varphi(t)+b_i = t-D_i$, $i = 1, 2,..., n$, is proposed by Horn [295]. The case of $\varphi(t) = t$ is discussed by Baker [193]. In the case $d_i = 0$, $i = 1, 2,..., n$, and under the additional condition of scheduling without violating deadlines, this problem is considered in [36, 417].

The problems of minimizing the total cost under additional constraints on usage of the machine are studied in [8, 131].

Polynomial-time algorithms for multicriteria single-machine problems, as well as single-criteria problems with additional constraints such as processing without violating the due dates are discussed in [30, 32, 69, 71, 159, 256, 272, 274, 291, 431, 434].

The papers [84, 86, 161, 387, 428] are devoted to stochastic counterparts of the problem of optimal single-machine processing. The problems of optimal scheduling with non-monotonic objective functions are discussed in [250, 311, 358, 401].

The algorithm for minimizing the makespan when $M \geq 1$, $t_i = 1$, $i = 1, 2,..., n$, and the reduction graph of the precedence relation is a forest (Section 5.2) is proposed by Hu [168]. In [11, 298], simpler proofs than that in [168] are given for the optimality of the schedule obtained by this algorithm. The proof given in Section 5.3 is due to Sethi [404]. Other algorithms with the same running time are developed in [40, 240], see also [438]. The $O(n^2)$ algorithm for solving the above problem for an arbitrary reduction graph of the precedence relation and $M = 2$ (Sections 5.4-5.6) is proposed by Coffman and Graham [234]. (The graph in Fig. 5.4 is given in [403], that in Fig. 5.7 is taken from [404]). Earlier, an $O(n^3)$ algorithm was known [263, 264]. An $O(n^2)$ algorithm by Garey and Johnson [272] also solves this problem (see Section 7.3). In [271], a polynomial-time algorithm is proposed for solving the problem for $G = (N, \varnothing)$ with resource constraints, while [329] studies the case of $M \geq 2$, $d_i \geq 0$, $G = (N, \varnothing)$ under additional condition of processing without violating the deadlines (d_i are integers, $i = 1, 2,..., n$).

The problem of finding a preemptive time-optimal schedule for $M \geq 1$, $t_{iL} = t_i$, $d_i = 0$, $i = 1, 2,..., n$, $L = 1, 2,..., M$, (Section 6) is solved by McNaughton [356] ($G = (N, \varnothing)$, the running time is $O(n)$), by Muntz and Coffman [370, 369] (either graph G is tree-like or $M = 2$, the running time is $O(n^2)$). The algorithms the same running times but different from those in [369, 370] and based on results obtained in [216, 272] are due to Lawler [338] (if $M = 2$, the machines may have different speeds), see also [389, 390]. A more efficient algorithm (the running time is $O(n\log M)$) presented in [284] solves the problem if the graph G is a forest of outtrees. The same paper also describes an $O(nM)$ algorithm for the case $d_i \geq 0$, $i = 1, 2,..., n$, $G = (N, \emptyset)$. The proof of Theorem 6.1 is given by Muntz and Coffman in [370], who also introduce the concept of a machine-sharing schedule. In [125], minimizing the number of preemptions in a time-optimal schedule (G being an intree) is considered.

The algorithms for finding a deadline-feasible schedule (given equal job processing times) discussed in Section 7 are proposed in [216] for the case if G is an intree (Section 7.2) and in [272] for $M = 2$ (Section 7.3). In both cases, it is assumed that $d_i = 0$, $i = 1, 2,..., n$. In [274], an $O(n^3)$ algorithm is given for solving the problem for $M = 2$, $t_i = 1$, $d_i \geq 0$, d_i being integers, $i = 1, 2,..., n$.

Minimizing L_{max} for $M > 1$ is considered in [29, 216, 272, 274, 295, 324, 338, 339, 353, 396, 413, 414]. For the case of $d_i = 0$, $t_i = 1$, $i = 1, 2,..., n$, provided that preemption is not allowed (Section 8.2) the problem is solved by Brucker et al. [216] (G is an intree, the running time is $O(n\log n)$) and by Garey and Johnson [272] (G in an arbitrary graph and $M = 2$; the running time is $O(n^2)$). For $M = 2$, d_i being integers, $t_i = 1$, $i = 1, 2,..., n$, and an arbitrary circuit-free graph G, the problem is solved in [274] (the running time is $O(n^3\log n)$). For $t_i = t$, $d_i \geq 0$, $i = 1, 2,..., n$, $M > 1$, an $O(n^3\log^2 n)$ algorithm is given in [414]. Horn [295] develops an $O(n^2)$ algorithm for solving the problem for $G = (N, \emptyset)$, $d_i = 0$, $M > 1$, $t_i > 0$, $i = 1, 2,..., n$, preemption is allowed. The algorithm (Section 8.3) for solving the problem for $G = (N, \emptyset)$, $M > 1$, provided that all d_i and t_i are integers and preemption is allowed, runs in $O(n^3 \min\{n^2, \log n + \log(\max\{t_i | i \in N\})\})$ time and is proposed by Labetoulle et al. [324]. This algorithm can be extended to the case of $M = 2$, $t_{iL} = a_L t_i$, $i = 1, 2,..., n$, $L = 1, 2,..., M$, [324]. For the case of $M > 1$, $t_{iL} = a_L t_i$, $i = 1, 2,..., n$, $L = 1, 2,..., M$, preemption is allowed, an $O(Mn\log n + M^2 n)$ algorithm is given in [324, 396] for $d_i = 0$, $i = 1, 2,..., n$. Martel [353] proposes a polynomial-time algorithm for arbitrary d_i. In [338], Lawler extends the results obtained for $t_i = 1$, $i = 1, 2,..., n$, to the case of arbitrary t_i provided that preemption is allowed ($M > 1$, $d_i = 0$, G is an intree, the running time is $O(n^2)$; $M = 2$, $t_{iL} = a_L t_i$, G is arbitrary, the running time is $O(n^2)$ or $O(n^6)$ if $d_i = 0$ or $d_i \geq 0$, respectively). Reducing the problem of minimizing L_{max} (for unrelated machines and preemptive processing) to a linear programming problem (Section 9.7) is given in [339]. If a job is allowed to be processed on several machines at a time, the problem is solved in [29] (for M identical machines, $d_i \geq 0$, $i = 1, 2,..., n$, preemption is allowed, the running time is $O(n^4 M)$).

Reducing a scheduling problem with the objective function (9.1) ($\varphi_i(t) = t$, $i = 1, 2,...,$

n) to the transportation problem (Section 9.2) is proposed by Horn [294] and Bruno et al. [220]. An $O(n^3)$ algorithm for solving the later problem is given in [219]; see also [218]. The algorithm for optimal (with respect to function (9.1)) scheduling for $\varphi_i(t) = t$ and $t_{iL} = a_L t_i$, $i = 1, 2,..., n$, $L = 1, 2,..., M$, (Section 9.3) is proposed by Conway et al. [78], see also [218, 296].

Reducing the problem of minimizing the objective function (9.1) ($t_{iL} = a_L$, $i = 1, 2,..., n$, $L = 1, 2,..., M$) to the transportation problem (Section 9.4) is given in [290]. In a special case of $t_i = 1$, $i = 1, 2,..., n$, $L = 1, 2,..., M$, the solution is obtained by Lawler [331].

An $O(n\log n)$ algorithm for minimizing the weighted number of late jobs for $t_i = 1$, $i = 1, 2,..., n$, $L = 1, 2,..., M$, (Section 9.5) is given in [334]. The same paper also considers the case when the number of available machines is specified in each unit time interval.

The scheduling problems to minimize (9.1) for $t_{iL} = a_L t_i$, $i = 1, 2,..., n$, $L = 1, 2,..., M$, are considered in [282, 337] (preemption is allowed). In [282], an $O(n\log n + Mn)$ algorithm is given for $\varphi_i(t) = \alpha_i t_i$, $i = 1, 2,..., n$, assuming that the weights α_i and the processing times t_i are agreeable ($t_i < t_j$ implies $\alpha_i \geq \alpha_j$). The problem of minimizing the number of late jobs if a_L are time-dependent is considered in [337]. The proposed algorithms run in $O(n^4)$ time for $M = 2$ and in $O(n^{3M-3})$ time for $M \geq 3$.

Reducing the scheduling problem with the objective function (9.2) to the transportation problem ($t_{iL} = a_L$, $i = 1, 2,..., n$, $L = 1, 2,..., M$, no preemption is allowed) considered in Section 9.4 is given in [290]. In the preemptive case, Lawler and Labetoulle [339] reduce this problem for $\varphi_i(t) = t$, $i = 1, 2,..., n$, to a linear programming problem (Section 9.6). The procedure for finding an optimal schedule from a solution of the linear programming problem if based on the results of [285].

The algorithms for time-optimal preemptive scheduling on uniform parallel machines are proposed in [297, 328, 338] ($M = 2$, an ordered set of jobs; the running time is $O(n^2)$), in [286] ($d_i = 0$, $i = 1, 2,..., n$, the running time is $O(n + M\log M)$) and in [324, 396] ($d_i \geq 0$, $i = 1, 2,..., n$, the running time is $O(Mn\log n + M^2 n)$).

The problems of finding schedules that are either time-optimal or feasible with respect to deadlines under resource constraints are discussed in [105, 211, 271]. The algorithms for finding time-optimal schedules are described in [271] ($M = 2$, $t_i = 1$, $i = 1, 2,..., n$, q types of resources, the running time is $O(qn^2 + n^{5/2})$), in [211] ($M = 2$, $t_{iL} = a_L$, $i = 1, 2,..., n$, $L = 1, 2,..., M$, $q = 1$, the running time is $O(n\log n)$; and also, $t_{iL} = a_L$, $i = 1, 2,..., n$, $L = 1, 2,..., M$, $q = 1$ and one unit of the resource is required for the processing of each job (the running time is $O(n^3)$). These algorithms can be used to find schedules that are feasible with respect to deadlines when $D_i = D$, $i = 1, 2,..., n$.

Polynomially solvable optimal scheduling problems with some constraints on the grouping of the jobs are considered in [135, 136, 372].

Section 9.8 is based on paper [156] by Tanaev. A meaningful formulation of the problem is given by Rothkopf [386]. Case (a) is considered in [357], and case (b) in [388].

10.2. We now consider the main results which have not been reflected in the Russian edition of the book.

An excellent survey of recent results in scheduling theory can be found in [115*]. Developments in some specific areas are reviewed in [7*, 10*, 11*, 12*, 38*, 66*, 70*, 114*, 117*-119*].

The following polynomial-time algorithms for solving traditional scheduling problems have been developed recently.

Frederickson [50*] proposes an $O(n)$ algorithm for minimizing the maximum lateness on a single machine with $t_i = 1$, $d_i > 0$, $i = 1, 2,..., n$.

Garey et al. [57*] give an $O(n\log n)$ algorithm for finding a feasible schedule with respect to the given release dates d_i and deadlines D_i for the single machine problem with $t_i = t$, $i = 1, 2,..., n$. The binary search over the possible values of L_{max} yields a polynomial-time algorithm for minimizing the maximum lateness.

Monma [130*] proposes a linear-time algorithm minimizing the maximum lateness on a single machine under precedence constraints assuming that $d_i = 0$, $t_i = 1$, $i = 1, 2,..., n$.

Rinnooy Kan [324] describes an $O(n\log n)$ procedure for the single machine problem to minimize the total tardiness, provided that the release dates are integers and the processing times are unit.

Lawler [112*] proposes an $O(n\log n)$ algorithm for minimizing the number of late jobs on a single machine if the release dates and due dates are similarly ordered.

Gordon and Baranova [8*, 68*] extend polynomially solvable cases of the minimum weighted number of late jobs problem (see Sections 4.3–4.4) to cover the problem in which specified jobs have to be completed on time and the release dates and due dates are similarly ordered. This generalizes the results in [407] obtained for the case of $d_i = 0$, $\alpha_i = 1$, $i = 1, 2,..., n$.

Monma [130*] describes an $O(n)$ algorithm for minimizing the number of late jobs on a single machine in the case of $d_i = 0$, $t_i = 1$, $i = 1, 2,..., n$.

Scheduling unit-length jobs with $d_i = 0$, $i = 1, 2,..., n$ on uniform parallel machines with a minisum or minimax criterion and arbitrary non-decreasing cost functions is considered in [39*], see also [115*]. The problem of minimizing the total cost is solved in $O(n^3)$ by reducing to the nxn weighted bipartite matching problem. The problem of minimizing the maximal cost is solved in $O(n^2)$ time using a generalization of the algorithm for the corresponding single-machine problem. The problems of minimizing $\sum_{i \in N} z_i$ or L_{max} are solved in $O(n\log n)$ time by matching the kth smallest due date with $\bar{t}_k^{\min}$. Here $\bar{t}_1^{\min} < \bar{t}_2^{\min} <... < \bar{t}_n^{\min}$ denote the n earliest possible completion times. These values can be obtained in $O(n\log M)$ time by arranging a priority queue with the completion times a_1, a_2, ..., a_M and then, in a general step, by removing the smallest completion time from the queue and, if this time is ka_L, inserting $(k+1)a_L$ into the queue. A similar approach is used for the problem of minimizing $\bar{t}_{max}$ with different d_i (matching the kth smallest release date with $\bar{t}_k^{\min}$), as well as for that of minimizing $\sum_{i \in N} \alpha_i \bar{t}_i$ (matching the kth largest weight with $\bar{t}_k^{\min}$). The problem of minimizing $\sum_{i \in N} \alpha_i u_i$ is solved in $O(n\log n)$ time by scanning the n earliest possible completion times from the largest to the smallest. Among unscheduled jobs which can be completed on time (if any), a job with the largest weight is chosen to start processing.

McCormick and Pinedo [127*] generalize the $O(n\log n+Mn)$ algorithm by Gonzalez [282] for the preemptive scheduling on uniform machines to minimize $w\bar{t}_{max}+ \sum_{i \in N} \bar{t}_i$ with an arbitrary weight $w \geq 0$.

Federgruen and Groenevelt [48*] propose an $O(tn^3)$ algorithm for preemptive scheduling n jobs with given release dates on M uniform parallel machines of t, $t \leq M$, distinct speeds to minimize L_{max}.

Monma [130*] presents a linear-time algorithm for scheduling unit-length jobs on parallel identical machines under precedence constraints of the form of an intree, the objective is L_{max}. Garey et al. [58*] show that this problem with the objective $\bar{t}_{max}$ can be solved in polynomial time if the reduction graph is an opposing forest (the disjoint union of an inforest and an outforest) and the number of machines is fixed (if the number of machines is variable the problem is NP-hard). Möhring [129*] shows that the problem can be solved in polynomial time by dynamic programming if the width of the reduction graph is bounded.

Gabow and Tarjan [56*] propose a linear-time algorithm for minimizing $\bar{t}_{max}$ on two identical machines under arbitrary precedence constraints assuming that $t_i = 1$, $i = 1, 2,..., n$. If the reduction graph is a tree, the problem can be solved in $O(n\log n)$ time for $t_i \in \{1, 2\}$ [134*] and in $O(n^2\log n)$ time for $t_i \in \{1, 3\}$ [43*].

Using results of symmetric functions study, Tanaev has found some properties which allow to describe a class of polynomially solvable scheduling problems [154*, 155*]

Below, we consider some extensions of the traditional scheduling problems that seem to be of particular interest and lead to new efficient algorithms.

Hochbaum and Shamir [80*] consider the high multiplicity problems in which the jobs can be partitioned into relatively few groups (or types), and in each group all the jobs are identical, i.e. they have the same set of parameters (the due dates, the weights etc.). The number of jobs of an individual type is called the multiplicity of that type. These problems may also be interpreted as classical scheduling problems, with a different kind of the objective. Each type is considered as a superjob of length equal to its multiplicity and the goal is to arrange a preemptive scheduling of the superjobs. The objective function takes into account the contribution of each unit (rather then that of a superjob) to the total cost.

The running times of the algorithms described in [80*] are polynomial with respect to the number k of groups rather than in the total number of jobs. Polynomial algorithms are proposed for the single-machine problem with p_i identical unit-time jobs of type i, $i = 1, 2, \ldots, k$, to minimize (1) the weighted number of late jobs (the running time is $O(k\log k)$; preemptive solution); (2) the total weighted tardiness with agreeable weights ($O(k\log k)$); (3) the maximum and total weighted completion time ($O(k\log k)$); (4) the maximum weighted lateness or tardiness ($O(k\log^2 k)$); and (5) the total weighted tardiness (reduced to a quadratic integer transportation model solvable in polynomial time; preemptive solution).

Batching and lot-sizing problems as combinations of sequencing and partitioning problems have recently become of great interest. Batching is considered as the decision of whether or not to schedule similar jobs contiguously. On the other hand, lot-sizing refers to the decision on when and how to split a production lot of identical items into sublots. Recent applications of these problems can be found in flexible manufacturing systems which provide the possibility to economically process jobs in small batches. Batching of similar jobs is mainly done to avoid set-up times or set-up costs. The review of the results concerning this type of problems is recently given by Potts and Van Wassenhove [140*]. See also [2*, 24*, 28*, 30*, 31*, 96*, 97*, 104*, 156*].

Using the approach of Lawler and Moore [342], Monma and Potts [131*] develop dynamic programming algorithms for single machine batching problem to minimize the maximum lateness, the total weighted completion time and the number or late jobs. In all this cases this approach yields the algorithms that are polynomial with respect to the number of jobs, but exponential with respect to the number of batches. Thus, the problems are efficiently solvable even with sequence dependent setup times, assuming that the number of batches is fixed.

Coffman et al. [31*] describe an $O(\sqrt{n})$algorithm for the single machine scheduling to minimize $\sum_{i \in N} \bar{t}_i$ in the case of two types of batches, sequence independent setup times and equal processing times in each batch.

Albers and Brucker [2*] propose an $O(n\log n)$ algorithm for minimizing the total weighted flow time $\sum_{i \in N} \alpha_i \bar{t}_i$ on a single machine, assuming that the processing times of jobs are equal, the setup times are both sequence and batch independent and the flow time of a job is determined by the completion time of the last scheduled job in a batch (all jobs in a batch are supposed to have the same flow time). They also give an $O(n^2)$ algorithm for the problem with $\alpha_i = 1$, $i = 1, 2, \ldots, n$, under arbitrary precedence constraints.

Cheng and Kahlbacher [28*], Cheng and Gordon [24*], Cheng, Gordon and Kovalyov [26*] present polynomial-time algorithms for some batch delivery problems in which the jobs in each batch have to delivered to the customer together, and the batch delivery cost is included into the objective function.

Another popular research topic in the recent scheduling literature is the optimal due date assignment and scheduling [6*, 16*-22*, 25*, 27*, 29*, 64*, 65*, 67*, 135*, 145*, 160*]. The due dates in these problems are not given in advance and have to be assigned during decision making. Polynomial-algorithms for the optimal due date assignment and scheduling are proposed for some single machine problems with minimax lateness [21*, 22*, 25*, 64*, 65*, 67*], minisum lateness [18*-20*, 135*, 144*, 160*], and minimal square lateness objectives [16*]. Extensive surveys of scheduling research involving due-date determination decision have been presented by Cheng and Gupta [27*] and Baker and Scudder [7*].

Scheduling theory is one of the areas that are likely to benefit from advances in parallel computers. Ribero [142*] and Kindervater and Lenstra [91*] present detailed reviews of parallelism in combinatorial optimization, see also [90*, 66*]. The complexity theory for parallel computation explains the speedups possible due to the introduction of parallelism. Within the class $\mathcal{P}$, this leads to a distinction between "very easy" problems that are solvable in polylogarithmic parallel time, and "not so easy" ones for which a speedup due to parallelism is unlikely (they are P-complete under log-space transformations). Well-solvable problems belong to the class $\mathcal{NC}$ which contains all problems solvable in polylog parallel time using only a polynomial number of processors. We refer to Johnson [86*] and Cook [32*] for further details.

Scheduling problems of the class $\mathcal{NC}$ are considered by Dekel and Sahni [35*-37*], Gordon [62*, 63*], Helmbold and Mayr [78*, 79*]. Single machine problems with release dates to minimize a minimax objective (either preemption is allowed or the processing times are unit) are considered in [36*, 62*, 63*]. Single machine problems to minimize either the number of late jobs or the weighted number of late jobs (unit processing times) are considered in [35*–37*]. Helmbold and Mayr [78*, 79*] study the problem of minimizing the makespan $\overline{t}_{max}$ for the jobs of unit length on two machines under precedence constraints. The problems to minimize either $\overline{t}_{max}$ (preemption is allowed) or L_{max} (the release dates may be different and the processing times are unit) for parallel identical machines are considered in [35*–37*]. The problem of minimizing $\overline{t}_{max}$ for parallel uniform machines (preemption is allowed) is considered in [125*].

CHAPTER 3

PRIORITY-GENERATING FUNCTIONS. ORDERED SETS OF JOBS

As mentioned in the previous chapters, a number of scheduling theory problems can be formulated in terms of optimizing functions over sets of permutations of the elements of a given finite set N. In particular, among such problems are those of finding optimal single-machine schedules for a finite set of jobs, provided that preemption is not allowed and that at most one job is processed at a time.

This chapter considers problems of optimizing functions over some subsets $\mathcal{P}$ of the set $\hat{\mathcal{P}}$ of all permutations of the elements of set N. Special classes of functions are distinguished and methods for their optimization are described under various assumptions on the structure of set $\mathcal{P}$. Attention is paid to analyzing the situation where N is a partially ordered set and $\mathcal{P}$ is a set of all permutations maintaining the order defined over N.

The concept of a priority-generating function is introduced in Section 1. In that section, a number of combinatorial extremal problems are presented which can be reduced to optimizing a priority-generating function over an appropriate set $\mathcal{P}$. Section 2 describes specific transformations of graph G of reduction of a precedence relation defined over set N. These transformations are the basis of the methods for optimizing priority-generating functions over a set of permutations maintaining the order defined over N. Sections 3 and 4 consider the cases in which graph G is tree-like and series-parallel, respectively. A general case is studied in Sections 5 and 6. In Section 7 the concept of a 1-priority-generating function is introduced, and methods for optimizing such functions are described.

1. Priority-Generating Functions

Let $\hat{\mathcal{P}}$ be a set of all permutations $\pi_r = (i_1, i_2, \ldots, i_r)$, $r = 0, 1, \ldots, n$, of the elements of a set $N = \{1, 2, \ldots, n\}$. Here r is *the length of a permutation* π_r, $\{\pi_0\} = \emptyset$. If $\mathcal{P} \subseteq \hat{\mathcal{P}}$, then let $Q[\mathcal{P}]$ denote a set of all those permutations $\pi^{(q)} \in \hat{\mathcal{P}}$, $\pi^{(q)} \neq \pi_0$, for which there exist permutations $\pi \in \mathcal{P}$ and $\pi^{(1)}, \pi^{(2)} \in \hat{\mathcal{P}}$ such that $\pi = (\pi^{(1)}, \pi^{(q)}, \pi^{(2)})$.

A function $F(\pi)$ is defined over a set $\mathcal{P}' \subseteq \hat{\mathcal{P}}$. For $\mathcal{P} \subseteq \mathcal{P}'$, let there exist a function $\omega(\pi)$ defined over the set $Q[\mathcal{P}]$ and having the following property. For any permutations $\pi' = (\pi^{(1)}, \pi^{(a)}, \pi^{(b)}, \pi^{(2)})$ and $\pi'' = (\pi^{(1)}, \pi^{(b)}, \pi^{(a)}, \pi^{(2)})$ belonging to $\mathcal{P}$ the condition $\omega(\pi^{(a)}) \geq \omega(\pi^{(b)})$ implies that $F(\pi') \leq F(\pi'')$. In this case, $F(\pi)$ is called *a priority-generating function over set* $\mathcal{P}$, and $\omega(\pi)$ is called *its priority function*. The value of $\omega(\pi)$ is called *the priority* of permutation π.

Note that if the function $F(\pi)$ is priority-generating over some set, it is also priority-generating over any of its subsets.

Priority-generating functions play an important role in scheduling theory. Many spectacular results are obtained while analyzing situations in which some priority-generating function is to be optimized over a certain set $\mathcal{P}$ of permutations. As a rule, a subset of the set $\hat{\mathcal{P}}_n$ is chosen as $\mathcal{P}$. Here $\hat{\mathcal{P}}_n$ is a subset of $\mathcal{P}$ consisting of all permutations of $\hat{\mathcal{P}}$ having the length n.

In the following, we consider the problem of minimizing $F(\pi)$. To maximize $F(\pi)$, it is enough to take $-\omega(\pi)$ as a priority function and to use an algorithm for minimizing $F(\pi)$.

We consider several problems which can be reduced to minimizing a priority-generating function. The following notation is used throughout this section. For a real number λ_i associated with an element $i \in N$ denote $\lambda(\pi) = \sum_{i \in \{\pi\}} \lambda_i$ where $\pi \in \hat{\mathcal{P}}$.

1.1. The jobs of a set $N = \{1, 2, \ldots, n\}$ are processed on a single machine starting at time $d = 0$. For each job i, the processing time $t_i > 0$ and the function $\varphi_i(t)$ of the cost to be "paid" for having that job completed at time t are given. Preemption is not allowed, and at most one job is processed at a time.

Let the function

$$F(\pi) = \sum_{k=1}^{r} \varphi_{i_k}\left(\sum_{j=1}^{k} t_{i_j}\right) \tag{1.1}$$

be defined over the set $\hat{\mathcal{P}}$ of permutations of the elements of set N, where $\pi = (i_1, i_2, \ldots,$

$i_r) \in \hat{\mathcal{P}}$ and $F(\pi_0) = 0$. It is required to find a feasible (in a certain sense) job processing sequence (i.e., a permutation π_n of $\mathcal{P} \subseteq \hat{\mathcal{P}}_n$), for which the total processing cost $F(\pi_n)$ is minimal.

We introduce the function

$$\Phi(\pi, C) = \sum_{k=1}^{r} \varphi_{i_k}\left[C + \sum_{j=1}^{k} t_{i_j}\right] \tag{1.2}$$

where C is a real number. It is obvious that $F(\pi) = \Phi(\pi, 0)$. We have

$$\begin{aligned} F(\pi^{(1)}, \pi^{(a)}, \pi^{(b)}, \pi^{(2)}) &= \Phi(\pi^{(1)}, 0) + \Phi(\pi^{(a)}, t(\pi^{(1)})) + \\ &+ \Phi(\pi^{(b)}, t(\pi^{(1)}, \pi^{(a)})) + \Phi(\pi^{(2)}, t(\pi^{(1)}, \pi^{(a)}, \pi^{(b)})); \\ F(\pi^{(1)}, \pi^{(b)}, \pi^{(a)}, \pi^{(2)}) &= \Phi(\pi^{(1)}, 0) + \Phi(\pi^{(b)}, t(\pi^{(1)})) + \\ &+ \Phi(\pi^{(a)}, t(\pi^{(1)}, \pi^{(b)})) + \Phi(\pi^{(2)}, t(\pi^{(1)}, \pi^{(b)}, \pi^{(a)})). \end{aligned}$$

Since $t(\pi^{(1)}, \pi^{(a)}, \pi^{(b)}) = t(\pi^{(1)}, \pi^{(b)}, \pi^{(a)})$, it follows that the inequality $F(\pi^{(1)}, \pi^{(a)}, \pi^{(b)}, \pi^{(2)}) \le F(\pi^{(1)}, \pi^{(b)}, \pi^{(a)}, \pi^{(2)})$ holds if and only if

$$\begin{aligned} &\Phi(\pi^{(a)}, t(\pi^{(1)})) + \Phi(\pi^{(b)}, t(\pi^{(1)}, \pi^{(a)})) \\ &\le \Phi(\pi^{(b)}, t(\pi^{(1)})) + \Phi(\pi^{(a)}, t(\pi^{(1)}, \pi^{(b)})). \end{aligned} \tag{1.3}$$

We now consider some special cases of function (1.1).

(a) Let $\varphi_i(t) = \alpha_i t + \beta_i$, where α_i, β_i are real numbers, $i = 1, 2, \ldots, n$. Then relation (1.3) reads $\alpha(\pi^{(b)})t(\pi^{(a)}) \le \alpha(\pi^{(a)})t(\pi^{(b)})$. Since $t_i > 0$, $i = 1, 2, \ldots, n$, we may define $\omega(\pi^{(a)}) = \alpha(\pi^{(a)})/t(\pi^{(a)})$ and $\omega(\pi^{(b)}) = \alpha(\pi^{(b)})/t(\pi^{(b)})$. Hence, if the cost functions are linear, function (1.1) is priority-generating over set $\hat{\mathcal{P}}$ and its priority function is

$$\omega(\pi) = \sum_{i \in \{\pi\}} \alpha_i / \sum_{i \in \{\pi\}} t_i. \tag{1.4}$$

(b) Let $\varphi_i(t) = \alpha_i \exp(\gamma t) + \beta_i$, $i = 1, 2, \ldots, n$, $\gamma \ne 0$. Then relation (1.3) reads $(\exp(\gamma t(\pi^{(a)})) - 1)(F(\pi^{(b)}) - \beta(\pi^{(b)})) \le (\exp(\gamma t(\pi^{(b)})) - 1)(F(\pi^{(a)}) - \beta(\pi^{(a)}))$. Since $t_i > 0$, $i = 1, 2, \ldots, n$, we may define $\omega(\pi^{(a)}) = (F(\pi^{(a)}) - \beta(\pi^{(a)}))/(\exp(\gamma t(\pi^{(a)})) - 1)$ and $\omega(\pi^{(b)}) = (F(\pi^{(b)}) - \beta(\pi^{(b)}))/(\exp(\gamma t(\pi^{(b)})) - 1)$. Hence, for exponential cost functions (with the same coefficient at the exponent), function (1.1) is priority-generating over set $\hat{\mathcal{P}}$ with the priority function

$$\omega(\pi) = \left[F(\pi) - \sum_{i \in \{\pi\}} \beta_i\right] \Big/ \left[\exp\left[\gamma \sum_{i \in \{\pi\}} t_i\right] - 1\right]. \tag{1.5}$$

(c) Let $\varphi_i(t) = \varphi(t)$, $i = 1, 2, \ldots, n$, where $\varphi(t)$ is a non-decreasing function for $t \ge 0$. We show that in this case, function (1.2) is not, generally speaking, priority-generating over either $\hat{\mathcal{P}}$ or $\hat{\mathcal{P}}_n$. In fact, let $\varphi(t) = t^2/3$, $N = \{1, 2, 3, 4, 5\}$, $t_1 = 10$,

$t_2 = 1$, $t_3 = 2$, $t_4 = 7$, $t_5 = 5$. Assume that $\pi^{(a)} = (3, 4)$, $\pi^{(b)} = (5)$. If the priority function $\omega(\pi)$ existed, then the relation of the form $F(\pi^{(1)}, \pi^{(a)}, \pi^{(b)}, \pi^{(2)}) \le F(\pi^{(1)}, \pi^{(b)}, \pi^{(a)}, \pi^{(2)})$ would hold irrespective of what are chosen as permutations $\pi^{(1)}$, $\pi^{(2)}$. However, in the situation under consideration, we have $F(1, 3, 4, 5, 2) = 602 < F(1, 5, 3, 4, 2) = 605$, $F(2, 3, 4, 5, 1) = 320 > F(2, 5, 3, 4, 1) = 317$.

1.2. The jobs of a set $N = \{1, 2,..., n\}$ are processed on a single machine starting at time $d = 0$. Preemption is not allowed, and at most one job is processed at a time. The processing time of a job i depends on its starting time $\overset{0}{t_i}$ and is equal to $t_i = \alpha_i \overset{0}{t_i} + \beta_i$, $\alpha_i > 0$, $\beta_i > 0$, $i = 1, 2,..., n$. It is required to find a job processing sequence $\pi_n \in P \subseteq \hat{P}_n$ which minimizes the total flow time.

Let the function

$$F(\pi) = \sum_{k=1}^{r} t_{i_k}, \ \pi = (i_1, i_2,..., i_r) \in \hat{P}, \ F(\pi_0) = 0 \tag{1.6}$$

be defined over set $\hat{P}$ of permutations of the elements of set N.

It is clear that $F(\pi_n)$ is the total flow time of the jobs of set N processed according to the sequence π_n. We show that this function is priority-generating over set $\hat{P}$ with the priority function

$$\omega(\pi) = \Psi(\pi)/F(\pi), \tag{1.7}$$

where

$$\Psi(\pi) = \sum_{k=1}^{r} \alpha_{i_k}(1+\overset{0}{\tau}_{i_k}), \ \overset{0}{\tau}_{i_1} = 0, \ \overset{0}{\tau}_{i_k} = \sum_{j=1}^{k-1} \alpha_{i_j}(1+\overset{0}{\tau}_{i_j}).$$

Let $\Phi(\pi, C)$ denote the total flow time of the jobs of the set $\{\pi\}$ processed according to the sequence π, provided that the processing of the first job starts at time C, i.e., $\overset{0}{t}_{i_1} = C$. It is easy to verify that

$$\Phi(\pi,C) = F(\pi)+C\Psi(\pi). \tag{1.8}$$

The relation $F(\pi^{(1)}, \pi^{(a)}, \pi^{(b)}, \pi^{(2)}) \le F(\pi^{(1)}, \pi^{(b)}, \pi^{(a)}, \pi^{(2)})$ holds if and only if $\Phi(\pi^{(a)}, F(\pi^{(1)}))+\Phi(\pi^{(b)}, F(\pi^{(1)}, \pi^{(a)}))+\Phi(\pi^{(2)}, F(\pi^{(1)}, \pi^{(a)}, \pi^{(b)})) \le \Phi(\pi^{(t)}, F(\pi^{(1)}))+ \Phi(\pi^{(a)}, F(\pi^{(1)}, \pi^{(b)}))+\Phi(\pi^{(2)}, F(\pi^{(1)}, \pi^{(b)}, \pi^{(a)}))$. Due to relation (1.8), the later inequality is equivalent to $(1+\Psi(\pi^{(2)}))(\Psi(\pi^{(a)})F(\pi^{(b)})-\Psi(\pi^{(b)})F(\pi^{(a)})) \ge 0$. Since $\Psi(\pi^{(2)}) > 0$, it follows that (1.7) is the desired priority function.

1.3. We now consider the problem of minimizing a linear form over a set of permutations. Two vectors $(\alpha_1, \alpha_2, \dots, \alpha_n)$ and $(\beta_1, \beta_2, \dots, \beta_n)$ with real components are given. It is required to find such a permutation $\pi_n = (i_1, i_2, \dots, i_n) \in \mathcal{P} \subseteq \hat{\mathcal{P}}_n$ which minimizes the function

$$F(\pi) = \sum_{k=1}^{r} \alpha_k \beta_{i_k} \tag{1.9}$$

for $r = n$. Here $\pi = (i_1, i_2, \dots, i_r)$, $F(\pi_0) = 0$.

For some sets $\mathcal{P} \subseteq \hat{\mathcal{P}}_n$, fast algorithms for finding an optimal permutation are known (e.g., when either $\mathcal{P} = \hat{\mathcal{P}}_n$ or $\mathcal{P}$ is the set of even (or odd) permutations). We show that, in a general case, function (1.9) is not priority-generating over $\hat{\mathcal{P}}_n$. Let $\pi^{(1)}$ be a permutation of length ν, $\pi^{(a)} = (i_1, i_2, \dots, i_q)$, $\pi^{(b)} = (j_1, j_2, \dots, j_s)$. The relation $F(\pi^{(1)}, \pi^{(a)}, \pi^{(b)}, \pi^{(2)}) \le F(\pi^{(1)}, \pi^{(b)}, \pi^{(a)}, \pi^{(2)})$ holds if and only if

$$\sum_{k=1}^{s} (\alpha_{\nu+q+k} - \alpha_{\nu+k})\beta_{j_k} \le \sum_{k=1}^{q} (\alpha_{\nu+s+k} - \alpha_{\nu+k})\beta_{j_k}. \tag{1.10}$$

It is obvious that inequality (1.10) depends on the length ν of permutation $\pi^{(1)}$, which contradicts the definition of a priority-generating function.

We consider two special cases, in which function (1.9) can be proved to be priority-generating.

(a) Let $\alpha_i = \alpha_1 + (i-1)h$, $i = 1, 2, \dots, n$. In this case, relation (1.10) reads $qh \sum_{i\in\{\pi^{(b)}\}} \beta_i \le sh \sum_{i\in\{\pi^{(a)}\}} \beta_i$. Hence, the priority function exists and is of the form

$$\omega(\pi) = \frac{h}{r} \sum_{i\in\{\pi\}} \beta_i, \tag{1.11}$$

where r is the length of a permutation π. Thus, in this case, function (1.9) is priority-generating over $\hat{\mathcal{P}}$.

(b) Let $\alpha_i = \alpha_i h^{i-1}$, $i = 1, 2, \dots, n$, $h > 0$. In this case, function (1.9) is also priority-generating over $\hat{\mathcal{P}}$. In fact, relation (1.10) reads $\alpha_1(h^q - 1)\sum_{k=1}^{s} \beta_{j_k} h^{k-1} \le \alpha_1(h^s - 1)\sum_{k=1}^{q} \beta_{j_k} h^{k-1}$ and, hence, the priority function exists and is of the form

$$\omega(\pi) = \frac{\alpha_1}{h^r - 1} \sum_{k=1}^{r} \beta_{i_k} h^{k-1}, \quad \pi = (i_1, i_2, \dots, i_r). \tag{1.12}$$

1.4. The function introduced below plays an important role in solving a number of optimal sequencing problems, some of which are presented in Sections 1.5 and 1.6.

Suppose that each element i of set N is associated with two real numbers α_i and β_i. Let the function

$$F(\pi) = \max\left\{ \sum_{k=1}^{u} \alpha_{i_k} + \beta_{i_u} \middle| 1 \leq u \leq n \right\}, \ \pi = (i_1, i_2, \ldots, i_r), \ F(\pi_0) = 0 \tag{1.13}$$

be defined over set $\hat{\mathcal{P}}$.

We show that function (1.13) is priority-generating over $\hat{\mathcal{P}}$. For any permutations π', π'' belonging to $\hat{\mathcal{P}}$ and such that $\{\pi'\} \cap \{\pi''\} = \emptyset$, we have

$$F(\pi', \pi'') = \max\{F(\pi'), \alpha(\pi') + F(\pi'')\}. \tag{1.14}$$

Let us establish the conditions under which the inequality

$$F(\pi^{(1)}, \pi^{(a)}, \pi^{(b)}, \pi^{(2)}) \leq F(\pi^{(1)}, \pi^{(b)}, \pi^{(a)}, \pi^{(2)}) \tag{1.15}$$

holds. Observe that, due to (1.14) the later inequality is equivalent to $\max\{F(\pi^{(1)}, \alpha(\pi^{(1)}) + F(\pi^{(a)}, \pi^{(b)}, \pi^{(2)})\} \leq \max\{F(\pi^{(1)}, \alpha(\pi^{(1)}) + F(\pi^{(b)}, \pi^{(a)}, \pi^{(2)})\}$.

This inequality holds if $F(\pi^{(a)}, \pi^{(b)}, \pi^{(2)}) \leq F(\pi^{(b)}, \pi^{(a)}, \pi^{(2)})$. Similarly, it can be shown that the latter inequality holds if $F(\pi^{(a)}, \pi^{(b)}) \leq F(\pi^{(b)}, \pi^{(a)})$ or, due to (1.14), if

$$\max\{F(\pi^{(a)}, \alpha(\pi^{(a)}) + F(\pi^{(b)})\} \leq \max\{F(\pi^{(b)}, \alpha(\pi^{(b)}) + F(\pi^{(a)})\}.$$

Subtracting $F(\pi^{(a)}) + F(\pi^{(b)})$ from both sides of this inequality yields

$$\min\{F(\pi^{(a)}, F(\pi^{(b)}) - \alpha(\pi^{(b)})\} \leq \min\{F(\pi^{(b)}, F(\pi^{(a)}) - \alpha(\pi^{(a)})\}. \tag{1.16}$$

To find a priority function, we need to prove the following auxiliary statement. Let x, y, w, z, W be real numbers, $W > \max\{|x|, |y|, |w|, |z|\}$. Then the inequality

$$\min\{x, y\} \leq \min\{w, z\} \tag{1.17}$$

holds if the following

$$\operatorname{sgn}(z - x)[W - \min\{x, z\}] \geq \operatorname{sgn}(y - w)[W - \min\{y, w\}] \tag{1.18}$$

is true.

In fact, three cases are possible: (1) $z - x > 0$, $y - w > 0$; (2) $z - x \geq 0$, $y - w \leq 0$; (3) $z - x < 0$, $y - w < 0$.

In case (1), inequality (1.18) reduces to the inequality $x \leq w$. Thus, $x \leq w$ and $x < z$, hence, inequality (1.17) holds satisfied for any y. In case (2), we have $x \leq z$ and $y \leq w$. If $x \leq y$, then $x \leq \min\{w, z\}$ and inequality (1.17) holds. If $x > y$, then $y \leq \min\{w, z\}$, and inequality (1.17) also holds. In case (3), inequality (1.18) reduces to $y \leq z$, hence, $y \leq \min\{w, z\}$, and inequality (1.17) holds. Thus, for inequality (1.16) and, hence, for inequality (1.15) to be true, it is sufficient (but not necessary) that

$$\operatorname{sgn}(-\alpha(\pi^{(a)}))[W - \min\{F(\pi^{(a)}), F(\pi^{(a)}) - \alpha(\pi^{(a)})\}] \geq$$

$$\geq \operatorname{sgn}(-\alpha(\pi^{(b)}))[W - \min\{F(\pi^{(b)}), F(\pi^{(b)}) - \alpha(\pi^{(b)})\}],$$

where $W \geq \sum_{i=1}^{n} (|\alpha_i| + |\beta_i|)$.

Hence, function (1.13) is priority-generating over set $\hat{\mathcal{P}}$, and its priority function is of the form

$$\omega(\pi) = \operatorname{sgn}\left[-\sum_{i\in\{\pi\}} \alpha_i\right]\left[W - F(\pi) + \max\left\{0, \sum_{i\in\{\pi\}} \alpha_i\right\}\right]. \tag{1.19}$$

1.5. At time $d = 0$, a number of requests enter a system which provides information recording, storage, and output. Processing a request $i \in N$ implies either recording t_i information units in storage (if $t_i > 0$) or extracting t_i information units from storage (if $t_i < 0$). It is required to choose such a sequence π_n of some set $\hat{\mathcal{P}} \subseteq \hat{\mathcal{P}}_n$ of feasible sequences which minimizes the maximum information storage volume.

Let the function

$$F(\pi) = \max\left\{\sum_{k=1}^{u} t_{i_k} \,\middle|\, 1 \leq u \leq r\right\} \tag{1.20}$$

be defined over set $\hat{\mathcal{P}}$, where $\pi = (i_1, i_2, \dots, i_r)$, $F(\pi_0) = 0$. For $r = n$, the quantity $F(\pi)+C$ is equal to the maximum volume of information to be kept in storage at a time, provided that the requests are processed according to the sequence π. Here C is the storage volume at time $d = 0$.

Since function (1.20) is a special case of function (1.13) with $\alpha_i = t_i$ and $\beta_i = 0$, it is priority-generating over $\hat{\mathcal{P}}$ with the priority function

$$\omega(\pi) = \operatorname{sgn}\left[-\sum_{i\in\{\pi\}} t_i\right]\left[W - F(\pi) + \max\left\{0, \sum_{i\in\{\pi\}} t_i\right\}\right] \tag{1.21}$$

where $W \geq \sum_{i=1}^{n} |t_i|$.

1.6. The jobs of a set $N = \{1, 2, \dots, n\}$ enter a two-machine processing system. A job $i \in N$ enters the system at time $d_i \geq 0$, and is processed on the first machine during $t_{1i} > 0$ time units and then on the second machine during $t_{2i} > 0$ time units. Each machine processes the jobs according to the same sequence with no preemption and at most one job at a time. The processing a job i on the second machine may start no earlier than time $t_i^0 + \delta_i$. Here t_i^0 denotes the starting time of job i on the first machine, $\delta_i \geq 0$. If $\delta_i \geq t_{1i}$, then job i cannot be processed simultaneously on both machines. If $\delta_i < t_{1i}$, the simultaneous processing of job i on both machines is allowed. If $\delta_i > t_{1i}$, then at least $\delta_i - t_{1i}$ time units must pass after the processing of job i on the first machine is completed until this job can start on the second machine.

The release dates d_i are assumed to satisfy the conditions

$$d_i \le d_l + t_{1l},\ i = 1, 2,\dots, n,\ l = 1, 2,\dots, n,\ i \ne l, \tag{1.22}$$

while the values of δ_i satisfy the conditions

$$\delta_i \ge t_{1i} - t_{2i},\ i = 1, 2,\dots, n. \tag{1.23}$$

It follows from (1.23) that the makespan is determined by the completion times of all jobs on the second machine.

For $N' \subseteq N$, $|N'| = r$, let $F(\pi)$ denote the smallest value of the makespan of the jobs of set N', provided that these jobs are processed according to the sequence $\pi = (i_1, i_2,\dots, i_r)$. It is required to find a permutation π_n of some given set $\mathcal{P} \subseteq \hat{\mathcal{P}}_n$ which minimizes the function $F(\pi_n)$.

If the jobs enter the system simultaneously (i.e., $d_i = 0$, $i = 1, 2,\dots, n$), then it is easy to verify by induction with respect to r that

$$F(\pi) = \sum_{k=1}^{r} t_{2i_k} + \max\left\{ \sum_{k=1}^{u} (t_{1i_k} - t_{2i_k}) + \delta_{i_u} + t_{2i_u} - t_{1i_u} \,\middle|\, 1 \le u \le r \right\}. \tag{1.24}$$

Assuming $\alpha_i = t_{1i} - t_{2i}$, $\beta_i = \delta_i + t_{2i} - t_{1i}$, we derive that the value of $F(\pi)$ differs from that of function (1.13) only by the constant $\sum_{k=1}^{r} t_{2i_k}$. Hence, in the case $d_i = 0$, $i = 1, 2,\dots, n$, function $F(\pi)$ is priority-generating over $\hat{\mathcal{P}}$, and its priority function is of the form

$$\omega(\pi) = \operatorname{sgn}\left[\sum_{i \in \{\pi\}} (t_{2i} - t_{1i}) \right] \left[W - F(\pi) + \max\left\{ \sum_{i \in \{\pi\}} t_{1i}, \sum_{i \in \{\pi\}} t_{2i} \right\} \right], \tag{1.25}$$

where $W \ge \sum_{i=1}^{n} (t_{1i} + t_{2i} + \delta_i)$.

If the jobs do not enter the system simultaneously, then due to condition (1.22), we have $F(\pi) = d_{i_1} + A$, where A is the right-hand side of relation (1.24). In this case, function $F(\pi)$ is not, in general, priority-generating over $\hat{\mathcal{P}}$. In fact, let $N = \{1, 2, 3, 4\}$, $t_{11} = 6$, $t_{12} = 4$, $t_{13} = 1$, $t_{14} = 6$, $t_{21} = 4$, $t_{22} = t_{23} = 2$, $t_{24} = 5$, $\delta_1 = 6$, $\delta_2 = 4$, $\delta_3 = 1$, $\delta_4 = 6$, $d_1 = 2$, $d_2 = d_3 = d_4 = 1$. It follows that $F(1, 2, 3) = 16 < F(2, 1, 3) = 17$, but $F(1, 2, 4) = 23 > F(2, 1, 4) = 22$.

Divide set $\hat{\mathcal{P}}$ into n pairwise disjoint non-empty subsets $\hat{\mathcal{P}}(j)$, $j = 1, 2,\dots, n$. Here $\hat{\mathcal{P}}(j)$ is the set of those and only those permutations $\pi = (i_1, i_2,\dots, i_r)$ of $\hat{\mathcal{P}}$, in which $i_1 = j$. Since for all permutations $\pi \in \hat{\mathcal{P}}(j)$, we have $d_{i_1} = d_j$, if follows that function $F(\pi)$ is priority-generating over set $\hat{\mathcal{P}}(j)$. Thus, if the jobs do not enter the system simultaneously, function $F(\pi)$ is priority-generating over each subset $\hat{\mathcal{P}}(j)$ of set $\hat{\mathcal{P}}$, but is not priority-generating over $\hat{\mathcal{P}}$.

1.7. Let $G = (N, U)$ be a directed circuit-free graph. The number $t_i > 0$ corresponds to each vertex $i \in N$, and the number w_{ij} is associated with each arc $(i, j) \in U$. The vertices of the graph are located on the interval $[0, \sum_{i \in N} t_i]$ in the following way. Each vertex i occupies the interval whose length is equal to t_i, and the intervals corresponding to different vertices must not intersect. Given such an allocation, the length of an arc $(i, j) \in U$ is determined by the coordinate difference $(x_j - x_i)$ of the right ends of the intervals corresponding to the vertices j and i. The length of arc an (i, j) may appear to be negative if the vertex j is located on the left of the vertex i. The vertices of graph G have to be allocated in such a way that the length of each arc $(i, j) \in U$ is positive and the total "weighted" length $\sum_{(i,j) \in U} w_{ij}(x_j - x_i)$ is minimal.

It is obvious that the required allocation of the vertices of graph G is specified by a permutation $\pi = (i_1, i_2, \ldots, i_n)$ of the elements of set N.

We show that the function

$$F(\pi_n) = \sum_{(i,j) \in U} w_{ij}(x_j - x_i) \tag{1.26}$$

is priority-generating over the set $\hat{\mathcal{P}}_n$.

Define $w_{ij} = 0$ for all $(i, j) \notin U$. Then $F(\pi_n) = \sum_{i \in N} \sum_{j \in N} w_{ij}(x_j - x_i) = \sum_{i \in N} \sum_{j \in N} w_{ij} x_j - \sum_{i \in N} \sum_{j \in N} w_{ji} x_j = \sum_{j \in N} \big(\sum_{i \in N} (w_{ij} - w_{ji}) \big) x_j$. Thus, we have

$$F(\pi_n) = \sum_{k=1}^{n} \Big[\sum_{l \in N} (w_{l i_k} - w_{i_k l}) \Big] x_{i_k}. \tag{1.27}$$

Since $x_{i_k} = \sum_{p=1}^{k} t_{i_p}$, defining $\alpha_i = \sum_{l \in N} (w_{li} - w_{il})$ implies that function (1.26) coincides with function (1.1) with $\varphi_i(t) = \alpha_i t$, $i = 1, 2, \ldots, n$. Hence, function (1.26) is priority-generating over $\hat{\mathcal{P}}_n$, and its priority function is of the form

$$\omega(\pi) = \sum_{i \in \{\pi\}} \sum_{l \in N} (w_{li} - w_{il}) / \sum_{i \in \{\pi\}} t_i. \tag{1.28}$$

1.8. This section gives an example of a function which is not priority-generating over either the set $\hat{\mathcal{P}}$ or the set $\hat{\mathcal{P}}_n$, although, it appears to be priority-generating on some special subset $\mathcal{P}_n$ of the set $\hat{\mathcal{P}}_n$.

The jobs of a set $N = \{1, 2, \ldots, n\}$ starting at $d = 0$ are processed on a single machine. The real number w_{ij} corresponds to each ordered pair (i, j) of jobs. The processing time of a job i is equal to t_i. The jobs are processed without preemption and at most one job at a time.

Let the function

$$F(\pi) = \sum_{1 \le l < k \le n} w_{i_l i_k}(\bar{t}_{i_k} - \bar{t}_{i_l}) \tag{1.29}$$

be defined over the set $\hat{\mathcal{P}}_n$ where $\pi = (i_1, i_2, \ldots, i_n) \in \hat{\mathcal{P}}_n$, $\bar{t}_{i_p} = \sum_{s=1}^{p} t_{i_s}$.

Minimizing function (1.29) over the set $\hat{\mathcal{P}}_n$ or over some subset $\mathcal{P}_n \subseteq \hat{\mathcal{P}}_n$ reflects a desire to obtain a certain grouping of jobs and to reduce (or increase) the length of time intervals between the processing of individual jobs. Usually, a set of permutations of $\hat{\mathcal{P}}_n$ that are feasible with respect to a given precedence relation over N is chosen as $\mathcal{P}_n$.

We show that, in a general case, function (1.29) is not priority-generating over $\hat{\mathcal{P}}_n$. Let $N = \{1, 2, 3, 4\}$, $t_i = 1$, $i = 1, 2, 3, 4$, $w_{12} = 10$, $w_{32} = 3$, $w_{34} = 1$, and all remaining numbers w_{ij} are equal to zero. Assume that $\pi^{(a)} = (1)$, $\pi^{(b)} = (3)$, then $F(2, \pi^{(a)}, \pi^{(b)}, 4) = 1 < F(2, \pi^{(b)}, \pi^{(a)}, 4) = 2$, $F(\pi^{(a)}, \pi^{(b)}, 2, 4) = 25 > F(\tau^{(b)}, \pi^{(a)}, 2, 4) = 19$. Hence, in the case under consideration, a priority function $\omega(\pi)$ does not exist.

Let the numbers w_{ij} satisfy the following condition: there exists a directed circuit-free graph $G = (N, U)$ such that $w_{ij} \neq 0$ implies $(i, j) \in U$. In this case, denote function (1.29) by $F_G(\pi)$. It is easy to note that in the above example the numbers w_{ij} satisfy that condition. Hence, the function $F_G(\pi)$ is also not priority-generating over $\hat{\mathcal{P}}_n$.

Let $\mathcal{P}_n(G)$ denote the set of all permutations that are feasible with respect to the precedence relation defined over N and given by a graph G. Functions $F_G(\pi)$ and (1.26) coincide over $\mathcal{P}_n(G)$. Since function (1.26) is priority-generating over $\hat{\mathcal{P}}_n$, it is also priority-generating over $\mathcal{P}_n(G)$. Hence, it follows that the function $F_G(\pi)$ is also priority-generating over set $\mathcal{P}_n(G)$.

1.9. To conclude this section, note that in minimizing priority-generating functions the values of $\omega(\pi)$ often have to be calculated, provided that the values of $\omega(\pi^{(1)})$ and $\omega(\pi^{(2)})$ have been calculated and that the permutation π is of the form: $\pi = (\pi^{(1)}, \pi^{(2)})$. In this case, it is possible to reduce the volume of computations essentially by using information obtained while computing $\omega(\pi^{(1)})$ and $\omega(\pi^{(2)})$. We illustrate this by considering several of the examples given above. Let $\pi^{(1)}, \pi^{(2)} \in \hat{\mathcal{P}}$, $\{\pi^{(1)}\} \cap \{\pi^{(2)}\} = \emptyset$ and $\pi = (\pi^{(1)}, \pi^{(2)})$.

(a) For priority function (1.4) we have

$$\omega(\pi^{(1)}, \pi^{(2)}) = (\alpha(\pi^{(1)}) + \alpha(\pi^{(2)}))/(t(\pi^{(1)}) + t(\pi^{(2)})). \tag{1.30}$$

(b) For exponential penalty functions under the conditions of Section 1.1, (b) we have $F(\pi^{(1)}, \pi^{(2)}) = F(\pi^{(1)}) + \exp(\gamma t(\pi^{(1)}))(F(\pi^{(2)}) - \beta(\pi^{(2)})) + \beta(\pi^{(2)})$. Hence,

$$\omega(\pi^{(1)}, \pi^{(2)}) = \frac{F(\pi^{(1)} - \beta(\pi^{(1)}) + \exp(\gamma t(\pi^{(1)}))(F(\pi^{(2)}) - \beta(\pi^{(2)})}{\exp(\gamma(t(\pi^{(1)}) + t(\pi^{(2)}))) - 1}. \tag{1.31}$$

(c) For function (1.6), it follows from relation (1.8) that $F(\pi^{(1)}, \pi^{(2)}) = F(\pi^{(1)}) + F(\pi^{(1)})\Psi(\pi^{(2)}) + F(\pi^{(2)})$. Similarly, $\Psi(\pi^{(1)}, \pi^{(2)}) = \Psi(\pi^{(1)}) + \Psi(\pi^{(1)})\Psi(\pi^{(2)}) + \Psi(\pi^{(2)})$. Hence, we obtain

$$\omega(\pi^{(1)}, \pi^{(2)}) = \frac{\Psi(\pi^{(1)}) + \Psi(\pi^{(1)})\Psi(\pi^{(2)}) + \Psi(\pi^{(2)})}{F(\pi^{(1)}) + F(\pi^{(1)})\Psi(\pi^{(2)}) + F(\pi^{(2)})}. \tag{1.32}$$

(d) For function (1.19), due to relation (1.14), we obtain

$$\begin{aligned}\omega(\pi^{(1)}, \pi^{(2)}) &= \operatorname{sgn}(-\alpha(\pi^{(1)}) - \alpha(\pi^{(2)}))[W - \max\{F(\pi^{(1)}),\\ &\alpha(\pi^{(1)}) + F(\pi^{(2)})\} + \max\{0, \alpha(\pi^{(1)}) + \alpha(\pi^{(2)})\}]\end{aligned} \tag{1.33}$$

where $W \geq \sum_{i \in N}(|\alpha_i| + |\beta_i|)$.

A similar expression for $\omega(\pi^{(1)}, \pi^{(2)})$ can also be obtained in the remaining cases. In any case, computing $\omega(\pi^{(1)}, \pi^{(2)})$ using information obtained while calculating $\omega(\pi^{(1)})$ and $\omega(\pi^{(2)})$ involves performing a certain number of operations independent of the length of the permutations $\pi^{(1)}$ and $\pi^{(2)}$. Thus, for the priority function determined by relation (1.7), calculation of $\omega(\pi^{(1)})$ and $\omega(\pi^{(2)})$ determines the values of $\Psi(\pi^{(1)})$, $F(\pi^{(1)})$, $\Psi(\pi^{(2)})$, and $F(\pi^{(2)})$. The use of relation (1.31) allows us to obtain the value of $\omega(\pi^{(1)}, \pi^{(2)})$ by performing just seven arithmetic operations.

2. Elimination Conditions

2.1. Let a precedence relation $\longrightarrow$ be defined over set $N = \{1, 2, \ldots, n\}$ and $G = (N, U)$ be the reduction graph of this relation. A permutation $\pi = (i_1, i_2, \ldots, i_r) \in \hat{\mathcal{P}}$ is called feasible (with respect to $\longrightarrow$, or, equivalently, with respect to G), if the condition $i_k \longrightarrow i_l$ implies $k < l$. Let $\mathcal{P}(G)$ denote the set of all feasible permutations, and $\mathcal{P}_n(G)$ denote the set of all feasible permutations of the length n.

In the following, attention is paid to the developing of methods to optimize priority-generating functions over a set of feasible permutations under various assumptions on the precedence relation structure (i.e., on the form of graph G).

We introduce the following operations on directed circuit-free graphs $\Gamma = (X, Y)$ with no transitive arcs.

The operation of identifying vertices x and y of a graph $\Gamma = (X, Y)$ such that $(x, y) \in Y$ involves replacing these two vertices by a single vertex followed by removing the arc (x, y). In this case, all the arcs that enter or leave either x or y are replaced by those that either enter or leave the new vertex, respectively. All transitive arcs are removed from the obtained graph.

Suppose that a graph $\Gamma = (X, Y)$ contains neither path from x to y nor from y to x. *The operation of including an arc* (x, y) involves substituting the graph Γ for the graph obtained from $\Gamma' = (X, Y \cup (x, y))$ by removing all its transitive arcs.

In the following, these operations are to be successively and repeatedly applied to graph $G = (N, U)$. While identifying two vertices $i, j \in N$ connected by the arc $(j, i) \in U$, the permutation $(j, i) \in \hat{\mathcal{P}}$ is associated with the new vertex. Let $G' = (N', U')$ be a graph obtained from G as a result of multiple applications of the operations of identifying vertices and including arcs. If a permutation $\pi^{(i')}$ corresponds to vertex i' of this graph and a permutation $\pi^{(j')}$ is associated with vertex j', and $(j', i') \in U'$, then after having identified the vertices j' and i', the permutation $\pi = (\pi^{(j')}, \pi^{(i')})$ corresponds to the new vertex.

A permutation $\pi = (i_1, i_2, \ldots, i_r) \in \hat{\mathcal{P}}$, corresponding to a vertex obtained as a result of identifying vertices is called *a composite element* and is denoted by $\pi = [i_1, i_2, \ldots, i_r]$. We do not distinguish between the vertices and the corresponding elements.

It is easy to verify that graph $G' = (N', U')$ defines a strict order over set N' of composite elements. To denote this order, the notation $\xrightarrow{G'}$ is used, and if none of the relations $i \xrightarrow{G'} j$ and $j \xrightarrow{G'} i$ holds for $i, j \in N'$, then $i \overset{G'}{\sim} j$ is used. A permutation of the elements of set N' that is feasible with respect to G' is at the same time a permutation of the elements of set N that is feasible with respect to G.

Since a composite element is a permutation of set $\hat{\mathcal{P}}$, it is possible to define *priorities of composite elements*. In what follows, we write $\omega(i_1, i_2, \ldots, i_r)$ rather than $\omega[i_1, i_2, \ldots, i_r]$.

2.2. *Example.* Consider the graph G shown in Fig. 2.1a. Using the operation of identifying vertices 3 and 4 yields the graph G_1' (see Fig. 2.1b). The dashed line in this figure shows the transitive arc that has been removed. The graph G_2' in Fig.2.1c is obtained from the graph G_1' by including the arc (5, 7). Figures 2.1d and 2.1e show graphs G_3' and G_4' obtained from G_2' by identifying the vertices 5 and 7 (the graph G_3'), followed by identifying the vertices [3, 4] and [5, 7] (the graph G_4'). The elements [3, 4], [5, 7] and [3, 4, 5, 7] are composite. The permutations (1, 2, [3, 4], [5, 7], 6) = (1, 2, 3, 4, 5,

7, 6), (1, [3, 4], 2, [5, 7], 6) = (1, 3, 4, 2, 5, 7, 6), (1, [3, 4], [5, 7], 2, 6) = (1, 3, 4, 5, 7, 2, 6) form the set of all permutations of length $n = 7$ that are feasible with respect to the graph G_3', while the set $\mathcal{P}_7(G_4')$ consists of two permutations (1, 2, [3, 4, 5, 7], 6) = (1, 2, 3, 4, 5, 7, 6) and (1, [3, 4, 5, 7], 2, 6) = (1, 3, 4, 5, 7, 2, 6).

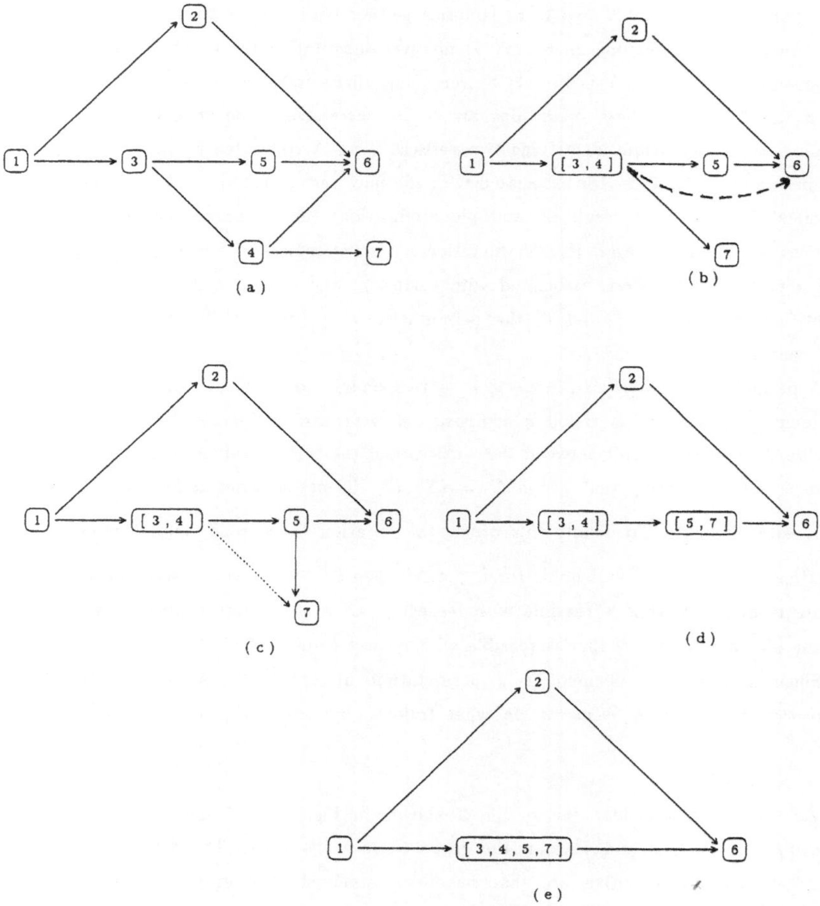

Fig. 2.1

2.3. While solving the problems of optimizing the priority-generating functions over the set $\mathcal{P}_n(G)$, the operations of identifying the vertices of graph G and including arcs to

G reduce the search considerably. We now consider the conditions, under which the above operations guarantee that a set $\mathcal{P}^0 \subseteq \mathcal{P}_n(G)$ containing at least one optimal permutation can be found.

First, we prove the following widely used statement.

Lemma 2.1. *Let a function $F(\pi)$ be priority-generating over a set $\mathcal{P}$, and permutations $\pi = (\pi^{(1)}, \pi^{(a)}, \pi^{(2)}, \pi^{(b)}, \pi^{(3)})$, $\pi' = (\pi^{(1)}, \pi^{(2)}, \pi^{(a)}, \pi^{(b)}, \pi^{(3)})$ and $\pi'' = (\pi^{(1)}, \pi^{(a)}, \pi^{(b)}, \pi^{(2)}, \pi^{(3)})$ belong to $\mathcal{P}$. If $\omega(\pi^{(a)}) \le \omega(\pi^{(b)})$, then either $F(\pi') \le F(\pi)$ or $F(\pi'') \le F(\pi)$.*

Proof. Two cases are possible: $\omega(\pi^{(2)}) \ge \omega(\pi^{(a)})$ and $\omega(\pi^{(2)}) < \omega(\pi^{(a)})$. The definition of a priority-generating function implies that, in the first case, $F(\pi') \le F(\pi)$, and in the second $F(\pi'') \le F(\pi)$.

As above, the sets of those and only those elements $j \in N$, for which $i \xrightarrow{G} j$, $j \xrightarrow{G} i$ and $i \overset{G}{\sim} j$, are denoted by $A_G(i)$, $B_G(i)$, and $E_G(i)$, respectively. Similarly, $A_G^0(i)$ and $B_G^0(i)$ denote the sets of those and only those elements $j \in N$ for which $i \overset{G}{\rightarrowtail} j$ and $j \overset{G}{\rightarrowtail} i$, respectively. Given $s, t \in N$, denote $\overline{B}_G(s, t) = B_G(s) \backslash (B_G(t) \cup t)$ and $\overline{A}_G(s, t) = A_G(t) \backslash (A_G(s) \cup s)$.

Some of the above notations are shown schematically in Fig. 2.2 (the index G is omitted): (a) $B^0(s) = t$; (b) $s \sim t$, and (c) $A^0(t) = s$.

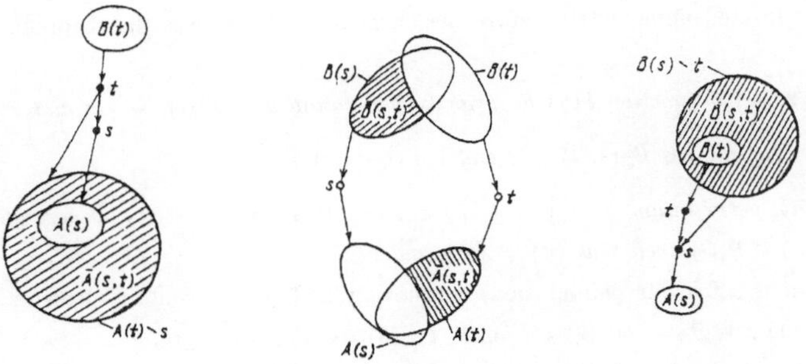

(a) (b) (c)

Fig. 2.2

In the following, the functions $F(\pi)$ are priority-generating over set $\mathcal{P}(G)$.

Theorem 2.1. *Let a function $F(\pi)$ be priority-generating over $\mathcal{P}(G)$, $B_G^0(s) = t$ and*

$$\omega(s) \geq \omega(i) \text{ for all } i \in \overline{A}_G(s, t) \cup t. \tag{2.1}$$

Then, for any permutation $\pi = (..., t, \tilde{\pi}, s,...) \in \mathcal{P}(G)$, there exists a permutation $\pi^0 = (..., t, s,...) \in \mathcal{P}(G)$ such that $\{\pi\} = \{\pi^0\}$ and $F(\pi^0) \leq F(\pi)$.

Proof. Since π is a feasible permutation, it follows that $\{\tilde{\pi}\}$ may contain elements of just two types: (1) $i \in \overline{A}_G(s, t)$, and (2) $l \in E_G(t)$. Let i' be a type 1 element of $\{\tilde{\pi}\}$ nearest to s in π, and between i' and s in π there is a permutation of the type 2 elements denoted by $\pi^{(l)}$. For all elements $l \in \{\pi^{(l)}\}$, the relation $l \sim i'$ holds. In fact, $l \longrightarrow i'$ is impossible due to the feasibility of permutation π, while $t \longrightarrow l$ would follow from $i' \longrightarrow l$ due to the transitivity of relation $\longrightarrow$, which is impossible since $l \in E_G(t)$. It follows from (2.1) that $\omega(s) \geq \omega(i')$. Therefore, due to Lemma 2.1, permutation π can be transformed into a feasible permutation $\pi^{(1)}$, by interchanging either $\pi^{(l)}$ and s or i' and $\pi^{(l)}$. Note that, in $\pi^{(1)}$, the element i' is placed immediately before the element s and $F(\pi') \leq F(\pi)$. Since $s \sim i'$ and $\omega(s) \geq \omega(i')$, permutation $\pi^{(1)}$ can be transformed into a feasible permutation $\pi^{(2)}$ such that $F(\pi^{(2)}) \leq F(\pi^{(1)})$ by interchanging i' and s.

Having applied the above procedure sufficiently many times, we can transform $\pi^{(2)}$ into a feasible permutation $\pi^{(3)}$ such that $F(\pi^{(3)}) \leq F(\pi^{(2)})$ and between t and s in $\pi^{(3)}$ there are no other elements besides, possibly, type 2 elements. In $\pi^{(3)}$, let the permutation of elements located between t and s again be denoted by $\pi^{(l)}$. Relation (2.1) implies that $\omega(s) \geq \omega(t)$. Due to Lemma 2.1, $\pi^{(3)}$ can be transformed into a feasible permutation π^0 such that $F(\pi^0) \leq F(\pi^{(3)})$ by interchanging either t and $\pi^{(l)}$ or $\pi^{(l)}$ and s. In permutation π^0, the element t is located immediately before the element s. This proves the theorem.

Theorem 2.2. *Let a function $F(\pi)$ be priority-generating over $\mathcal{P}(G)$, $s \overset{G}{\sim} t$ and*

$$\omega(j) \geq \omega(i) \text{ for all } j \in \overline{B}_G(s, t) \cup s \text{ and } i \in \overline{A}_G(s, t) \cup t. \tag{2.2}$$

Then, for any permutation $\pi = (..., t, \tilde{\pi}, s,...) \in \mathcal{P}(G)$, there exists a permutation $\pi^0 = (..., s, t,...) \in \mathcal{P}(G)$ such that $\{\pi\} = \{\pi^0\}$ and $F(\pi^0) \leq F(\pi)$.

Proof. Since π is a feasible permutation, it follows that $\{\tilde{\pi}\}$ may contain elements only of three types: (1) $j \in \overline{B}_G(s, t)$; (2) $i \in \overline{A}_G(s, t)$; (3) $l \in E_G(s) \cap E_G(t)$.

Let j' be a type 1 element of $\{\pi\}$ nearest to t in π, and i' be a type 2 element located between t and j' in π and be the nearest to j' among all such elements. If, in π, some elements are located between i' and j', then these may be only of type 3. The permutation of the elements between i' and j' is denoted by $\pi^{(l)}$. Since, due to (2.2), we have $\omega(i') \leq \omega(j')$ and for all $l \in \{\pi^{(l)}\}$ relations $l \sim i'$ and $l \sim j'$ hold, it follows that permutation π (due to Lemma 2.1) can be transformed into a feasible permutation $\pi^{(1)}$ in which the element i' is located immediately before the element j' and $F(\pi^{(1)}) \leq F(\pi)$. The definitions of $\overline{B}_G(s, t)$ and $\overline{A}_G(s, t)$ imply that if j is a type 1 element and i is a type 2 element, then either $j \longrightarrow i$ or $j \sim i$. Since π is a feasible permutation, it follows that $i' \sim j'$. Hence, $\pi^{(1)}$ can be transformed into a feasible permutation $\pi^{(2)}$, in which the element j' is located before the element i' and $F(\pi^{(2)}) \leq F(\pi^{(1)})$. Having applied the described procedure sufficiently many times, we obtain a feasible permutation $\pi^{(3)}$ such that $F(\pi^{(3)}) \leq F(\pi^{(2)})$, and there is no type 2 element between t and j'. If there is a sequence of type 3 elements between them, then again denote it by $\pi^{(l)}$. Since $\omega(t) \leq \omega(j')$, by Lemma 2.1 we can obtain a feasible permutation, in which the element t is located immediately before j'. Taking into account that $t \sim j'$, it is then possible to obtain a feasible permutation $\pi^{(4)}$ in which j' is located on the left of t and $F(\pi^{(4)}) \leq F(\pi^{(3)})$. Similarly, in $\pi^{(4)}$, it is possible to exclude all type 1 elements from the permutation between t and s and to obtain a feasible permutation $\pi^{(5)}$, in which between t and s there may exist only elements of types 2 and 3, and $F(\pi^{(5)}) \leq F(\pi^{(4)})$.

Let $\pi^{(5)} = (..., t, \tilde{\pi}', s,...)$ and i'' be a type 2 element of $\{\tilde{\pi}'\}$ nearest to s. Since $\omega(i'') \leq \omega(s)$ and $i'' \sim s$, by the same procedure as in the proof of Theorem 2.1, $\pi^{(5)}$ can be transformed into a feasible permutation $\pi^{(6)}$ such that the element s is located on the left of element i'' and $F(\pi^{(6)}) \leq F(\pi^{(5)})$. Similarly, in $\pi^{(6)}$, it is possible to exclude all type 2 elements from the permutation between the elements s and t, and, by Lemma 2.1, to obtain a permutation $\pi^{(7)}$ such that the element t is located immediately before the element s and $F(\pi^{(7)}) \leq F(\pi^{(6)})$.

It is now clear that $\pi^{(7)}$ can be transformed into the desired permutation π^0. This proves the theorem.

2.4. Theorems 2.1 and 2.2 are formulated regarding the original set N ordered by relation $\longrightarrow$, or, equivalently, regarding the original graph G. It is easy to see that these statements still hold for a set N', a graph G' and a relation $\xrightarrow{G'}$. In the latter case, the elements s and t are, in general, composite ones, i.e., they may be some permutations of $\mathcal{P}(G)$.

In the proofs of Theorems 2.1 and 2.2, function $F(\pi)$ has been assumed to be priority-generating over $\mathcal{P}(G)$. In many scheduling problems, it is required to find a feasible permutation of the length n, delivering an extremum to function $F(\pi)$. In such situations, it suffices to demand that function $F(\pi)$ is priority-generating over $\mathcal{P}_n(G)$, and to assume that in these theorems the permutations π, π^0 belong to $\mathcal{P}_n(G)$.

Theorems 2.1 and 2.2 can be given a simple graph-theoretical interpretation. Satisfying the conditions of these theorems guarantees that using the operation of identifying vertices t and s (Theorem 2.1) or the operation of including arc (s, t) to graph G (Theorem 2.2) enables one to obtain a new graph G^0 with the following properties: $\mathcal{P}_n(G^0) \subseteq \mathcal{P}_n(G)$ and for any permutation $\pi \in \mathcal{P}_n(G)$ there exists a permutation $\pi^0 \in \mathcal{P}_n(G^0)$ such that $F(\pi^0) \leq F(\pi)$.

The operation of identifying vertices t and s satisfying the conditions of Theorem 2.1 is called *a transformation* I (the notation I-$[t, s]$). The operation of including arc (s, t) under the conditions of Theorem 2.2 is called *a transformation* II (the notation II-(s, t)). If, in graph G, there exists a pair of vertices s and t satisfying either the conditions of Theorem 2.1 or those of Theorem 2.2, then we say that *transformation* I *or transformation* II, respectively, *may be applied to graph* G or that transformation (I-$[t, s]$) or II-(s, t)) is *feasible* for graph G.

Corollary 2.1. *If function $F(\pi)$ is priority-generating over $\mathcal{P}_n(G)$ and graph G' is obtained from graph G by performing sequence of transformations I or II, then*

$$\min\{F(\pi) \mid \pi \in \mathcal{P}_n(G')\} = \min\{F(\pi) \mid \pi \in \mathcal{P}_n(G)\}.$$

This directly follows from Theorems 2.1 and 2.2.

If graph G' is obtained from the graph G by performing a sequence L of transformations I and II, then L is said *to transform graph G into graph G'*. If a sequence L_1 transforms graph G into graph G' and a sequence L_2 transforms graph G' into graph G'', then the sequence $L = (L_1, L_2)$ transforms graph G into graph G'' according to the scheme $G \rightarrow G' \rightarrow G''$. Finally, if in each of the transformations of the sequence L only such s and t are involved for which $\{s\} \subset N_1$, $\{t\} \subset N_1$, $N_1 \subseteq N$, then *sequence L* is said *to act on set N_1*.

3. Tree-like Order

Let a function $F(\pi)$ be priority-generating over $\mathcal{P}_n(G)$. This section considers the problem of finding a permutation $\pi_n^* \in \mathcal{P}_n(G)$ which minimizes $F(\pi)$, provided that G is *tree-like*. A permutation π_n^* is called *optimal*.

3.1. Let a graph $G' = (N', U')$ be obtained from graph $G = (N, U)$ by performing a sequence of transformations I or II (see Section 2 of this chapter). If graph G' is a chain, then it obviously specifies the only feasible permutation π_n which is optimal due to Corollary 2.1. For transforming the initial graph G into a chain G', the concept of an ω-chain is of great importance.

Construct a chain $C = (i_1, i_2, \ldots, i_m)$, whose vertices are all or some elements of set N'. Chain C is called an *ω-chain* if the permutation $(i_1, i_2, \ldots, i_m)$ is feasible with respect to G', $\omega(i_k) \geq \omega(i_{k+1})$, $k = 1, 2, \ldots, m-1$, and the equality $\omega(i_k) = \omega(i_{k+1})$ implies $i_k \overset{G'}{\sim} i_{k+1}$. A chain consisting of one vertex is an ω-chain by definition.

Lemma 3.1. *If all connected components of graph G are ω-chains, then there exists a sequence of transformations II converting G into a single ω-chain.*

Proof. Assuming that the statement holds for all graphs G with at most l, $l \geq 1$, connected components, we prove this is also true for a graph with $l+1$ components.

Let $C_1 = (i_1, i_2, \ldots, i_{m_1})$, $C_2 = (j_1, j_2, \ldots, j_{m_2})$ be connected components of graph G and $\omega(i_1) \geq \omega(j_1)$. Since C_1 and C_2 are ω-chains, we have that $\omega(i_1) \geq \omega(j_k)$, $k = 2, 3, \ldots, m_2$. Hence, transformation II-(i_1, j_1) can be applied to G. Note that in the case $\omega(i_1) = \omega(j_1)$, any of the arcs (i_1, j_1) or (j_1, i_1) can be included in G. Then, compare the values of $\omega(i_2)$ and $\omega(j_1)$ (here it is assumed that the arc (i_1, j_1) is included in the previous step). If $\omega(i_2) \geq \omega(j_1)$, then transformation II-(i_2, j_1) can be applied to the graph obtained from G as a result of the previous transformation; if $\omega(i_2) \leq \omega(j_1)$, then transformation II-(j_1, i_2) can be used. As a result of applying at most m_1+m_2-1 such steps, a pair of ω-chains C_1 and C_2 is transformed into the single chain $C' = (i_1', i_2', \ldots, i_{m_1+m_2}')$. By construction, the permutation $(i_1', i_2', \ldots, i_{m_1+m_2}')$ is feasible with respect to G, the vertices in C' are sorted in non-increasing order of the priorities of the corresponding elements, and $\omega(i_k') = \omega(i_{k+1}')$ only if i_k' and i_{k+1}' belong to different initial chains. Thus, C' is an ω-chain, and the graph obtained from G by the described transformations has l connected components. This proves the lemma.

Under the conditions of Lemma 3.1, *in order to find a desired ω-chain, i.e., an optimal permutation π_n^*, it suffices to sort the elements of set N (i.e., the vertices of graph G) in non-increasing order of their priorities.*

Lemma 3.2. *If all connected components of graph G are chains, then there exists a sequence of transformations I converting each chain into an ω-chain.*

Proof. Let $C = (i_1, i_2, \ldots, i_m)$ be a connected component of graph G, and

$\omega(i_k) \le \omega(i_{k+1})$. Apply transformation I-$[i_k, i_{k+1}]$ to graph G. As a result, the chain C is transformed to the chain $C' = (i'_1, i'_2,..., i'_{m-1})$. If in C' there are such vertices i'_l and i'_{l+1} that $\omega(i'_l) \le \omega(i'_{l+1})$, then apply transformation I again. It is clear that, in order to convert C into an ω-chain, it suffices to apply transformation I at most $m-1$ times to graph G. This proves the lemma.

3.2. We now consider the situation in which graph G is either an outtree or an intree.

Theorem 3.1. *If the graph G is an outtree (an intree), then there exists a sequence of transformations I and II converting G into an ω-chain.*

Proof. Let graph $G = (N, U)$ be an outtree. The proof of the theorem is by induction with respect to the number of pairs of non-comparable elements in N. If all elements in N are pairwise comparable, then G is a chain and the theorem follows from Lemma 3.2.

Let the theorem hold for all outtrees G such that there exist at most m pairs of non-comparable elements in set N, $m \ge 0$. We show that the theorem also holds for any outtree G containing $m+1$ pairs of non-comparable elements.

By assumption, N has at least one pair of non-comparable elements. Therefore, there exists a vertex i^0 such that $A^0(i^0) = \{i_1, i_2,..., i_l\}$, $l \ge 2$, and for any vertex $i \in A(i^0)$ the relation $|A(i^0)| \le 1$ holds. A subgraph G^0 of the graph G induced by the set $A(i^0)$ of vertices consists of l connected components, each of which is a chain of the form $(i_k, j_1^{(k)},..., j_{\nu_k}^{(k)})$, $1 \le k \le l$.

Let s and t be vertices of graph G^0. Since $B_G(i_1) = B_G(i_2) = ... = B_G(i_l) = B_G(i^0) \cup i^0$, the conditions of Theorems 2.1 and 2.2 either are satisfied or not satisfied for the graphs G and G^0 simultaneously. Hence, transformations I-$[t, s]$ or II-(s, t) can be applied to graph G if and only if these are feasible for graph G^0.

Lemmas 3.2 and 3.1. imply the existence of a sequence L of transformations I, II converting graph G^0 into an ω-chain. Applying the sequence L to graph G yields a new graph $G' = (N', U')$ such that set N' contains at most m pairs of non-comparable elements (with respect to the order defined by graph G'). It is obvious that G' is an outtree.

The proof of the other part of the theorem (i.e., G is an intree) is essentially the same, the only difference is that the symbols A and B must be interchanged. This remark completes the proof.

3.3. Let G be a tree-like graph, and $G_1 = (N_1, U_1)$ be one of its connected components and $s, t \in N_1$. It is obvious that any of transformations I-$[t, s]$ or II-(s, t) can be applied to G if and only if it can be applied to G_1. This observation and Theorem 3.1 imply that there exists a sequence of transformations I and II that converts graph G into a graph G' with each connected component being an ω-chain. As follows from Lemma 3.1, to transform graph G' into a single ω-chain it suffices to sort its vertices in non-increasing order of the priorities of the corresponding elements. The obtained ω-chain specifies an optimal (due to Corollary 2.1) permutation π_n^* of the elements of set N.

Based on the proof of Theorem 3.1, it is easy to construct a procedure for transforming the components of graph G (outtrees and intrees) into ω-chains.

(a) Let $G = (N, U)$ be *an outtree*, $|N| = n$. *The procedure of transforming* G *into an* ω*-chain* involves a sequence of transitions from one outtree to another, each time reducing the number of vertices. Transformations are to be performed until a single-vertex graph is obtained. In each step, the vertices of the current outtree are associated with some ω-chains. In the first step, the vertices of graph G are chosen as such chains.

Find, in G, a vertex i^0 (called *a supporting vertex*), with all direct successors being terminal vertices. Let the ω-chains $C_1, C_2,..., C_l$ correspond to these successors. Due to Lemma 3.1, the chains $C_1, C_2,..., C_l$ can be replaced by a single chain C_0'. To find C_0', it suffices to sort the vertices of the chains $C_1, C_2,..., C_l$ in non-increasing order of the priorities of the corresponding elements. The chain C_0' is called *the union of the* ω*-chains* $C_1, C_2,..., C_l$, while these ω-chains are said to be *united.*

Insert the vertex i^0 into C_0' from the left and transform the obtained chain C_0'' into an ω-chain. Let $\{i_1, i_2,.., i_\nu\}$ be the set of vertices of chain C_0'. To transform C_0'' into an ω-chain it suffices to apply transformation I at most ν times: if $\omega(i^0) \leq \omega(i_{k_0}) = \max\{\omega(i_k) \,|\, k = 1, 2,..., \nu\}$, unite i^0 and i_{k_0} into the composite element $[i^0, i_{k_0}]$. Then compare $\omega(i^0, i_{k_0})$ and $\max\{\omega(i_k) \,|\, k = 1, 2,..., \nu, k \neq k_0\}$, and, if necessary, the next two elements, and so on. Let C_0 denote the obtained ω-chain.

Remove from G all successors of the vertex i^0, and associate the ω-chain C_0 with i^0. In the obtained tree $G^{(1)}$ there are at most $n-1$ vertices.

Applying described transformations to $G^{(1)}$, we obtain some outtree $G^{(2)}$, and so on, until a graph $G^{(h)}$ consisting of a single vertex is obtained. The chain corresponding to this vertex is the desired ω-chain.

(b) *The procedure for converting an intree into an* ω*-chain* is essentially the same as that for an outtree. In each step, a vertex i^0 is chosen as the supporting vertex if all its direct predecessors have no predecessors in the tree obtained in the previous step. The chain C_0'' is found by inserting the vertex i^0 into the ω-chain C_0' from the right. To transform C_0'' into an ω-chain C_0, compare $\omega(i^0)$ and $\omega(i_{k_0}) = \min\{\omega(i_k) \,|\, k = 1, 2,..., \nu\}$.

The composite element $[i_{k_0}, i^0]$ is to be formed, provided that $\omega(i^0) \geq \omega(i_{k_0})$.

3.4. To implement the procedure for transforming outtrees and intrees into ω-chains it is possible to use balanced 2-3-trees for data representation (see Section 2 of Chapter 1). Such data representation allows finding an optimal permutation π_n^* in at most $O(n\log n)$ time.

Let a perfect pseudo-order relation $\Longrightarrow$ be defined over set $Q[\mathcal{P}_n(G)]$ (see Section 1 of this chapter) in the following way: $\pi^{(1)} \Longrightarrow \pi^{(2)}$ for any two permutations $\pi^{(1)}, \pi^{(2)} \in Q[\mathcal{P}_n(G)]$ if and only if $\omega(\pi^{(1)}) \geq \omega(\pi^{(2)})$. It is easy to check that, in this case, the relation $\Longrightarrow$ is, in fact, a perfect quasi-order relation.

When implementing the procedure for transforming a tree (either an outtree or an intree) into an ω-chain, the ω-chains appearing in this process are represented by balanced 2-3-trees. All such 2-3-trees are represented by the same table. To refer to a particular ω-chain it suffices to refer to the number of the root of the corresponding balanced 2-3-tree.

Any ω-chain is specified by the permutation of the numbers of its vertices sorted in non-increasing order of their priorities. Representing an ω-chain by the balanced 2-3-tree, with the labels either v_{min} or v_{max} corresponding to an intermediate vertex v, this chain may be reconstructed in at most $O(n' \log n')$ time, where n' is the length of the chain. The value of the priority function corresponding to a given label of a vertex v is called *the value of this label.*

Consider the implementation of the procedure for transforming an outtree $G_1 = (N_1, U_1)$ into an ω-chain. Without loss of generality, the vertices of G_1 can be assumed to be numbered by the integers $1, 2, \ldots, n_1$, $n_1 = |N_1|$, in the following way. The root of G_1 has number 1. If α_ν denotes the number of vertices belonging to the νth rank of tree G_1, then the second-rank vertices are numbered $2, 3, \ldots, \alpha_2+1$; the third-rank vertices are numbered $\alpha_2+2, \alpha_2+3, \ldots, \alpha_2+\alpha_3+1$, etc. While numbering the vertices of each current rank, the direct successors of a vertex with a minimum number are given numbers first, followed by the direct successors of a vertex having the next number, etc.

Graph G_1 is represented by a table consisting of four rows and n_1 columns. The first row contains the numbers of the vertices of G_1. The kth cell of the second row contains the number of the immediate predecessor of vertex k; while the kth cell of the fourth row indicates the minimal and maximal numbers of the direct successors of the kth vertex. The kth cell of the third row contains the number of the root of the balanced 2-3-tree representing the ω-chain corresponding to the kth vertex of the graph.

In the following, a table representing either the graph G_1 or the current graph $G_1^{(s)}$, $1 \leq s \leq h$, is called Table 1; while a table representing the balanced 2-3-trees is called Table 2. The columns n_1+1, $n_1+2,...,$ of the third and fourth rows of Table 2 contain the labels and the values of the labels; the cells 1, 2,..., n_1 of the third and/or the fourth row, contain the values of $\omega(\pi^{(C)})$. Here C is the ω-chain corresponding to a given vertex of the graph G_1, and $\pi^{(C)}$ is the feasible (with respect to the chain C) permutation of all elements of the set N involved in C.

Note that the implementation of the procedure for transforming an outtree into an ω-chain does not require the third row of Table 2 to be filled, since, in this case, only the label v_{max} is used. Similarly, the fourth row of Table 2 can be skipped in the case of an intree.

In the first step of transforming G_1 into an ω-chain, each vertex of G_1 is considered as an ω-chain; therefore, we start with Table 2 having only the first row filled (with the numbers 1, 2,..., $2n_1-1$).

The process of transforming graph G_1 into an ω-chain includes the implementation of the following subroutines: find the next supporting vertex and the corresponding ω-chains to be united; unite several ω-chains into a single ω-chain; insert a supporting vertex into an ω-chain and transform the resulting chain into an ω-chain; remove the chains united in some step from the graph, and associate a chain C_0 with a supporting vertex i^0. The implementation of these subroutines is considered below.

Let a supporting vertex i^0 be chosen, then using the cell i^0 of the fourth row of Table 1, find the numbers i_1, $i_1+1,...,$ i_1+l of terminal vertices which are direct successors of vertex i^0. The cells i_1, $i_1+1,...,$ i_1+l of the third row of Table 1 contain the numbers of the root of the balanced 2-3-trees representing the ω-chains which correspond to the vertices i_1, $i_1+1,...,$ i_1+l. Let the found ω-chains be united into the ω-chain C_0', and let the chain C_0'' be found, which, in turn, is transformed into the ω-chain C_0. Removing the vertices i_1, $i_1+1,...,$ i_1+l from the current graph and replacing the vertex i^0 by the vertex associated with the chain C_0 can be done in the following way. Remove the contents of the cells i_1, $i_1+1,...,$ i_1+l in all four rows of Table 1, as well as that of the cell i^0 of the fourth row; replace the content of the cell i^0 of the third row by the number of the root of the balanced 2-3-tree representing the chain C_0.

Thus, while transforming outtree G_1 into an ω-chain, all subroutines for finding the ω-chains to be united, removing the vertices corresponding to these chains from the current graph, and associating a chain C_0 with a supporting vertex i^0 require at most $O(n_1)$ time.

The chosen way of numbering the vertices of graph G_1 and removing the vertices from the current graph allows a simple implementation of the search for the next supporting vertex. In fact, the last filled cell of the first row of Table 1 contains the number of a terminal vertex of the current graph $G_1^{(s)}$, and this vertex belongs to the last level of $G_1^{(s)}$. Hence, the immediate predecessor of this vertex (its number is given in the corresponding cell of the second row) can be chosen as a supporting vertex. While transforming G_1 into an ω-chain, the search for all supporting vertices takes at most $O(n_1)$ time.

Uniting two ω-chains into a single ω-chain can be done by uniting the corresponding 2-3-trees. As shown in Section 2 of Chapter 1, this takes at most $O(\log n')$ time, where n' is the largest length of the chains to be united. Hence, while transforming G_1 into an ω-chain, uniting all the ω-chains takes at most $O(n_1 \log n_1)$ time.

The vertex i^0 can be inserted into ω-chain C_0' and the obtained chain C_0'' can be transformed into an ω-chain simultaneously. Let v be the root of the 2-3-tree representing the chain C_0'. Compare $\omega(i^0)$ and the value $\omega(v_{max})$ of the label v_{max}. If $\omega(i^0) \le \omega(v_{max})$, then the cell v of the fourth row of Table 2 contains the number of a vertex i_{k_0} which is the label v_{max}. Unite i^0 and i_{k_0} into the composite element $[i^0, i_{k_0}]$. For the chosen way of representing the data, it is enough to know only the priority of the composite element. Therefore, associate the value of $\omega(i^0, i_{k_0})$ with the element i^0. To do this, replace the content of the cell i^0 of the fourth row of Table 2 by $\omega(i^0, i_{k_0})$. Remove the vertex i_{k_0} from the ω-chain C_0' using the procedure for deleting an element from a set represented by a balanced 2-3-tree (see Section 2.6 of Chapter 1). The composite element $[i^0, i_{k_0}]$ itself is required only for finding the permutation π_n^*, and is stored separately. Again, let v denote the root of the balanced 2-3-tree representing the ω-chain obtained from C_0' after removing the vertex i_{k_0}. Compare the new value of $\omega(v_{max})$ with the new value of $\omega(i^0)$ (i.e., compare the contents of the corresponding cells of the fourth row of Table 2). If $\omega(i^0) \le \omega(v_{max})$, then a new composite element is to be formed. If $\omega(i^0) > \omega(v_{max})$, then the vertex i^0 is included in the 2-3-tree with the root v using the procedure for uniting two sets represented by balanced 2-3-trees (see Sections 2.4 and 2.5 of Chapter 1).

Let n' be the length of C_0'; then forming each new composite element takes at most $O(\log n')$ time, and inserting i^0 into the 2-3-tree with the root v also takes $O(\log n')$ time. It is clear that while transforming G_1 into an ω-chain, new composite elements may be formed at most $n_1 - 1$ times, and the procedure for including i^0 to C_0' is to be performed at most $n_1 - 1$ times as well. Hence, the running time of all procedures for including a supporting vertex in an ω-chain followed by transforming the resulting chain into an

ω-chain does not exceed $O(n_1 \log n_1)$.

Having completed the transformation of G_1 into graph $G_1^{(h)}$ consisting of a single vertex, the ω-chain corresponding to $G_1^{(h)}$ has to be recovered using Table 2. This takes at most $O(n_1 \log n_1)$ time.

Thus, the procedure for converting an outtree into an ω-chain can be run in $O(n_1 \log n_1)$ time.

While performing the procedure for transforming an intree $G_1 = (N_1, U_1)$ into an ω-chain, the vertices of the tree are numbered by the integers 1, 2,..., n_1, starting with the root. To do this, change the orientation of all arcs and number the vertices as in the case of an outtree. The table representing an intree differs from that for an outtree by the second and the fourth rows. Here, the second row contains the numbers of the direct successors, while the fourth row contains minimal and maximal numbers of the direct predecessors. A supporting vertex is chosen based on the last filled cell of the table representing the current graph: the content of the corresponding cell of the second row is the number of the supporting vertex. The procedure for transforming an intree into an ω-chain also requires at most $O(n_1 \log n_1)$ time.

Having transformed all connected components of a tree-like graph $G = (N, U)$ into ω-chains, the desired permutation π_n^* can be recovered in at most $O(n \log n)$ time.

Thus, *the running time* of the algorithm for finding an optimal permutation in the case of a tree-like graph G does not exceed $O(n \log n)$. This estimate does not involve the time required for calculating the priorities of composite elements. Note, however, that this time is a constant for all priority-generating functions considered in Section 1 of this chapter.

3.5. *Example.* Consider the problem of minimizing the information storage volume (see Section 1.5 of this chapter), provided that tree-like precedence constraints are defined over a set of requests.

Table 3.1

i	1	2	3	4	5	6	7	8	9	10
t_i	10	15	20	5	-30	8	12	-22	6	8

The reduction graph G of precedence relation $\rightarrow$ is shown in Fig. 3.1, and the values of the parameters t_i are listed in Table 3.1. At time $t = 0$, the storage contains 100 information units.

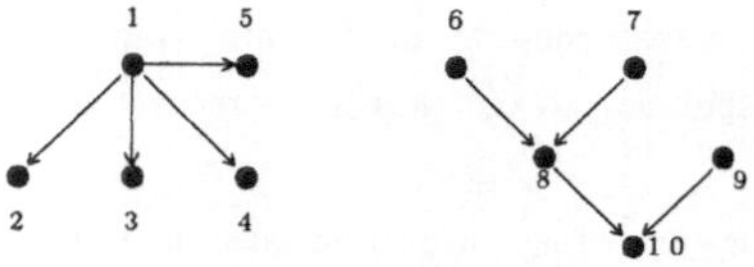

Fig. 3.1

Using formula (1.22), calculate the priorities of the elements of set $N = \{1, 2, \dots, 10\}$ (assuming $W = 140$): $\omega(1) = \omega(2) = \omega(3) = \omega(4) = \omega(6) = \omega(7) = \omega(9) = \omega(10) = -140$, $\omega(5) = 170$, $\omega(8) = 162$.

Let $G_1 = (N_1, U_1)$ denote the connected component of G being an outtree, and $G_2 = (N_2, U_2)$ denote the connected component being an intree; $N_1 = \{1, 2, \dots, 5\}$, $N_2 = \{6, 7, \dots, 10\}$. Transform G_1 into an ω-chain. For the graph G_1, construct Tables 1 (see Table 3.2) and 2 (see Table 3.3).

Table 3.2

I	The number of a vertex	1	2	3	4	5
II	The number of the direct predecessor		1	1	1	1
III	The number of the root of a 2-3-tree	1	2	3	4	5
IV	The numbers of direct successors	2, 5				

Table 3.3

I	The number of a vertex in a balanced 2-3-tree	1	...	4	5	6	...	9
II	The number of the direct predecessor							
III	v_{min} (the value of a label)							
IV	v_{max} (the value of a label)	(-140)	...	(-140)	(170)			
V	The numbers of direct successors							

In Table 1, find a supporting vertex and a set of ω-chains to be united in the first step. Vertex 1 is taken as supporting, and 2, 3, 4, and 5 are the numbers of the roots of the balanced 2-3-trees representing the chains to be united in this step. Table 3.4 (the third row is omitted) represents Table 2 after completing the procedure for uniting ω-chains 2, 3, 4, and 5 into ω-chain C_1'. The chain C_1' is represented by the balanced 2-3-tree with the root 8. Since $\omega(1) < \omega(8_{max}) = \omega(5)$, form the composite element [1, 5]. After removing vertex 5 from the ω-chain C_1', the obtained ω-chain is given by the balanced

2-3-tree with the root 6, $6_{max} = 2$. Since $\omega(1, 5) = 130 > \omega(6_{max}) = -140$, no more new composite elements are required.

Table 3.4

I	1	2	3	4	5	6	7	8	9
II		6	6	7	7	8	8		
IV						2	5	5	
	(-140)	(-140)	(-140)	(-140)	(170)	(-140)	(170)	(170)	
V						2, 3	4, 5	6, 7	

Now, unite the balanced 2-3-tree with the root 6 and the balanced 2-3-tree representing the element [1, 5]. Tables 1 and 2 obtained as a result of applying the above procedures are given by Tables 3.5 and 3.6, respectively. The empty columns in Table 1 are omitted.

Table 3.5

I	1
II	
III	
IV	8

Table 3.6

I	1	2	3	4	5	6	7	8	9
II		6	6	7		8	8		
IV						3	1	1	
	(140)	(-140)	(-140)	(-140)		(-140)	(130)	(130)	
V						2, 3	4, 5	6, 7	

The graph $G_1^{(1)}$ obtained after performing the first step consists of a single vertex associated with the ω-chain $C(G_1)$. This ω-chain is represented by the balanced 2-3-tree with the root 8. Recovering this chain yields: $C(G_1) = ([1, 5], 2, 3, 4)$.

Table 3.7

The initial number of a vertex	6	7	8	9	10
The new number of a vertex	5	4	3	2	1

Let us now convert G_2 into an ω-chain. Renumber the vertices of G_2 according to

Table 3.7.

For the graph G_2, construct Tables 1 (see Table 3.8) and 2 (see Table 3.9).

Table 3.8

I	The number of a vertex	1	2	3	4	5
II	The number of the direct successor		1	1	3	3
III	The number of the root of a 2-3-tree	1	2	3	4	5
IV	The numbers of direct predecessors	2, 3		4, 5		

Table 3.9

I	1	2	3	4	5	6	7	8	9
II									
III	(-140)	(-140)	(162)	(-140)	(-140)				
V									

In the first step, vertex 3 is supporting and the ω-chains to be united are given by the balanced 2-3-tree with the roots 4 and 5. Tables 3.10 and 3.11 correspond to Tables 1 and 2, respectively, after performing the first step of the procedure for transforming G_2 into an ω-chain (empty columns of Table 1 are omitted). While performing the first step, the composite element [4, 5, 3] is formed, $\omega(4, 5, 3) = 120$.

Table 3.10

I	1	2	3
II		1	1
III	1	2	4
IV	2, 3		

Table 3.11

I	1	2	3	4	5	...	9
II							
III	(-140)	(-140)		(120)			
V							

In the second step, vertex 1 is supporting, and the ω-chains to be united are represented by the balanced 2-3-trees with the roots 2 and 4. Tables 3.12 and 3.13 correspond to Tables 1 and 2, respectively, after performing the second step. In this step, the composite element [2, 1] is formed, and $\omega(2, 1) = -140$. The graph $G_2^{(2)}$ obtained after the two described steps of the procedure for transforming the graph G_2 into an ω-chain consists of one vertex ($G_2^{(2)}$ is given by Table 3.12). Recovering the obtained ω-chain yields: $C(G_2)$ = ([4, 5, 3], [2, 1]). In the initial numbering, this chain is of the form: $C(G_2)$ = ([7, 6, 8], [9, 10]).

Table 3.12

I	1
II	
III	6
IV	

Table 3.13

I	1	2	3	4	5	6	7	8	9
II		6		6		2			
III		(-140)		(120)		(-140)			
V						2, 4			

Thus, the graphs G_1 and G_2 have been transformed into the ω-chains $C(G_1)$ and $C(G_2)$, respectively. Unite these chains into one ω-chain $C(G)$ by sorting the vertices of $C(G_1)$ and $C(G_2)$ in non-increasing order of the priorities of the corresponding composite elements: $C(G)$ = ([1, 5], [7, 6, 8], 2, 3, 4, [9, 10]). Thus, the permutation $\pi_{10}^* = (1, 5, 7, 6, 8, 2, 3, 4, 9, 10)$ is optimal.

4. Series-Parallel Order

This section considers the situations in which the reduction graph G of a precedence relation $\longrightarrow$ either is series-parallel or may be converted into a series-parallel graph by performing some sequence of transformations I and II.

The concept of a series-parallel graph, as well as related concepts such as a decomposition tree $T(G)$ of an arbitrary graph G, operations of a series (notation s) and parallel (notation p) composition of graphs were introduced in Section 1 of Chapter 1. That section also presented a scheme for constructing a tree $T(G)$ and a procedure for reconstructing graph G by its decomposition tree. Recall that graph G which can be given either in the form of series ($G = G_1 s G_2$) or parallel ($G = G_1 p G_2$) composition of two graphs

G_1 and G_2, is called decomposable (in the opposite case, non-decomposable), and the graphs G_1 and G_2 are the decomposition components of graph G. If, in turn, graphs G_1 or G_2 can be presented in the form of either series or parallel composition of some graphs G_3 and G_4, then the latter graphs are also called decomposition components of G. A decomposition component of graph G corresponds to each terminal vertex in tree $T(G)$, and the operations of series or parallel composition are associated with intermediate vertices. The terminal vertices of *a complete* decomposition tree correspond to non-decomposable graphs.

In the following, no distinction is made between terminal vertices of tree $T(G)$ and the corresponding decomposition components of graph G, as well as between the operational vertices of $T(G)$ and the corresponding composition operations.

4.1. Consider some properties of graphs and their decomposition trees, as well as the relations between some operations over graph $G = (N, U)$ and its decomposition components.

Lemma 4.1. *Let $G_1 = (N_1, U_1)$ be a decomposition component of a graph G. Then for any elements i^0, $j^0 \in N_1$ and $i \in N \backslash N_1$, exactly one pair of the following relations holds: (1) $i \sim i^0$, $i \sim j^0$, or (2) $i \longrightarrow i^0$, $i \longrightarrow j^0$, or (3) $i^0 \longrightarrow i$, $j^0 \longrightarrow i$.*

Proof. Since G_1 is a decomposition component of G, there must exist a decomposition tree $T(G)$ in which some terminal vertex corresponds to graph G_1, while i is a vertex of another graph G_2 associated with another terminal vertex of tree $T(G)$. In $T(G)$, find an operational vertex O_1 of the highest rank such that there are paths from O_1 to each vertex G_1 and vertex G_2.

Implement the procedure for reconstructing graph G by $T(G)$ up to the moment when the vertex O_1 happens to be adjacent to two terminal vertices G' and G''. If i is a vertex of G', then i^0, j^0 are vertices of graph G''. If i is a vertex of G'', then i^0, j^0 are vertices of G'. In fact, otherwise (i.e., if i, i^0, j^0 were vertices of exactly one of these two graphs) there would exist an operational vertex O_2 from which two paths go to vertices G_1 and G_2, and whose rank is higher than that of O_1). If the operation of parallel composition corresponds to vertex O_1, then $i \sim i^0$ and $i \sim j^0$. Let the operation of series composition correspond to O_1. If i is a vertex of graph G', then the definition of operation of series composition implies that $i \longrightarrow i^0$ and $i \longrightarrow j^0$. If i is a vertex of G'', then $i^0 \longrightarrow i$ and $j^0 \longrightarrow i$. This proves the lemma.

Lemma 4.2. *Let $T(G)$ be a decomposition tree of graph G and $G_1 = (N_1, U_1)$ be a terminal vertex of $T(G)$, i^0, $j^0 \in N_1$, $i^0 \sim j^0$. If G' and G_1' are the graphs obtained from G and G_1*

by including the arc (i^0, j^0), *then the tree* T' *obtained from* $T(G)$ *by replacing the vertex* G_1 *with the vertex* G_1' *is a decomposition tree of the graph* G'.

Proof. Let $G'' = (N, U'')$ be a graph such that T' is its decomposition tree. We show that the graphs G' and G'' coincide. Graph G is the reduction graph of the precedence relation $\longrightarrow$, therefore, graph G' is circuit-free; moreover, by construction, it has no transitive arcs. Hence, G' can be considered as the reduction graph of some precedence relation denoted by $\overset{G'}{\longrightarrow}$. Graph G_1' has no circuits and transitive arcs, and neither does graph G''. Hence, G'' can be viewed as the reduction graph of some precedence relation $\overset{G''}{\longrightarrow}$. To prove that the graphs G' and G'' are the same, it suffices to show that for any $i, j \in N$ the relation $i \overset{G'}{\longrightarrow} j$ holds if and only if the relation $i \overset{G''}{\longrightarrow} j$ holds.

Let $i \in N \backslash N_1$, $j \in N$. The relation $i \overset{G'}{\longrightarrow} j$ holds if and only if $i \longrightarrow j$. The sufficiency is obvious. Suppose that $i \overset{G'}{\longrightarrow} j$ but $i \sim j$, i.e., in G', there exists a path from vertex i to vertex j but there is no such a path in graph G. Observe that G' differs from G by the only arc (i^0, j^0). Hence, in G' a path from i to j must contain the arc (i^0, j^0), and for the relation $\longrightarrow$ defined by graph G the following two conditions must hold: $i \longrightarrow i^0$ and $i \sim j^0$. By Lemma 4.1, the latter is impossible, since i^0, j^0 are vertices of graph G_1, and G_1 is a decomposition component of graph G. Similarly, it can be proved that $j \overset{G'}{\longrightarrow} i$ if and only if $j \longrightarrow i$.

Let $i \in N_1$ and $j \in N$. The relation $i \overset{G'}{\longrightarrow} j$ holds if and only if either $i \longrightarrow j$ or $i \longrightarrow i^0$ and $j^0 \longrightarrow j$. The sufficiency is obvious. If $j \in N \backslash N_1$, then the necessity follows from the previous considerations. If $j \in N_1$, then, assuming that there is no path from i to j in G, we derive that such a path in G' must contain the arc (i^0, j^0). Similarly, $j \overset{G'}{\longrightarrow} i$ if and only if either $j \longrightarrow i$ or $j \longrightarrow i^0$ and $j^0 \longrightarrow i$.

Consider graph G''. Let $i \in N \backslash N_1$, $j \in N$. The relation $i \overset{G''}{\longrightarrow} j$ (or $j \overset{G''}{\longrightarrow} i$) holds if and only if $i \longrightarrow j$ ($j \longrightarrow i$, respectively). The sufficiency is obvious, and the necessity can be proved in a similar way as in the case of graph G'. Analogously, for $i \in N_1$, $j \in N$, the relation $i \overset{G''}{\longrightarrow} j$ ($j \overset{G''}{\longrightarrow} i$) holds if and only if either $i \longrightarrow j$ (or $j \longrightarrow i$) or $i \longrightarrow i^0$ and $j^0 \longrightarrow j$ (or $j \longrightarrow i^0$ and $j^0 \longrightarrow i$, respectively).

Thus, $i \overset{G'}{\longrightarrow} j$ if and only if $i \overset{G''}{\longrightarrow} j$. This proves the lemma.

Lemma 4.3. *Let* $T(G)$ *be a decomposition tree of a graph* G, $G_1 = (N_1, U_1)$ *be a terminal* *vertex of $T(G)$, and i^0, $j^0 \in N_1$, $j^0 \rightarrowtail i^0$. If G' and G_1' are the graphs obtained from G and G_1 by identifying the vertices j^0 and i^0, then the tree T' obtained from $T(G)$ by replacing the vertex G_1 by the vertex G_1' is a decomposition tree of the graph G'.*

Proof. Let i' denote the vertex obtained by identifying the vertices j^0 and i^0. Assume that $G' = (N', U')$ and $G_1' = (N_1', U_1')$. Let $G'' = (N', U'')$ be a graph such that T' is its decomposition tree. We show that the graphs G' and G'' coincide. The graphs G' and G'' are circuit-free and do not contain transitive arcs; they therefore, define the precedence relations $\xrightarrow{G'}$ and $\xrightarrow{G''}$, respectively. To prove that graphs G' and G'' are the same it suffices to show that for any $i, j \in N$, the relation $i \xrightarrow{G'} j$ holds if and only if the relation $i \xrightarrow{G''} j$ holds.

Since graph G' is obtained from graph G by identifying vertices j^0 and i^0, it follows that, for any $i, j \in N \backslash i'$, the relation $i \xrightarrow{G'} j$ is valid if and only if $i \rightarrow j$. Moreover, $i \xrightarrow{G'} i'$ (or $i' \xrightarrow{G'} i$) holds for all $i \in N \backslash i'$ if and only if either $i \rightarrow i^0$ or $i \rightarrow j^0$ (or is either $i^0 \rightarrow i$ or $j^0 \rightarrow i$).

Lemma 4.1 and the procedure for constructing T' imply that for any $i \in N \backslash i'$ and $j \in N_1'$, the relation $i \xrightarrow{G''} j$ ($j \xrightarrow{G''} i$) holds if and only if $i \rightarrow j$ (or $j \rightarrow i$). If $j = i'$, then $i \xrightarrow{G''} i'$ ($i' \xrightarrow{G''} i$) if and only if either $i \rightarrow i^0$ or $i \rightarrow j^0$ (or if either $i^0 \rightarrow i$ or $j^0 \rightarrow i$). If $i, j \in N \backslash N_1'$, then $i \xrightarrow{G''} j$ if and only if $i \rightarrow j$.

Thus, the relation $i \xrightarrow{G'} j$ holds if and only if the relation $i \xrightarrow{G''} j$ holds, $i, j \in N'$. This proves the lemma.

Lemma 4.4. *Let $G_1 = (N_1, U_1)$ be a decomposition component of a graph G, and i^0, $j^0 \in N_1$. Transformation I-$[j^0, i^0]$ or II-$[i^0, j^0]$ can be applied to graph G if and only if it can be applied to graph G_1.*

Proof. The possibility of applying transformations I-$[j^0, i^0]$ and II-$[i^0, j^0]$ depends on conditions (2.1) and (2.2), respectively (with $s = i^0$ and $t = j^0$). Lemma 4.1 implies that for any element $i \in N \backslash N_1$ exactly one of the following relations holds: $i \in E(i^0) \cap E(j^0)$, or $i \in B(i^0) \cap B(j^0)$, or $i \in A(i^0) \cap A(j^0)$. In any case, $i \notin \overline{A}(i^0, j^0)$ and $i \notin \overline{B}(i^0, j^0)$, hence, conditions (2.1) and (2.2) are satisfied or not satisfied simultaneously for graphs G and G_1. The lemma is proved.

Corollary 4.1. *Let $G_1 = (N_1, U_1)$ be a decomposition component of a graph G. A sequence*

L of transformations I and II acting on a set N_1 can be applied to graph G if and only if it can be applied to graph G_1.

This statement directly follows from the last three lemmas.

Corollary 4.2. *Let $T(G)$ be a decomposition tree of a graph G, a graph $G_1 = (N_1, U_1)$ be a terminal vertex of $T(G)$, L be a sequence of transformations I and II acting on the set N_1. If L transforms graphs G and G_1 into graphs G' and G_1', respectively, then a decomposition tree $T(G')$ of graph G' can be obtained form $T(G)$ by replacing the vertex G_1 by the vertex G_1'.*

This statement follows from Corollary 4.1 and from Lemmas 4.2 and 4.3.

Theorem 4.1. *Let $\{G_1, G_2, \dots, G_m\}$ be a set of terminal vertices of a decomposition tree $T(G)$ of a graph G. If for each of the graphs $G_1, G_2, \dots, G_m$ there exists a sequence of transformations I and II which transforms a graph into a chain, then for graph G there exists a sequence of transformations I and II which transforms G into an ω-chain.*

Proof. Let $L_1, L_2, \dots, L_m$ be sequences of transformations I and II which transform the graphs $G_1, G_2, \dots, G_m$ into chains $C_1, C_2, \dots, C_m$, respectively. Due to Lemma 3.2, C_1, $C_2, \dots, C_m$ can be considered to be ω-chains. Let $G^{(1)}$ denote the graph obtained from graph G by applying the sequence $L = (L_1, L_2, \dots, L_m)$ of transformations I and II. The existence of the graph $G^{(1)}$ is guaranteed by Corollary 4.1. As follows from Corollary 4.2, a tree T_1 obtained from decomposition tree $T(G)$ by replacing the vertices $G_1, G_2, \dots, G_m$ by vertices $C_1, C_2, \dots, C_m$ is a decomposition tree of the graph $G^{(1)}$.

Let us reconstruct graph $G^{(1)}$ by its decomposition tree T_1, while simultaneously making some transformations of $G^{(1)}$. Let O be an operational vertex of tree T_1 adjacent to two terminal vertices C_{l_1} and C_{l_2}. Construct a decomposition tree T_1' of the graph $G^{(1)}$ by removing the vertices C_{l_1} and C_{l_2} from T_1 and by replacing the vertex O by the vertex G' where either $G' = C_{l_1} s C_{l_2}$ if O is the operation of series composition, or $G' = C_{l_1} p C_{l_2}$ if O is the operation of parallel composition. Lemmas 3.1 and 3.2 imply that in any of these cases there exists a sequence L_1' of transformations I and II which transforms graph G', into some ω-chain C'. In fact, if $G' = C_{l_1} s C_{l_2}$, then G' is a chain; if $G' = C_{l_1} p C_{l_2}$, then G' consists of two connected components, each of which is an ω-chain. In the decomposition tree T_1', replace the vertex G' by the vertex C' and denote the constructed decomposition tree by T_2. Due to Corollary 4.2, the tree T_2 is a decomposition tree of a graph $G^{(2)}$, which is the result of applying the sequence L_1' of transformations to graph $G^{(1)}$.

The decomposition tree T_2 has $m-1$ terminal vertices. Having applied the described procedure $m-2$ times, we obtain a graph $G^{(m-1)}$ and its decomposition tree T_{m-1}. The tree T_{m-1} has two terminal vertices associated with some ω-chains. Performing the operation of composition corresponding to the root of tree T_{m-1} results in a graph G''. For graph G'', as well as for graph G', there exists a sequence L'_{m-1} of transformations I and II which transforms G'' into an ω-chain C''.

It is obvious that the chain C'' is obtained from graph $G^{(1)}$ as a result of performing the sequence $L' = (L'_1, L'_2, \ldots, L'_{m-1})$ of transformations I and II. Thus, the sequence $L^0 = (L, L')$ transforms graph G into an ω-chain C''. This proves the theorem.

If graph G is series-parallel, then each terminal vertex of its complete decomposition tree is a single-vertex graph; therefore, the following statement holds.

Corollary 4.3. *For any series-parallel graph, there exists a sequence of transformations I and II which transforms the graph into an ω-chain.*

4.2. Based the Theorem 4.1, we now describe *an algorithm for transforming a series-parallel graph G into a chain*, assuming that graph G is represented by its complete decomposition tree.

The algorithm for transforming graph G into a chain consists of $n-1$ steps. In each step, the algorithm passes from one series-parallel graph to another. In the first $n-2$ steps, these graphs are represented by their decomposition trees, and as a result of performing one step we pass to a tree having one terminal vertex less than the previous one. Some ω-chains correspond to the terminal vertices of decomposition trees. The complete decomposition tree $T(G)$ of the graph G is considered as the initial decomposition tree T_1. The vertices of graph G are terminal vertices of tree T_1 (recall that a single-vertex chain is, at the same time, an ω-chain).

Let T_r be a decomposition tree obtained after having performed the first $r-1$ steps, $1 \le r \le n-1$. Tree T_r has $n-r+1$ terminal vertices. In T_r, choose an operational vertex O adjacent to two terminal vertices $C_{l_1} = (i_1, i_2, \ldots, i_{\nu_1})$ and $C_{l_2} = (j_1, j_2, \ldots, j_{\nu_2})$. By analogy with Section 3 of this chapter, the vertex O is called ***supporting***.

If O is the operation of parallel composition, then unite the chains C_{l_1} and C_{l_2} into a single ω-chain C. To construct chain C, it suffices to sort the vertices of the ω-chains C_{l_1} and C_{l_2} in non-increasing order of their priorities. If O is the operation of series composition, then form the chain $C' = (i_1, i_2, \ldots, i_{\nu_1}, j_1, j_2, \ldots, j_{\nu_2})$. If $\omega(i_{\nu_1}) > \omega(j_1)$, the chain C' is an ω-chain (denote it by C). Otherwise, transform C' to the ω-chain C as follows. Unite i_{ν_1} and j_1 into the composite element $i^0 = [i_{\nu_1}, j_1]$. If

$\omega(i_{\nu_1-1}) \le \omega(i^0)$, then unite i_{ν_1-1} and i^0 into the composite element $[i_{\nu_1-1}, i^0]$ again denoted by i^0. If $\omega(i_{\nu_1-1}) > \omega(i^0)$ and $\omega(i^0) \le \omega(j_2)$, form the composite element $[i^0, j_2]$ again denoted by i^0. The transformation of C' into an ω-chain results in the chain C which has one of the following forms: $C = (i^0)$, $C = (i_1, i_2,..., i_k, i^0)$, $C = (i^0, j_l,..., j_{\nu_2})$, $C = (i_1, i_2,..., i_k, i^0, j_l,..., j_{\nu_2})$, where $\omega(i_k) > \omega(i^0)$ and $\omega(i^0) > \omega(j_l)$.

Remove the vertices C_{l_1} and C_{l_2} from the decomposition tree T_r, and replace the supporting vertex O by the vertex C. The resulting decomposition tree T_{r+1} has $n-r$ terminal vertices, each of which is an ω-chain. Performing $n-2$ steps yields a decomposition tree T_{n-1} with two terminal vertices $C^{(1)}$, $C^{(2)}$ and one operational vertex O. If O is the operation of parallel composition, then sorting the vertices of the chains $C^{(1)}$ and $C^{(2)}$ in non-increasing of their priorities yields a desired chain C. If O is an operation of series composition, then a desired chain is $C = (C^{(1)}, C^{(2)})$. In the latter case, the chain C, in turn, can be transformed, if necessary, into an ω-chain by the procedure described above.

4.3. When minimizing a priority-generating function $F(\pi)$ over set $\mathcal{P}_n(G)$, the constructed chain C specifies an optimal permutation π_n^*. We show that using balanced 2-3-tree to represent ω-chains allows permutation π_n^* to be found in at most $O(n\log n)$ time, provided a series-parallel graph G is given by its complete decomposition tree $T(G)$.

Define a perfect pseudo-order relation $\Longrightarrow$ over set $Q[\mathcal{P}_n(G)]$ in a similar way to that used in Section 3 of this chapter: $\pi^{(1)} \Longrightarrow \pi^{(2)}$ for any $\pi^{(1)}$ and $\pi^{(2)}$ of $Q[\mathcal{P}_n(G)]$ if and only if $\omega(\pi^{(1)}) \ge \omega(\pi^{(2)})$.

Let us number the vertices of the complete decomposition tree $T(G)$ of graph G in the following way. Remove all terminal vertices from $T(G)$, and, in the resulting tree, number the vertices by the integers $n+1$, $n+2$,..., $2n-1$ starting with the root, as in the case of an ordinary outtree (see Section 3 of this chapter). The elements of set $N = \{1, 2,..., n\}$ are associated with the terminal vertices of $T(G)$; therefore, these vertices may be considered to be numbered by the integers 1, 2,..., n.

The decomposition tree $T(G)$ is represented by a table consisting of 5 rows and $2n-1$ columns. The first row of this table contains the numbers of the vertices of the tree $T(G)$; the kth cell of the second row contains the number of the direct predecessor of vertex k. The kth cell of the third row contains the number of the root of the balanced 2-3-tree representing the ω-chain associated with the kth vertex of tree $T(G)$. The fourth row contains the numbers of direct successors. The kth cell of the fifth row contains an index of the operation of composition (either s or p) corresponding to the kth operational vertex of the tree $T(G)$.

In what follows, a table representing a decomposition tree is called Table 1, and a table representing balanced 2-3-trees is called Table 2.

To start, place the integers 1, 2,..., n into the first n cells of the third row of Table 1; the cells $n+1$, $n+2$,..., $2n-1$ of this row remain empty. The way of filling the remainder rows of Table 1 is quite obvious. The cells of the third and fourth rows of Table 2 contain the labels and their values (see Section 3 of this chapter).

Due to the chosen way of numbering the operational vertices of $T(G)$, it follows that the vertex numbered $2n-k$ can be taken as supporting in the kth step of the algorithm.

To run the algorithm, one has to be able to implement the following procedures: find the next supporting vertex; remove two terminal vertices from the current decomposition tree and associate an ω-chain C with the current supporting vertex; unite two ω-chains into one ω-chain (if the index p corresponds to the supporting vertex); transform a series composition of two ω-chains into a single ω-chain if the index s corresponds to the supporting vertex). The implementation of all these procedures except the last is discussed in detail in Section 3 of this chapter, and in all $n-1$ steps of the algorithm they can be implemented in at most $O(n\log n)$ time. Consider the last procedure among those mentioned above.

Let $C' = C_1 s C_2$, $C_1 = (i_1, i_2,..., i_{\nu_1})$, $C_2 = (j_1, j_2,..., j_{\nu_2})$, and $v^{(1)}$ and $v^{(2)}$ are the roots of the balanced 2-3-trees representing the ω-chains C_1 and C_2, respectively. Compare $\omega(i_{\nu_1})$ and $\omega(j_1)$ (it is obvious that $\omega(i_{\nu_1}) = \omega(v^{(1)}_{min})$, $\omega(j_1) = (v^{(2)}_{max})$). If $\omega(i_{\nu_1}) > \omega(j_1)$, then C' is an ω-chain, and, in this case, it suffices to unite the balanced 2-3-trees with the roots $v^{(1)}$ and $v^{(2)}$. If $\omega(i_{\nu_1}) \le \omega(j_1)$, then unite i_{ν_1} and j_1 into the composite element $[i_{\nu_1}, j_1]$. To implement such uniting, it suffices to remove the vertices i_{ν_1} and j_1 from the ω-chains C_1 and C_2, to remove the contents of cells j_1 of the third and fourth rows of Table 2, and to replace the contents of cells i_{ν_1} of these rows by the value $\omega(i_{\nu_1}, j_1)$. Let C_1' and C_2' be the chains obtained from C_1 and C_2 by removing the vertices i_{ν_1} and j_1, respectively. Again, let $v^{(1)}$ and $v^{(2)}$ denote the roots of the balanced 2-3-tree representing the chains C_1' and C_2'.

Compare $\omega(i^0)$ and $\omega(v^{(1)}_{min})$ (here $i^0 = [i_{\nu_1}, j_1]$). If $\omega(v^{(1)}_{min}) \le \omega(i^0)$, then form a new composite element by uniting the element which is the label $\omega(v^{(1)}_{min})$ and the element i^0. Otherwise, compare $\omega(i^0)$ and $\omega(v^{(2)}_{max})$. If $\omega(i^0) \le \omega(v^{(2)}_{max})$, then unite i^0 and the element which is the label $v^{(2)}_{max}$ and go to further comparisons.

The process of forming the new composite elements is completed if one of the following situations is achieved: (1) the composite element i^0 includes all vertices of the chains

C_1 and C_2; (2) i^0 includes all vertices of C_1 and $\omega(i^0) > \omega(v_{max}^{(2)})$; or (3) i^0 includes all vertices of C_2 and $\omega(v_{min}^{(1)}) > \omega(i^0)$; (4) $\omega(v_{min}^{(1)}) > \omega(i^0) > \omega(v_{max}^{(2)})$. In the first case, the element i^0 is the desired ω-chain C. In the second and third cases, the element i^0 must be inserted into the balanced 2-3-tree with the root $v^{(2)}$ or $v^{(1)}$, respectively. In the fourth case, it is necessary to unite trees with the roots $v^{(1)}$ and $v^{(2)}$ and to inset the element i^0 into the obtained balanced 2-3-tree.

Each of the procedures for removing a vertex from an ω-chain represented by a balanced 2-3-tree and for uniting such trees takes at most $O(n\log n)$ time. While transforming a series-parallel graph G into a chain, new composite elements are to be formed at most $n-1$ times. Therefore, all procedures for transforming a series composition of two ω-chains into a single ω-chain require at most $O(n\log n)$ time.

Thus, *the running time of the algorithm* for finding an optimal permutation π_n^*, provided that graph G is series-parallel and is given by its complete decomposition tree $T(G)$, does not exceed $O(n\log n)$. This estimate does not take into account the time required for calculating the priorities of the composite elements to be formed. Note, however, that this time is constant for all priority-generating functions analyzed in Section 1 of this chapter.

4.4. *Example.* Consider the problem of minimizing the sum of linear penalty functions (see Section 1.1(a) of this chapter), assuming that the precedence relation is defined over the set of jobs, and its reduction graph is series-parallel.

The reduction graph G of the precedence relation and its decomposition tree $T(G)$ are shown in Fig. 4.1. The job parameters α_i, β_i and t_i are listed in Table 4.1.

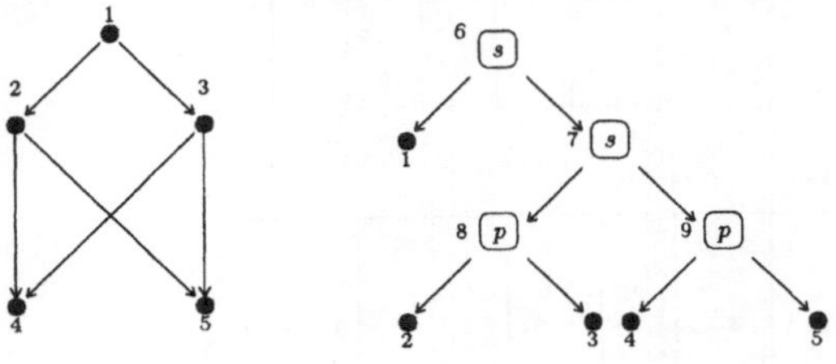

Fig. 4.1

Table 4.1

i	1	2	3	4	5
α_i	12	6	-20	12	8
β_i	5	8	16	-1	7
t_i	1	2	5	3	8

Calculate the priorities of the elements of set $N = \{1, 2, 3, 4, 5\}$ by formula (1.4): $\omega(1) = 12$, $\omega(2) = 3$, $\omega(3) = -4$, $\omega(4) = 4$, $\omega(5) = 1$. Construct Table 1 (see Table 4.2) and Table 2 (see Table 4.3).

Table 4.2

I	The number of a vertex	1	2	3	4	5	6	7	8	9
II	The number of the direct predecessor	6	8	8	9	9		6	7	7
III	The number of the root of a 2-3-tree	1	2	3	4	5				
IV	The numbers of direct successors						1, 7	8, 9	2, 3	4, 5
V	Index of the operation						*s*	*s*	*p*	*p*

Table 4.3

I	1	2	3	4	5	6	7	8	9
II									
III	(12)	(3)	(-4)	(4)	(1)				
IV	(12)	(3)	(-4)	(4)	(1)				
V									

In the first step, vertex 9 is chosen as supporting, and in the second step, vertex 8 is chosen as supporting. Tables 4.4 and 4.5 present Tables 1 and 2, respectively, after two steps of the algorithm have been performed. The empty columns of Table 1 are omitted.

Table 4.4

I	1	6	7	8	9
II	6		6	7	7
III	1			6	7
IV		1, 7	8, 9		
V		*s*	*s*		

Table 4.5

I	1	2	3	4	5	6	7	8	9
II		6	6	7	7				
III	(12)	(3)	(-4)	(4)	(1)	3(-4)	5(1)		
IV	(12)	(3)	(-4)	(4)	(1)	2(3)	4(4)		
V						2, 3	4, 5		

In the third step, vertex 7 is chosen as supporting. Tables 4.6 and 4.7 represent Tables 1 and 2, respectively, after the third step has been performed. In the third step, the composite element [3, 4, 5] is formed, $\omega(3, 4, 5) = 0$. The number of an element of N

occupying the first position in the composite element is used as i^0.

Table 4.6

I	1	6	7
II	6		6
III	1		
IV		1, 7	
V		s	s

Table 4.7

I	1	2	3	4	5	6	7	8	9
II	1	6	6						
III	(12)	(3)	(0)			3(0)			
IV	(12)	(3)	(0)			2(3)			
V						2, 3			

Vertex 6 is supporting in the fourth (the last) step. Table 4.8 represents Table 2 after this step has been performed.

Table 4.8

I	1	2	3	4	5	6	7	8	9
II	6	6	6						
III	(12)	(3)	(0)			3(0)			
IV						1(12)			
V						1, 2, 3			

In the fourth step, the ω-chain C is constructed and is represented by the balanced 2-3-tree with root 6 (Table 4.8). Reconstructing this ω-chain yields $C = (1, 2, 3)$. Since the composite element [3, 4, 5] is denoted by 3, we derive that $C = (1, 2, [3, 4, 5])$. Thus, the permutation $\pi_5^* = (1, 2, 3, 4, 5)$ is optimal.

4.5. Let $\{G_1, G_2, \ldots, G_m\}$ be a set of terminal vertices of the complete decomposition tree $T(G)$ of a graph G, and, for each graph G_i of this set there exists a sequence L_i of transformations which converts it into a series-parallel graph. Corollaries 4.1 and 4.2 imply that the sequence $L = (L_1, L_2, \ldots, L_m)$ transforms graph G into a series-parallel graph G'.

In this case, by applying transformations I and II graph G can be converted into a chain.

In fact, construct the tree $T(G)$ and transform the decomposition components G_i, $i = 1$,

2,..., m, of graph G into series-parallel graphs G_i', respectively. For each graph G_i', construct its complete decomposition tree $T(G_i')$. In the tree $T(G)$, replace the vertices G_i, $i = 1, 2,..., m$, by the decomposition trees $T(G_i')$ (this can be done by removing a vertex G_i' from $T(G)$ followed by replacing the arc entering this vertex by the arc entering the root of the tree $T(G_i')$). This yields the complete decomposition tree $T(G')$ of graph G'. Further, run the algorithm described in Section 4.3.

5. General Case

The problem of finding a permutation π_n^* that minimizes a priority-generating function $F(\pi)$ over the set $\mathcal{P}_n(G)$ in the case of an arbitrary graph G is NP-hard. This follows directly from the fact that the problem of minimizing function (1.1) over $\mathcal{P}_n(G)$ (see Section 1 of this chapter) is NP-hard in the case of linear penalties (see Section 5 of Chapter 4).

Applying transformations I and II to a graph G allows a graph G' to be found such that $\mathcal{P}_n(G') \subseteq \mathcal{P}_n(G)$, and the set $\mathcal{P}_n(G')$ contains at least one optimal permutation. In some cases, such as when the graph G is series-parallel, it is possible to find a sequence of transformations I and II which transforms G into a chain, thus yielding an optimal permutation π_n^*. In a general case, such a sequence may not exist and applying transformations I and II only reduces the search for π_n^*.

This section presents an algorithm for finding a sequence L^0 of transformations I and II which converts a graph G into a so-called *deadlock* graph. The running time of this algorithm does not exceed $O(n^4)$. In Section 6, the situations are analyzed in which a deadlock graph obtained from G by performing a sequence L^0 of transformations I and II is a chain.

5.1. A graph is called *deadlock* if neither of transformations I or II can be applied to it. A sequence L^0 of transformations I and II which converts a graph $G = (N, U)$ into a deadlock graph is called *a deadlock sequence* for G.

To describe *an algorithm for transforming the graph G into a deadlock graph* G^0, the following notation is used: $\max \overline{A}(i, j) = \max\{\omega(k) \mid k \in \overline{A}(i,j) \cup j\}$; $\min \overline{B}(i, j) = \min\{\omega(k) \mid k \in \overline{B}(i, j) \cup i\}$; $M(G) = \|m_{ij}\|$ and $M(\overline{G}) = \|\overline{m}_{ij}\|$ are the adjacency matrices of a graph G and its transitive closure $\overline{G}$, respectively.

The algorithm runs according to the following *scheme*.

(a) Form a list of the elements of set N.

(b) Scanning the list, apply all possible transformations II to the graph. If no transformation II can be applied to the current graph, return to the beginning of the list and go to (c).

(c) Choosing the next element i in the current list, check whether transformation I-$[j, i]$, where $j \rightarrowtail\!\!\!\rightarrow i$, can be applied to the current graph. If this transformation is feasible, then it is performed. Modify the list, return to its beginning, and go to (b). If transformation I-$[j, i]$ is non-feasible for the current graph, then take the next element in the list.

This process is completed when neither of transformations I or II can be applied to the current graph, i.e., in performing (c) no transformation has been performed.

5.2. Let matrices $M(G)$ and $M(\overline{G})$ be given, and the priorities of the elements of set N be calculated. The algorithm for converting G into a deadlock graph consists of two stages: auxiliary and main.

At *the auxiliary stage*, the following procedures are to be performed.

(a) For each $i \in N$, find the set $B^0(i)$. If the condition $|B^0(i)| = 1$ is satisfied, then compare $\omega(i)$ and $\omega(j)$, where $j \rightarrowtail\!\!\!\rightarrow i$. If $\omega(i) \geq \omega(j)$, then find the set $\overline{A}(i, j)$ and compute $\max \overline{A}(i, j)$. If $\omega(i) < \omega(j)$, then take the next element $i \in N$.

(b) For each $i \in N$, find all such $j \in E(i)$ that $\omega(i) \geq \omega(j)$. For these j, find the sets $\overline{B}(i, j)$ and $\overline{A}(i, j)$ and compute $\min \overline{B}(i, j)$ and $\max \overline{A}(i, j)$.

For the auxiliary stage of the algorithm, *the data representation* is as follows.

Form a $3 \times n$ table (Table 1) and a $2 \times n^2$ table (Table 2).

The first row of Table 1 contains the values of $\omega(i)$. The ith cell of the second row contains the number of the element j if $|B^0(i)| = 1$ and $j \rightarrowtail\!\!\!\rightarrow i$. Otherwise (i.e., if $|B^0(i)| \neq 1$), this cell remains empty. If $|B^0(i)| = 1$, then the ith cell of the third row contains the value of $\max \overline{A}(i, j)$ where $j \rightarrowtail\!\!\!\rightarrow i$. If $|B^0(i)| \neq 1$, then the ith cell of the third row remains empty.

Table 2 contains the values of $\min \overline{B}(i, j)$ (the first row) and $\max \overline{A}(i, j)$ (the second row). These values are placed in the column numbered $n(i-1)+j$. It is obvious that the cells of a column numbered by $n(i-1)+i$ are always empty.

The main stage of this algorithm involves the following procedures.

(a) Arrange the list of the elements of set N by sorting them arbitrarily (e.g., in non-decreasing order of their priorities).

(b) For each element i in the list, starting with the first, perform the following. For each $j \in E(i)$ check whether min $\overline{B}(i, j) \geq$ max $\overline{A}(i, j)$. For all $j \in E(i)$ that satisfy this condition, perform transformation II-(i, j), and take the next element in the list. If either the inequality min $\overline{B}(i, j) <$ max $\overline{A}(i, j)$ holds for all $j \in E(i)$ or $E(i) = \emptyset$, then also take the next element in the list. If during the current scan of the list no transformation II has been performed, go to (c).

(c) For each element i in the list, starting with the first, perform the following. If $|B^0(i)| = 1$ and the condition $\omega(i) \geq$ max $\overline{A}(i, j)$ is satisfied for $j \rightarrowtail i$, then perform transformation I-$[j, i]$. Remove the elements i and j from the list, insert the element $[j, i]$ into the list, return to the beginning of the modified list, and go to (b). If, otherwise, either $|B^0(i)| \neq 1$ or $\omega(i) <$ max $\overline{A}(i, j)$, then take the next element in the list.

This process is completed if while performing (c), no transformation I that is feasible for the current graph is found.

We now estimate the running time of the algorithm and present some details of the implementation of the procedures to be performed at the main and auxiliary stages.

For each element in the list, checking all conditions takes at most $O(n)$ time. Therefore, before performing the first transformation I or II all conditions can be checked by scanning the list in at most $O(n^2)$ time. The list is to be scanned at most $n(n+3)/2$ times. In fact, transformation I can be applied at most $n-1$ times, while transformation II can be applied at most $n(n-1)/2$ times since matrix $\overline{M}(G)$ contains at most $n(n-1)/2$ non-zero entries (recall that G is a circuit-free graph), and as a result of performing one transformation II, at least one new non-zero element is included in that matrix. Moreover, n "failure" checks are possible while performing (b), and one such check while performing (c).

Thus, while transforming G into a deadlock graph, all conditions can be checked in at most $O(n^4)$ time.

Let us estimate the running time of the auxiliary-stage procedures described in (a) and (b).

The set $B^0(i)$ can be found by matrix $M(G)$: an element $j \in B^0(i)$ if and only if $m_{ij} = 1$. Finding all elements $i \in N$ such that $|B^0(i)| = 1$ takes at most $O(n^2)$ time. For each $i \in N$, finding the set $\overline{A}(i, j)$ requires at most $O(n)$ time. In fact, $\overline{A}(i, j) = A(j)\backslash(A(i) \cup i)$ and, hence, an element $k \in N$ belongs to the set $\overline{A}(i, j)$ if and only if both $\bar{m}_{jk} = 1$ and $\bar{m}_{ik} = 0$. Computing max $\overline{A}(i, j)$ for all $i \in N$ such that $|B^0(i)| = 1$ takes at most $O(n^2)$ time. Thus, the procedure described in (a) can be implemented in at most $O(n^2)$ time.

For any $i \in N$, the set $E(i)$ can be found by matrix $M(\overline{G})$ (an element $j \in E(i)$ if and only if $\overline{m}_{ij} = \overline{m}_{ji} = 0$) and this takes at most $O(n)$ time. Finding the sets $\overline{A}(i, j)$ and $\overline{B}(i, j)$ for a fixed pair of elements i, j requires at most $O(n)$ time, and this amounts to $O(n^3)$ time for all pairs. Thus, the procedure described in (b) requires at most $O(n^3)$ time.

We now consider the implementation of transformations I and II and estimate their running times. Let G be a graph to which a transformation I or II is applied, and G' be the graph obtained from G by this transformation.

The implementation of transformation I. Let $|B^0(i)| = 1$ and $\omega(i) \geq \max \overline{A}_G(i, j)$, where $j \rightarrowtail i$. Replace each element of the ith row and ith column of the matrices $M(G)$ and $M(\overline{G})$ by -1, which indicates that vertex i has been removed from the graph. Let M' denote the matrix obtained from $M(G)$. To construct the matrix $M(G')$, it is necessary to replace some zeros in the row j of matrix M' by unities. If a vertex k is a direct successor of vertex i in graph G but it is not a successor of vertex j in the graph with the adjacency matrix M', then replace the kth element of the row j of matrix M' by unity. Since finding all successors of vertex j by matrix M' requires at most $O(n^2)$ time (see Section 1.4 of Chapter 1) and each of the remained procedures takes at most $O(n)$ time, we conclude that matrices $M(G')$ and $M(\overline{G}')$ can be found in at most $O(n^2)$ time.

In Table 1, replace the jth cell of the first row by $\omega(j, i)$, and delete the contents of all cells of the ith column. For all $k \in A_G^0(i)$, find the sets $B_{G'}^0(k)$. If both relations $|B_{G'}^0(k)| = 1$ and $|B_G^0(k)| \neq 1$ hold, then compute $\max \overline{A}_{G'}(k, j)$, where $j \rightarrowtail k$, replace the kth cell of the second row of Table 1 by the index j, and replace the kth cell of the third row by $\max \overline{A}_{G'}(k, j)$. Since $|A_G^0(i)| < n$, this procedure requires at most $O(n^2)$ time.

For each $k \in B_G(j) \cup A_G(j) \backslash i$ and for all found sets $\overline{A}(k, l)$, $\overline{B}(k, l)$, $\overline{A}(l, k)$, $\overline{B}(l, k)$, compute $\max \overline{A}_{G'}(k, l)$, $\min \overline{B}_{G'}(k, l)$, $\max \overline{A}_{G'}(l, k)$, $\min \overline{B}_{G'}(l, k)$, and use these values to replace the contents of the corresponding cells of Tables 1 and 2. To find the cell to be corrected, it suffices to perform the following. In Table 1, check the contents of the kth cell of the second row (if the cell is not empty it is to be corrected). Find those cells of the second row which contain the number k (thereby, the numbers l of the cells containing $\max \overline{A}_G(l, k)$ are found). In Table 2, scan the cells with the numbers of the form $n(k-1)+l$ and $n(l-1)+k$, $l = 1, 2, \ldots, n$. If a cell is not empty, it is to be corrected.

Since $|B_G(j) \cup A_G(j)| < n$, and the total number of the sets $\overline{A}(k, l)$, $\overline{B}(k, l)$, $\overline{A}(l, k)$, $\overline{B}(l, k)$ does not exceed $4(n-1)$, it follows that the latter procedure requires at most $O(n^3)$ time. Transformation I is to be performed at most $n-1$ times; therefore, performing all transformations I takes at most $O(n^4)$ time.

The implementation of transformation II. Let $i \sim j$ and $\min \overline{B}_G(i, j) \geq \max \overline{A}_G(i, j)$. Then, in the matrix $M(G)$, define $m_{ij} = 1$, and, in the matrix $M(\overline{G})$, define $\overline{m}_{kl} = 1$ for all pairs k and l such that $k \in B_G(i) \cup i$, $l \in A_G(j) \cup j$. Besides, if in the initial matrix $M(\overline{G})$ for some of the above pairs k, l the element $\overline{m}_{kl} = 1$, then in $M(G)$ define $m_{kl} = 0$ (in this case, after the arc (i, j) has been included in the graph, the arc (k, l) becomes transitive). This procedure takes at most $O(n^2)$ time.

For all $k \in \overline{B}_G(i, j) \cup i$, and $l \in E_{G'}(k)$, find the sets $\overline{A}_{G'}(k, l)$ if $\omega(k) \geq \omega(l)$ and the sets $\overline{A}_{G'}(l, k)$ if $\omega(k) \leq \omega(l)$. Compute $\max \overline{A}_{G'}(k, l)$ and $\max \overline{A}_{G'}(l, k)$ for all found sets and use these values to replace the contents of the corresponding cells of Table 2. For all $k \in \overline{A}_G(i, j) \cup j$ and $l \in E_{G'}(k)$, find the sets $\overline{B}_{G'}(k, l)$ if $\omega(k) \geq \omega(l)$, and the sets $\overline{B}_{G'}(l, k)$ if $\omega(k) \leq \omega(l)$. Compute $\min \overline{B}_{G'}(k, l)$, $\min \overline{B}_{G'}(l, k)$ and correct Table 2. Besides, for $k \in \overline{A}_G(i, j) \cup j$ such that $|B^0_{G'}(k)| = 1$ but $|B^0_G(k)| \neq 1$ and for such l that $l \overset{G'}{\rightarrowtail} k$ and $\omega(k) \geq \omega(l)$, place the index l in the kth cell of the second row of Table 1; find the sets $\overline{A}_{G'}(k, l)$, compute $\max \overline{A}_{G'}(k, l)$, and place this value into the kth cell of the third row of Table 1. If $|B^0_{G'}(j)| \neq 1$ and $|B^0_G(j)| = 1$, delete the contents of cell l of the second and third rows of Table 1.

The procedures for finding the sets mentioned above and for computing $\max \overline{A}_{G'}(k, l)$, $\min \overline{B}_{G'}(k, l)$, $\max \overline{A}_{G'}(l, k)$, $\min \overline{B}_{G'}(l, k)$ require at most $O(n^2(|\overline{B}_G(i, j)| + |\overline{A}_G(i, j)|))$ time. On the other hand, as a result of performing transformation II-(i, j), at least $|\overline{B}_G(i, j)| + |\overline{A}_G(i, j)| + 1$ new unit entries are to be included in matrix $M(\overline{G})$. Hence, the addition of one new unit element to $M(\overline{G})$ takes at most $O(n^2)$ time. Since the number of unit entries in the matrix $M(\overline{G})$ may not exceed $n(n-1)/2$, all transformations II can be implemented in at most $O(n^4)$ time.

Thus, *the running time* of the algorithm for transforming a graph G into a deadlock graph does not exceed $O(n^4)$.

In the following, this algorithm is called the *D-algorithm.*

5.3. *Example.* Let the numbers $\alpha_1 = 4$, $\alpha_2 = 2$, $\alpha_3 = \alpha_4 = 3$, $\alpha_5 = 1$, $\alpha_6 = 7$, $\alpha_7 = 5$ be associated with the elements of set $N = \{1, 2, 3, 4, 5, 6, 7\}$. The precedence relation $\rightarrow$ is defined over N, its reduction graph G being shown in Fig. 5.1a. A priority-generating function $F(\pi)$ is defined over set $\hat{\mathcal{P}}$ and its priority function $\omega(\pi)$ is defined over set $Q[\hat{\mathcal{P}}] = \hat{\mathcal{P}}$ as follows: $\omega(\pi_r) = \sum_{i \in \{\pi\}} \alpha_i / r$ if $\pi_r \neq (2, 3)$ and $\omega(2, 3) = 6$.

We use the D-algorithm to find a permutation π_7^* that minimizes the function $F(\pi)$ over set $\hat{\mathcal{P}}_7(G)$.

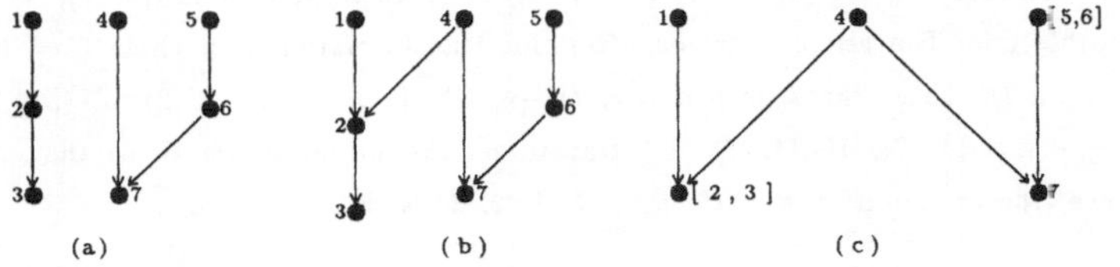

Fig. 5.1

Construct matrices $M(G)$ and $M(\overline{G})$, and calculate the priorities of the elements of set N:

$$M(G) = \begin{Vmatrix} 0&1&0&0&0&0&0 \\ 0&0&1&0&0&0&0 \\ 0&0&0&0&0&0&0 \\ 0&0&0&0&0&0&1 \\ 0&0&0&0&0&1&0 \\ 0&0&0&0&0&0&1 \\ 0&0&0&0&0&0&0 \end{Vmatrix}, \quad M(\overline{G}) = \begin{Vmatrix} 0&1&1&0&0&0&0 \\ 0&0&1&0&0&0&0 \\ 0&0&0&0&0&0&0 \\ 0&0&0&0&0&0&1 \\ 0&0&0&0&0&1&1 \\ 0&0&0&0&0&0&1 \\ 0&0&0&0&0&0&0 \end{Vmatrix},$$

$\omega(1) = 4$, $\omega(2) = 2$, $\omega(3) = \omega(4) = 3$, $\omega(5) = 1$, $\omega(6) = 7$, $\omega(7) = 5$.

Arrange the list $S_1 = (5, 2, 3, 4, 1, 7, 6)$ of the elements of set N by sorting them in non-decreasing order of their priorities. Transformation II-(4, 2) is feasible for the graph G. Having performed this, we obtain the graph $G^{(1)}$ (see Fig. 5.1b), for which no transformations II can be applied. Transformation I-[2, 3] is the first transformation (with respect to the list S_1) that is feasible for $G^{(1)}$. Having performed this transformation, we obtain the graph $G^{(2)}$, for which only one transformation may be applied; this transformation is I-[5, 6]. It is easy to check that the resulting graph $G^{(3)}$ (see Fig. 5.1c) is a deadlock graph. Thus, the sequence L_1 = (II-(4, 2), I-[2, 3], I-[5, 6]) converts G into the deadlock graph $G^{(3)}$ which is not a chain. Obviously, $\mathcal{P}_7(G^{(3)}) \subset \mathcal{P}_7(G)$, $|\mathcal{P}_7(G)| = 105$, $|\mathcal{P}_7(G^{(3)})| = 16$.

Let us return to the initial graph G and arrange another list $S_2 = (6, 7, 1, 3, 4, 2, 5)$ of the elements of set N. In this list, the elements are sorted in non-increasing order of their priorities. Transformation I-[5, 6] is the first feasible transformation for the graph $G^{(1)}$ with respect to the list S_2 (here $\omega(5, 6) = 4$). Having performed the transformation, we obtain the list S_2 = (7, 1, [5, 6], 3, 4, 2) (we maintain the elements sorted in non-increasing order of their priorities). Transformation II-(7, 2) is the first transformation that is feasible with respect to the obtained list. The next feasible transformation is II-([5, 6], 1), and then II-([5, 6], 4). We return to the beginning of the list and conclude that transformation I-[4, 7] is feasible ($\omega(4, 7) = 4$). The corrected list is of the form S_2 = ([4, 7], 1, [5, 6], 3, 2). Transformation II-([4, 7], 1) is the next feasible one. The graph obtained after this transformation is a chain C = ([5, 6], [4, 7], 1, 2, 3). Thus, the sequence L = (II-(4, 2), I-[5, 6], II-(7, 2), II-([5, 6], 1), II-([5, 6], 4), I-[4, 7], II-([4, 7], 1)) transforms the initial graph G to the chain C and, hence, the permutation π = (5, 6, 4, 7, 1, 2, 3) is the desired one.

5.4. The example considered above implies that, in general, several deadlock sequences of transformations I and II exist for a graph G. Some of them may transform G into a deadlock graph that is not a chain (sequence L_1 in the example above), while the others may transform G into a chain (sequence L_2). Obtaining this or that sequence depends on the order in which the elements of set N are to be scanned, i.e., on the initial list of the elements and on the way this list is corrected after performing each transformation I. In a general case, a sequence of transformations I and II which transforms the initial graph into a chain need not exist.

A priority function $\omega(\pi)$ is called *auto-bounded* if for any permutations $\pi^{(a)}$, $\pi^{(b)}$, $\pi^{(c)}$, $\pi^{(d)}$ of such $Q[\mathcal{P}]$ that the permutation $(\pi^{(a)}, \pi^{(b)})$ belongs to $Q[\mathcal{P}]$, $\omega(\pi^{(c)}) \le \omega(\pi^{(a)}) \le \omega(\pi^{(d)})$ and $\omega(\pi^{(c)}) \le \omega(\pi^{(b)}) \le \omega(\pi^{(d)})$, the condition $\omega(\pi^{(c)}) \le \omega(\pi^{(a)}, \pi^{(b)}) \le \omega(\pi^{(d)})$ holds, and, moreover, if $\omega(\pi^{(a)}) = \omega(\pi^{(b)}) = c$, then $\omega(\pi^{(a)}, \pi^{(b)}) = c$ holds.

It is obvious that a function $\omega(\pi)$ is auto-bounded if the condition

$$\min\{\omega(\pi^{(a)}), \omega(\pi^{(b)})\} \le \omega(\pi^{(a)}, \pi^{(b)}) \le \max\{\omega(\pi^{(a)}), \omega(\pi^{(b)})\} \tag{5.1}$$

holds for any permutations $\pi^{(a)}$, $\pi^{(b)}$ of $Q[\mathcal{P}]$ such that $(\pi^{(a)}, \pi^{(b)}) \in Q[\mathcal{P}]$.

As shown below (see Section 6 of this chapter), the condition for a priority function to be auto-bounded is sufficient for transforming a series-parallel graph G into a chain by a sequence of transformations I and II which is deadlock for G. Note that, in the example considered above, the function $\omega(\pi)$ is not auto-bounded, while the graph in Fig. 5.1a is series-parallel.

Some other examples of priority-generating functions $F(\pi)$ with non-auto-bounded priority functions can be given. For a set $N = \{1, 2,..., n\}$, $n \ge 4$, let the numbers α_i and β_i be associated with each element $i \in N$. The set $\mathcal{P}$ consists of all permutations of length $r \ge 4$ being of the form $\pi = (\pi', \pi'')$, where $\{\pi'\} = \{1, 2, 3\}$, $\{\pi''\} \subseteq N \backslash \{\pi'\}$. The function $F(\pi) = \max\left\{\sum_{k=1}^{u} \alpha_{i_k} + \beta_{i_u} \,\middle|\, 1 \le u \le r\right\}$ is defined over set $\mathcal{P}$, where $\pi = (i_1, i_2,..., i_r)$, $r \ge 4$. Let $\alpha_1 = -1$, $\alpha_2 = \alpha_3 = 1$, $\beta_1 = \beta_2 = 1$, $\beta_3 = 2$, $\alpha_i \ge 5$, $\beta_i \ge 0$, $i = 4, 5,...,$ n. Define the function $\omega(\pi)$ over the set $Q[\mathcal{P}]$ as follows: $\omega(\pi) = \mathrm{sgn}\left(-\sum_{i\in\{\pi\}} \alpha_i\right) \left(W - F(\pi) +\right.$

$\max\left\{0, \sum_{i\in\{\pi\}} \alpha_i\right\}$ for all $\pi \in Q[\mathcal{P}]$ different from the permutation (2, 3), and assume $\omega(2, 3) = W+1$ where $W \geq \sum_{i=1}^{n} (|\alpha_i| + |\beta_i|)$. It follows from Section 1.4 of this chapter that the function $\omega(\pi)$ is a priority function for $F(\pi)$. We have $\omega(1) = W$, $\omega(2) = -W+1$, $\omega(3) = -W+2$, and thus $\omega(1) > \omega(2)$, $\omega(1) > \omega(3)$, but $\omega(2, 3) > \omega(1)$, i.e., the function $\omega(\pi)$ is not auto-bounded.

Note that, for a given priority-generating function $F(\pi)$, finding a priority function which is not auto-bounded takes some effort. At the same time, all known priority functions (see Section 1, Chapter 3) are auto-bounded.

Consider, for example, function (1.7). It is of the form $\omega(\pi) = \Psi(\pi)/F(\pi)$. Let $\pi^{(a)}$, $\pi^{(b)}$ be permutations of $Q[\mathcal{P}]$ such that $\pi = (\pi^{(a)}, \pi^{(b)}) \in Q[\mathcal{P}]$ and $\omega(\pi^{(a)}) \leq \omega(\pi^{(b)})$.

We show that $\omega(\pi^{(a)}) \leq \omega(\pi) \leq \omega(\pi^{(b)})$. It follows from relation (1.31) (see Section 1.9 of this chapter) that

$$\omega(\pi) = \frac{\Psi(\pi^{(a)})+\Psi(\pi^{(a)})\Psi(\pi^{(b)})+\Psi(\pi^{(b)})}{F(\pi^{(a)})+F(\pi^{(a)})\Psi(\pi^{(b)})+F(\pi^{(b)})}.$$

Since $F(\pi^{(a)}) > 0$, it follows from $\omega(\pi^{(a)}) \leq \omega(\pi^{(b)})$ that $\Psi(\pi^{(b)}) \geq F(\pi^{(b)})\Psi(\pi^{(a)})/F(\pi^{(a)})$. Hence,

$$\omega(\pi) \geq \frac{\Psi(\pi^{(a)})+\Psi(\pi^{(a)})\Psi(\pi^{(b)})+F(\pi^{(b)})\Psi(\pi^{(a)})/F(\pi^{(a)})}{F(\pi^{(a)})+F(\pi^{(a)})\Psi(\pi^{(b)})+F(\pi^{(b)})} = \Psi(\pi^{(a)})/F(\pi^{(a)}) = \omega(\pi^{(a)}).$$

On the other hand, $\Psi(\pi^{(a)}) > 0$, therefore, $F(\pi^{(a)}) \geq \Psi(\pi^{(a)})F(\pi^{(b)})/\Psi(\pi^{(b)})$. Hence,

$$\omega(\pi) = \frac{\Psi(\pi^{(a)})+\Psi(\pi^{(a)})\Psi(\pi^{(b)})+\Psi(\pi^{(b)})}{\Psi(\pi^{(a)})F(\pi^{(b)})\,/\,\Psi(\pi^{(b)})+\Psi(\pi^{(a)})F(\pi^{(b)})+F(\pi^{(b)})} = \Psi(\pi^{(b)})/F(\pi^{(b)}) = \omega(\pi^{(b)}).$$

Thus, $\omega(\pi^{(a)}) \leq \omega(\pi) \leq \omega(\pi^{(b)})$ and function (1.7) is auto-bounded.

Similarly, we may check that the other priority functions constructed in Sections 1.1–1.7 of this chapter are auto-bounded.

Theorem 5.1. *If a function $F(\pi)$ is priority-generating over set $\mathcal{P}$, then there exists its auto-bounded priority function over the set $Q[\mathcal{P}]$.*

Proof. To simplify the notation, we assume that all permutations π for which the values of $F(\pi)$ or $\omega(\pi)$ are calculated, belong to the set $\mathcal{P}$ or to the set $Q[\mathcal{P}]$, respectively. Let $\omega(\pi)$ be a priority function for $F(\pi)$ and there exists such a permutation $\pi^{(1)} = (\pi^{(a)}, \pi^{(b)})$ that $\omega(\pi^{(1)}) > \max\{\omega(\pi^{(a)}), \omega(\pi^{(b)})\}$. If $\pi^{(2)}$ is such a permutation that $\pi^{(1)} \cap \pi^{(2)} = \emptyset$ and either $\max\{\omega(\pi^{(a)}), \omega(\pi^{(b)})\} \leq \omega(\pi^{(2)}) < \omega(\pi^{(1)})$ or $\max\{\omega(\pi^{(a)}), \omega(\pi^{(b)})\} < \omega(\pi^{(2)}) \leq \omega(\pi^{(1)})$, then as follows from the definition of a priority-generating function, we have $F(\pi^{(1)}, \pi^{(2)}) \le F(\pi^{(2)}, \pi^{(1)}) = F(\pi^{(2)}, \pi^{(a)}, \pi^{(b)}) \le F(\pi^{(a)}, \pi^{(2)}, \pi^{(b)}) \le F(\pi^{(a)}, \pi^{(b)}, \pi^{(2)})$. Hence, we obtain

$$F(\pi^{(a)}, \pi^{(b)}, \pi^{(2)}) = F(\pi^{(a)}, \pi^{(2)}, \pi^{(b)}) = F(\pi^{(2)}, \pi^{(a)}, \pi^{(b)}). \tag{5.2}$$

Define $\omega'(\pi^{(1)}) = \max\{\omega(\pi^{(a)}), \omega(\pi^{(b)})\}$ and $\omega'(\pi) = \omega(\pi)$ for all permutations π of $Q[\mathcal{P}]$ different from $\pi^{(1)}$. Relation (5.2) implies that the constructed function $\omega'(\pi)$ is a priority function for $F(\pi)$.

Similarly, if $\omega(\pi^{(1)}) < \min\{\omega(\pi^{(a)}), \omega(\pi^{(b)})\}$, define $\omega'(\pi^{(1)}) = \min\{\omega(\pi^{(a)}), \omega(\pi^{(b)})\}$, and $\omega'(\pi) = \omega(\pi)$ for $\pi \ne \pi^{(1)}$. The constructed function $\omega'(\pi)$ is, as $\omega(\pi)$, a priority function for $F(\pi)$.

We construct a function $\omega(\pi)$, starting with unit length permutations, which satisfy conditions (5.1). Define $\omega'(i) = \omega(i)$ for all elements $i \in N$. Let $\pi = (i, j)$. Define $\omega'(\pi) = \omega(\pi)$ for all π satisfying the condition $\min\{\omega'(i), \omega'(j)\} \le \omega(\pi) \le \max\{\omega'(i), \omega'(j)\}$. Then, for all $\pi = (i, j)$ such that $\omega(\pi) > \max\{\omega'(i), \omega'(j)\}$, define $\omega'(\pi) = \max\{\omega'(i), \omega'(j)\}$. After this, for all $\pi = (i, j)$ such that $\omega(\pi) < \min\{\omega'(i), \omega'(j)\}$, define $\omega'(\pi) = \min\{\omega'(i), \omega'(j)\}$.

Let a new priority function be constructed for all permutations of the length m, $2 \le m < n$. We construct $\omega'(\pi)$ for permutations of the length $m+1$ as follows. Define $\omega'(\pi) = \omega(\pi)$ if $\min\{\omega'(\pi^{(a)}), \omega'(\pi^{(b)})\} \le \omega(\pi) \le \max\{\omega'(\pi^{(a)}), \omega'(\pi^{(b)})\}$ for all permutations $\pi^{(a)}, \pi^{(b)}$ of $Q[\mathcal{P}]$ such that $\pi = (\pi^{(a)}, \pi^{(b)})$. Then find all permutations $\pi_{m+1} \in Q[\mathcal{P}]$ such that there exist permutations $\pi^{(a)}, \pi^{(b)} \in Q[\mathcal{P}]$ satisfying the conditions $\pi = (\pi^{(a)}, \pi^{(b)})$, $\omega(\pi) > \max\{\omega'(\pi^{(a)}), \omega'(\pi^{(b)})\}$. Define $\omega'(\pi) = \min\{\max\{\omega'(\pi^{(a)}), \omega'(\pi^{(b)})\} \mid (\pi^{(a)}, \pi^{(b)}) = \pi, \omega(\pi) > \max\{\omega'(\pi^{(a)}), \omega'(\pi^{(b)})\}\}$.

Now, for all $\pi_{m+1} \in Q[\mathcal{P}]$ such that there exist permutations $\pi^{(a)}, \pi^{(b)} \in Q[\mathcal{P}]$ satisfying the conditions $\pi = (\pi^{(a)}, \pi^{(b)})$, $\omega(\pi) < \min\{\omega'(\pi^{(a)}), \omega'(\pi^{(b)})\}$, define $\omega'(\pi) = \max\{\min\{\omega'(\pi^{(a)}), \omega'(\pi^{(b)})\} \mid (\pi^{(a)}, \pi^{(b)}) = \pi, \omega(\pi) < \min\{\omega'(\pi^{(a)}), \omega'(\pi^{(b)})\}\}$.

The described process of constructing function $\omega'(\pi)$ in fact can be implemented. Suppose that function $\omega'(\pi)$ satisfies condition (5.1) for all π_r, $r \le m$, and $\pi = (\pi^{(a)}, \pi^{(b)})$ is a permutation of the length $m+1$ such that $\omega(\pi) > \max\{\omega'(\pi^{(a)}), \omega'(\pi^{(b)})\}$. We show that in this case, there are no permutations $\pi^{(c)}, \pi^{(d)}$ such that $\pi = (\pi^{(c)}, \pi^{(d)})$ and $\omega'(\pi) < \min\{\omega'(\pi^{(c)}), \omega'(\pi^{(d)})\}$, where $\omega'(\pi) = \max\{\omega'(\pi^{(a)}), \omega'(\pi^{(b)})\}$. Suppose that these permutations exist. The following variants are possible:

1) $\omega'(\pi^{(a)}) \ge \omega'(\pi^{(b)})$, $\omega'(\pi^{(c)}) \ge \omega'(\pi^{(d)})$;

2) $\omega'(\pi^{(a)}) \ge \omega'(\pi^{(b)})$, $\omega'(\pi^{(c)}) < \omega'(\pi^{(d)})$;

3) $\omega'(\pi^{(a)}) < \omega'(\pi^{(b)})$, $\omega'(\pi^{(c)}) \ge \omega'(\pi^{(d)})$;

4) $\omega'(\pi^{(a)}) < \omega'(\pi^{(b)})$, $\omega'(\pi^{(c)}) < \omega'(\pi^{(d)})$;

For the first variant, we have

$$\omega'(\pi^{(b)}) \leq \omega'(\pi^{(a)}) = \omega'(\pi) < \omega'(\pi^{(d)}) \leq \omega'(\pi^{(c)}). \quad (5.3)$$

Let $\pi^{(c)} = (\pi^{(a)}, \tilde{\pi})$ then $\pi^{(b)} = (\tilde{\pi}, \pi^{(d)})$. Since condition (5.1) is satisfied for function $\omega'(\pi)$ for all permutations of length $r \leq m$, it follows from (5.3) that $\omega'(\tilde{\pi}) > \omega'(\pi^{(a)})$. Hence, $\omega'(\pi^{(b)}) = \omega'(\tilde{\pi}, \pi^{(d)}) > \omega'(\pi^{(a)})$, which contradicts the conditions of first variant. Let $\pi^{(a)} = (\pi^{(c)}, \tilde{\pi})$ then $\pi^{(d)} = (\tilde{\pi}, \pi^{(b)})$. From (5.3), we have $\omega'(\pi^{(a)}) < \omega'(\pi^{(c)})$ and, hence, $\omega'(\pi^{(a)}) \geq \omega'(\tilde{\pi})$. On the other hand, $\omega'(\pi^{(b)}) < \omega'(\pi^{(d)})$, therefore, $\omega'(\tilde{\pi}) \geq \omega'(\pi^{(d)})$. Thus, $\omega'(\pi^{(a)}) \geq \omega'(\tilde{\pi}) \geq \omega'(\pi^{(d)})$ which contradicts (5.3).

For the third variant, we have

$$\omega'(\pi^{(a)}) < \omega'(\pi^{(b)}) = \omega'(\pi) < \omega'(\pi^{(d)}) \leq \omega'(\pi^{(c)}). \quad (5.4)$$

If $\pi^{(c)} = (\pi^{(a)}, \tilde{\pi})$ then $\pi^{(b)} = (\tilde{\pi}, \pi^{(d)})$ and it follows from (5.4) that $\omega'(\pi^{(b)}) \geq \omega'(\pi^{(a)})$ and $\omega(\pi^{(b)}) \geq \omega(\tilde{\pi})$. Hence, $\omega'(\pi^{(b)}) \geq \omega'(\pi^{(c)})$, which contradicts (5.4). Let $\pi^{(a)} = (\pi^{(c)}, \tilde{\pi})$, $\pi^{(d)} = (\tilde{\pi}, \pi^{(b)})$, then it follows from (5.4) that $\omega'(\tilde{\pi}) \geq \omega'(\pi^{(d)}) > \omega'(\pi^{(b)})$ and $\omega'(\pi^{(a)}) = \omega'(\pi^{(c)}, \tilde{\pi}) > \omega'(\pi^{(b)})$, which contradicts the conditions of the third variant.

Similarly, it can be shown that the conditions of the second and the fourth variants also lead to a contradiction.

The above considerations imply that, given a function $\omega(\pi)$, and manipulating as described above, it is possible to construct a new priority function $\omega'(\pi)$ for a function $F(\pi)$ which is auto-bounded. This proves the theorem.

It follows from the proof of Theorem 5.1 that *for any function that is priority-generating over a set $\mathcal{P}$ there exists its priority function defined over the set $\mathcal{Q}[\mathcal{P}]$ satisfying condition* (5.1).

6. Convergence Conditions

It is shown in this section that if a priority function is auto-bounded, then any deadlock sequence of transformations I and II transforms a series-parallel graph G into a chain (see Theorem 6.3). Thereby, using the D-algorithm described in Section 5 of this chapter guarantees that an optimal permutation can be found at least for series-parallel graphs.

Theorem 6.3 is proved by induction with respect to the number of vertices of graph G, based on the fact that any series-parallel graph may be represented as a series or a parallel composition of two graphs G_1 and G_2, each of which is in turn series-parallel. Besides, the proof uses the fact (established by Theorems 6.1 and 6.2) that any sequence of transformations I and II which is deadlock for graph G transforms that graph into a chain if and only if any sequences of transformations I and II which are deadlock for the graphs G_1 and G_2, respectively, transform G_1 and G_2 into chains.

To conclude this section, it is shown that using interdependence of graphs G and priority functions it is possible to describe an essentially more general class of situations for which the D-algorithm also guarantees that an optimal permutation will be found.

6.1. Let L be some sequence of transformations I and II of a graph G. Let $L^{(\nu)}$ denote a subsequence of sequence L consisting of the first ν transformations, and $G^{(\nu)}$ denote the graph obtained from the original graph G by sequence of transformations $L^{(\nu)}$.

A sequence L of transformations I and II is called feasible for graph G if the $(\nu+1)$th transformation in the sequence L is feasible for graph $G^{(\nu)}$, $\nu = 0, 1, \ldots, l-1$. Here l is the number of transformations in the sequence L and $G^{(0)} = G$.

A graph G is called *reducible* (with respect to a given priority function) if any sequence of transformations I and II which is deadlock for G transforms it into a chain.

Theorem 6.1. *Let $G = G_1 s G_2$. Graph G is reducible if and only if graphs G_1 and G_2 are reducible.*

Proof. Necessity. Let there exist a sequence L_1 of transformations I and II which transforms graph $G_1 = (N_1, U_1)$ into a deadlock graph $G_1' = (N_1', U_1')$ which is not a chain. We show that, in this case, there exists a sequence of transformations I and II which transforms graph $G = (N, U)$ into a deadlock graph which is not a chain either. Let $G_2' = (N_2', U_2')$ denote a deadlock graph obtained from G_2 as a result of applying some sequence L_2 of transformations I and II.

Suppose that graph G_1' cannot be represented as $G_1' = G_1'' s G_1'''$, where G_1''' is a chain. Construct a sequence $L = (L_1, L_2)$. Due to Corollary 4.1 (see Section 4 of this chapter), sequence L is feasible for graph G. Let G' denote a graph into which L transforms graph G. It is easy to see that $G' = G_1' s G_2'$, and sequence L is deadlock for graph G.

Suppose now that $G_1' = G_1'' s G_1'''$, where G_1''' is a chain of the maximal length. Then $G' = G_1'' s G_2''$, where $G_2'' = G_1''' s G_2'$. Let L_3 denote a sequence of transformations that is deadlock for G_2'' (it is possible to show that L_3 consists of transformations I only). Construct a sequence $L' = (L, L_3)$. If L_3 transforms G_2'' into a graph G_2''', and L' transforms graph G' into a graph G'', then $G'' = G_1''sG_2'''$, where G_1'' is a deadlock graph and is not a chain. Graph G_1'' cannot be represented as series composition of two graphs such that the second of them is a chain. Hence, G'' is a deadlock graph and L' is the desired sequence.

Similar considerations can also be given if there is a sequence of transformations I and II which transforms G_2 into a deadlock graph that is not a chain.

Sufficiency. Suppose that there exists a sequence L which transforms G into a deadlock graph G' that is not a chain. Since $G = G_1sG_2$, it follows that graph G' can be represented as $G' = G_1'sG_2'$, where $G_1' = (N_1', U_1')$, and the composite elements of set N_1' are formed from the elements of set N_1. Suppose that G_1' is not a chain. We show that, in this case, there exists a sequence L_1 of transformations I and II that transforms graph G_1 into a deadlock graph that is not a chain.

If all elements of set N_1 are included into composite elements of set N_1', then transformations of sequence L can be divided into two groups: transformations acting on set N_1 and transformations acting on set N_2. Having ordered transformations of the first group in the same order as they appear in sequence L, we obtain sequence L_1 transforming G_1 into G_1'.

If some elements $i \in N_1$ are included in composite elements of set N_2', then this is possible only when all arcs of the form (k, i) are included (as a result of applying transformations II), where k is a terminal vertex of graph G_1. It is easy to check that all elements of set N_2' are either of the form $\pi^{(2)}$, where $\{\pi^{(2)}\} \subset N_2$, or of the form $[\pi^{(1)}, \pi^{(2)}]$, where $\{\pi^{(1)}\} \subset N_1$, $\{\pi^{(2)}\} \subseteq N_2$. Therefore, having removed from the sequence L all transformations I which form the composite elements $[\pi^{(1)}, \pi^{(2)}]$ as well as all transformations acting on set N_2, we obtain a sequence L_1' which transforms graph G_1 into graph $G_1'' = G_1'sG_1'''$, where G_1''' is a chain. If L_1'' denotes a sequence which transforms G_1''' into a deadlock graph, then it is obvious that $L_1 = (L_1', L_1'')$ is the desired sequence.

If G_1' is a chain, then G' may be represented as $G' = G_1^0sG_2^0$, where graph $G_2^0 = (N_2^0, U_2^0)$ is not a chain and the composite elements of set N_2^0 are formed from the elements of set N_2. In this case, using the considerations similar to those given above, it is possible to prove the existence of a sequence of transformations I and II which transforms G_2 into a deadlock graph that is not a chain. This proves the theorem.

6.2. We now consider the situation when $G = G_1pG_2$ and prove the statement similar to Theorem 6.1. First, we prove several auxiliary statements.

Lemma 6.1. *Let G' be a graph obtained from graph G by applying some sequence of transformations I and II, and i^0 be a composite element corresponding to some vertex of graph G'. If a priority function is auto-bounded, then $\omega(\pi^{(a)}) \leq \omega(\pi^{(b)})$ for any permutations $\pi^{(a)}$, $\pi^{(b)}$ such that $i^0 = [\pi^{(a)}, \pi^{(b)}]$.*

Proof. For $|\{i^0\}| = 2$, the lemma is obvious. Suppose that the lemma holds for all composite elements i^0 such that $|\{i^0\}| \leq m$, $m \geq 2$. Let $|\{i^0\}| = m+1$. If i^0 is obtained by transformation I$-[\pi', \pi'']$, then $\omega(\pi') \leq \omega(\pi'')$. Let $\pi'' = [\tilde{\pi}, \pi^{(b)}]$, then $\pi^{(a)} = (\pi', \tilde{\pi})$. Since $|\{\pi''\}| \leq m$, we have $\omega(\tilde{\pi}) \leq \omega(\pi^{(b)})$ and, since the priority function is auto-bounded, it follows that $\omega(\pi') \leq \omega(\pi^{(b)})$. Hence, $\omega(\pi^{(a)}) \leq \omega(\pi^{(b)})$. Let $\pi' = [\pi^{(a)}, \tilde{\pi}]$, $\pi^{(b)} = (\tilde{\pi}, \pi'')$, then $\omega(\pi^{(a)}) \leq \omega(\tilde{\pi})$ and $\omega(\pi^{(a)}) \leq \omega(\pi')$. Moreover, $\omega(\pi') \leq \omega(\pi'')$, therefore, $\omega(\pi^{(a)}) \leq \omega(\pi'')$. Thus, $\omega(\pi^{(a)}) \leq \omega(\pi^{(b)})$. The lemma is proved.

Let $N^{(\nu)}$ be the set of vertices of a graph $G^{(\nu)}$ obtained from G by applying sequence $L^{(\nu)}$ of transformations I and II. If sequence L of transformations I and II transforms graph G into a graph G', and i' is a composite element corresponding to some vertex of G', then let $\{i'\}^{(\nu)}$ denote a set of all composite elements of set $N^{(\nu)}$ that are incorporated into the element i'.

Let $G_1 = (N_1, U_1)$, $G_2 = (N_2, U_2)$ and $G = G_1pG_2$. Transformation I$-[i, j]$ or II$-(i, j)$ is said to be *mixed* if $(\{i\} \cup \{j\}) \cap N_1 \neq \varnothing$ and $(\{i\} \cup \{j\}) \cap N_2 \neq \varnothing$. Otherwise, a transformation is called *uniform.* A composite element i such that $\{i\} \cap N_1 \neq \varnothing$ and $\{i\} \cap N_2 \neq \varnothing$ is called *mixed.* If $\{i\} \subseteq N_1$ or $\{i\} \subseteq N_2$, then i is called a *uniform* composite element.

Lemma 6.2. *Let a sequence L of transformations I and II, in which all transformations I are uniform, transform a graph $G = G_1pG_2$ into a graph $G' = (N', U')$, and a priority function be auto-bounded. If i', j' are such elements of the set N' that $i' \overset{G'}{\rightarrowtail} j'$ and either $\{i'\} \subseteq N_1$, $\{j'\} \subseteq N_2$ or $\{i'\} \subseteq N_2$, $\{j'\} \subseteq N_1$, then $\omega(i) \geq \omega(j)$ for all $i \in B_{G'}(i') \cup i'$, $j \in A_{G'}(j') \cup j'$.*

Proof. Without loss of generality, consider the case $\{i'\} \subseteq N_1$, $\{j'\} \subseteq N_2$. The proof is by induction with respect to the number $\varphi(L)$ of mixed transformations in L. Assume that II$-(i^0, j^0)$ is the unique mixed transformation in L, $\{i^0\} \subseteq N_1$, $j^0 \subseteq N_2$, and this transformation is placed in L in the $(\nu+1)$th position, $\nu \geq 0$. Then $\omega(i) \geq \omega(j)$ for all $i \in B_{G^{(\nu)}}(i^0) \cup i^0$ and $j \in A_{G^{(\nu)}}(j^0) \cup j^0$. Suppose that $i' = [\pi^{(a)}, i^0, \pi^{(b)}]$, $j' = [\pi^{(c)},$

j^0, $\pi^{(d)}]$. The set $\{\pi^{(a)}\}^{(\nu)}$ can be divided into two subsets. The first of these sets contains elements of $B_{G^{(\nu)}}(i^0)$ (the priority of any of them is at least $\max\{\omega(j) \mid j \in A_{G^{(\nu)}}(j^0) \cup j^0\}$). The second set contains the composite elements i formed from the elements of set $N^{(\nu)}$ and such that $\{i\}^{(\nu)} \cap B_{G^{(\nu)}}(i^0) = \emptyset$, but $\{i\}^{(\nu)} \subseteq B_{G^{(\nu+\rho)}}(i^0)$ for some $\rho \geq 1$. In the latter case, it is obvious that $\omega(i) \geq \omega(i^0)$. Since the priority function is auto-bounded, it follows that $\omega(\pi^{(a)}, i^0) \geq \omega(j^0, \pi^{(d)})$, and Lemma 6.1 implies that $\omega(\pi^{(a)}, i^0) \leq \omega(\pi^{(b)})$ and $\omega(\pi^{(c)}) \leq \omega(j^0, \pi^{(d)})$. Thus, $\omega(\pi^{(a)}, i^0, \pi^{(b)}) \geq \omega(\pi^{(c)}, j^0, \pi^{(d)})$.

Suppose that the lemma holds for all such sequences L that $\varphi(L) \leq m$, $m \geq 1$. We prove that this also holds for $\varphi(L) = m+1$. Let in L the $(m+1)$th mixed transformation occupy the $(\nu+1)$th place. By the induction assumption, the lemma holds for graph $G^{(\nu)}$. Let the $(\nu+1)$th transformation in L be of the form II$-(i^0, j^0)$, and assume $\{i^0\} \subseteq N_1$, $j^0 \subseteq N_2$. Then it is obvious that $\min\{\omega(i) \mid i \in \overline{B}_{G^{(\nu)}}(i^0, j^0) \cup i^0\} \geq \max\{\omega(j) \mid j \in \overline{A}_{G^{(\nu)}}(i^0, j^0) \cup j^0\}$. Let $i \subseteq B_{G^{(\nu)}}(i^0) \cap B_{G^{(\nu)}}(j^0)$ and either $\{i\} \subset N_1$ or $\{i\} \subset N_2$. Then the inequality $\omega(i) \geq \max\{\omega(j) \mid j \in A_{G^{(\nu)}}(j^0) \cup j^0\}$ follows either from the induction assumption (in the former case) or from both the induction assumption and the inequality $\omega(i) \geq \omega(i^0)$ (in the latter case). Similarly, if $j \subseteq A_{G^{(\nu)}}(i^0) \cap A_{G^{(\nu)}}(j^0)$, then $\omega(j) \leq \min\{\omega(i) \mid i \in B_{G^{(\nu)}}(i^0) \cup i^0\}$. Thus, $\min\{\omega(i) \mid i \in B_{G^{(\nu)}}(i^0) \cup i^0\} \geq \max\{\omega(j) \mid j \in A_{G^{(\nu)}}(j^0) \cup j^0\}$.

If $i' = [\pi^{(a)}, i^0, \pi^{(b)}]$, $j' = [\pi^{(c)}, j^0, \pi^{(d)}]$, then, using the same argument as in the case $\varphi(L) = 1$, it can be easily proved that the inequality $\min\{\omega(i) \mid i \in B_{G'}(i') \cup i'\} \geq \max\{\omega(j) \mid j \in A_{G'}(j') \cup j'\}$ holds. This proves the lemma.

Let i^0 be a mixed composite element. Let $\tilde{\pi}^{(1)}(i^0)$ (or $\tilde{\pi}^{(2)}(i^0)$) denote a permutation obtained from i^0 after removing all elements of N_2 (or N_1).

Lemma 6.3. *Let a sequence L of transformations I and II containing exactly one mixed transformation I transform a graph $G = G_1pG_2$ into a graph G', and a priority function be auto-bounded. If i^0 is a mixed composite element corresponding to some vertex of graph G', then $\omega(i^0) = \omega(\tilde{\pi}^{(1)}(i^0)) = \omega(\tilde{\pi}^{(2)}(i^0))$.*

Proof. Suppose that the mixed transformation I occupies the $(\nu+1)$th position in the sequence L, and that this transformation is of the form I$-[i', j']$. Without loss of generality, assume that $\{i'\} \subseteq N_1$, $\{j'\} \subseteq N_2$. Due to Lemma 6.2, we have $\omega(i') \geq \omega(j')$. On the other hand, $\omega(i') \leq \omega(j')$ and, hence, $\omega(i') = \omega(j')$. Since the priority function is auto-bounded, it follows that $\omega(i', j') = \omega(i') = \omega(j')$. It is obvious that $[i', j'] = i^0$, $i' = \tilde{\pi}^{(1)}(i^0)$, $j' = \tilde{\pi}^{(2)}(i^0)$. The lemma is proved.

Lemma 6.4. *Let a sequence L of transformation I and II transform a graph $G = G_1pG_2$ into a graph G', and a priority function is auto-bounded. If i^0 is a mixed composite element corresponding to some vertex of graph G', then $\omega(i) \geq \omega(i^0) \geq \omega(j)$ for all $i \in B_{G'}(i^0)$, $j \in A_{G'}(i^0)$.*

Proof. The proof is by induction with respect to the number $\varphi(L)$ of mixed transformations I in sequence L. For $\varphi(L) = 1$ the lemma follows from Lemmas 6.2 and 6.3.

Let the lemma hold for all sequences L such that $\varphi(L) \leq m$, $m \geq 1$. We show that this also holds for $\varphi(L) = m+1$. Suppose that the $(m+1)$th mixed transformation I occupies the $(\nu+1)$th position in L, and that this transformation is of the form $\text{I}-[i', j']$. Then $L^{(\nu)}$ contains exactly m mixed transformations I.

If i' is a mixed composite element, then by the induction assumption it follows that $\min\{\omega(i) \mid i \in B_{G^{(\nu)}}(i')\} \geq \omega(i') \geq \max\{\omega(j) \mid j \in A_{G^{(\nu)}}(i')\}$ and, hence, $\omega(i') \geq \omega(j')$. The latter inequality also holds if j' is a mixed composite element. On the other hand, $\omega(j') \geq \omega(i')$. This and the fact that the priority function is auto-bounded imply that $\omega(i') = \omega(j') = \omega(i^0)$, where $i^0 = [i', j']$. Thus, $\min\{\omega(i) \mid i \in B_{G^{(\nu+1)}}(i^0)\} \geq \omega(i^0) \geq \max\{\omega(j) \mid j \in A_{G^{(\nu+1)}}(i^0)\}$, since $B_{G^{(\nu+1)}}(i^0) = B_{G^{(\nu)}}(i')$, and $A_{G^{(\nu+1)}}(i^0) = A_{G^{(\nu)}}(i')\backslash j'$. Using arguments similar to those in the proof of Lemma 6.2, it is easy to show that $\min\{\omega(i) \mid i \in B_{G'}(i^0)\} \geq \omega(i^0) \geq \max\{\omega(j) \mid j \in A_{G'}(i^0)\}$. It is also clear that the lemma holds for all mixed elements i^0 obtained by mixed transformations I belonging to $L^{(\nu)}$.

Let i' and j' be uniform composite elements. Without loss of generality, assume that $\{i'\} \subseteq N_1$, and $\{j'\} \subseteq N_2$. Find such k, $0 \leq k < \nu$, that $i \overset{G^{(k)}}{\sim} j$ for all $i \in \{i'\}^{(k)}$ and $j \in \{j'\}^{(k)}$, and the $(k+1)$th transformation in the sequence L is of the form $\text{II}-(i'', j'')$, where $i'' \in \{i'\}^{(k)}$ and $j'' \in \{j'\}^{(k)}$. Then, by induction with respect to k, it is not difficult to show that $\omega(i) \geq \omega(j)$ for all $i \in B_{G^{(k)}}(i'') \cup i''$ and $j \in A_{G^{(k)}}(j'') \cup j''$. Using arguments similar to those in the proof of Lemma 6.2, it is easy to show that $\omega(i) \geq \omega(j)$ for all $i \in B_{G^{(\nu)}}(i') \cup i'$ and $j \in A_{G^{(\nu)}}(j') \cup j'$. This implies (since $\omega(j') \geq \omega(i')$) that $\omega(i') = \omega(j') = \omega(i^0)$ and $\min\{\omega(i) \mid i \in B_{G^{(\nu+1)}}(i^0)\} \geq \omega(i^0) \geq \max\{\omega(j) \mid j \in A_{G^{(\nu+1)}}(i^0)\}$. The lemma is proved.

Lemma 6.5. *Let a sequence L of transformations I and II transform a graph $G = G_1pG_2$ into a graph G', and a priority function be auto-bounded. If i^0 is a mixed composite element corresponding to some vertex of graph G', then $\omega(i^0) = \omega(\tilde{\pi}^{(1)}(i^0)) = \omega(\tilde{\pi}^{(2)}(i^0))$.*

Proof. The proof is by induction with respect to the number $\varphi(L)$ of mixed transformations I in the sequence L. If $\varphi(L) = 1$, then the lemma follows from Lemma 6.3.

Suppose that the lemma holds for all such sequences L that $\varphi(L) \leq m$, $m \geq 1$. We prove this holds for $\varphi(L) = m+1$. Suppose that the $(m+1)$th mixed transformation I occupies the $(\nu+1)$th position in sequence L is of the form i $-[i', j']$. Then we may assume that all mixed elements different from $\pi = [i', j']$ are obtained by applying the sequence $L^{(\nu)}$ of transformations I and II, and the lemma holds due to the induction assumption.

Consider the composite element $\pi = [i', j']$. If i' and j' are uniform elements, then, as in the case of the proof of Lemma 6.4, it is easy to show that $\omega(i') = \omega(j')$. Since $\tilde{\pi}^{(1)}(i^0) = i'$, $\tilde{\pi}^{(2)}(i^0) = j'$ and the priority function is auto-bounded, it follows that $\omega(i^0) = \omega(\tilde{\pi}^{(1)}(i^0)) = \omega(\tilde{\pi}^{(2)}(i^0))$.

If i' (or j') is a mixed composite element then Lemma 6.4 implies that $\omega(i') \geq \omega(j')$, since $j' \in A_{G^{(\nu)}}(i')$. Therefore, $\omega(i') = \omega(j')$. By the induction assumption, $\omega(\tilde{\pi}^{(1)}(i')) = \omega(\tilde{\pi}^{(2)}(i')) = \omega(i')$, $\omega(j') = \omega(\tilde{\pi}^{(1)}(j')) = \omega(\tilde{\pi}^{(2)}(j'))$. Since $\omega(i') = \omega(j')$, we obtain that $\omega(i^0) = \omega(\tilde{\pi}^{(1)}(i^0)) = \omega(\tilde{\pi}^{(2)}(i^0))$. The lemma is proved.

The proof of Lemma 6.5 also implies the following statement.

Corollary 6.1. *If the conditions of lemma* 6.5 *are satisfied and* I $-[i, j]$ *is a mixed transformation of a sequence* L, *then* $\omega(i) = \omega(j)$.

Two feasible sequences L_1 and L_2 of transformations I and II of graph G are called *equivalent* if each of them transforms G into the same graph G'.

Let L be a feasible sequence of transformations I and II of graph G. Let $\tilde{L}_\alpha$ (or $\tilde{L}_\beta$) denote a sequence obtained from L by deleting all mixed (or uniform) transformations. Also, define $\tilde{L} = (\tilde{L}_\alpha, \tilde{L}_\beta)$.

Lemma 6.6. *If a priority function is auto-bounded, then a sequence* $\tilde{L}$ *of transformations I and II is feasible for graph* G *and is equivalent to sequence* L.

Proof. The proof is by induction with respect to the number $\varphi(L)$ of mixed transformations in L. If $\varphi(L) = 1$, then the mixed transformation is transformation II. In this case, the lemma is obvious. Suppose that the lemma holds for all sequences L such that $\varphi(L) \leq m$, $m \geq 1$. We prove this holds for $\varphi(L) = m+1$.

Let l be the number of transformations in sequence L. If a mixed transformation occupies the lth position in L, then the lemma holds. In fact, let $L^{(l-1)}$ transform graph G into a graph G', then by the induction assumption, $\tilde{L}^{(l-1)}$ also transforms G to G', and the same transformation occupies the last position in both L and $\tilde{L}$.

Let the last mixed transformation occupy the $(\nu+1)$th position, $\nu < l-1$, in L. If this is transformation II, then it is obvious that L is equivalent to the sequence L' resulted from L by transferring this transformation from the $(\nu+1)$th to the lth place. Suppose that a transformation I $-[i', j']$ occupies the $(\nu+1)$th position in L. Lemma 6.5 implies that $\omega(i') = \omega(j') = \omega(i', j')$. Moreover, the element $[i', j']$ does not participate in transformations occupying the positions $\nu+2, \nu+3, \ldots, l$ in L (otherwise, at least one of those transformations is mixed). Hence, in this case the sequences L and L' are equivalent. The induction assumption implies that L and $\tilde{L}$ are equivalent as well. The lemma is proved.

Theorem 6.2. *Let $G = G_1pG_2$ and a priority function be auto-bounded. Graph G is reducible if and only if the graphs G_1 and G_2 are reducible.*

Proof. Necessity. Suppose that there exists a sequence L_1 of transformations I and II which transforms graph $G_1 = (N_1, U_1)$ into a deadlock graph $G_1' = (N_1', U_1')$ that is not a chain. We show that, in this case, there exists a sequence of transformations I and II which transforms graph $G = (N, U)$ into a deadlock graph that is not a chain.

Let $G_2' = (N_2', U_2')$ denote a graph obtained from G_2 by applying an arbitrary deadlock sequence L_2 of transformations I and II.

Let the sequence $L = (L_1, L_2)$ transform graph G into a graph G'. Then $G' = G_1'pG_2'$. If $G' = (N', U')$ is a deadlock graph, then L is the desired sequence. If graph G' is not deadlock, then Lemma 4.4 implies that none of transformations I can be applied to G'. Let L_3 denote a sequence of all transformations II which can be applied to G'. Any transformation of the sequence L_3 must be either of the form II $-(i, j)$ or of the form II $-(j, i)$, where $i \in N_1'$, $j \in N_2'$. Let sequence L_3 transform graph G' into some graph $G'' = (N', U'')$.

If transformation I $-[i^0, j^0]$ can be applied to graph G'', then $(i^0, j^0) \in U''\backslash U'$ and $\omega(i^0) = \omega(j^0)$. In fact, $\omega(i^0) \geq \omega(j^0)$ since the arc (i^0, j^0) is formed as a result of transformation II. On the other hand, since transformation I $-[i^0, j^0]$ is feasible for G'', we have $\omega(i^0) \leq \omega(j^0)$. Since the priority function is auto-bounded, it follows that $\omega(i^0, j^0) = \omega(i^0) = \omega(j^0)$. Therefore, none of transformations II can be applied to a graph obtained from G'' by transformation I $-[i^0, j^0]$. This also holds for a graph G''' obtained from G'' by applying all feasible transformations I. Thus, graph G''' is deadlock. Besides, if $s \overset{G''}{\sim} t$, then $s' \overset{G'''}{\sim} t'$, where s' and t' are such elements that $\{s\} \subseteq \{s'\}$ and $\{t\} \subseteq \{t'\}$. Since G_1' is not a chain, there exist elements s and t in N_1' such that $s \overset{G_1'}{\sim} t$. Hence, $s \overset{G''}{\sim} t$.

Sufficiency. Let there exist a sequence L of transformations I and II that transforms G into a deadlock graph that is not a chain. We show that, in this case, there is a sequence of transformations which is deadlock for G_1 (or for G_2) and which transforms G_1 (or G_2) into a graph that is not a chain.

Transform sequence L into the sequence $\tilde{L} = (\tilde{L}_\alpha, \tilde{L}_\beta)$ (see Lemma 6.6). Let $\text{I}-[i', j']$ be the first transformation in the sequence $\tilde{L}_\beta$ such that $\{i'\} \cap N_1 \neq \emptyset$ and $\{j'\} \cap N_1 \neq \emptyset$. Suppose that this transformation occupies the $(\nu+1)$th position, $\nu \geq 1$, in $\tilde{L}_\beta$. The definition of the $(\nu+1)$th transformation in $\tilde{L}_\beta$ implies that either $i' = [\tilde{\pi}^{(1)}(i'), \tilde{\pi}^{(2)}(i')]$ or $i' = [\tilde{\pi}^{(2)}(i'), \tilde{\pi}^{(1)}(i')]$. Similarly, either $j' = [\tilde{\pi}^{(1)}(j'), \tilde{\pi}^{(2)}(j')]$ or $j' = [\tilde{\pi}^{(2)}(j'), \tilde{\pi}^{(1)}(j')]$.

Suppose that $j' = [\tilde{\pi}^{(1)}(j'), \tilde{\pi}^{(2)}(j')]$. In the sequence $\tilde{L}_\beta^{(\nu)}$, find such transformation $\text{I}-[i^0, j^0]$ that $\{i^0\} \subseteq \{\tilde{\pi}^{(1)}(i')\}$, $\{j^0\} \subseteq \{\tilde{\pi}^{(2)}(i')\}$. It is clear (see the definition of the $(\nu+1)$th transformation) that $i^0 = \tilde{\pi}^{(1)}(i')$. Due to Corollary 6.1, we have $\omega(i^0) = \omega(j^0)$. Let transformation $\text{I}-[i^0, j^0]$ occupy the position μ_1 in sequence $\tilde{L}_\beta^{(\nu)}$, and the transformation which forms the element i' occupy the position μ_2. Delete the μ th and μ_2th transformations from $\tilde{L}_\beta^{(\nu)}$ and modify successively the transformations placed in the positions $\mu_1+1,\ldots, \mu_2-1$ in the following way. Let the current transformation to be modified be of the form $\text{I}-[i'', j'']$, where $\{[i^0, j^0]\} \subseteq \{i''\}$, $\{j''\} \subset \{\tilde{\pi}^{(2)}(i')\}$. Replace this transformation by $\text{I}-[\tilde{\pi}^{(2)}(i''), j'']$. It is clear that $i'' = [i^0, j^0, j''']$, where $\{j'''\} \subset \{\tilde{\pi}^{(2)}(i')\}$. Due to Lemma 6.5 and Corollary 6.1, we have $\omega(i'') = \omega(\tilde{\pi}^{(2)}(i'')) = \omega(i^0, j^0) = \omega(i^0) = \omega(j^0) = \omega(j'')$. If the next transformation is of the form $\text{II}-[i'', j'']$, where $\{[i^0, j^0]\} \subseteq \{i''\}$ (or $\{[i^0, j^0]\} \subseteq \{j''\}$), then replace it by the pair of transformations $\text{II}-(i^0, j'')$ and $\text{II}-(\tilde{\pi}^{(2)}(i''), j'')$ (or by $\text{II}-(i'', i^0)$ and $\text{II}-(i'', \tilde{\pi}^{(2)}(j''))$, respectively).

If $i' = [\tilde{\pi}^{(2)}(i'), \tilde{\pi}^{(1)}(i')]$, then the μ_1th transformation in $\tilde{L}_\beta^{(\nu)}$ is of the form $\text{I}-[i^0, j^0]$, where $i^0 \subseteq \{\tilde{\pi}^{(2)}(i')\}$ and $j^0 = \tilde{\pi}^{(1)}(i')$. In this case, the transformations of the form $\text{I}-[i'', j'']$, where $\{i''\} \subset \{\tilde{\pi}^{(2)}(i')\}$, $\{[i^0, j^0]\} \subseteq \{j''\}$ are replaced by those of the form $\text{I}-[i'', \tilde{\pi}^{(2)}(j'')]$.

Similarly, we modify the part of the sequence $\tilde{L}_\beta^{(\nu)}$ related to constructing the element j'.

Delete all transformations of the form $\text{II}-[i, j]$, where either $\{i\} \subseteq \{\tilde{\pi}^{(2)}(i')\}$ or $\{i\} \subseteq \{\tilde{\pi}^{(2)}(j')\}$, and $\{j\} \subseteq \{\tilde{\pi}^{(1)}(j')\}$, from $\tilde{L}_\beta^{(\nu)}$. Note that if $i' = [\tilde{\pi}^{(2)}(i'), \tilde{\pi}^{(1)}(i')]$, then also delete the transformations of the form $\text{II}-(i, j)$, where either $\{i\} \subseteq \{\tilde{\pi}^{(1)}(i')\}$ or $\{i\} \subseteq \{\tilde{\pi}^{(1)}(j')\}$, and $\{j\} \subseteq \{\tilde{\pi}^{(2)}(j')\}$.

Let $\overline{L}_\beta^{(\nu)}$ denote the sequence obtained from $\tilde{L}_\beta^{(\nu)}$ in the described way. It is easy to check that applying transformations of the sequence $\overline{L}_\beta^{(\nu)}$ results in forming the composite elements $\tilde{\pi}^{(1)}(i')$, $\tilde{\pi}^{(2)}(i')$, $\tilde{\pi}^{(1)}(j')$, and $\tilde{\pi}^{(2)}(j')$. Lemma 6.5 and Corollary 6.1 imply

$\omega(\tilde{\pi}^{(1)}(i')) = \omega(\tilde{\pi}^{(2)}(i')) = \omega(i') = \omega(\tilde{\pi}^{(1)}(j')) = \omega(\tilde{\pi}^{(2)}(j')) = \omega(j') = \omega(i', j')$.

In the sequence $\tilde{L}_\beta^{(\nu+1)}$, replace its subsequence $\tilde{L}_\beta^{(\nu)}$ by the sequence $\overline{L}_\beta^{(\nu)}$, and replace the transformation in the $(\nu+1)$th position by three transformations: $\mathrm{I}-[\tilde{\pi}^{(1)}(i'), \tilde{\pi}^{(1)}(j')]$, $\mathrm{I}-[\tilde{\pi}^{(2)}(i'), \tilde{\pi}^{(2)}(j')]$, and $\mathrm{I}-[\overline{i}', \overline{j}']$, where $\overline{i}' = [\tilde{\pi}^{(1)}(i'), \tilde{\pi}^{(1)}(j')]$, $\overline{j}' = [\tilde{\pi}^{(2)}(i'), \tilde{\pi}^{(2)}(j')]$. It may happen that either $\tilde{\pi}^{(2)}(i') = \pi_0$ or $\tilde{\pi}^{(2)}(j') = \pi_0$, in which case, the second of the above transformations is dropped. Let $\overline{L}_\beta^{(\nu+1)}$ denote the resulting sequence. In general, the number of transformations in $\overline{L}_\beta^{(\nu+1)}$ may be different from $\nu+1$.

Let the sequence $\tilde{L}_\alpha$ transform graph G into graph G', and the sequence $\tilde{L}_\beta^{(\nu+1)}$ transform graph G' into graph G''. Then, by construction, the sequence $\overline{L}_\beta^{(\nu+1)}$ is feasible for graph G' and transforms that graph into the graph G''' which differs from G'' only in that the vertex, associated with the element $[i', j']$ in G'', is associated with the element $[\overline{i}', \overline{j}']$ in G'''. In this case, the equality $\omega(i', j') = \omega(\overline{i}', \overline{j}')$ holds.

In $\tilde{L}_\beta$, replace its subsequence $\tilde{L}_\beta^{(\nu+1)}$ by the sequence $\overline{L}_\beta^{(\nu+1)}$, and in the other transformations of $\tilde{L}_\beta$ replace the element $[i', j']$ by the element $[\overline{i}', \overline{j}']$. Let $\overline{L}_\beta$ denote the obtained sequence. In $\tilde{L}$, replace its subsequence $\tilde{L}_\beta$ by the sequence $\overline{L}_\beta$ and denote the obtained result by $\overline{L}$. It is obvious that, if $\tilde{L}$ transforms graph G into a deadlock graph G', then $\overline{L}$ transforms G into graph G'', which is isomorphic to G'. Moreover, there exists such an isomorphism that the priorities of the elements associated with the corresponding vertices are equal. Hence, the graph G'' is deadlock and is not a chain.

The sequence $\overline{L}$ contains one mixed transformation of the form $\mathrm{I}-[i', j']$, where $\{i'\} \cap N_1 \neq \varnothing$ and $\{j'\} \cap N_1 \neq \varnothing$ less than L. Using the described procedure sufficiently many times, we obtain some sequence L' which has no mixed transformations of the above form. Sequence L' transforms graph G into a deadlock graph G' that is not a chain. In a similar way, we can pass from L' to a sequence L'' which does not contain mixed transformations of the form $\mathrm{I}-[i', j']$, where $\{i'\} \cap N_2 \neq \varnothing$ and $\{j'\} \cap N_2 \neq \varnothing$. The sequence L'' transforms G into a deadlock graph G'' that is not a chain.

Transform L'' into the sequence $\tilde{L}''= (L_\alpha'', L_\beta'')$ (similar to the way L in which has been transformed into sequence $\tilde{L}$). Due to Lemma 6.6, sequence $\tilde{L}''$ transforms graph G into graph G''.

In sequence $\tilde{L}_\beta''$, all transformations are either of the form I-$[i', j']$ or of the form $\mathrm{II}-(i', j')$, where either $\{i'\} \subseteq N_1$, $\{j'\} \subseteq N_2$ or $\{i'\} \subseteq N_2$, $\{j'\} \subseteq N_1$. Lemma 6.5 and Corollary 6.1 imply that the sequence $\tilde{L}_\alpha''$ transforms each of the graphs G_1 and G_2 into some deadlock graphs G_1' and G_2', respectively. If both of these graphs are chains (more precisely, ω-chains), then it is easy to check that any sequence of transformations I and II which is deadlock for graph $G_1'pG_1'$ must transform that graph into a chain, although, G''

is not a chain. Hence, at least, one of deadlock graphs G_1' and G_2' is not a chain. Suppose that G_1' is not a chain.

Having deleted all transformations acting on set N_2 from $\tilde{L}_\alpha''$, we obtain a sequence which transforms G_1 to G_1'. This proves the theorem.

Recall that in Theorem 6.1, unlike Theorem 6.2, a priority function need not be auto-bounded.

Theorem 6.3. *For an auto-bounded priority function, any series-parallel graph G is reducible.*

Proof. The proof is by induction with respect to the number n of vertices of graph G. For $n = 2$ the theorem is obvious. Suppose that this holds for all $n \leq m$, $m \geq 2$. Let $n = m+1$. Since G is a series-parallel graph, it follows that either $G = G_1sG_2$ or $G = G_1pG_2$, where graphs $G_1 = (N_1, U_1)$, $G_2 = G(N_2, U_2)$ are series-parallel. Observe that $|N_1| \leq m$, $|N_2| \leq m$.

If $G = G_1sG_2$, then by the induction assumption graph G_1 is reducible. This also holds for graph G_2. Theorem 6.1 implies that, in this case, any sequence of transformations I and II which is deadlock for graph G transforms that graph into a chain. Similarly, if $G = G_1pG_2$, then the theorem follows from Theorem 6.2. The theorem is proved.

6.3. Theorem 6.3 implies that *the D-algorithm guarantees that an optimal permutation can be found for any auto-bounded priority function and any series – parallel graph G.* At the same time, it is easy to give examples in which graph G is not series-parallel but for some specific priority function the D-algorithm transforms that graph into a chain. Imposing certain constraints on a pair "priority function – graph G", we may describe more general (compared with series-parallel graphs) classes of "solvable" situations. Consider one of such classes.

Let the priority function $\omega(\pi)$ and graphs $G_1 = (N_1, U_1)$, $G_2 = (N_2, U_2)$ be given such that $N_1 \cap N_2 = \emptyset$ and $N_1 \cup N_2 = N$. Consider the graph $G^0 = (N, U^0)$, which is a subgraph of the graph $G' = (N, U_1 \cup U_2 \cup N_1 \times N_2)$ such that $U_1 \cup U_2 \subseteq U^0$ and if $(i, j) \in N_1 \times N_2$ but $i \overset{G^0}{\sim} j$, then $\omega(i) > \omega(j)$. Graph $G = (N, U)$ is called *an ω-series-composition* of graphs G_1 and G_2 (the notation $G = G_1s_\omega G_2$) if that graph can be constructed from the graph G^0 by removing all its transitive arcs belonging to the set $N_1 \times N_2$.

Graph G is said to be obtained as a result of *the operation of an ω – series composition* of graphs G_1 and G_2 if $G = G_1s_\omega G_2$.

Let G^t denote the graph obtained from graph G as a result of the successive removal of all transitive arcs G.

The graph G is called *an ω-series-parallel* graph if the graph G^t can be constructed by successive application of the operations of ω-series and parallel compositions of single-vertex graphs $G_i = (i, \varnothing)$, $i = 1, 2, \ldots, n$. By definition, a single-vertex graph is ω-series-parallel.

It is easy to verify that for any priority function, any series-parallel graph is at the same time ω-series-parallel.

If the graphs $G_1, G_2, \ldots, G_m$ are such that graph G can be obtained from these graphs as a result of the successive implementation of $m-1$ operations of ω-series and parallel composition, then these graphs are called *components of an ω-decomposition* of graph G.

Lemma 6.7. *Let $G_1 = (N_1, U_1)$ be a component of an ω-decomposition of graph G and $i^0, j^0 \in N_1$. Transformation $I-[j^0, i^0]$ or $II-(i^0, j^0)$ can be applied to graph G if and only if it can be applied to graph G_1.*

Proof. If $i \notin N_1$, then, it is easy to check that exactly one of the following situations may happen: (a) $i \sim i^0$, $i \sim j^0$; (b) $i \rightarrow i^0$, $i \rightarrow j^0$; (c) $i^0 \rightarrow i$, $j^0 \rightarrow i$; (d) $i \rightarrow i^0$, $i \sim j^0$ and $\omega(i) > \omega(j^0)$; (e) $i \sim i^0$, $i \rightarrow j^0$ and $\omega(i) > \omega(i^0)$; (f) $i^0 \rightarrow i$, $j^0 \sim i$ and $\omega(j^0) > \omega(i)$; (g) $i \sim i^0$, $j^0 \rightarrow i$ and $\omega(i^0) > \omega(i)$. Hence, the lemma holds.

Lemma 6.8. *Let a sequence of transformations* II *transform graph G into a graph G'. If transformation* $\text{II}-(i^0, j^0)$ *can be applied to G, and $\omega(i^0) > \omega(j^0)$ then either $i^0 \overset{G'}{\rightarrow} j^0$ or transformation* $\text{II}-(i^0, j^0)$ *can be applied to G'.*

Proof. The relation $j^0 \overset{G'}{\rightarrow} i^0$ may not be valid since $i^0 \overset{G}{\sim} j^0$ and $\omega(i^0) > \omega(j^0)$. Moreover, if $l \overset{G'}{\rightarrow} i^0$ and $l \overset{G}{\sim} i^0$, then the definition of transformation II implies that $\omega(l) \geq \omega(i^0)$. Similarly, if $j^0 \overset{G'}{\rightarrow} l$ and $j^0 \overset{G}{\sim} l$, then $\omega(j^0) \geq \omega(l)$. Therefore, the relation $\min \overline{B}_G(i^0, j^0) \geq \max \overline{A}_G(i^0, j^0)$ yields $\min \overline{B}_{G'}(i^0, j^0) \geq \max \overline{A}_{G'}(i^0, j^0)$ if $i^0 \overset{G'}{\sim} j^0$. Hence, transformation $\text{II}-(i^0, j^0)$ is feasible for the graph G'. The lemma is proved.

Theorem 6.4. *For a given auto-bounded priority function $\omega(\pi)$, let a graph G be ω-series-parallel. Then the D–algorithm transforms G into a chain.*

Proof. Let $G = G_1 s_\omega G_2 = (N, U)$, $G_3 = G_1 s G_2 = (N, U_3)$ and an arc $(i, j) \in U_3$ but $(i, j) \notin U$. Let $G^0 = (N, U^0)$ denote a graph resulting from G by applying all possible transformations II. We show that $(i, j) \in U^0$.

Suppose that $i^0 \in \overline{B}_G(i, j)$, $\omega(i^0) < \omega(i)$ and $\omega(i^0) = \min \overline{B}_G(i, j) < \max \overline{A}_G(i, j)$. Then transformation II – (i^0, j) can be applied to graph G. In fact, $\min \overline{B}_G(i^0, j) = \omega(i^0)$ and for any $j^0 \in A_G(j) \cup j$ at least one of the following relations is valid: $i^0 \xrightarrow{G} j^0$ or $\omega(i^0) > \omega(j^0)$. Hence, $\omega(i^0) > \max \overline{A}_G(i^0, j)$. Lemma 6.8 implies that $i^0 \xrightarrow{G^0} j$.

After the arc (i^0, j) is included in the graph, all other arcs of the form (i', j), where $i' \in \overline{B}_G(i, j)$ and $\omega(i') < \omega(i)$, may also be included successively. Then transformation II – (i, j) can be applied to the obtained graph. Lemma 6.8 guarantees that all the above-mentioned arcs are in $\overline{G}^0$ ($\overline{G}^0$ is the transitive closure of graph G^0).

Thus, if $G = G_1 s_\omega G_2$, then for all pairs of the elements i, j such that the arc $(i, j) \in U_3$ but $(i, j) \notin U$, we have that the relation $i \xrightarrow{G^0} j$ is valid. It may therefore be considered that graph G^0 is obtained from the graph $G_3 = G_1 s G_2$ as a result of applying some sequence of transformations II.

Let G be an ω-series-parallel graph, G_1 be such a component of its ω-decomposition that $G_1 = G_1' s_\omega G_1''$, and i be a vertex of the graph G_1', j be a vertex of G_1'' and $i \sim j$. It follows from the above that there exists a sequence of transformations II which transforms G_1 into a graph G_1^0 such that $i \xrightarrow{G_1^0} j$. Transformation II – (i, j) can be applied to G_1 if and only if it may be applied to G (see Lemma 6.7). This and Lemma 6.8 imply $i \xrightarrow{G^0} j$.

Hence, the arcs which are included in graph G as a result of the implementation of all feasible transformations II, supplement that graph up to a series-parallel graph, and graph G^0 can be considered as a graph obtained from a series-parallel graph as a result of applying some sequence of transformations II.

By implementing Step (c) of the D-algorithm, the graph G^0 can be transformed to some deadlock graph G'. Theorem 6.3 implies that G' is a chain. The theorem is proved.

6.4. There exist situations in which graph G is not ω-series-parallel but the D-algorithm transforms that graph into a chain. Consider the following example.

Let $N = \{1, 2, 3, 4, 5\}$ and the numbers $a_1 = 7$, $a_2 = 4$, $a_3 = 6$, $a_4 = 12$, $a_5 = 10$ be associated with the elements of this set. The precedence relation with the reduction graph $G = (N, U)$, where $U = \{(1, 2), (2, 4), (3, 4), (3, 5)\}$, is defined over set N. The priority function $\omega(\pi) = \sum_{i \in \{\pi\}} a_i / r$, where $\pi = (i_1, i_2, \ldots, i_r)$, is defined over the set $\hat{\mathcal{P}} \backslash \pi_0$.

It is easy to check that G is not an ω-series-parallel graph. It is also easy to verify that the sequence $L =$ (II – (3, 2), I – [2, 4], I – [3, 5], II – ([3, 5], 1), I – [1, [2, 4]]) is constructed by the D-algorithm and transforms the graph into the chain C = ([3, 5], [1, 2, 4]).

7. 1-Priority-Generating Functions

So-called 1-priority-generating functions can be viewed as a natural extension of priority-generating functions. In a number of cases, due to the properties of these functions, efficient algorithms can be developed to optimize them.

7.1. Let $\mathcal{P} \subseteq \mathcal{P}' \subseteq \hat{\mathcal{P}}$, $\mathcal{Q}^{(1)}[\mathcal{P}] = N \cap \mathcal{Q}[\mathcal{P}]$ and $F(\pi)$ be a function defined over set $\mathcal{P}'$.

A function $F(\pi)$ is called 1-*priority-generating over set* $\mathcal{P}$ if there exists a function $\omega^{(1)}(i)$ defined over set $\mathcal{Q}^{(1)}[\mathcal{P}]$ and having the following property: for any elements j, l of $\mathcal{Q}^{(1)}[\mathcal{P}]$ and for any permutations $\pi' = (\pi^{(1)}, j, l, \pi^{(2)})$ and $\pi'' = (\pi^{(1)}, l, j, \pi^{(2)})$ belonging to $\mathcal{P}$, the condition $\omega^{(1)}(j) \geq \omega^{(1)}(l)$ implies $F(\pi') \leq F(\pi'')$. Function $\omega^{(1)}(i)$ is called *a* 1-*priority function*, and the value of $\omega^{(1)}(i)$ is called *the priority* of element i.

It follows from the definition that *any priority-generating function over* $\mathcal{P}$ *is at the same time* 1-*priority-generating over* $\mathcal{P}$. In general, the opposite need not hold.

We present some examples of functions which are 1-priority-generating over set $\mathcal{P}$, but not priority-generating over this set.

(a) Let $F(\pi)$ be function (1.1) (see Section 1 of this chapter), under the condition that $\varphi_i(t) = \varphi(t)+\beta_i$. Here $\varphi(t)$ is a monotonic function. Defining $\pi^{(a)} = j$, $\pi^{(b)} = l$, and using relation (1.3), we obtain

$$\varphi(t_j+t(\pi^{(1)})) \leq \varphi(t_l+t(\pi^{(1)})). \tag{7.1}$$

If $\varphi(t)$ is a non-decreasing function, then to satisfy (7.1) it is sufficient that $t_j \leq t_l$. Define $\omega^{(1)}(i) = -t_i$. Then to satisfy the inequality $F(\pi^{(1)}, j, l, \pi^{(2)}) \leq F(\pi^{(1)}, l, j, \pi^{(2)})$ it is sufficient that $\omega^{(1)}(j) \geq \omega^{(1)}(l)$. Hence, function (1.1) for $\varphi_i(t) = \varphi(t)+\beta_i$, $i = 1, 2,..., n$, where $\varphi(t)$ is the non-decreasing function is 1-priority-generating over $\hat{\mathcal{P}}$ and its 1-priority function is

$$\omega^{(1)}(i) = -t_i.$$

If $\varphi(t)$ is a non-increasing function, then the 1-priority function is $\omega^{(1)}(i) = t_i$.

Recall that, in the case under consideration, function (1.1) is not, in general, priority-generating over $\hat{\mathcal{P}}$.

(b) In Section 1.4 it was shown that, in general, function (1.9) is not priority-generating over $\hat{\mathcal{P}}_n$. We show that this function (which is of the form $F(\pi) = \sum_{k=1}^{r} \alpha_k \beta_{i_k}$, $\pi = (i_1, i_2, \ldots, i_r)$) is 1-priority-generating over $\hat{\mathcal{P}}$ for $\alpha_{i+1} \geq \alpha_i$, $i = 1, 2, \ldots, n-1$. Defining $\pi^{(a)} = j$, $\pi^{(b)} = l$ in relation (1.10) we obtain $(\alpha_{\nu+2} - \alpha_{\nu+1})\beta_l \leq (\alpha_{\nu+2} - \alpha_{\nu+1})\beta_j$. Since $\alpha_{i+1} \geq \alpha_i$, $i = 1, 2, \ldots, n-1$, it follows that

$$\omega^{(1)}(i) = \beta_i$$

is an 1-priority function for function (1.9).

(c) The jobs of a set $N = \{1, 2, \ldots, n\}$ starting at time $d = 0$ are processed successively and continuously on a single machine. The processing time of a job i depends on its starting time t_i^0 and is equal to $t_i = \varphi(t_i^0) + \beta_i$, where $\varphi(t)$ is a non-decreasing and non-negative function for $t \geq 0$, and $\beta_i > 0$, $i = 1, 2, \ldots, n$. It is required to find, in a given set $\mathcal{P} \subseteq \hat{\mathcal{P}}_n$, a sequence π_n^* of jobs which minimizes the total processing time.

Over set $\hat{\mathcal{P}}$, define the function

$$F(\pi) = \sum_{k=1}^{r} t_{i_k}, \tag{7.2}$$

where $\pi = (i_1, i_2, \ldots, i_r) \in \hat{\mathcal{P}}$, $F(\pi_0) = 0$. It is obvious that $F(\pi_n)$ represents the total processing time of the jobs of set N processed according to the sequence π_n.

We now establish the conditions under which the inequality $F(\pi^{(1)}, j, l, \pi^{(2)}) \leq F(\pi^{(1)}, l, j, \pi^{(2)})$ holds. To satisfy this inequality it is sufficient that $F(\pi^{(1)}, j, l) \leq F(\pi^{(1)}, l, j)$, which is equivalent to

$$\varphi(F(\pi^{(1)}) + \varphi(F(\pi^{(1)})) + \beta_j) \leq \varphi(F(\pi^{(1)}) + \varphi(F(\pi^{(1)})) + \beta_l). \tag{7.3}$$

Since $\varphi(t)$ is a non-decreasing function, the last inequality holds if $\beta_j \leq \beta_l$. Hence, the function $\omega^{(1)}(i) = -\beta_i$ is an 1-priority function for $F(\pi)$, and $F(\pi)$ is 1-priority-generating over $\hat{\mathcal{P}}$.

We show that, in general, function $F(\pi)$ is not priority-generating over $\hat{\mathcal{P}}$. In fact, let $N = \{1, 2, 3, 4\}$, $\beta_1 = \beta_2 = 1$, $\beta_3 = 9$, $\beta_4 = 4$, $\varphi(t) = t^2$, $\pi^{(a)} = (2, 3)$, $\pi^{(b)} = 4$. Consider two variants: $\pi^{(1)} = 1$, $\pi^{(2)} = \pi_0$ and $\pi^{(1)} = \pi^{(2)} = \pi_0$. In the first case, we have $F(1, 2, 3, 4) = 464 < F(1, 4, 2, 3) = 471$, while in the second case, we have $F(2, 3, 4) = 134 > F(4, 2, 3) = 65$.

(d) For the previous problem, let function $\varphi(t)$ be non-decreasing, non-negative and satisfy the condition $\Delta t \geq |\Delta\varphi(t)|$ for $t \geq 0$, where $\Delta\varphi(t) = \varphi(t + \Delta t) - \varphi(t)$.

The condition $\Delta t \geq |\Delta\varphi(t)|$ implies that the inequality $F(\pi^{(1)}, j, l, \pi^{(2)}) \leq F(\pi^{(1)}, l, j, \pi^{(2)})$ holds if relation (7.3) holds. Since $\varphi(t)$ is a non-increasing function, inequality (7.3) holds if $\beta_j \geq \beta_l$.

Thus, in this case, function (7.2) is 1-priority-generating with the 1-priority-function $\omega^{(1)}(i) = \beta_i$.

7.2. If a function $F(\pi)$ is 1-priority-generating over set $\mathcal{P}$, then for the search for an optimal permutation π^* over $\mathcal{P}$ (minimizing $F(\pi)$ over $\mathcal{P}$), the following obvious procedure is widely used. Let $P(l, j)$ be a set of all permutations of $\mathcal{P}$ of the form $(\pi^{(1)}, l, j, \pi^{(2)})$, for each of which in $\mathcal{P}$ there exists a permutation of the form $(\pi^{(1)}, j, l, \pi^{(2)})$. If $\omega^{(1)}(j) \geq \omega^{(1)}(l)$, then the set $\mathcal{P}\backslash P(l, j)$ contains at least one optimal permutation. Hence, while searching for π^*, the set of permutations $P(l, j)$ may be skipped.

In particular, this implies the following statement.

Theorem 7.1. *If a function $F(\pi)$ is 1-priority- generating over set $\hat{\mathcal{P}}_n$, then the permutation in which the elements are sorted in non-increasing order of their priorities minimizes $F(\pi)$ over $\hat{\mathcal{P}}_n$.*

In fact, let $\omega^{(1)}(i_{k_1}) = \max\{\omega^{(1)}(i) \mid i \in N\}$ and $P(j)$ be a set of all permutations $\pi = (i_1, i_2,..., i_n) \in \hat{\mathcal{P}}_n$ such that $i_1 = j$. Then any permutation π that does not belong to the set $P(i_{k_1})$ belongs to some set $P(l, i_{k_1})$ and, hence, the set $P(i_{k_1})$ contains an optimal permutation (minimizing $F(\pi)$ over $\hat{\mathcal{P}}_n$). Similarly, an optimal permutation can be found among those permutations of $P(i_{k_1})$, in which the second position is occupied by an element i_{k_2} such that $\omega^{(1)}(i_{k_2}) = \max\{\omega^{(1)}(i) \mid i \in N\backslash i_{k_1}\}$, etc.

As a result of this successive reduction of the search region, a permutation $\pi^* = (i_{k_1}, i_{k_2},..., i_{k_n})$ is obtained such that $F(\pi^*) \leq F(\pi)$, $\pi \in \hat{\mathcal{P}}_n$, and $\omega^{(1)}(i_{k_j}) \geq \omega^{(1)}(i_{k_{j+1}})$, $j = 1, 2,..., n-1$.

7.3. To conclude this section, we consider a function which is not 1-priority-generating over $\hat{\mathcal{P}}_n$ but has an 1-priority function over some special subset $\mathcal{P} \subset \hat{\mathcal{P}}_n$.

Let $N = N_1 \cup N_2$ and $N_1 \cap N_2 \neq \emptyset$. Associate real numbers α_i and $t_i > 0$ with each element $i \in N$. The function $\varphi_i^{(1)}(t) = \alpha_i t$ corresponds to each element $i \in N_1$ and the function $\varphi_i^{(2)}(t) = \alpha_i \exp(\gamma t)$, $\gamma \neq 0$, corresponds to each element $i \in N_2$. Let the function

$$F(\pi) = \sum_{k=1}^{n} \varphi_{i_k}^{(\nu)}\left[\sum_{j=1}^{k} t_{i_j}\right] \tag{7.4}$$

be defined over set $\hat{\mathcal{P}}_n$, where $\pi = (i_1, i_2,..., i_n)$, and $\nu = 1$ if $i_k \in N_1$, $\nu = 2$ if $i_k \in N_2$.

We show that, in a general case, function (7.4) is not 1-priority-generating over $\hat{\mathcal{P}}_n$. Let $N = \{1, 2, 3\}$, $N_1 = \{1, 2\}$, $N_2 = \{3\}$, $\alpha_1 = 1$, $\alpha_2 = 3$, $\alpha_3 = 1/16$; $t_1 = 5$, $t_2 = 1$, $t_3 = 4$; $\gamma = \ln 2$. Then $F(2, 3, 1) = 15 < F(3, 2, 1) = 26$ but $F(1, 2, 3) = 87 > F(1, 3, 2) =$

67.

Let $\widetilde{\mathcal{P}}_n$ denote the set of all permutations of the form $(\pi^{(1)}, \pi^{(2)})$ and $(\pi^{(2)}, \pi^{(1)})$, where $\{\pi^{(\nu)}\} = N_\nu$, $\nu = 1, 2$. Function (7.4) is 1-priority-generating over set $\widetilde{\mathcal{P}}_n$. Using the results of Section 1.1 of this chapter (see Items (a) and (b)), we may conclude that, in the case under consideration, the function

$$\omega^{(1)}(i) = \begin{cases} \alpha_i / t_i & \text{for } i \in N_1, \\ \alpha_i \exp(\gamma t_i)/(\exp(\gamma t_i) - 1) & \text{for } i \in N_2 \end{cases}$$

is an 1-priority function for $F(\pi)$.

8. Bibliography and Review

The concept of a priority-generating function was introduced by Shafransky [173, 51] and then defined more precisely together with Tanaev. Later on, similar concepts were introduced independently by Burdyuk and Reva [19, 28], as well as by Monma and Sidney [365] (under some extra restrictions for a function). These restrictions have been then removed by Monma [364], thereby resulting in a concept equivalent to the one introduced in [173].

The problem of minimizing the functions from Items (a) and (c) of Section 1.1 over set $\hat{\mathcal{P}}_n$ is considered by Smith [417] and by Tanaev [153] and Rothkopf [386], respectively. These papers propose algorithms with the running time of $O(n\log n)$. Priority functions (1.4) and (1.5) for the above objective functions are constructed by Horn [293] and by Gordon and Tanaev [48], respectively. The problem of minimizing function (1.6) over set $\hat{\mathcal{P}}_n$ is a special case of the so-called parametric scheduling problem formulated by Mel'nikov [108]. The problem of minimizing linear form (1.9) over set $\mathcal{P} \subset \hat{\mathcal{P}}_n$ is formulated by Suprunenko [149]. In particular, it is shown in [149] that a number of known extremal problems over permutations may be reduced to the above problem. Suprunenko, Aizenshtat, Lepeshinsky, Metel'sky, Kuntsevich, Kravchuk have carried out interesting research on minimizing function (1.9) over various subsets $\mathcal{P} \subset \hat{\mathcal{P}}_n$. A review of the obtained results can be found in [150]. Later on, these studies have been continued by Suprunenko, Metel'sky and Sarvanov [151, 109, 137, 138, 139]. An $O(n\log n)$ algorithm for minimizing function (1.9) over set $\hat{\mathcal{P}}_n$ is proposed by Hardy et al. [164]. The problem of minimizing function (1.13) is formulated and studied by Tanaev in [152] where an $O(n\log n)$ algorithm for minimizing this function over set $\hat{\mathcal{P}}_n$ is proposed. The problem of minimizing function (1.20) over set $\mathcal{P}_n(G)$ is formulated in [187]. A number of papers are devoted to studying and solving the problem presented in Section 1.6. Livshits and Rublinetsky [100] prove this problem to be NP-hard (even for $\delta_i = t_{1i}$). For $d_i = 0$, $\delta_i = t_{1i}$, the problem is solved by Johnson [59] and Bellman [204]. The algorithm by Johnson is of $O(n\log n)$ running time. The solution given by Bellman is based on the dynamic programming method. Under the above-mentioned conditions, the problem is known as the $2\times n$ flow shop (or Bellman-Johnson) problem. More general cases are considered in [91, 126, 152, 309, 359, 371, 420]. Reducing the optimal linear arrangement problem for a directed graph G to that of minimizing the function from Item (a) Section 1.1 over set $\mathcal{P}_n(G)$ is made by Adolphson and Hu [189] (see also [336]). The problem of minimizing function (1.29) over set $\hat{\mathcal{P}}_n$ is considered by Elmaghraby [251] and Nikitin [123]; special cases of this problem are considered by Burkov and Sokolov [26]; see also [21, 22]. Priority functions (1.7), (1.11), (1.12) and (1.19) are found by Shafransky [172]; the results obtained in [152] have been essentially used for constructing function (1.19).

Tuzikov shows that the function $F(\pi) = \max\{\varphi_{i_k}(\bar{t}_{i_k}) \mid 1 \le k \le r\}$ where $\pi = (i_1, i_2, \ldots, i_r)$, $\bar{t}_{i_k} = \sum_{l=1}^{k} t_{i_l}$, is priority-generating over set $\hat{\mathcal{P}}_n$ in the following cases: 1) $\varphi_i(t) = \alpha t + \beta_i$; 2) $\varphi_i(t) = \alpha_i \exp(\gamma t)$; 3) $\varphi_i(t) = \varphi(t - D_i)$, where $\varphi(x)$ is a non-decreasing function, $D_i \ge 0$. The corresponding priority functions are of the form: $\omega(\pi) = F(\pi) - \varphi\big(\sum_{i\in\{\pi\}} t_i\big)$, $\omega(\pi) = F(\pi)/\exp \gamma\big(\sum_{i\in\{\pi\}} t_i\big)$ and $\omega(\pi) = \max\{\bar{t}_{i_k} - D_{i_k} \mid 1 \le k \le r\} - \sum_{i\in\{\pi\}} t_i$, respectively. This result has not been published previously.

Algorithms for minimizing the function from Section 1.1, provided that each connected component of graph G is a chain, have been proposed independently by Shkurba et al. [185] and by Conway et al. [78]. The running time of both algorithms is $O(n\log n)$. An algorithm for minimizing the function from Item (a) of Section 1.1, provided that each connected component of G is an outtree is proposed by Horn [293]. An algorithm for minimizing the function from Item (b) of Section 1.1, provided that G is a tree-like graph is due to Gordon and Tanaev [48]. The running time of both algorithms is $O(n^2)$. An $O(n\log n)$ algorithm the optimal linear arrangement problem for a directed graph G, assuming that each connected component of G is an outtree, is developed by Adolphson and Hu [189]. Kurisu [322] gives an $O(n\log n)$ algorithm for solving the $2\times n$ flow shop problem under chain-like precedence constraints. Here, the precedence constraints (defined by the relation $\rightarrow$) have the following meaning. If $i \rightarrow j$, then a machine may not start processing job j until the processing of job i on that machine is completed. If it is assumed that the processing of job j may not start on any machine unless job i is completed on all machines, the problem is proved to be NP-hard [345]. If G is an arbitrary circuit-free graph, Kurisu [323] describes a number of rules for reducing enumeration. An algorithm for minimizing an arbitrary priority-generating function over set $\mathcal{P}_n(G)$, provided that G is tree-like, is due to Shafransky [173]. The running time of that algorithm is $O(n^2)$.

Lawler [333] gives an $O(n\log n)$ algorithm for minimizing the function from Item (a) of Section 1.1 over set $\mathcal{P}_n(G)$, assuming that graph G is series-parallel. This paper uses the results obtained by Sidney [408]. In [408], a decomposition scheme is proposed to solve the problem of minimizing the above function over set $\mathcal{P}_n(G)$ for an arbitrary circuit-free graph G. Later, some results by Sidney were extended to the function from Item (b) of Section 1.1; see [279, 280]. An algorithm for minimizing the functions from Items (a) and (b) of Section 1.1 over set $\mathcal{P}_n(G)$ in the situation when graph G is of a somewhat more general form than a series-parallel graph, is proposed by Zinder [68]. An algorithm for minimizing an arbitrary priority-generating function over set $\mathcal{P}_n(G)$, assuming that graph G is series-parallel, is due to Gordon and Shafransky [51-54]; the running time of the algorithm is $O(n^2)$. Later, a similar approach was proposed by Monma and Sidney [365]. Besides, algorithms for minimizing some specific priority-generating functions over set $\mathcal{P}_n(G)$ for series-parallel graph G are described in [90, 187, 362, 410].

The papers [50] and [364] present examples illustrating that for the functions from Items (a) and (b) of Sections 1.1 and graphs G not being series-parallel, set $\mathcal{P}_n(G)$ may contain non-optimal permutations which cannot be "improved" by the transposition of any neighboring groups of elements.

Sections 2-6 are based on the results obtained by Shafransky and partly published in [174-177]. A transformation which is a prototype of transformation I was introduced by Adolphson [188] while studying the problem of minimizing the function from Item (a) of Section 1.1 over set $\mathcal{P}_n(G)$ for an arbitrary circuit-free graph G. Studying the so-called problem of minimizing the cost of the project performance under precedence constraints, Garey [266] introduced a transformation which can be viewed as a special case of the one described by Adolphson. An $O(n^2)$ algorithm for the case of a tree-like graph G is described in [266]. The same problem with no precedence constraints is solved by Livshits [97]; the running time of his algorithm is $O(n\log n)$. In [97], it is proved that the objective function in the above problem is in fact 1-priority-generating and the 1-priority function is found. As shown in [158], the objective function in that problem is a special case of the function from Item (b) of Section 1.1. In [55], the results by Adolphson are extended to the case of an arbitrary priority-generating function, and in [364], his transformations of graph G are generalized.

An algorithm close to the D-algorithm [174] is proposed by Burdyuk and Reva [19] (with no estimates of the running time and studying the properties); see also [107]. Structures occupying intermediate positions between series-parallel and ω-series-parallel graphs are constructed by Zinder [68] and Reva [129].

The concept of a 1-priority-generating function is one of the possible formalizations of the so-called interchange technique [158]. The function from Item (a) of Section 7.1 is considered by Tanaev [153, 154], and the function from Item (b) of Section 7.1 by Mel'nikov and Shafransky [108]. These papers also suggest the corresponding priority functions.

The function of the form $F(\pi) = \sum_{k=1}^{n} f(i_k)\varphi_k(i_1, i_2,..., i_k)$ where $\pi = (i_1, i_2,..., i_n)$ is considered by Rau [381]. He shows that this function is 1-priority-generating under the following conditions: (1) $f(i_k) > 0$, $k = 1, 2,..., n$; (2) $\varphi_k(i_1, i_2,..., i_{k-1}, i_k) = \varphi_k(i_1', i_2',..., i_{k-1}', i_k)$, if $\{i_1, i_2,..., i_{k-1}\} = \{i_1', i_2',..., i_{k-1}'\}$; 3) there exist the functions Φ, $\Phi_1 \equiv 1$, $\Phi_2,..., \Phi_n \geq 0$ such that $\varphi_k(i_1, i_2,..., i_{k-1}, i_k) - \varphi_{k+1}(i_1', i_2',..., j,..., i_{k-1}', i_k) = \Phi(j)\Phi_k(i_1, i_2,..., i_{k-1})$, if $\{i_1, i_2,..., i_{k-1}\} = \{i_1', i_2',..., i_{k-1}'\}$. The 1-priority function is of the form $\Phi(i)/f(i)$.

Lawler and Sivazlian [343] consider the problem of minimizing the function

$$F(\pi) = \sum_{k=1}^{n} \int_{\bar{t}_{i_k} - t_{i_k}}^{\bar{t}_{i_k}} \varphi_{i_k}(x)dx,$$

where $\pi = (i_1, i_2,..., i_n)$ and $\bar{t}_{i_k} = \sum_{l=1}^{k} t_{i_l}$. If $\varphi_i(x) = \alpha_i\varphi(x) + \beta_i$ and $\varphi(x)$ is a monotonic function over the interval $[0, \sum_{i=1}^{n} t_i]$, then $F(\pi)$ is a 1-priority-generating function. In this case, the 1-priority function is of the form $\omega^{(1)}(i) = \alpha_i$ if $\varphi(x)$ is a non-decreasing function and $\omega^{(1)}(i) = -\alpha_i$, if $\varphi(x)$ is a non-increasing function.

Kladov and Livshits [76] proved (in some other terms) that function (1.1) is 1-priority-generating (assuming that the functions $\varphi_i(x)$ are strictly increasing and sufficiently smooth) if and only if either $\varphi_i(x) = \alpha_i x + \beta_i$, $i = 1, 2,..., n$, or $\varphi_i(x) = \alpha_i \exp(\gamma x) + \beta_i$, $i = 1, 2,..., n$, or $\varphi_i(x) = \varphi(x) + \beta_i$, $i = 1, 2,..., n$. Zinder [68] shows (under the same assumptions) that function (1.1) is priority-generating over set $\hat{\mathcal{P}}_n$ if and only if $\varphi_i(x) = \alpha_i x + \beta_i$, $i = 1, 2,..., n$, or $\varphi_i(x) = \alpha_i \exp(\gamma x) + \beta_i$, $i = 1, 2,..., n$.

The issue of extending spheres of the effective use of the interchange technique is discussed by Shkurba [184], Burdyuk [18], Livshits [99], Khenkin [167]; see also [398, 399].

CHAPTER 4
NP-HARD PROBLEMS

This chapter establishes the NP-hardiness of a number of scheduling problems. To prove that a given Problem B is NP-hard, we use the following scheme. The decision Problem B' corresponding to Problem B is formulated, and a Problem A is shown to be polynomially reducible to B' where A is one of the standard problems, i.e., a decision problem known to be NP-complete. If Problem A is NP-complete in the strong sense, then sometimes it is shown to be pseudopolynomially reducible to Problem B'.

The following standard problems are chosen: the partition problem (Section 1), the 3-partition problem (Section 2), the vertex covering problem (Section 3), the clique problem (Section 4) and the linear arrangement problem (Section 5).

For most of the problems proved to be NP-hard, polynomially solvable special cases are presented.

Along with the usual notation such as t_i for the processing time of a job i and D_i for its due date, this chapter uses expressions of the form $t(i)$ and $D(i)$ to denote the same parameters. Similarly, together with the notation d_i, α_i, L_i, z_i, u_i etc., the notation $d(i)$, $\alpha(i)$, $L(i)$, $z(i)$, $u(i)$ is used.

1. Reducibility of the Partition Problem

In this section, the partition problem is used as a standard problem for proving the NP-hardness of some scheduling problems.

The *partition* problem can be formulated as follows. Given a set $N^0 = \{1, 2,..., n_0\}$, each element $i \in N^0$ is associated with a positive integer γ_i such that $\sum_{i \in N^0} \gamma_i = 2A$, does there exist a partition of set N^0 into two subsets N_1^0 and N_2^0 such that $A_1 = A_2$? Here $A_k = \sum_{i \in N_k^0} \gamma_i$ for $N_k^0 \subset N^0$.

In the binary alphabet, the input length of a partition problem belongs to the interval $[c_1 n_0 \log\gamma', c_2 n_0 \log\gamma'']$, where $\gamma' = \min\{\gamma_i | i \in N^0\}$, $\gamma'' = \max\{\gamma_i | i \in N^0\}$, and $c_1 \le c_2$ are positive constants.

The partition problem is NP-complete but not in the strong sense (a pseudopolynomial algorithm for solving this problem is known).

1.1. This section considers the following problems.

Problem 1.1 The jobs of a set $N = \{1, 2,..., n\}$ enter a system consisting of two identical parallel machines at time $d = 0$. A job $i \in N$ can be processed on any of the machines during $t_i > 0$ time units. Preemption is not allowed. It is required to find a schedule s^* which minimizes the function $F(s)$ in the following cases:

(a) $F(s) = \bar{t}_{max}(s) = \max\{\bar{t}_i(s) | i \in N\}$, where $\bar{t}_i(s)$ is the completion time of a job i in a schedule s;

(b) $F(s) = \bar{t}_{max}(s) \sum_{i \in N} \bar{t}_i(s)$;

(c) $F(s) = \sum_{i \in N} \alpha_i \bar{t}_i(s)$, where α_i is a non-negative real number associated with job $i \in N$.

Problem 1.2. A processing system consists of a single machine. A job i of a given set $N = \{1, 2, .., n\}$ enters the system at time $d_i \ge 0$, its processing time is $t_i > 0$. Preemption is not allowed. Each job $i \in N$ is associated with a non-negative number α_i and the deadline $D_i \ge 0$, by which it is desirable to complete processing. It is required to find a schedule s^* which minimizes the function $F(s)$ in the following cases:

(a) $F(s) = L_{max}(s) = \max\{L_i(s) | i \in N\}$, where $L_i(s) = \bar{t}_i(s) - D_i$;

(b) $F(s) = \sum_{i \in N} \alpha_i u_i(s)$, $d_i = 0$, $i = 1, 2,..., n$; here $u_i(s) = 0$ if $\bar{t}_i(s) \le D_i$, and $u_i(s) = 1$ if $\bar{t}_i(s) > D_i$.

Problem 1.3. The jobs of a set $N = \{1, 2,..., n\}$ enter a system consisting of $M \le n$ identical parallel machines at time $d = 0$. Each job i is processed during $t_i > 0$ time units on any machine with no preemption. It is required to find the smallest number M^* of machines which provides the completion of the processing of all jobs by a given deadline $D \ge \max\{t_i | i \in N\}$.

In the following, these problems are shown to be NP-hard.

1.2. For Problem 1.1(a) the value of $F(s)$ is specified by a distribution of the jobs among the machines, i.e., by partitioning set N into two subsets N_1 and N_2.

The following decision problem corresponds to Problem 1.1(a): determine whether there exists a schedule s^0 for processing the jobs of set N such that $\bar{t}_{max}(s^0) \le y$ for a given y.

The partition problem is reduced to this decision problem in polynomial time. In fact, define $n = n_0$, $t_i = \gamma_i$, $i = 1, 2, \ldots, n$, $y = A$. It is obvious that a schedule s^0 with $\bar{t}_{max}(s^0) \le y$ exists if and only if for the partition problem there exists a partition of set N^0 into two subsets N_1^0 and N_2^0 that $A_1 = A_2$, i.e., if and only if the partition problem has a solution. The described reduction can be implemented in $O(n_0)$ time.

Thus, Problem 1.1(a) is NP-hard.

Note that if preemption is allowed this problem can be solved in $O(n)$ time for any number $M \ge 2$ of processing machines (see Section 6.2 of Chapter 2).

1.3. Consider Problem 1.1(b). The corresponding decision problem is as follows: determine whether there exists a schedule s^0 such that $\bar{t}_{max}(s^0) \sum_{i \in N} \bar{t}_i(s^0) \le y$ for a given y. We show that the partition problem reduces to the latter problem in polynomial time.

Define

$$n = 2n_0,\ t_i = 2A+\gamma_i,\ t_{n_0+i} = 2Ai,\ i = 1, 2, \ldots, n_0;$$

$$y = (\frac{2}{3}\ An_0(n_0+1)(n_0+2) + \sum_{i=0}^{n_0}(n_0-i+1)\gamma_i)A(n_0(n_0+1)+1).$$

We show that for the constructed problem a schedule s^0 exists if and only if there exists a partition of set N^0 into subsets N_1^0 and N_2^0 such that $A_1 = A_2$.

It is clear that the value of $F(s)$ is specified by both the distribution of the jobs among the machines and the processing sequences for the jobs assigned to a machine.

We may consider only the situation in which each of the jobs i and $i+n_0$, $i = 1, 2, \ldots, n_0$, occupies the ith position in the processing sequence either on the first or the second machine. In fact, in any such a schedule s, the value $\sum_{i=1}^{n} \bar{t}_i(s)$ attains its minimum equal to $\frac{2}{3}An_0(n_0+1)(n_0+2) + \sum_{i=0}^{n_0}(n_0-i+1)\gamma_i$ (see Section 9.3 of Chapter 2). Moreover, it can be easily verified that it is possible to transform any schedule $\tilde{s}$ which does not satisfy the above condition into a schedule s' which satisfies this condition and such that the inequality $\bar{t}_{max}(s') \le \bar{t}_{max}(\tilde{s})$ holds.

Since the given condition fixes the order of job processing, the only question that remains to be answered is which of the two machines a job is processed on. In this case, the value of the function $\overline{t}_{max}(s)\sum_{i\in N}\overline{t}_i(s)$ is specified only by $\overline{t}_{max}(s)$. In turn, a schedule s^0 with $\overline{t}_{max}(s^0) \leq A(n_0(n_0+1)+1)$ exists if and only if the partition problem has a solution.

The implementation of the described reduction of the partition problem to the decision problem under consideration requires at most $O(n_0)$ time.

Thereby, Problem 1.1(b) is thus NP-hard.

1.4. A schedule s in Problem 1.1(c) is specified by a pair of permutations $\pi^{(1)}$ and $\pi^{(2)}$ which specify the processing sequence of the jobs of set N on each of the machines. It is obvious that $N = \{\pi^{(1)}\} \cup \{\pi^{(2)}\}$ and $\{\pi^{(1)}\} \cap \{\pi^{(2)}\} = \varnothing$.

The corresponding decision problem is as follows: determine whether there exists a schedule s^0 such that $\sum_{i\in N}\alpha_i\overline{t}_i(s^0) \leq y$ for a given number y. We show that the partition problem reduces to this decision problem in polynomial time.

Define $n = n_0$; $t_i = \gamma_i$, $\alpha_i = \gamma_i$, $i = 1, 2,..., n$; $y = A^2+\frac{1}{2}\sum_{i\in N^0}\gamma_i^2$.

Let a schedule s be defined by a pair of the permutations $\pi^{(1)} = (i_1, i_2,..., i_{n_1})$ and $\pi^{(2)} = (j_1, j_2,..., j_{n_2})$. Compute $\sum_{i\in N}\alpha_i\overline{t}_i(s)$. Define $N_1^0 = \{\pi^{(1)}\}$, $N_2^0 = \{\pi^{(2)}\}$. It is clear that $N_1^0 \cup N_2^0 = N^0$ and

$$\sum_{i\in N_1^0}\alpha_i\overline{t}_i(s) = \sum_{k=1}^{n_1}\gamma_{i_k}\sum_{l=1}^{k}\gamma_{i_l} = \sum_{i\in N_1^0}\gamma_i^2 + \sum_{1\leq l<k\leq n_1}\gamma_{i_k}\gamma_{i_l} = \frac{1}{2}\,(A_1^2+\sum_{i\in N_1^0}\gamma_i^2).$$

Similarly, $\sum_{i\in N_2^0}\alpha_i\overline{t}_i(s) = \frac{1}{2}\,(A_2^2+\sum_{i\in N_2^0}\gamma_i^2)$.

Hence,

$$\sum_{i\in N}\alpha_i\overline{t}_i(s) = \frac{1}{2}\,(A_1^2+A_2^2) + \frac{1}{2}\sum_{i\in N^0}\gamma_i^2.$$

Since $A_1+A_2 = 2A$, the value $A_1^2+A_2^2$ attains its minimum equal to $2A^2$ at $A_1 = A_2$. Therefore, a schedule s^0 for which $\sum_{i\in N}\alpha_i\overline{t}_i(s^0) \leq A^2+\frac{1}{2}\sum_{i\in N^0}\gamma_i^2$ exists if and only if the partition problem has a solution. The described reduction can be implemented in $O(n_0)$ time. Thus, Problem 1.1(c) is NP-hard.

Note that if $\alpha_i = 1$, $i = 1, 2,..., n$, then this problem is solvable in $O(n\log n)$ time (see Section 9.3 of Chapter 2). Moreover, the corresponding algorithm is designed for finding an optimal schedule in a more complex situation of $M \geq 2$ uniform machines.

1.5. We show that Problem 1.2(a) is NP-hard. The corresponding decision problem is as follows: determine whether there exists a schedule s^0 for single-machine processing of the jobs of set N such that $L_{max}(s^0) \le y$ for a given y.

Let us describe a polynomial reduction of the partition problem to the formulated decision problem.

Define $n = n_0+1$; $t_i = \gamma_i$, $d_i = 0$, $D_i = 2A+1$, $i = 1, 2,..., n_0$; $t_n = 1$, $d_n = A$, $D_n = A+1$; $y = 0$.

In Problem 1.2(a), a schedule is specified by a permutation $\pi = (i_1, i_2,..., i_n)$ of the elements of set N. Note that the starting time of a job i_k is $t^0_{i_k} = \max\{d_{i_k}, \overline{t}_{i_{k-1}}\}$, $k = 2, 3,..., n$, $t^0_{i_1} = d_{i_1}$.

Let a permutation π be of the form $\pi = (\pi^{(1)}, n, \pi^{(2)})$. Define $N^0_1 = \{\pi^{(1)}\}$, $N^0_2 = \{\pi^{(2)}\}$. It is clear that $N^0_1 \cup N^0_2 = N^0$. Therefore,

$$L_{max}(s) = \max\{\max\{A_1, A\} + 1-(A+1), \max\{A_1, A\}+1+A_2-(2A+1)\}$$

$$= \max\{\max\{A_1-A, 0\}, \max\{0, A-A\}\}$$

$$= \max\{\max\{A_1-A_2)/2, 0\}, \max\{(A_2-A_1)/2, 0\}\} = |A_1-A_2|/2.$$

Hence, $L_{max}(s) \le y = 0$ if and only if there exists a partition of set N^0 into two subsets N^0_1 and N^0_2 such that $A_1 = A_2$. The implementation of the described reduction requires at most $O(n_0)$ time. Thus, Problem 1.2(a) is NP-hard.

Note that if $d_i = 0$, $i = 1, 2,..., n$, Problem 1.2(a) can be solved in $O(n\log n)$ time (see Section 3.3 of Chapter 2). Besides, Problem 1.2 is polynomially solvable if $d_i = 0$, $i = 1, 2,..., n$, and $F(s) = \max\{\varphi_i(\overline{t}_i(s)) \mid i \in N\}$, where $\varphi_i(t)$ are non-decreasing functions. Moreover, precedence constraints may be defined over set N, and an optimal schedule must be feasible with respect to these constraints.

1.6. As in the previous problem, in Problem 1.2(b) a schedule s is determined by a permutation π of the elements of set N.

The decision problem corresponding to Problem 1.2(b) is as follows: determine whether there exists a schedule s^0 such that $\sum_{i \in N} \alpha_i u_i(s^0) \le y$ for a given y.

We show that the partition problem reduces to this decision problem in polynomial time.

Define $n = n_0+1$; $t_i = \gamma_i$, $\alpha_i = \gamma_i$, $D_i = 2A$, $i = 1, 2,..., n_0$; $t_n = 2A$, $\alpha_n = 2A$, $D_n = 3A$; $y = A$. It is clear that the described transformations can be done in at most $O(n_0)$ time.

A schedule s^0 such that $\sum_{i \in N} \alpha_i u_i(s^0) \le A$ exists if and only if the partition problem has a solution. In fact, let a permutation π that specifies a schedule s be of the form

$\pi = (\pi^{(1)}, n, \pi^{(2)})$. Define $N_1^0 = \{\pi^{(1)}\}$, $N_2^0 = \{\pi^{(2)}\}$. Then $N_1^0 \cup N_2^0 = N^0$ and $\sum_{i \in N} \alpha_i u_i(s) = 2Au_n(s) + A_2$. It is easy to check that $u_n(s) = 0$ if and only if $A_1 \leq A$. Therefore, if permutation π is such that $A_1 > A$, then $\sum_{i \in N} \alpha_i u_i(s) > A$. If $A_1 \leq A$, then $\sum_{i \in N} \alpha_i u_i(s) = A_2$ and, hence, the inequality $\sum_{i \in N} \alpha_i u_i(s) \leq A$ holds if and only if $A_1 = A_2 = A$. Thus, Problem 1.2(b) is NP-hard.

Problem 1.2(b) can be solved in $O(n\log n)$ time in the following situations (see Sections 4.3(a) and 4.3(b) of Chapter 2, respectively):

(1) $\alpha_i = 1$, $i = 1, 2,..., n$;

(2) for all i, $j \in N$ such that $t_i < t_j$ the inequality $\alpha_i \geq \alpha_j$ holds.

Also, note that if the jobs do not enter the processing system simultaneously (i.e., $d_i \geq 0$, $i = 1, 2,..., n$), Problem 1.2(b), remaining NP-hard in a general case, can be solved in $O(n^2)$ time when $\alpha_i = 1$, $i = 1, 2,..., n$, and for all i, $j \in N$ such that $d_i < d_j$, the inequality $D_i \leq D_j$ holds (see Item (c) of Section 4.3 of Chapter 2).

1.7. We show that Problem 1.3 is also NP-hard. The corresponding decision problem is as follows: determine whether there exists a number M^0 such that $M^0 \leq y$ for a given y, and in Problem 1.3 there exists a schedule s for which $\bar{t}_{max}(s) \leq D$.

We show that the partition problem reduces to the formulated decision problem in polynomial time. Define $n = n_0$, $t_i = \gamma_i$, $i = 1, 2,..., n$; $D = A$; $y = 2$. It is obvious that two machines can complete the processing of all n jobs by the deadline A if and only if the partition problem has a solution. The implementation of this reduction takes $O(n_0)$ time.

Problem 1.3, being NP-hard in the non-preemptive case, becomes trivial if preemption is allowed. It is easy to check that in latter case $M^* = \lceil \sum_{i \in N} t_i/D \rceil$, where $\lceil x \rceil$ is the smallest integer such that $x \leq \lceil x \rceil$.

1.8. Let $F(x_1, x_2,..., x_n)$ be a real function and s be some schedule for processing the jobs of set N. Denote $z_i(s) = \max\{0, \bar{t}_i(s) - D_i\}$ and $z_{max}(s) = \max\{z_i(s) \mid i \in N\}$.

Remark 1.1. Let A, B, C, and E be decision problems corresponding to the optimization problems that differ only in their objectives $L_{max}(s)$, $z_{max}(s)$, $\sum_{i \in N} z_i(s)$ and $\sum_{i \in N} u_i(s)$, respectively. Then there exist both polynomial and pseudopolynomial reductions of Problem A to Problems B, C, and E.

To see this, let N^0 and N denote the sets of jobs in Problems A and B, respectively, $|N^0| = n_0$. Let D_i^0 be the due dates in Problem A. Verify whether there exists a schedule s^0

such that $L_{max}(s^0) \leq y_0$ for a given y_0. For Problem B, define $N = N^0$, $D_i = D_i^0 + y_0$, $i = 1, 2, \ldots, n_0$, and $y = 0$. It is easy to check that a schedule s^0 exists if and only if in Problem B there exists a schedule s such that $z_{max}(s) \leq y$. It is evident that the described reduction is both polynomial and pseudopolynomial. Reductions of Problem A to Problems C and E can be constructed in a similar way.

Remark 1.2. In Problem Q, let the objective function be of the form $F(\bar{t}_1(s), \bar{t}_2(s), \ldots, \bar{t}_n(s))$, and Problems R and V differ from Problem Q only by the objective functions which have the form $F(L_1(s), L_2(s), \ldots, L_n(s))$ and $F(z_1(s), z_2(s), \ldots, z_n(s))$, respectively. Moreover, let in Problems R and V for all jobs $i \in N$ we have $D_i = D \geq 0$. Then there exist polynomial and pseudopolynomial reductions of Problem Q to Problems R and V.

In fact, by defining $D = 0$, we obtain $L_i(s) = z_i(s) = \bar{t}_i(s)$ for any schedule s, and, hence, $F(\bar{t}_1(s), \bar{t}_2(s), \ldots, \bar{t}_n(s)) = F(L_1(s), L_2(s), \ldots, L_n(s)) = F(z_1(s), z_2(s), \ldots, z_n(s))$.

The above considerations imply the following statement.

Remark 1.3. Suppose that in Problem Q (see Remark 1.2 above) we have $F(\bar{t}_1(s), \bar{t}_2(s), \ldots, \bar{t}_n(s)) = \bar{t}_{max}(s)$ and in Problems C and E (see Remark 1.1 above) we have $D_i = D \geq 0$ for all jobs $i \in N$. Then there exist both polynomial and pseudopolynomial reductions of Problem Q' to each Problem C or Problem E. Here Q' is the decision problem corresponding to Problem Q.

Remark 1.4. Let Problem H be as follows. The jobs of a set N are processed on a single machine. The jobs enter the system simultaneously and must be processed with no preemption. A precedence relation with the reduction graph G is defined over set N. It is required to determine whether there exists a schedule s (that is feasible with respect to G) such that $F(\bar{t}_1(s), \bar{t}_2(s), \ldots, \bar{t}_n(s)) \leq y$ for a given y. Function $F(x_1, x_2, \ldots, x_n)$ is assumed to be non-decreasing with respect to x_i for $x_i > 0$, $i = 1, 2, \ldots, n$.

If Problem $\tilde{H}$ is the preemptive counterpart of Problem H, then there exist both polynomial and pseudopolynomial reductions of Problem H to Problem $\tilde{H}$.

In fact, if in Problem H a required schedule exists, then it may be also taken as the desired one in Problem $\tilde{H}$. On the other hand, Theorem 1.1 (see Section 1 of Chapter 2) implies that for any feasible schedule s in Problem $\tilde{H}$ there exists a feasible schedule s' in Problem H such that $F(s') \leq F(s)$.

Remark 1.5. Let Problem K be as follows. The jobs of a set N enter simultaneously the processing system consisting of $M \geq 2$ identical machines. Each job is processed with no preemption. It is required to determine whether there exists a schedule s such that $F(s) \leq y$ for a given y. The function $F(x)$ is assumed to be e-quasiconcave for $x_i > 0$,

$i = 1, 2,\ldots, n$, (see Section 1.3 of Chapter 2).

If Problem $\tilde{K}$ is the preemptive counterpart of Problem K, then there exist both polynomial and pseudopolynomial reductions of Problem K to Problem $\tilde{K}$. This follows directly from Theorem 1.2 (see Section 1 of Chapter 2).

1.9. Due to Remarks 1.2 and 1.3, the NP-hardness of Problems 1.1(a) and 1.1.(b) imply that Problem 1.1 is also NP-hard in the following cases:

(d) $F(s) = z_{max}(s)$, $D_i = D$, $i = 1, 2,\ldots, n$;

(e) $F(s) = \sum_{i \in N} z_i(s)$, $D_i = D$, $i = 1, 2,\ldots, n$;

(f) $F(s) = \sum_{i \in N} u_i(s)$, $D_i = D$, $i = 1, 2,\ldots, n$;

(g) $F(s) = L_{max}(s)$;

(h) $F(s) = z_{max}(s) \sum_{i \in N} z_i(s)$, $D_i = D$, $i = 1, 2,\ldots, n$;

(i) $F(s) = L_{max}(s) \sum_{i \in N} L_i(s)$, $D_i = D$, $i = 1, 2,\ldots, n$.

It follows from Remark 1.1 and the NP-hardness of Problem 1.2(a) that Problem 1.2 is also NP-hard in the following cases:

(c) $F(s) = z_{max}(s)$;

(d) $F(s) = \sum_{i \in N} u_i(s)$.

Remark 1.4 (or Remark 1.5) implies that Problem 1.2(b) (or Problem 1.1(c), respectively) also remains NP-hard in the preemptive case.

2. Reducibility of the 3-Partition Problem

In this section, we prove some scheduling problems to be NP-hard using the 3-partition problem as standard. Recall that the 3-*partition* problem is formulated as follows: given a set $N^0 = \{1, 2,\ldots, 3n_0\}$, a positive integer δ, and a positive integer γ_i associated with $i \in N^0$ such that $\delta/4 < \gamma_i < \delta/2$ and $\sum_{i \in N^0} \gamma_i = n_0\delta$, does there exists a partition of set N^0 into n_0 three-element subsets N_j^0 such that $\sum_{i \in N_j^0} \gamma_i = \delta$, $j = 1, 2,\ldots, n_0$?

The 3-partition problem is NP-complete in the strong sense. The length of its input encoded in the unary alphabet is equal to $O(\delta n_0)$, while for the binary alphabet the length is $O(n_0 \log\delta)$.

2.1. In this section, the following scheduling problems are considered.

Problem 2.1. The jobs of a set $N = \{1, 2, \ldots, n\}$ enter the processing system consisting of three identical parallel machines at time $d = 0$. Each job may be processed on any of the machines during one time unit. Preemption is not allowed. At any time, the processing of a job $i \in N$ requires r_i units of some resource. The total amount of the resource available at each time is equal to R. It is required to find a schedule s^* which minimizes the function $F(s)$ in the following cases:

(a) $F(s) = \bar{t}_{max}(s) = \max\{\bar{t}_i(s) \mid i \in N\}$, where $\bar{t}_i(s)$ is the completion time of job i in schedule s;

(b) $F(s) = \sum_{i \in N} \bar{t}_i(s)$.

Problem 2.2. The jobs of a set $N = \{1, 2, \ldots, n\}$ are to be processed on a single machine. A job $i \in N$ becomes available not earlier than at time $d_i \geq 0$, and its processing time is $t_i > 0$ time units. Unless stated otherwise, preemption is not allowed. Each job $i \in N$ is associated with a number $\alpha_i \geq 0$. The due date $D_i \geq 0$, by which it is desirable to complete job i is given for each $i \in N$. A precedence relation is defined over set N such that each connected component of the reduction graph $G = (N, U)$ is a chain. It is required to find a schedule s^* that is feasible with respect to G and minimizes the function $F(s)$ in the following cases:

(a) $F(s) = \sum_{i \in N} \bar{t}_i(s)$, $G = (N, \varnothing)$;

(b) $F(s) = \sum_{i \in N} \alpha_i \bar{t}_i(s)$, $G = (N, \varnothing)$ and preemption is allowed;

(c) $F(s) = \sum_{i \in N} \alpha_i \bar{t}_i(s)$, $t_i = 1$, $i = 1, 2, \ldots, n$;

(d) $F(s) = \sum_{i \in N} \alpha_i z_i(s)$, $d_i = 0$, $i = 1, 2, \ldots, n$, $G = (N, \varnothing)$; here $z_i(s) = \max\{0, \bar{t}_i(s) - D_i\}$;

(e) $F(s) = \sum_{i \in N} \alpha_i z_i(s)$, $d_i = 0$, $t_i = 1$, $i = 1, 2, \ldots, n$;

(f) $F(s) = \sum_{i \in N} \alpha_i \bar{t}_i(s)$; $d_i = 0$, $i = 1, 2, \ldots, n$, $G = (N, \varnothing)$; a schedule s is assumed to be feasible if $\bar{t}_i(s) \leq D_i$, $i = 1, 2, \ldots, n$;

(g) $F(s) = \sum_{i \in N} \alpha_i \bar{t}_i(s)$; $d_i = 0$, $t_i = 1$, $i = 1, 2, \ldots, n$; as in case (f), a feasible schedule s must satisfy the condition $\bar{t}_i(s) \leq D_i$, $i = 1, 2, \ldots, n$;

(h) $F(s) = \sum_{i \in N} u_i(s)$; $d_i = 0$, $t_i = 1$, $i = 1, 2, \ldots, n$; here $u_i(s) = 1$ if $\bar{t}_i(s) > D_i$ and $u_i(s) = 0$, if $\bar{t}_i(s) \leq D_i$.

Problem 2.3. The jobs of a set $N = \{1, 2, \ldots, n\}$ enter a processing system consisting of two identical parallel machines at time $d = 0$. A job $i \in N$ may be processed on any of the machines, and this processing takes t_i time units. Preemption is not allowed. A precedence relation is defined over set N such that each connected component of the reduction graph $G = (N, U)$ is an intree. It is required to find a schedule s^* which is feasible with respect to G and minimizes the function $F(s) = \sum_{i \in N} \bar{t}_i(s)$.

Problem 2.4. The jobs of a set $N = \{1, 2,..., n\}$ enter a processing system consisting of two identical parallel machines at time $d = 0$. A precedence relation is defined over set N such that each connected component of the reduction graph $G = (N, U)$ is a chain. Each job $i \in N$ may be processed on any machine with no preemption. All processing times are unit. At each time, the processing of a job i requires r_i units of some resource, and $r_i \in \{0, 1\}$, $i = 1, 2,..., n$. At each time no more than one unit of the resource is available. It is required to find a schedule s^* that is feasible with respect to G, satisfies the resource constraints, and minimizes the function $F(s)$ in the following cases:

(a) $F(s) = \bar{t}_{max}(s)$;

(b) $F(s) = \sum_{i \in N} \bar{t}_i(s)$.

2.2. We start by proving a statement that is useful for showing some scheduling problems to be NP-hard.

Let us consider the following class of problems.

The jobs of a set $N = \{1, 2,..., n\}$ enter a single-machine processing system at time $d = 0$. The machine can process no more than one job at a time and must operate without idle time. The processing time of a job $i \in N$ is equal to t_i time units. Each job $i \in N$ is associated with a non-decreasing function $\varphi_i(t)$ and the due date $D_i \geq 0$, by which it is desirable to complete this job i. A precedence relation is defined over set N, and $G = (N, U)$ is its reduction graph. Moreover, a non-decreasing function $\varphi(t)$ is given such that $\varphi(0) = 0$. It is required to find a schedule s^* that is feasible with respect to G and minimizes the function $F(s) = \sum_{i \in N} \varphi[\varphi_i(\bar{t}_i(s))]$.

A graph G' is said to be obtained from graph G *by substituting a chain* $C = (i'_1, i'_2,..., i'_r)$, $r \geq 1$, if G' may be obtained from G by replacing some its vertex i by the chain C so that all arcs entering i (leaving i) are replaced by those entering i'_1 (leaving i'_r, respectively).

A graph G' is said to be obtained from graph G *by substituting chains* if G' is obtained from G by replacing each of its vertices by some chain (specific for each vertex).

For an extremal problem H, let H' denote the corresponding decision problem.

Let A and B be the problems of the described class, and G and G' be the reduction graphs of the precedence relations defined over the sets of jobs of these problems, respectively, and suppose that G' is obtained from G by substituting chains.

Lemma 2.1. *Suppose that in Problems A and B we have* $\varphi_i(t) = \alpha_i t$, $\alpha_i \geq 0$, *and that these problems differ from each other only in that in Problem A the processing times* t_i *are positive integers while in Problem B all processing times are unit. Moreover, suppose that, if in Problem A we have* $\alpha_i \in \{\lambda_1, \lambda_2, ..., \lambda_k\}$, *then in Problem B we have* $\alpha_i \in \{0, \lambda_1, \lambda_2, ..., \lambda_k\}$. *Also, assume that Problems C and E differ from Problems A and B, respectively, in that in both C and E we have* $\varphi_i(t) = \alpha_i \max\{0, t - D_i\}$. *Then there exists a pseudopolynomial reduction of Problem A′ to Problem B′, as well as that of Problem C′ to Problem E′.*

Proof. Let us construct a pseudopolynomial reduction of Problem A' to Problem B'. For Problem B', define set N' of jobs as follows. Associate each job $i \in N$ with t_i jobs $i^{(1)}, i^{(2)}, ..., i^{(t_i)}$, assuming $t(i^{(k)}) = 1$, $k = 1, 2, ..., t_i$; $\alpha(i^{(k)}) = 0$, $D(i^{(k)}) = \sum_{i \in N} t_i$, $k = 1, 2, ..., t_i - 1$; $\alpha(i^{(t_i)}) = \alpha_i$, $D(i^{(t_i)}) = D_i$. Define the precedence relation over the set N' assuming that (a) $i^{(t_i)} \to j^{(1)}$ if and only if $i \to j$ and (b) $i^{(k)} \to i^{(k+1)}$ for all $i \in N$, $k = 1, 2, ..., t_i - 1$. Let the reduction graph of this relation be denoted by G'. Note that graph G' is obtained from graph G by substituting chains.

Suppose that there exists a schedule s' for processing the jobs of set N which is feasible with respect to G and such that $\sum_{i \in N} \varphi(\alpha_i \bar{t}_i(s')) \leq y$. Then it is obvious that in the constructed Problem B' there exists a schedule s'' that is feasible with respect to G' and such that $\sum_{i^{(k)} \in N'} \varphi[\alpha(i^{(k)}) \bar{t}_{i^{(k)}}(s'')] \leq y$.

Suppose now that there exists a schedule s'' for processing the jobs of set N' which is feasible with respect to G' and satisfies the condition $\sum_{i^{(k)} \in N'} \varphi[\alpha(i^{(k)}) \bar{t}_{i^{(k)}}(s'')] \leq y$. We show that this implies that there exists a schedule s' for processing the jobs of set N that is feasible with respect to G and such that $\sum_{i \in N} \varphi(\alpha_i \bar{t}_i(s')) \leq y$.

Let there exist a job $i \in N$ such that in schedule s'' the relation $\bar{t}_{i^{(k+1)}}(s'') = \bar{t}_{i^{(k)}}(s'') + 1 + c$ holds for some k, $1 \leq k \leq t_i - 1$, and $c > 0$. Transform s'' into a schedule s''' in which the processing of job $i^{(k)}$ starts c time units later than in schedule s'', and each of the jobs processed in schedule s'' in the time interval $(\bar{t}_{i^{(k)}}(s''), \bar{t}_{i^{(k)}}(s'') + c]$ is to be processed in s''' one time unit earlier. It is easy to verify that s''' is feasible with respect to G' and

$$\sum_{i^{(k)}\in N'} \varphi[\alpha(i^{(k)})\overline{t}_{i^{(k)}}(s''')] \leq \sum_{i^{(k)}\in N'} \varphi[\alpha(i^{(k)})\overline{t}_{i^{(k)}}(s'')].$$

Using the described transformations sufficiently many times (no more than t_i), we can transform s'' into a schedule s^0 which is feasible with respect to G' and such that

$$\sum_{i^{(k)}\in N'} \varphi[\alpha(i^{(k)})\overline{t}_{i^{(k)}}(s^0)] \leq \sum_{i^{(k)}\in N'} \varphi[\alpha(i^{(k)})\overline{t}_{i^{(k)}}(s'')]$$

and $\overline{t}_{i^{(k+1)}}(s^0) = \overline{t}_{i^{(k)}}(s^0)+1$ for all $i \in N$, $k = 1, 2,..., t_i-1$. Schedule s^0 specifies a schedule s' for processing the jobs of set N that is feasible with respect to G and such that $\sum_{i\in N}\varphi(\alpha_i\overline{t}_i(s')) = \sum_{i^{(k)}\in N'}\varphi[\alpha(i^{(k)})\overline{t}_{i^{(k)}}(s^0)] \leq y$.

The described reduction can be implemented in $O(\sum_{i\in N} t_i)$ time. The input length of Problem A' in the binary and unary alphabets is at most

$$c_1\left[\sum_{i=1}^{n}(\log t_i+\log\alpha_i+\log D_i) + \sum_{i=1}^{n}\log i\right]+c_1',$$

and

$$c_2\left[\sum_{i=1}^{n}(t_i+\alpha_i+D_i)+n^2\right]+c_2',$$

respectively, while that of Problem B' is at most

$$c_3\left[\sum_{i=1}^{n}(\log\alpha_i+\log D_i) + \sum_{i=1}^{n}\log(t_i)\right]+c_3',$$

and

$$c_4\left[\sum_{i=1}^{n}(\alpha_i+D_i) + \left[\sum_{i=1}^{n}t_i^2\right]\right]+c_4',$$

respectively. Here c_1, c_2, c_3, c_4, c_1', c_2', c_3', c_4' are some constants, the first four being positive.

The polynomials $p'(x) = cx^2$ and $p''(x) = c'x$, where c and c' are some positive constants can be taken as polynomials p' and p'' (see the definition of pseudopolynomial reduction in Chapter 1). Thus, the described reduction of Problem A' to Problem B' is pseudopolynomial.

A pseudopolynomial reduction of Problem C' to Problem E' can be constructed in a similar way. The only difference is that instead of $\alpha(i^{(k)}) = 0$, $k = 1, 2,..., t_i-1$, we now assume $\alpha(i^{(k)}) = \alpha_i$, $k = 1, 2,..., t_i$. Note that in any schedule s for processing the jobs of set N' we have $z_{i^{(k)}}(s) = 0$ for all $i \in N$ and $k = 1, 2,..., t_i-1$. This proves the lemma.

Remark 2.1. The above considerations imply that in Problem A' a schedule s' which is feasible with respect to G and satisfies the conditions $\overline{t}_i(s') \leq D_i$ exists if and only if in the constructed Problem B' there exists schedule s'' that is feasible with respect to G' and such that $\bar{t}_{i(k)}(s'') \le D_{i(k)}$, $i = 1, 2,..., n$, $k = 1, 2,..., t_i$.

Corollary 2.1. *If Problem A' (or Problem C') is NP-hard in the strong sense, then Problem B (or Problem D, respectively) is NP-hard in the strong sense as well.*

This directly follows from Theorem 3.2 (see Section 3 of Chapter 1).

2.3. Since the 3-partition problem is *NP*-hard in the strong sense, it follows that, in order to prove that any of the problems in Section 2.1 is *NP*-hard, it suffices to construct a pseudopolynomial reduction of the 3-partition problem to the corresponding decision problem.

The following decision problem corresponds to Problem 2.1(a): determine whether there exists a schedule s^0 for processing the jobs of set N such that at most R resource units are to be consumed at any time and $\bar{t}_{max}(s^0) \le y$ for a given y.

We now construct a polynomial reduction of the 3-partition problem to the formulated decision problem.

Define $n = 3n_0$, $r_i = \gamma_i$, $i = 1, 2,..., n$; $R = \delta$, $y = n_0$.

Let the 3-partition problem have a solution. Then each of the subsets N_j^0 forming a partition of set N^0 specifies a triplet of jobs to be processed in the time interval $[j-1, j]$. These three jobs can be distributed over the machines arbitrarily. It is clear that in schedule s^0 obtained this way, all jobs of set N are completed by time n_0 and δ resource units are to be consumed at any time.

On the other hand, if for the constructed scheduling decision problem there exists a schedule s^0 such that $\bar{t}_{max}(s^0) \le n_0$, then this implies that exactly three jobs are processed at any time. Each triplet of jobs processed in the time interval $[j-1, j]$ specifies a subset N_j^0 of the required 3-partition of set N^0.

It is easy to verify that the described reduction requires $O(n_0)$ time, i.e., this reduction is both polynomial and pseudopolynomial.

Since the 3-partition problem is *NP*-hard in the strong sense, it follows that Problem 2.1(a) is *NP*-hard in the strong sense as well.

Note that if in Problem 2.1(a) the processing system consists of two machines, then the corresponding problem is solvable in $O(n\log n)$ time [211].

2.4. The decision problem corresponding to Problem 2.1(b) is as follows: determine whether there exists a schedule s^0 for processing the jobs of set N such that

$\sum_{i \in N} \bar{t}_i(s^0) \leq y$ for a given y.

Define $n = 3n_0$, $r_i = \gamma_i$, $i = 1, 2,..., n$; $R = \delta$, $y = \frac{3}{2}n_0(n_0+1)$. It is easy to check that the described transformation of the 3-partition problem into the formulated decision problem takes at most $O(n_0)$ time.

If the 3-partition problem has a solution, then a partition of set N^0 specifies a partition of set N into n_0 subsets each consisting of three jobs such that exactly δ resource units are to be consumed at any time in the processing of each triplet of jobs. If s^0 is a schedule in which in each time interval $[j-1, j]$, $j = 1, 2,..., n_0$, the jth triplet of jobs is processed, then $\sum_{i \in N} \bar{t}_i(s^0) = \frac{3}{2}n_0(n_0+1)$.

Suppose that there exists a resource-feasible schedule s^0 such that $\sum_{i \in N} \bar{t}_i(s^0) \leq \frac{3}{2}n_0(n_0+1)$. It is easy to verify that the latter inequality implies that at each time moment exactly three jobs are processed. Since $\sum_{i \in N} \gamma_i = n_0\delta$, we conclude that schedule s^0 specifies a solution of the 3-partition problem. Thus, Problem 2.1(b) is NP-hard in the strong sense.

2.5. The following decision problem corresponds to Problem 2.2(a): determine whether there exists a schedule s^0 for processing the jobs of set N such that $\sum_{i \in N} \bar{t}_i(s^0) \leq y$ for a given y.

We now construct a pseudopolynomial reduction of the 3-partition problem to the formulated decision problem.

Define $n = 3n_0+(n_0+1)(n_0+\delta)^3$. Assume that set N contains jobs of two types: main and auxiliary. The main jobs i have the parameters $t_i = \gamma_i$, $d_i = 0$, $i = 1, 2,..., 3n_0$. The set of the auxiliary jobs $J_l^{(k)}$ consists of n_0+1 groups with the parameters

$$t(J_l^{(k)}) = 1/(n_0+\delta)^3,\ d(J_l^{(k)}) = (\delta+1)(k-1),\ k = 1, 2,..., n_0,\ l = 1, 2,..., (n_0+\delta)^3;$$

$$t(J_l^{(n_0+1)}) = 1,\ d(J_l^{(n_0+1)}) = n_0(\delta+1),\ l = 1, 2,..., (n_0+\delta)^3.$$

Define

$$y = 3(\delta+1)n_0(n_0+1)/2+n_0(n_0+\delta)^3(1+1/(n_0+\delta)^3)/2$$
$$+(n_0-1)(\delta+1))+(2n_0(\delta+1)+(n_0+\delta)^3+1)((n_0+\delta)^3)/2.$$

If the 3-partition problem has a solution, then the processing of each of the auxiliary jobs $J_l^{(k)}$ may be completed by time $t = (k-1)(\delta+1)+l/(n_0+\delta)^3$, $k = 1, 2,..., n_0$, $l = 1, 2,..., (n_0+\delta)^3$ and each of the jobs $J_l^{(n_0+1)}$ can be completed by time $t = n_0(\delta+1)+l$, $l = 1, 2,..., (n_0+\delta)^3$. This may be done in the manner shown in Fig. 2.1, where a shaded rectangle corresponds to an auxiliary job, and a non-shaded rectangle represents a main job.

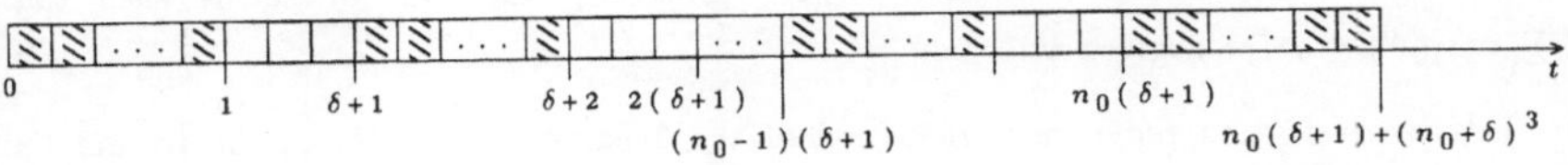

Fig. 2.1

Let s^0 be a schedule corresponding to the situation in Fig. 2.1. In this schedule, each triplet of the main jobs processed immediately after the kth group of auxiliary jobs is completed at time $t = k(\delta+1)$. Therefore, $\sum_{i\in N'} \overline{t}_i(s^0) \leq \frac{3}{2}(\delta+1)n_0(n_0+1)$, where N' denotes the set of all main jobs. Hence,

$$\sum_{i\in N} \overline{t}_i(s^0) \leq \frac{3}{2}(\delta+1)n_0(n_0+1) + \sum_{k=1}^{n_0} \sum_{l=1}^{(n_0+\delta)^3} ((k-1)(\delta+1)$$

$$+ l/(n_0+\delta)^3) + \sum_{l=1}^{(n_0+\delta)^3} (n_0(\delta+1)+l) = y.$$

Suppose now that the 3-partition problem has no solution. Since γ_i are positive integers and

$$\sum_{l=1}^{(n_0+\delta)^3} t(J_l^{(k)}) = 1, \; k = 1, 2,..., n_0,$$

we conclude that in any schedule s either the processing of at least one group of auxiliary jobs starts at least one time unit later than in the schedule shown in Fig. 2.1 or at least one main job is processed either after the (n_0+1)th group of auxiliary jobs or in the time interval between the processing of two jobs of this group. In any case, since $(n_0+\delta)^3 > \frac{3}{2}(\delta+1)n_0(n_0+1)$, we have

$$\sum_{i\in N} \overline{t}_i(s) > (n_0+\delta)^3 + \sum_{k=1}^{n_0} \sum_{l=1}^{(n_0+\delta)^3} ((k-1)(\delta+1) + l/(n_0+\delta)^3) + \sum_{l=1}^{(n_0+\delta)^3} (n_0(\delta+1)+l) > y,$$

Hence, a required schedule s^0 exists if and only if the 3-partition problem has a solution. The described reduction can be implemented in at most $O(n_0(n_0+\delta)^3)$ time. Since the 3-partition problem is *NP*-hard in the strong sense, we conclude that Problem 2.2(a) is *NP*-hard in the strong sense as well.

If in Problem 2.2(a) we assume $d_i = 0$, $i = 1, 2,..., n$, then the problem reduces to one of minimizing a priority-generating function over a set $\hat{P}_n$ of all permutations of the elements of set N. In this case, the problem becomes solvable in $O(n\log n)$ time (see Section 7 of Chapter 3).

2.6. The following decision problem corresponds to Problem 2.2(b): determine whether there exists a (preemptive) schedule s^0 for processing the jobs of set N such that $\sum_{i \in N} \alpha_i \bar{t}_i(s^0) \leq y$ for a given y.

Let us construct a polynomial reduction of the 3-partition problem to the formulated decision problem.

Define $n = 4n_0 - 1$; $d_i = 0$, $t_i = \alpha_i = \gamma_i$, $i = 1, 2, \ldots, 3n_0$; $d_i = (i - 3n_0)(\delta + 1) - 1$, $t_i = 1$, $\alpha_i = \delta$, $i = 3n_0 + 1, 3n_0 + 2, \ldots, 4n_0 - 1$; $y = \sum_{1 \leq l \leq k \leq 3n_0} \gamma_l \gamma_k + \delta(\delta + 2) n_0(n_0 - 1)/2$.

First, we show that a schedule s^0 can be found in a class of schedules in which the processing of each job starts at time d_i, $i = 3n_0 + 1, 3n_0 + 2, \ldots, 4n_0 - 1$. If the condition $t_i = 1$ were substituted for $t_i = 0$, $i = 3n_0 + 1, 3n_0 + 2, \ldots, 4n_0 - 1$, and the objective function were $F(s) = \sum_{i=1}^{3n_0} \alpha_i \bar{t}_i(s)$, then for any non-preemptive schedule s the equality $F(s) = \sum_{1 \leq l \leq k \leq 3n_0} \gamma_l \gamma_k$ would be valid. Let this sum be denoted by ν. Besides, let N' denote the set $\{1, 2, \ldots, 3n_0\}$, and $N_k(s)$ denote a set of jobs in N' which are completed in schedule s later than $3n_0 + k$, $k = 1, 2, \ldots, n_0 - 1$. For a schedule s, the starting time of a job $3n_0 + k$, $k = 1, 2, \ldots, n_0 - 1$, is denoted by $t_k^0(s)$. It is easy to verify that

$$\sum_{i \in N} \alpha_i \bar{t}_i(s) = \nu + \sum_{k=1}^{n_0 - 1} \sum_{j \in N_k(s)} \gamma_j + \delta \sum_{k=1}^{n_0 - 1} (t_k^0(s) + 1).$$

Suppose that in a schedule s some job $i \in N \backslash N'$ starts at time $d_i + r_1$, where $r_1 \geq 1$ and there exists a job $j \in N'$ which starts at time $d_i - r_2$, where $r_2 \geq 0$ and is processed in the time interval $[d_i, d_i + r_1]$. Transform schedule s into a schedule s' in the following way. Define $t_i^0(s') = d_i$ if $t_i^0(s) = d_i + r_1$, and either start processing job j at time $d_i + 1$ (if $r_2 = 0$) or interrupt processing job j (if $r_2 > 0$) resuming that processing at time $d_i + 1$. It is obvious that $F(s) - F(s') \geq r_1\delta - \gamma_j > 0$. Hence, it follows that the search for schedule s^0 can be restricted to consideration of such schedules s that $t_k^0(s) = d_{3n_0 + k}$, $k = 1, 2, \ldots, n_0 - 1$. Note that in any such a schedule we have

$$F(s) = \nu + \delta(\delta + 1) \sum_{k=1}^{n_0 - 1} k + \sum_{k=1}^{n_0 - 1} \sum_{j \in N_k(s)} \gamma_j = \nu + \mu + \sum_{k=1}^{n_0 - 1} \sum_{j \in N_k(s)} \gamma_j,$$

where $\mu = \delta(\delta + 1) n_0(n_0 - 1)/2$. Thus, for any schedule s from the described class, the value of $F(s)$ is determined by $\sum_{k=1}^{n_0 - 1} \sum_{j \in N_k(s)} \gamma_j$.

Suppose that there exists a partition of set N^0 into n_0 of three-element subsets N_j^0 such that $\sum_{j \in N_j^0} \gamma_i = \delta$. Without loss of generality, we may assume that $N_j^0 = \{3j - 2, 3j - 1, 3j\}$, $j = 1, 2, \ldots, n_0$. Then it is easy to see that for a schedule s' defined by the permutation $\pi' = (1, 2, 3, 3n_0 + 1, \ldots, 3j - 2, 3j - 1, 3j, 3n_0 + j, \ldots, 3n_0 - 5, 3n_0 - 4, 3n_0 - 3, 4n_0 - 1, 3n_0 - 2, 3n_0 - 1, 3n_0)$, each job $i \in N/N'$ starts at time d_i and

$$\sum_{i\in N}\alpha_i\bar{t}_i(s') = \nu+\mu+\sum_{k=1}^{n_0-1}\delta(n_0-k) = \nu+\mu+\delta n_0(n_0-1)/2 = y.$$

Suppose now that no required partition of set N^0 exists and that in a schedule s' the processing of a job $i \in N\backslash N'$ starts at a time d_i while the jobs of set N' are processed according to a permutation $\pi = (i_1, i_2,..., i_{3n_0})$. Then there must exist such an index k that $\gamma_{i_{3k-2}}+\gamma_{i_{3k-1}}+\gamma_{i_{3k}} = \beta \neq \delta$. Let k be the smallest index satisfying this condition. If $\beta < \delta$, then $\sum_{j\in N_k(s')}\gamma_j\gamma_j = (n_0-k)\delta+(\delta-\beta)$. If $\beta > \delta$, then in schedule s' job i_{3k} is processed with preemption and $\sum_{j\in N_k(s')}\gamma_j = (n_0-k)\delta+(\delta-\beta)+\gamma_{i_{3k}}$. On the other hand, for any j, $j = 1, 2,..., n_0-1$, we derive that $d_{3n_0+j} = j(\delta+1)-1$ and that in schedule s' only those jobs of set N' may be completed by time d_{3n_0+j} whose total processing time does not exceed $j(\delta+1)-1-(j-1) = j\delta$. This implies that the inequality $\sum_{i\in N_j(s')}\gamma_i \geq (n_0-j)\delta$ holds for any j, $j = 1, 2,..., n_0-1$.

Thus, either $F(s') \geq \nu+\mu+\sum_{j=1}^{n_0-1}(n_0-j)\delta+(\delta-\beta)$ or $F(s') \geq \nu+\mu+\sum_{j=1}^{n_0}(n_0-j)\delta+\gamma_{i_{3k}}$. In any case, $F(s') > y$ due to $0 < \beta-\delta < \gamma_{i_{3k}}$.

The described reduction can be implemented in $O(n_0^2)$ time. This reduction is both polynomial and pseudopolynomial. Thus, Problem 2.2(b) is *NP*-hard in the strong sense.

2.7. In this section, we construct a reduction of Problem 2.2(b) to Problem 2.2(c). Let N' denote the set of jobs in Problem 2.2(c). This set is formed as follows. Each $i \in N$ is associated with t_i jobs $i^{(1)}, i^{(2)},..., i^{(t_i)}$ (without loss of generality, t_i are assumed here to be integers), define $\alpha(i^{(k)}) = 0$, $k = 1, 2,..., t_i-1$; $\alpha(i^{(t_i)}) = \alpha_i$; $t(i^{(k)}) = 1$, $d(i^{(k)}) = d_i$, $k = 1, 2,..., t_i$. The precedence relation $\rightarrow$ is defined over N' by its reduction $i^{(k)} \rightarrow i^{(k+1)}$, $k = 1, 2,..., t_i-1$, $i = 1, 2,..., n$.

Let s' be a schedule for processing the jobs of set N' determined by a permutation π of the elements of set N'. It is obvious that any such a permutation that is feasible with respect to graph G specifies a schedule s for processing the jobs of set N in Problem 2.2(b), and $F(s) = F'(s')$, where $F'(s') = \sum_{i^{(k)}\in N'}\alpha(i^{(k)})\bar{t}_{i^{(k)}}(s')$. On the other hand, a schedule s for processing the jobs of set N specifies a permutation π of the elements of N'.

The described reduction can be implemented in $O\left(\sum_{i\in N}t_i\right)$ time, so this reduction is pseudopolynomial. Since Problem 2.2(b) is *NP*-hard in the strong sense, it follows that Problem 2.2(c) is *NP*-hard in the strong sense as well.

Note that if $G = (N, \varnothing)$, Problem 2.2(c) is solvable in $O(n^3)$ time (see Section 4.5 of

Chapter 2). The same algorithm (with the same running time) solves Problem 2.2, provided that $G = (N, \emptyset)$, $t_i = 1$, $i = 1, 2,\ldots, n$, and $F(s) = \sum_{i\in N}\varphi_i(\bar{t}_i(s))$, where $\varphi_i(x)$ are non-decreasing functions for all $i \in N$.

2.8. The following decision problem corresponds to Problem 2.2(d): determine whether there exists a schedule s^0 for processing the jobs of set N such that $\sum_{i\in N}\alpha_i z_i(s^0) \le y$ for a given y.

We construct a polynomial reduction of the 3-partition problem to the formulated decision problem.

The set N of jobs is formed as follows. Define $n = 4n_0$ and assume that set $N = \{1, 2,\ldots, n\}$ contains jobs of the following two types: I_k, $k = 1, 2,\ldots, n_0$, and J_k, $k = 1, 2,\ldots, 3n_0$.

Define $t(I_k) = a = \delta^2 n_0^2$, $\alpha(I_k) = \delta(\delta+a)n_0(n_0+1)/2+1$, $D(I_k) = ak+\delta(k-1)$, $k = 1, 2,\ldots, n_0$; $t(J_k) = \alpha(J_k) = \gamma_k$, $D(J_k) = 0$, $k = 1, 2,\ldots, 3n_0$; $y = \delta(\delta+a)n_0(n_0+1)/2$.

Let the 3-partition problem have a solution. Without loss of generality, it may be assumed that $\gamma_{3j-2}+\gamma_{3j-1}+\gamma_{3j} = \delta$, $j = 1, 2,\ldots, n_0$. Consider a schedule s^0 determined by the permutation $\pi^0 = (I_1, J_1, J_2, J_3, I_2, J_4, J_5, J_6, I_3,\ldots, I_{n_0}, J_{3n_0-2}, J_{3n_0-1}, J_{3n_0})$. It is easy to verify that $\bar{t}(I_k(s^0)) = D(I_k)$, and $\bar{t}(J_{3j-2}(s^0)) < \bar{t}(J_{3j-1}(s^0)) < \bar{t}(J_{3j}(s^0)) = aj+\delta j$, $j = 1, 2,\ldots, n_0$. Here $\bar{t}(I_k(s))$ and $\bar{t}(J_k(s))$ denote the completion times of jobs I_k and J_k in a schedule s, respectively. Since $\alpha(J_{3j-2})+\alpha(J_{3j-1})+\alpha(J_{3j}) = \delta$, $j = 1, 2,\ldots, n_0$, we have

$$\sum_{i\in N}\alpha_i z_i(s^0) < \sum_{k=1}^{n_0}\delta(a+\delta)j = \delta(a+\delta)n_0(n_0+1)/2 = y.$$

Suppose now that there exists a schedule s' for processing the jobs of set N such that $\sum_{i\in N}\alpha_i z_i(s') \le y$. Since δ and γ_k are positive integers and $\alpha(I_k) > y$, it follows that the condition $\bar{t}(I_k(s')) \le D(I_k)$, $k = 1, 2,\ldots, n_0$, must be satisfied. In fact, if $\bar{t}(I_k(s')) > D(I_k)$ for some k, then $z(I_k(s')) \ge 1$ and $\sum_{i\in N}\alpha_i z_i(s') \ge y+1$.

Let $N_k(s')$, $k = 1, 2,\ldots, n_0$, denote the set of all those jobs J_j, $j \in \{1, 2,\ldots, 3n_0\}$, which in schedule s' are completed after job I_k is completed; by definition, it is assumed that $N_{n_0+1}(s') = \emptyset$. Denote $A_k = \sum_{J_j\in N_k(s')}\alpha(J_j)$. Since $\bar{t}(I_k(s')) \le ak+\delta(k-1)$ and $\alpha(J_j) = t(J_j)$, it follows that the condition $A_k \ge \delta n_0 - \delta(k-1) = \delta(n_0-k+1)$ must be satisfied. It is obvious that for any job J_j processed in schedule s' after job I_k, the condition $\bar{t}(J_j(s')) > ak$ holds. Therefore,

$$\sum_{i\in N}\alpha_i z_i(s') > \sum_{k=1}^{n_0} ak(A_k - A_{k+1}) = a\sum_{k=1}^{n_0} A_k.$$

Suppose that there exists an index k' such that

$$A_{k'} = \delta(n_0-k'+1)+1.$$

Then it follows that

$$\sum_{i\in N}\alpha_i z_i(s') > a\delta\sum_{k=1}^{n_0}(n_0-k+1)+a = a\delta n_0(n_0+1)/2+a = y-\delta^2 n_0(n_0+1)/2-\delta^2 n_0^2 \geq y,$$

which contradicts the definition of schedule s'.

Hence, $A_k = \delta(n_0-k+1)$ and $A_k-A_{k+1} = \delta$. Since $\alpha(J_k) = t(J_k)$, $k = 1, 2,\ldots, 3n_0$, the total processing time of jobs J_j to be processed in schedule s' in the time interval between the completion time of job I_k and the starting time of job I_{k+1}, $k = 1, 2,\ldots, n_0-1$, is equal to δ. Thus, in each of these intervals exactly three jobs J_k must be processed.

Thus, schedule s' specifies the desired 3-partition of set N^0. The implementation of the described reduction requires $O(n_0)$ time. This reduction is both polynomial and pseudopolynomial.

Hence, Problem 2.2(d) is *NP*-hard in the strong sense.

If $t_i = 1$, $i = 1, 2,\ldots, n$, then Problem 2.2(d) is solvable in $O(n^3)$ time even when the jobs do not enter the processing system simultaneously (see Section 4.5 of Chapter 2).

Problem 2.2(e) is *NP*-hard in the strong sense. This follows from Lemma 2.1 and the fact that Problem 2.2(d) is *NP*-hard in the strong sense. If $G = (N, \emptyset)$, the algorithm mentioned above also solves Problem 2.2(e).

2.9. The following decision problem corresponds to Problem 2.2(f): determine whether there exists a schedule s^0 for processing the jobs of set N such that $\bar{t}_i(s^0) \leq D_i$, $i = 1, 2,\ldots, n$, and $\sum_{i\in N}\alpha_i\bar{t}_i(s^0) \leq y$ for a given y.

We construct a polynomial reduction of the 3-partition problem to the formulated decision problem.

The set N of jobs is formed as follows. Define $n = 4n_0$ and assume that set $N = \{1, 2,\ldots, n\}$ consists of jobs of two following types: I_k, $k = 1, 2,\ldots, n_0$, and J_k, $k = 1, 2,\ldots, 3n_0$.

Define $t(I_k) = a = \delta^2 n_0^2$, $\alpha(I_k) = 0$, $D(I_k) = ak+\delta(k-1)$, $k = 1, 2,\ldots, n_0$; $t(J_k) = \alpha(J_k) = \gamma_k$, $D(J_k) = an_0+\delta n_0$, $k = 1, 2,\ldots, 3n_0$; $y = \delta(\delta+a)n_0(n_0+1)/2$.

The proof is similar to the one presented in Section 2.8.

Suppose that the 3-partition problem has a solution and $\gamma_{3j-2}+\gamma_{3j-1}+\gamma_{3j} = \delta$, $j = 1, 2,\ldots, n_0$. Then, for schedule s^0 determined by the permutation $\pi = (I_1, J_1, J_2, J_3, I_2, J_4, J_5, J_6, I_3,\ldots, I_{n_0}, J_{3n_0-2}, J_{3n_0-1}, J_{3n_0})$, we have $\bar{t}(I_k(s^0)) = D(I_k)$, $k = 1, 2,\ldots,$

n_0, $\overline{t}(J_k(s^0)) \le D(J_k)$, $k = 1, 2, \ldots, 3n_0$, $\sum_{i \in N} \alpha_i \overline{t}_i(s^0) < \sum_{j=1}^{n_0} \delta(a+\delta)j = y$.

Suppose now that there exists a schedule s' for processing the jobs of set N such that $\overline{t}(I_k(s')) \le D(I_k)$ and $\sum_{i \in N} \alpha_i \overline{t}_i(s') \le y$. Introduce the numbers A_k, $k = 1, 2, \ldots, n_0$, as in Section 2.8. It then follows from $\overline{t}(I_k(s')) \le ak + \delta(k-1)$ that $A_k \ge \delta(n_0 - k + 1)$ and $\alpha_i \overline{t}_i(s') > \sum_{k=1}^{n_0} ak(A_k - A_{k+1}) = a \sum_{k=1}^{n_0} A_k$.

If an index k' that satisfies the condition $A_{k'} \ge \delta(n_0 - k + 1) + 1$ exists, we obtain $\sum_{i \in N} \alpha_i \overline{t}_i(s') > a\delta n_0(n_0+1)/2 + a \ge y$, which contradicts the definition of schedule s'. Hence, $A_k = \delta(n_0 - k + 1)$, $k = 1, 2, \ldots, n_0$, and $A_k - A_{k+1} = \delta$. Then it follows that schedule s' specifies a required partition of set N^0.

The implementation of the described reduction takes $O(n_0)$ times. Thus, Problem 2.2(f) is NP-hard in the strong sense.

Note that if $\alpha_i = 1$, $i = 1, 2, \ldots, n$, Problem 2.2(f) is solvable in $O(n^2)$ time [417, 36].

Problem 2.2(g) is NP-hard in the strong sense. This follows from Lemma 2.1 and from the fact that Problem 2.2(f) is NP-hard in the strong sense.

2.10. The following decision problem corresponds to Problem 2.2(h): determine whether there exists a schedule s^0 for processing the jobs of set N which is feasible with respect to G and such that $\sum_{i \in N} u_i(s^0) \le y$ for a given positive integer y. This decision problem is called Problem 2.2(h′).

To prove that Problem 2.2(h′) is NP-hard, we introduce two auxiliary problems and prove that these are NP-hard in the strong sense.

The first problem is called *the 3-set exact covering problem* and can be formulated as follows. Given a finite set $M^0 = \{1, 2, \ldots, 3m_0\}$ and a cover $\widetilde{M} = \{M_1, M_2, \ldots, M_m\}$ of this set by its three-element subsets ($m \ge m_0$), does $\widetilde{M}$ contain an exact cover of set M^0, i.e., such a subset $\widetilde{M}' = \{M_{j_1}, M_{j_2}, \ldots, M_{j_l}\} \subseteq \widetilde{M}$ that $l = m_0$ and $\bigcup_{k=1}^{l} M_{j_k} = M^0$?

We construct a polynomial reduction of the 3-partition problem to the 3-set exact covering problem. Define $M^0 = N^0$. A collection $\widetilde{M}$ is formed as follows. Construct all three-element subsets of set N^0 (their number is equal to $\binom{n_0}{3}$). For each such a subset, calculate the sum of the corresponding γ_i's. Those and only those subsets, for which this sum is equal to δ are to be included in collection $\widetilde{M}$.

It is obvious that the 3-partition problem has a solution if and only if the constructed collection $\tilde{M}$ contains a 3-set exact cover of set M^0.

The implementation of this reduction takes at most $O(n_0^3)$ time. Thus, the 3-set exact covering problem is NP-hard in the strong sense.

The second auxiliary problem to be discussed differs from Problem 2.2(h') in that the job processing times are positive, integer and, generally speaking, different numbers. This problem is called Problem 2.2(h″). We show that it is NP-hard in the strong sense.

We construct a polynomial reduction of the 3-set exact covering problem to Problem 2.2(h″).

For Problem 2.2(h″), form the set N of jobs as follows. Associate each subset $M_j \in \tilde{M}$, $j = 1, 2,..., m$, with job job J_j (these jobs are called jobs of the first type). Associate each element $i \in M^0$ with as many jobs $J_{i,j}$ as many times i can be found in the triplets of set $\tilde{M}$: job $J_{i,j}$ corresponds to element i if and only if $i \in M_j \in \tilde{M}$. The jobs $J_{i,j}$ are called jobs of the second type.

The precedence relation is defined over the constructed set N as follows. Let $M_j = \{i, i', i''\}$, where $i < i' < i''$. Define $J_j \rightarrow J_{i,j}$, $J_{i,j} \rightarrow J_{i',j}$, $J_{i',j} \rightarrow J_{i'',j}$ (here only the reduction of the defined precedence relation is presented).

Define $y = 3(m-m_0)$; $t(J_{i,j}) = mi$, $D(J_{i,j}) = m_0+mi(i+1)/2$, $t(J_j) = 1$, $D(J_j) = m+m\sum_{j=1}^{m}\sum_{i\in M_j} i$, $j = 1, 2,..., m$, $i \in M_j$.

It is clear that for any schedule s for processing the jobs of set N such that the machine has no intermediate idle time, we have $u_{J_j}(s) = 0$, $j = 1, 2,...,m$.

Suppose that the 3-set exact covering problem has a solution and $\tilde{M}' = \{M_{j_1}, M_{j_2},..., M_{j_l}\} \subseteq \tilde{M}$ is such a subset that $l = m_0$ and $\bigcup_{k=1}^{m_0} M_{j_k} = M^0$. Without loss of generality, assume that $\tilde{M} = \{M_1, M_2,..., M_{m_0}\}$.

Consider a schedule s^0 specified by the permutation $\pi^0 = (\pi^{(1)}, \pi^{(2)})$, where $\pi^{(1)} = (J_1, J_2,..., J_{m_0}, J_{1,j_1}, J_{2,j_2},..., J_{3m_0,j_{3m_0}})$, $j_k \in \{1, 2,..., m_0\}$, $k = 1, 2,..., 3m_0$, and $\pi^{(2)}$ is a permutation of the elements of set $N\backslash\{\pi^{(1)}\}$ that is feasible with respect to G. It is easy to verify that s^0 is feasible with respect to G and such that $\sum_{J_{i,j}\in N} u_{J_{i,j}}(s^0) = 3(m-m_0) = y$.

Suppose now that there exists a schedule s' for processing the jobs of set N which is feasible with respect to G and such that $\sum_{J_{i,j}\in N} u_{J_{i,j}}(s^0)\ (s') \leq y$. We show that, in this case, $\tilde{M}$ contains an exact cover of set M^0.

In schedule s, a job $J_{i,j}$ is said to be processed with no tardiness if $u_{J_{i,j}}(s) = 0$.

First, we prove the following statement.

If, in a schedule s feasible with respect to G, i jobs of the second type are processed with no tardiness by time $t = D(J_{i,j})$, then these are the jobs $J_{1,j_1}, J_{2,j_2}, \ldots, J_{i,j_i}$.

The proof is by induction with respect to i. For $i = 1$ and $i = 2$, the statement obviously holds.

Suppose that this holds for all $i \leq r$, where $r \geq 2$. We show that the statement holds for $i = r+1$.

Thus, in schedule s, $r+1$ jobs of the second type are processed with no tardiness by the time $t = D(J_{r+1,j}) = m_0+m(r+1)(r+2)/2$. Then, in schedule s, at least r jobs of the second type must be processed with no tardiness by the time $t = D(J_{r,j}) = m_0+mr(r+1)/2$. In fact, among the jobs of the second type, only the jobs $J_{k,j}$, $k \geq r+1$, can be processed without tardiness after $D(J_{r,j})$, while no more than one such a job may be processed in the time interval $[D(J_{r,i}), D(J_{r+1,j})]$ (since $t(J_{k,j}) \geq (r+1)m$ for $k \geq r+1$).

Due to the induction assumption, in schedule s, the jobs $J_{1,j_1}, J_{2,j_2}, \ldots, J_{r,j_r}$ must be completed by the time $t = D(J_{r,j})$ and (since s is feasible with respect to G) at least one job of the first type must be completed. The total processing time of all these jobs is at least $mr(r+1)/2+1$. Therefore, in schedule s, processing the $(r+1)$th job of the second type may start no earlier than time $t = mr(r+1)/2+1$ and must be completed by time $t = D(J_{r+1,j})$. The length of this time interval is $m_0+m(r+1)-1$. Since $m \geq m_0$ and for $k \geq r+1$ the inequality $t(J_{k,j}) \geq m(r+1)$ holds, it follows that only job $J_{r+1,j_{r+1}}$ may be processed in this time interval. This proves the required statement.

Schedule s' satisfies the condition $\sum_{J_{i,j} \in N} u_{J_{i,j}}(s') \leq y = 3(m-m_0)$. Hence, in schedule s', at least $3m_0$ jobs of the second type are processed with no tardiness. Moreover, the processing of these jobs must be completed by time $t = \max\{D(J_{i,j}) \mid J_{i,j} \in N\} = D(J_{3m_0,j})$. The statement proved above implies that, in schedule s', the jobs $J_{1,j_1}, J_{2,j_2}, \ldots, J_{3m_0,j_{3m_0}}$ are processed with no tardiness. Each element M_j of set $\tilde{M}$ is associated with exactly three different jobs of the second type, and a one-to-one correspondence exists between the elements of set M^0 and the jobs $J_{1,j_1}, J_{2,j_2}, \ldots, J_{3m_0,j_{3m_0}}$. Thus, the above jobs specify the desired 3-set exact cover $\tilde{M}' \subseteq \tilde{M}$ of set M^0.

The implementation of the described reduction requires $O(m)$ time. This reduction is both polynomial and pseudopolynomial, and Problem 2.2(h″) is, therefore, NP-hard in the strong sense.

Problem 2.2(h′) is NP-hard in the strong sense as well. This follows from Lemma 2.1 and NP-hardness in the strong sense of Problem 2.2(h″).

Thus, Problem 2.2(h) is NP-hard in the strong sense.

If $G = (N, \emptyset)$, Problem 2.2(h) is solvable in $O(n\log n)$ time (see Section 4.3(b) of

Chapter 2). That algorithm may be applied not only if the processing times of all jobs are unit but also if these times are arbitrary.

2.11. Let us consider Problem 2.3. The corresponding decision problem is as follows: determine whether there exists a schedule s^0 for processing the jobs of set N that is feasible with respect to G and such that $\sum_{i \in N} \bar{t}_i(s^0) \leq y$ for a given y.

We construct a pseudopolynomial reduction of the 3-partition problem to this decision problem.

Define $n = 20n_0^4\delta + 6n_0 + 2$ and denote $a = 4n_0^2\delta$, $b = 20n_0^4\delta$. Set $N = \{1, 2, \dots, n\}$ is assumed to contain the jobs of two types: main and auxiliary. The main jobs are denoted by i, $i = 1, 2, \dots, 3n_0$, and have the parameters $t_i = \gamma_i$. For the auxiliary jobs J_l, define $t(J_{2l-1}) = t(J_{2n_0+1+l}) = a$, $l = 1, 2, \dots, n_0+1$; $t(J_l) = 1$, $l = 3n_0+3, 3n_0+4, \dots, 3n_0+2+b$; $t(J_{2l}) = \delta$, $l = 1, 2, \dots, n_0$. The precedence relation $\rightarrow$ is defined over the constructed set N as follows: $J_{l-1} \rightarrow J_l$, $l = 2, 3, \dots, 2n_0+1$, $l = 3n_0+4, 3n_0+5, \dots, 3n_0+2+b$; $J_{2n_0+1+l} \rightarrow J_{2l}$, $l = 1, 2, \dots, n_0$; $J_{2n_0+1} \rightarrow J_{3n_0+3}$; $J_{3n_0+2} \rightarrow J_{3n_0+3}$; $i \rightarrow J_{3n_0+3}$, $i = 1, 2, \dots, 3n_0$. A subgraph of the graph $G = (N, U)$ of the reduction of the constructed precedence relation induced by the set of all auxiliary jobs is shown in Fig. 2.2. Each vertex in this figure is accompanied by the processing time of the corresponding job. It is easy to check that G is an intree.

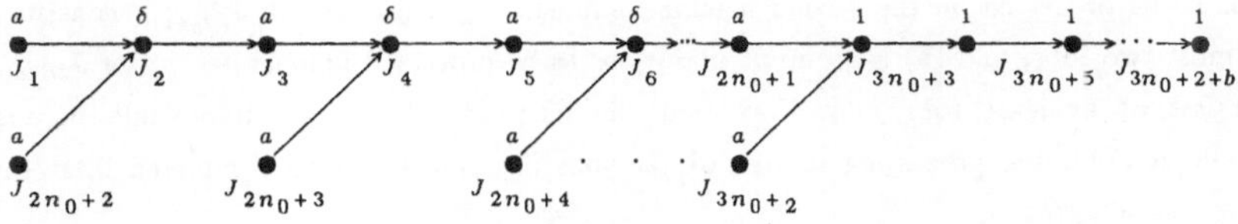

Fig. 2.2

Define

$$y = 2\sum_{l=1}^{n_0+1}(al+\delta(l-1)) + 4\sum_{l=1}^{n_0}(a+\delta)l + \sum_{l=1}^{b}((n_0+1)a+n_0\delta+l).$$

Without loss of generality, assume that the jobs J_l, $l = 1, 2, \dots, 2n_0+1$, $l = 3n_0+3, 3n_0+4, \dots, 3n_0+2+b$, are processed immediately one after another and that the completion time of job J_{3n_0+2+b} determines the makespan. In fact, if there is no schedule s that is feasible with respect to G satisfying the described conditions, then we have

$$\sum_{i \in N}\bar{t}_i(s) \geq \sum_{l=1}^{n_0+1}(al+\delta(l-1)) + \sum_{l=1}^{n_0}(a+\delta)l + \sum_{l=1}^{b}((n_0+1)a+n_0\delta+1+l)$$

$$= y - \sum_{l=1}^{n_0+1}(al+\delta(l-1)) - 3\sum_{l=1}^{n_0}(a+\delta)l + b > y.$$

Therefore, we may assume that: (a) the jobs mentioned above are processed on the same machine (e.g., on the first one); (b) this machine does not process other jobs; (c) the processing of a job J_{2n_0+1+l} is completed (on the second machine) by no later than time $t = al+\delta(l-1)$, $l = 1, 2, \ldots, n_0+1$, and the processing of all main jobs is finished by no later than time $t = a(n_0+1)+\delta n_0$. In the following, we consider only those schedules which satisfy all the conditions introduced above.

Suppose that the 3-partition problem has a solution. Then the set of main jobs may be partitioned into groups of three jobs each so that each triple of jobs may be processed on the second machine simultaneously with the processing of one of the jobs J_{2l}, $l = 1, 2, \ldots, n_0$, on the first machine. In any such schedule, the sum of the completion times of those main jobs which are included in the same triple does not exceed $3(a+\delta)l$. Hence, for any such a schedule s^0 we have

$$\sum_{i\in N}\bar{t}_i(s^0) \le 2\sum_{l=1}^{n_0+1}(al+\delta(l-1)) + 4\sum_{l=1}^{n_0}(a+\delta)l + \sum_{l=1}^{b}((n_0+1)a+n_0\delta+l) = y.$$

Suppose now that the 3-partition problem has no solution. In this case, in any schedule s that is feasible with respect to G and satisfying the introduced conditions, there exists at least one pair of indices l_1 and l_2 ($l_1 < l_2$) such that: (a) a group of main jobs to be processed on the second machine immediately after the job $J_{2n_0+1+l_1}$ consists of at most two jobs, and (b) a group of main jobs to be processed immediately after $J_{2n_0+1+l_2}$ consists of at least four jobs. Note that the inequality $l_1 > l_2$ is impossible because, should it hold, the processing of one of the jobs J_{2n_0+1+l} would be completed later than time $t = al+\delta(l-1)$.

Let N_1 and N_2 denote the sets of jobs to be processed in schedule s on the first and second machines, respectively. We have

$$\sum_{i\in N_1}\bar{t}_i(s) = \sum_{l=1}^{n_0+1}(al+\delta(l-1)) + \sum_{l=1}^{n_0}(a+\delta)l + \sum_{l=1}^{b}((n_0+1)a+n_0\delta+l),$$

$$\sum_{i\in N_2}\bar{t}_i(s) > \sum_{l=1}^{n_0+1}al + 3\sum_{l=1}^{n_0}(a+1)l - al_1 + al_2 \ge \sum_{l=1}^{n_0+1}al + 3\sum_{l=1}^{n_0}(a+1)l + a.$$

The bound on $\sum_{i\in N_2}\bar{t}_i(s)$ is derived as follows. The completion times of the jobs J_{2n_0+1+l}, $l = 1, 2, \ldots, n_0+1$, are computed as if these jobs were processed immediately one after another, and the completion times of the main jobs of the group to be processed immediately after job J_{2n_0+1+l} were considered to be equal to $(a+l)$.

Since it may be assumed that $n_0 \geq 2$, we have

$$\sum_{i \in N} \bar{t}_i(s) = \sum_{i \in N_1} \bar{t}_i(s) + \sum_{i \in N_2} \bar{t}_i(s) > y - \sum_{l=1}^{n_0+1} \delta(l-1) - 3\sum_{l=1}^{n_0} (\delta-1)l + a$$

$$= y - \delta n_0(n_0+1)/2 - 3(\delta-1)n_0(n_0+1)/2 + a > y,$$

Hence, a schedule s^0 feasible with respect to G and such that $\sum_{i \in N} \bar{t}_i(s^0) \leq y$ exists if and only if the 3-partition problem has a solution. The implementation of the described reduction requires at most $O(\delta n_0^4)$ time. Since the 3-partition problem is NP-hard in the strong sense, it follows that Problem 2.3 is NP-hard is the strong sense as well.

If $G = (N, \varnothing)$, Problem 2.3 can be solved in $O(n\log n)$ time. This algorithm (see Section 9.3 of Chapter 2) is developed for solving a more general problem when a processing system consists of $M \geq 2$ uniform machines.

2.12. The decision problem corresponding to Problem 2.4(a) is as follows: determine whether there exists a schedule s^0 that is feasible with respect to G, provided that at each time moment no more than one resource unit is consumed and $\bar{t}_{max}(s^0) \leq y$ for a given y.

We show that the 3-partition problem reduces to the described decision problem in polynomial time.

Define $n = 4n_0\delta$. Assume that set N contains jobs of four types: I_l and I'_l, $l = 1, 2, ..., n_0\delta$, and $J_{i,k}$ and $J'_{i,k}$, $i = 1, 2,..., 3n_0$, $k = 1, 2,..., \gamma_i$. Define $r(I_l) = r(J_{i,k}) = 0$ for all jobs I_l and $J_{i,k}$, and $r(I'_l) = r(J'_{i,k}) = 1$, for all jobs I'_l and $J'_{i,k}$. Define $y = 2n_0\delta$.

The precedence relation $\longrightarrow$ is defined over set N as follows (only the reduction of the relation is given below):

$$I_{(j-1)\delta+l} \longrightarrow I_{(j-1)\delta+l+1},$$

$$I'_{(j-1)\delta+l} \longrightarrow I'_{(j-1)\delta+l+1},\ l = 1, 2,..., \delta-1,\ j = 1, 2,..., n_0;$$

$$I'_{j\delta} \longrightarrow I_{(j-1)\delta+1},\ j = 1, 2,..., n_0;$$

$$J_{i,k} \longrightarrow J_{i,k+1},\ J'_{i,k} \longrightarrow J'_{i,k+1},\ k = 1, 2,..., \gamma_i-1,\ i = 1, 2,..., 3n_0;$$

$$J_{i,\gamma_i} \longrightarrow J'_{i,1},\ i = 1, 2,..., 3n_0;\ I_{j\delta} \longrightarrow I'_{j\delta+1},\ j = 1, 2,..., n_0+1.$$

It is easy to verify that each element $i \in N^0$ is associated with a chain of jobs $J_{i,1} \longrightarrow J_{i,2} \longrightarrow ... \longrightarrow J_{i,\gamma_i} \longrightarrow J'_{i,1} \longrightarrow J'_{i,2} \longrightarrow ... \longrightarrow J'_{i,\gamma_i}$, and the reduction graph G consists of $3n_0+1$ chains, where $3n_0$ chains correspond to the elements of set N^0, while the elements I_l and I'_l, $l = 1, 2, \ldots, n_0\delta$, correspond to the vertices of the $(3n_0+1)$th chain.

Suppose that the 3-partition has a solution and that N_j^0 are the required three-element subsets of set N^0. Then a schedule s^0 can be constructed as follows. All jobs I_l and I'_l are assigned to the first machine and are processed according to the sequence that is feasible with respect to $\rightarrow$ (such a sequence is unique). Let $N_j^0 = \{i_1, i_2, i_3\}$. Then the jobs $J_{i_1,1}$, $J_{i_1,2}, \ldots$, $J_{i_1,\gamma_{i_1}}$, $J_{i_2,1}$, $J_{i_2,2}, \ldots$, $J_{i_2,\gamma_{i_2}}, J_{i_3,1}$, $J_{i_3,2}, \ldots$, $J_{i_3,\gamma_{i_3}}$ are assigned to be processed on the second machine in the same time interval as the jobs I'_l, $l = (j-1)\delta+1$, $(j-1)\delta+2, \ldots$, $j\delta$, and the jobs $J'_{i_1,1}$, $J'_{i_1,2}, \ldots$, $J'_{i_1,\gamma_{i_1}}$, $J'_{i_2,1}$, $J'_{i_2,2}, \ldots$, $J'_{i_2,\gamma_{i_2}}$, $J'_{i_3,1}$, $J'_{i_3,2}, \ldots$, $J'_{i_3,\gamma_{i_3}}$, in the same interval as the jobs I_l, $l = (j-1)\delta+1$, $(j-1)\delta+2, \ldots$, $j\delta$. It is obvious that $F(s^0) = 2n_0\delta$ and that schedule s^0 is feasible with respect to G and does not violate the resource constraints.

Suppose now that there exists a required schedule s^0. Without loss of generality, it may be assumed that in schedule s^0 the first machine processes all jobs I_l, I'_l and only those. In this case (since $F(s^0) = 2n_0\delta$), while job I'_l is processed on the first machine, the second machine must simultaneously process one of the jobs $J_{i,k}$. Similarly, while I_l is processed on the first machine, the second machine processes some job $J'_{i,k}$.

We show that, in schedule s^0, the sequence according to which the second machine processes the jobs $J_{i,k}$ and $J'_{i,k}$, defines the desired partition of set N^0. In fact, let the jobs $J_{i',1}$, $J_{i',2}, \ldots$, $J_{i',\gamma_{i'}}$, $J_{i'',1}$, $J_{i'',2}, \ldots$, $J_{i'',\gamma_{i''}}$, $J_{i''',1}$, $J_{i''',2}, \ldots$, $J_{i''',\gamma_{i'''}}$, and only these, be processed on the second machine, while simultaneously the jobs I'_l, $l = (j-1)\delta+1$, $(j-1)\delta+2, \ldots$, $j\delta$, are processed on the first machine. Then $\gamma_{i'}+\gamma_{i''}+\gamma_{i'''} = \delta$ and i', i'', i''' form one of the three-element groups of the desired partition. On the other hand, suppose that the above condition is violated, i.e., there exists such a j^0 that in the interval during which the jobs I'_l, $l = (j^0-1)\delta+1$, $(j^0-1)\delta+2, \ldots$, $j^0\delta$, are processed on the first machine, the second machine processes the jobs $J_{i^0,1}$, $J_{i^0,2}, \ldots$, $J_{i^0,\nu}$, where $\nu < \gamma_{i^0}$ (probably together with some other jobs), and does not process the job $J_{i^0,\nu+1}$. Let j^0 be the smallest such index. Then simultaneously with the processing of the jobs I_l, $l = (j^0-1)\delta+1$, $(j^0-1)\delta+2, \ldots$, $j^0\delta$, it is possible to process at most $\delta-\nu$ jobs $J'_{i,k}$ since $J_{i^0,\gamma_{i^0}} \rightarrow J'_{i^0,1}$, and the job $J_{i^0,\gamma_{i^0}}$ has not yet been processed. This contradicts the fact that, in schedule s^0, each job I_l is simultaneously processed with some job $J'_{i,k}$.

The implementation of the described reduction takes $O(n_0\delta)$ time. Since the 3-partition problem is NP-hard in the strong sense, we conclude that Problem 2.4(a) is also NP-hard in the strong sense.

Note that if $G = (N, \varnothing)$, Problem 2.4(a) is solvable in $O(n^3)$ time [211]. In this case, the processing system may consist of M uniform machines operating at different speeds. Problem 2.4(a) is solvable in $O(n^2)$ time if no resource constraints are imposed and the reduction graph G is an arbitrary circuit-free graph (see Section 5.5 of Chapter 2).

2.13. The decision problem corresponding to Problem 2.4(b) is as follows: determine whether there exists a schedule s^0 that is feasible with respect to G and satisfies the resource constraints such that $\sum_{i \in N} \bar{t}_i(s^0) \leq y$ for a given y.

To prove that the formulated problem is NP-hard, we use the reduction constructed in Section 2.12. The only change required here is: $y = 2n_0\delta(2n_0\delta+1)$.

If the desired partition of set N^0 into three-element subsets exists, then the schedule constructed in Section 2.12 may be taken as s^0. It is easy to verify that $\sum_{i \in N} \bar{t}_i(s^0) = 2n_0\delta(2n_0\delta+1)$.

Let us find a lower bound on $\sum_{i \in N} \bar{t}_i(s)$ for all schedules (including those that are not feasible with respect to G and/or to the resource constraints) in which both machines have no intermediate idle time. Let x denote the number of jobs to be processed on the first machine. Then the second machine processes $4n_0\delta - x$ jobs. It is easy to check that

$$\min_s \sum_{i \in N} \bar{t}_i(s) = \min\{x(x+1)/2+(4n_0\delta - x(4n_0\delta - x+1)/2 \,|\, 0 \leq x \leq 4n_0\delta\} = y,$$

where $x = 2n_0\delta$. Thus, for any schedule s (and therefore for any feasible schedule) we have $\sum_{i \in N} \bar{t}_i(s) = y$ if and only if each machine processes $2n_0\delta$ jobs and has no idle time. Otherwise, $\sum_{i \in N} \bar{t}_i(s) > y$.

Suppose that there exists a schedule s^0 that is feasible with respect to G and to the resource constraints such that $\sum_{i \in N} \bar{t}_i(s^0) = y$. Without loss of generality, the first machine may be assumed to process jobs I_l and I_l', $l = 1, 2, \ldots, n_0\delta$. Then the considerations presented in Section 2.12 imply that the sequence in which the jobs are processed on the second machine defines a desired partition of set N^0.

Thus, Problem 2.4(b) is NP-hard in the strong sense.

2.14. Due to Remarks 1.2 and 1.3 (see Section 1.8 of this chapter), the fact that Problems 2.1(a) and 2.4(a) are NP-hard in the strong sense implies that Problems 2.1 and 2.4 are NP-hard in the strong sense in the following cases:

(c) $F(s) = L_{max}(s)$;

(d) $F(s) = z_{max}(s)$; $D_i = D$, $i = 1, 2, \ldots, n$;

(e) $F(s) = \sum_{i \in N} z_i(s)$; $D_i = D$, $i = 1, 2,\dots, n$;

(f) $F(s) = \sum_{i \in N} u_i(s)$; $D_i = D$, $i = 1, 2,\dots, n$.

Similarly, Remark 1.2 and the fact that Problems 2.2(a)-(c) are *NP*-hard in the strong sense imply that Problem 2.2 is *NP*-hard in the strong sense in the following cases:

(i) $F(s) = \sum_{i \in N} z_i(s)$; $G = (N, \varnothing)$; $D_i = D$, $i = 1, 2,\dots, n$;

(j) $F(s) = \sum_{i \in N} \alpha_i z_i(s)$; $G = (N, \varnothing)$; $D_i = D$, $i = 1, 2,\dots, n$; and preemption is allowed;

(k) $F(s) = \sum_{i \in N} \alpha_i z_i(s)$; $t_i = 1$; $D_i = D$, $i = 1, 2,\dots, n$.

Remark 1.4 implies that Problems 2.2(d), 2.2(e), 2.2(h) also remain *NP*-hard in the strong sense if preemption is allowed.

2.15. *Remark* 2.2. Consider a class of problems for which all release times d_i and processing times t_i are integers. Let Problems A and B of this class differ only in that in Problem A either all $t_i = 1$ or $t_i \in \{0,1\}$ and no preemption is allowed, while in Problem B the values of t_i may be arbitrary integers and preemption is allowed only at integer times. Then it is obvious that there exist both polynomial and pseudopolynomial reductions of Problem A to Problem B.

This and the *NP*-hardness in the strong sense of Problems 2.1(a)-(f), 2.2(c), 2.2(e), 2.2(g), 2.2(h), 2.2(k), 2.4(a)-(f) imply that all these problems remain *NP*-hard in the strong sense if the processing times are not unit but arbitrary integers, provided that preemption is allowed only at integer times.

3. Reducibility of the Vertex Covering Problem

This section uses *the vertex covering problem* as a standard problem to prove the *NP*-hardness of scheduling problems.

A non-directed graph $\Gamma = (V, E)$ and a positive integer y_0 are given. A set of vertices $W \subseteq V$ is called *a vertex covering of graph* Γ if each edge of set E is incident to at least one vertex in W. Does there exist a vertex covering W^0 of graph Γ such that $|W^0| \le y_0$?

Let $V = \{1, 2,\dots, v\}$ and $e = |E|$. Then the input length of the vertex covering problem under binary encoding is in the interval $[c_1(v+e),\ c_2(v+e)\log v]$, while under unary encoding this is in the interval $[c_3(v^2+e),\ c_4 v(v+e)]$. Here c_1, c_2, c_3, c_4 are constants independent of v and e, $0 < c_1 \le c_2$, $0 < c_3 \le c_4$.

The vertex covering problem is *NP*-complete in the strong sense.

3.1. This section studies the following problems.

Problem 3.1. The jobs of a set $N = \{1, 2, ..., n\}$ enter a processing system consisting of two identical parallel machines at time $d = 0$. A job $i \in N$ can be processed on any machine, and this takes t_i time units. Preemption is not allowed. A precedence relation with the reduction graph $G = (N, U)$ is defined over set N. Each connected component of G is an outtree. It is required to find a schedule s^* for processing the jobs of set N that is feasible with respect to G and minimizes the function $F(s) = \sum_{i \in N} \bar{t}_i(s)$, where $\bar{t}_i(s)$ is the completion time of job i in schedule s.

Problem 3.2. The jobs of a set $N = \{1, 2, ..., n\}$ enter the processing system consisting of M identical parallel machines, $M \geq 2$, at time $d = 0$. The processing time of a job $i \in N$ is t_i time units. No preemption is allowed. Each job $i \in N$ is given the due date $D_i \geq 0$ by which it is desirable that processing is completed. A precedence relation is defined over set N. Each connected component of the reduction graph $G = (N, U)$ is an outtree. It is required to find a schedule s^* for processing the jobs of set N that is feasible with respect to G and minimizes the function $F(s) = L_{max}(s) = \max\{\bar{t}_i(s) - D_i \mid i \in N\}$ provided that $t_i = 1$, $i = 1, 2, ..., n$.

We show that both formulated problems are *NP*-hard in the strong sense.

3.2. The following decision problem corresponds to Problem 3.1: determine whether there exists a schedule s^0 for processing the jobs of set N that is feasible with respect to G and such that $\sum_{i \in N} \bar{t}_i(s^0) \leq y$ for a given y.

We show that the vertex covering problem reduces to the formulated decision problem in polynomial time.

The set N of jobs is defined as follows. Each vertex j of graph Γ is associated with the job V_j called a vertex job. Replace the edges of Γ by the arcs: an edge $[j_1, j_2] \in E$ is replaced by the arc (j_1, j_2) if $j_1 < j_2$ and by the arc (j_2, j_1) if $j_1 > j_2$. Let the arcs be numbered by the integers from 1 to e. The arc with the number k, $k = 1, 2, ..., e$, is associated with the job $E_k^{(1)}$, corresponding to the starting vertex of the arc, and with the job $E_k^{(2)}$ corresponding to the terminal vertex of the arc. These jobs are called edge jobs. Define the precedence relation $\rightarrow$ over the set of all vertex jobs and edge jobs as follows: $V_j \rightarrow E_k^{(1)}$ (or $V_j \rightarrow E_k^{(2)}$) if and only if vertex j is the starting vertex (or the terminal vertex) of the arc with the number k. Define $t(V_j) = 1$, $i = 1, 2, ..., v$; $t(E_k^{(1)}) = t(E_k^{(2)}) = 1+k$, $k = 1, 2, ..., e$.

Define $a = 16(v+e+1)^3$, and introduce $3v+6e+2a+a(v+2e+a)$ auxiliary jobs of four types:

J_r, $t(J_r) = 1$, $r = 1, 2, \ldots, 2v+2e+a$,

J'_r, $t(J'_r) = 0.5$, $r = 1, 2, \ldots, v+2e+a$,

$L_p^{(1)}$ and $L_p^{(2)}$, $t(L_p^{(1)}) = t(L_p^{(2)}) = 1+p$, $p = 1, 2, \ldots, e$,

$I_q^{(r)}$, $t(I_q^{(r)}) = 1/2a$, $q = 1, 2, \ldots, a$, $r = 1, 2, \ldots, v+2e+a$.

Define the precedence relation $\rightarrow$ over the set of the auxiliary jobs in the following way (only the reduction of this relation is given; see Fig. 3.1):

$J_{r-1} \rightarrow J_r$, $r = 2, 3, \ldots, 2y_0$, $r = 2y_0+e+1, 2y_0+e+2, \ldots, 2v+e$,
$r = 2v+2e+1, 2v+2e+2, \ldots, 2v+2e+a$;

$J_{2r-1} \rightarrow J'_r$, $r = 1, 2, \ldots, y_0$;

$J_{2r-e+1} \rightarrow J'_r$, $r = y_0+e+1, y_0+e+2, \ldots, v+e$;

$J_{v+r-1} \rightarrow J'_r$, $r = v+2e+1, v+2e+2, \ldots, v+2e+a$;

$L_p^{(1)} \rightarrow J_{2y_0+p}$, $L_p^{(1)} \rightarrow J'_{y_0+p}$, $J_{2y_0+p-1} \rightarrow L_p^{(1)}$, $p = 1, 2, \ldots, e$;

$L_p^{(2)} \rightarrow J_{2v+e+p}$, $L_p^{(2)} \rightarrow J'_{v+e+p}$, $J_{2v+e+p-1} \rightarrow L_p^{(2)}$, $p = 1, 2, \ldots, e$;

$J'_r \rightarrow I_1^{(r)}$, $r = 1, 2, \ldots, v+2e+a$;

$I_{q-1}^{(r)} \rightarrow I_q^{(r)}$, $q = 2, 3, \ldots, a$, $r = 1, 2, \ldots, v+2e+a$.

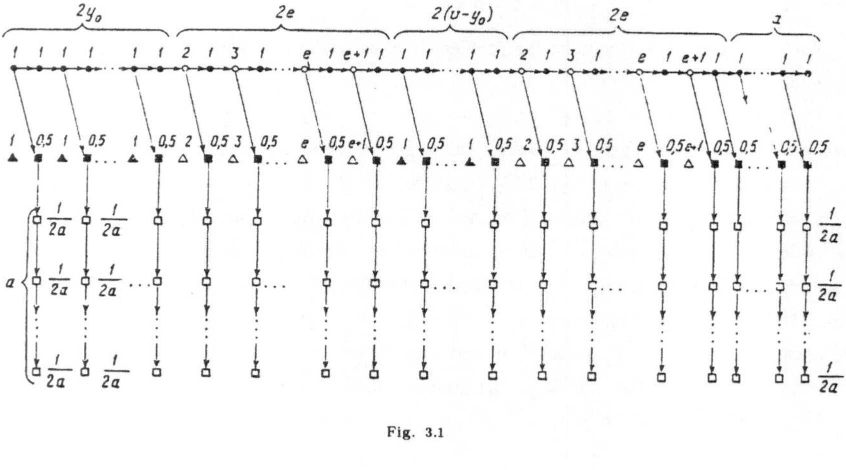

Fig. 3.1

It is easy to verify that each connected component of the reduction graph G of the constructed precedence relation is an outtree. Figure 3.1 displays a graph obtained from G by removing all arcs of the form $(V_j, E_k^{(1)})$ and $(V_j, E_k^{(2)})$. The following notation is used: ▲ - a vertex job; △ - an edge job; ● - a job J_r; ■ - a job J_r'; ○ - a job $L_p^{(1)}$ or $L_p^{(2)}$; □ - a job $I_q^{(r)}$. Images of the jobs are accompanied by their corresponding processing times.

Let set $N = \{1, 2,..., n\}$ consist of all vertex, edge and auxiliary jobs. Construct a schedule s^0 for processing the jobs of set N in the following way. The jobs J_r, $L_p^{(1)}$ and $L_p^{(2)}$, $r = 1, 2,..., 2v+2e+a$, $p = 1, 2,..., e$, and only these, are processed on the first machine one after another with no intermediate idle time according to a sequence that is feasible with respect to G (see Fig. 3.1). The remaining jobs are processed on the second machine. For job J_r', let its immediate predecessor (with respect to $\rightarrow$) be completed on the first machine at time τ_r. Assume that in schedule s^0 the processing of job J_r', $r = 1, 2,..., v+2e+a$, on the second machine starts at time τ_r. Denote the corresponding completion time by τ_r'. It is obvious that $\tau_r' = \tau_r+1/2$. Denote $\tau = \sum_{r=1}^{v+2e+a} \tau_r'$. Assume that in schedule s^0 the processing of job $I_1^{(r)}$, $r = 1, 2,..., v+2e+a$, on the second machine starts at time τ_r' and all jobs $I_q^{(r)}$, $q = 1, 2,..., a$, are processed by the second machine immediately one after another according to a sequence that is feasible with respect to G. In Fig. 3.1 the jobs placed in the same column are processed on the first machine (top row) and on the second machine (all other rows) in the same time interval. The vertex jobs and the edge jobs are processed on the second machine according to the sequence shown in Fig. 3.1.

Let graph Γ contain such a vertex covering W that $|W| \le y_0$. Then the sequence of processing of the vertex jobs and the edge jobs may be chosen to make schedule s^0 feasible with respect to G. In fact, let in the time interval $[0, 2y_0]$ the second machine process the vertex jobs corresponding to the vertices of W. If $|W| < y_0$, then in this interval it is possible to process $y_0 - |W|$ other vertex jobs. The definition of a vertex covering of graph Γ and that of the precedence relation over the set of the vertex jobs and the edge jobs imply that at time $t = 2y_0$ it is possible to start processing at least one of two jobs $E_k^{(1)}$ and $E_k^{(2)}$ for each $k = 1, 2,..., e$. These edge jobs are processed in the time interval $[2y_0, 2y_0+2e+e(e+1)/2]$ according to the sequence shown in Fig. 3.1. The rest of the vertex jobs are processed in the time interval $[2y_0+2e+e(e+1)/2, 2v+2e+e(e+1)/2]$ and the other edge jobs are processed in the interval $[2v+2e+e(e+1)/2, 2v+4e+2e(e+1)/2]$.

Define

$$y = 2b + a(2v + e(e+5)) + a(a+1)/2 + (a+1)\tau + (a+1)(v + 2e + a)/4,$$

where

$$b = 2(v+2e)(2v+e(e+5)).$$

Observe that $a > 2b$.

Let us estimate the value of $\sum_{i \in N} \bar{t}_i(s^0)$. The set $\{J_r, L_p^{(1)}, L_p^{(2)} \mid r = 1, 2, \ldots, 2v+2e+a, p = 1, 2, \ldots, e\}$ of jobs is partitioned into five subsets N_h, $h = 1, 2, \ldots, 5$, as shown in Fig. 3.1. Set N_1 contains $2y_0$ jobs J_r, $r = 1, 2, \ldots, 2y_0$, each of the sets N_2 and N_4 contains $2e$ elements; set N_3 consists of $2(v - y_0)$ jobs, and set N_5 consists of a elements J_r. In the expression $\sum_{h=1}^{4} \sum_{i \in N_h} \bar{t}_i(s^0)$, replace all $\bar{t}_i(s^0)$ by $2v + e(e+5)$ which is equal to the completion time of job J_{2v+2e} in schedule s^0. Then

$$\sum_{h=1}^{4} \sum_{i \in N_h} \bar{t}_i(s^0) < 2(v+2e)(2v+e(e+5)) = b.$$

It is not difficult to compute $\sum_{i \in N_5} \bar{t}_i(s^0) = a(2v+e(e+5)) + a(a+1)/2$. Let N_6 denote the set of the vertex and edge jobs, N_7 denote the set of jobs J_r', $r = 1, 2, \ldots, v+2e+a$, and N_8 denote the set of all jobs $I_q^{(r)}$. It is clear that $N = \bigcup_{h=1}^{8} N_h$, $\sum_{i \in N_6} \bar{t}_i(s^0) < b$, $\sum_{i \in N_7} \bar{t}_i(s^0) = \tau$. It is also easy to check that $\sum_{i \in N_8} \bar{t}_i(s^0) = a\tau + (a+1)(v+2e+a)/4$.

Thus, if graph Γ has a vertex covering W such that $|W| \le y_0$, then $\sum_{i \in N} \bar{t}_i(s^0) < y$.

Suppose that the inequality $|W| > y_0$ holds for any vertex covering W of graph Γ. Note that for any schedule s that is feasible with respect to G the relations $\sum_{i \in N_5} \bar{t}_i(s) \ge a(2v+e(e+5)) + a(a+1)/2$, $\sum_{i \in N_7} \bar{t}_i(s) \ge \tau$, $\sum_{i \in N_8} \bar{t}_i(s) \ge a\tau + (a+1)(v+2e+a)/4$ hold. Assume that in a schedule s, processing each job J_r' starts at time τ_r. Then, no more than y_0 vertex jobs may be completed by time $t = 2y_0$. Since $|W| > y$ for any vertex covering W of graph Γ, it follows that there exists at least one index k such that at time $t = 2y_0$ both jobs $E_k^{(1)}$ and $E_k^{(2)}$ have at least one non-completed predecessor. Hence, in any schedule s that is feasible with respect to G, at least one of the following situations arises: (a) there exists such a k that the processing of job J_k' starts no earlier than time $t = \tau_k + 1$; or (b) the processing of at least one of the edge jobs starts no earlier than time $t = 2v + e(e+5) + a - 1$. In any case, we have $\sum_{i \in N} \bar{t}_i(s) > \sum_{i \in N_5 \cup N_7 \cup N_8} \bar{t}_i(s) + a \ge y - 2b + a > y$ since $a > 2b$.

Thus, a schedule s^0 that is feasible with respect to G and such that $\sum_{i \in N} \bar{t}_i(s^0) \le y$ exists if and only if there exists a vertex covering of graph Γ containing at most y_0

vertices. The implementation of the described reduction requires at most $O((v+e)^6)$ time.

Hence, Problem 3.1 is *NP*-hard in the strong sense.

Note that, if $G = (N, \emptyset)$, Problem 3.1 is solvable in $O(n\log n)$ time (see Section 9.3 of Chapter 2).

3.3. The following decision problem corresponds to Problem 3.2: determine whether there exists such a schedule s^0 that is feasible with respect to G such that $L_{max}(s^0) \leq y$ for a given y.

We provide a polynomial reduction of the vertex covering problem to the formulated decision problem.

Define $M = (v+e)e+1$, $y = 0$.

The set N of jobs is defined as follows. Transform a non-directed graph Γ into the directed one as in Section 3.2. Associate a vertex j of the constructed graph with the vertex job V_j, $j = 1, 2,..., v$. Associate the arc with the number k with two groups of edge jobs $E^{(1)}_{k,p}$ and $E^{(2)}_{k,p}$, $k = 1, 2,..., e$, $p = 1, 2,..., e+k$. The first group of jobs corresponds to the starting vertex of an arc, while the second group corresponds to the terminal vertex of an arc. Besides, introduce two groups of auxiliary jobs: $Q^{(1)}_{k,q}$, $Q^{(2)}_{k,q}$, $k = 1, 2,..., e$, $q = 1, 2,..., (v+e)k$, and $J^{(h)}_r$, $h = 1, 2,..., 5e+4$, $r = 1, 2,..., r(h)$, where $r(1) = M-y_0$, $r(h) = M-e$, $h = 2, 3,..., e+2$, $h = 2e+5, 2e+6,..., 3e+5$; $r(h) = M-((v+e-1)(h-e-2)+e)$, $h = e+3, e+4,..., 2e+2$; $r(2e+3) = M$; $r(2e+4) = M-v+y_0$; $r(h) = M-4e+h-5$, $h = 3e+6, 3e+7,..., 4e+4$; $r(h) = M-(v+e)(h-4e-4)$, $h = 4e+5, 4e+6, ..., 5e+4$. All introduced jobs form the set $N = \{1, 2,..., n\}$.

Introduce the precedence relation $\rightarrow$ over the constructed set N as follows (only the reduction of this relation is given): $V_j \rightarrow E^{(1)}_{k,p}$ (or $V_j \rightarrow E^{(2)}_{k,p}$) if and only if j is the starting vertex (or the terminal vertex) of the arc with the number k;

$$E^{(1)}_{k,p-1} \rightarrow E^{(1)}_{k,p},\ E^{(2)}_{k,p-1} \rightarrow E^{(2)}_{k,p},\ p = 2, 3,..., e+k,\ k = 1, 2,..., e;$$

$$E^{(1)}_{k,e+k} \rightarrow Q^{(1)}_{k,q},\ E^{(2)}_{k,e+k} \rightarrow Q^{(2)}_{k,q},\ q = 1, 2,..., (v+e)k,\ k = 1, 2,..., e;$$

$$J^{(h-1)}_1 \rightarrow J^{(h)}_r,\ h = 2, 3,..., 5e+4,\ r = 1, 2,..., r(h).$$

Each connected component of the reduction graph G of the constructed precedence relation $\rightarrow$ is an outtree. Figure 3.2a gives an example of graph Γ and Figure 3.2b shows the corresponding graph G. The arcs $(J^{(h-1)}_1, J^{(h)}_r)$, $h = 9, 10, 11, 12$, $r = 2, 3,..., r(h)$, as well as $(E^{(1)}_{1,3}, Q^{(1)}_{3,q})$ $q = 1, 2,..., 5$, and $(E^{(2)}_{2,4}, Q^{(2)}_{4,q})$, $q = 1, 2,..., 10$, are not shown in Fig.3.2b. The following notation is used: ○ - a vertex job; △ - an edge job; □ - a job $Q^{(1)}_{k,q}$ or $Q^{(2)}_{k,q}$; ● - a job $J^{(h)}_r$.

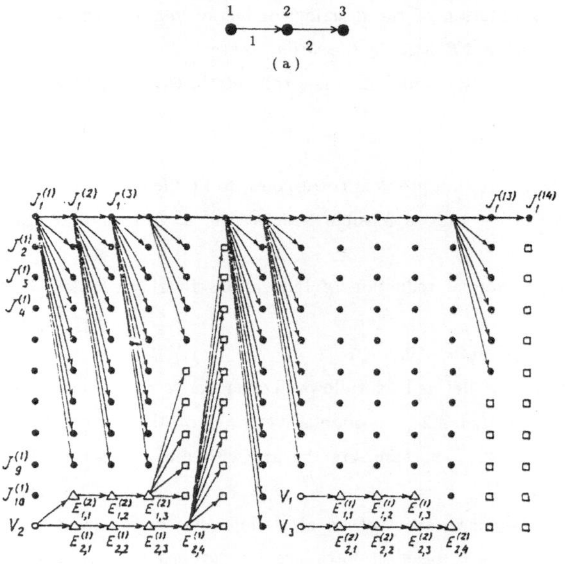

(b)

Fig.3.2.

Define $D(V_j) = 2e+4$, $j = 1, 2,\dots, v$; $D(E^{(1)}_{k,e+k}) = D(E^{(2)}_{k,e+k}) = 3e+k+4$, $k = 1, 2,\dots, e$; $D(E^{(1)}_{k,p}) = D(E^{(2)}_{k,p}) = 4(e+1)$, $p = 1, 2,\dots, e+k-1$, $k = 1, 2,\dots, e$; $D(Q^{(1)}_{k,q}) = D(Q^{(2)}_{k,q}) = 4e+k+4$, $q = 1, 2,\dots, (v+e)k$, $k = 1, 2,\dots, e$; $D(J^{(h)}_r) = h$, $h = 1, 2,\dots, 5e+4$, $r = 1, 2,\dots, r(h)$.

Let us establish the conditions under which there exists a schedule s^0 for processing the jobs of the constructed set N that is feasible with respect to G and such that $L_{max}(s^0) \le 0$.

Since each job $J^{(h)}_r$ has exactly $h-1$ predecessors and $D(J^{(h)}_r) = h$, it follows that if $L_{max}(s^0) \le 0$ then each job $J^{(h)}_r$ starts at time $t = h-1$. The latest due date in the formulated problem is equal to $5e+4$, while $|N| = (5e+4)M$. Therefore, if the inequality $L_{max}(s^0) \le 0$ holds, then at each time the processing system must process exactly M jobs.

Let graph Γ have a vertex covering W such that $|W| \le y_0$. In this case, it is possible to construct a schedule s^0 that is feasible with respect to G and satisfies the required conditions. In fact, in the time interval $[0, 1]$, it is possible to process y_0 vertex jobs, among which there are $|W|$ jobs corresponding to the vertex covering. Hence, at time $t = 1$ at least one of two jobs $E^{(1)}_{k,1}$ and $E^{(2)}_{k,1}$ may start processing, $k = 1, 2,\dots, e$. In the time interval [2, $2e+1$], it is possible to process e groups of edge jobs $E_{k,p}^{(\nu)}$, $p = 2, 3, \ldots, e+k$, $k = 1, 2, \ldots, e$, where $\nu = 1$ or $\nu = 2$ depending on which of the two jobs $E_{k,1}^{(1)}$ or $E_{k,1}^{(2)}$ was processed in the time interval [1, 2]. Moreover, in the time interval [1, $2e+1$], it is possible to process the jobs $Q_{k,q}^{(\nu)}$, $k = 1, 2, \ldots, e-1$, $q = 1, 2, \ldots, (v+e)k)$, where ν is defined as in the previous case, and the jobs $Q_{e,q}^{(\nu)}$, $q = 1, 2, \ldots, (v+e)e)$ can be assigned to be processed in the interval [$2e+1$, $2e+2$] for the same ν. The rest of the vertex jobs are processed in the time interval [$2e+3$, $2e+4$], the rest of the edge jobs are processed in the interval [$2e+4$, $4e+4$], and the rest of the jobs $Q_{k,q}^{(\nu)}$ are processed in increasing order of k in the interval [$4e+4$, $5e+4$]. As mentioned, the processing of each of the jobs $J_r^{(h)}$ starts at time $t = h-1$. A typical structure of schedule s^0 is shown schematically in Fig. 3.2b, where the jobs placed in the same row are processed on the same machine and those placed in the same column are processed in the same unit time interval.

It is not difficult to verify that $L_{max}(s^0) = 0$.

Suppose now that any vertex covering of graph Γ contains at least y_0+1 vertices. In this case, for any schedule for processing the jobs of set N that is feasible with respect to G, there exists at least one index k' such that both jobs $E_{k',1}^{(1)}$ and $E_{k',1}^{(2)}$ have at least one predecessor that is not completed by time $t = 1$. Therefore, in any schedule s that is feasible with respect to G, at most $M(2e+2)-y_0$ jobs may be completed by time $t = 2e+2$. In fact, the following cases are possible: (a) at time $t = 1$ less than e jobs $E_{k,1}^{(\nu)}$ may start processing; (b) at a time $t = 1$ it is possible to start processing at least e jobs $E_{k,1}^{(\nu)}$ but among them there is no such pair $E_{k,1}^{(1)}$, $E_{k,1}^{(2)}$ that $k < k'$ (for k' mentioned above); (c) at time $t = 1$, it is possible to start processing at least e jobs $E_{k,1}^{(\nu)}$ and among them there is such a pair $E_{k'',1}^{(1)}$, $E_{k'',1}^{(2)}$ of jobs that $k'' < k'$.

In case (a), at most $M(2e+2)-(e+k')-(v+e)k'+(v-y_0)$ jobs can be completed by time $t = 2e+2$ in any feasible schedule. In case (b), at most $M(2e+2)-(v+e)k'+(v-y_0)+(e-2)$ jobs can be completed. Consider case (c) in more detail. Unlike in schedule s^0, assume that instead of one of the jobs $E_{k',1}^{(1)}$, $E_{k',1}^{(2)}$ it is possible to start processing a job $E_{k'',1}^{(\nu)}$ at time $t = 1$. Since for a fixed ν the number of jobs $Q_{k'',q}^{(\nu)}$ is equal to k'' and $k'' < k'$, it follows that no more than $M(2e+2)-(v+e)k'+(v+e)k''+(e-2) < M(2e+2)-y_0$ jobs may be completed by time $t = 2e+2$. Hence, in the time interval [1, $2e+2$] there exists a unit subinterval in which less than M jobs are processed, i.e., one of the necessary conditions for the existence of the required schedule is violated.

Thus, a schedule s^0 that is feasible with respect to G and such that $L_{max}(s^0) \leq y$ exists if and only if graph Γ has a vertex covering containing at most y_0 vertices.

The implementation of the described reduction requires at most $O((v+e)e^2)$ time.

Since the vertex covering problem is *NP*-hard in the strong sense, Problem 3.2 is *NP*-hard in the strong sense as well.

Note that if each component of the graph G is an intree (not an outtree) Problem 3.2 becomes solvable in $O(n\log n)$ time (see Section 8.2 of Chapter 8).

3.4. It follows from Remark 1.1 (see Section 1 of this chapter) and from the fact that Problem 3.2 is *NP*-hard in the strong sense that this problem is also *NP*-hard in the strong sense if the objective function $F(s) = L_{max}(s)$ is replaced by:

(a) $F(s) = z_{max}(s)$;

(b) $F(s) = \sum_{i \in N} z_i(s)$;

(c) $F(s) = \sum_{i \in N} u_i(s)$;

Due to Remark 2.2 (see Section 2 of this chapter) and the fact that Problems 3.2 and 3.2(a)-(c) are *NP*-hard in the strong sense, we conclude that these problems remain *NP*-hard in the strong sense if the processing times are not unit but arbitrary integers, provided that preemption is allowed only at integer times.

4. Reducibility of the Clique Problem

This section uses the clique problem as a standard decision problem for proving the *NP*-hardness of some scheduling problems.

The *clique* problem is as follows. Given a non-directed graph $\Gamma = (V, E)$ and a positive integer y_0, does Γ contain such a complete subgraph (a clique) $\Gamma^0 = (V^0, E^0)$ that $|V^0| \le y_0$?

The input length of a clique problem in the binary alphabet is contained in the interval $[c_1(v+e), c_2(v+e)\log v]$, while the input length encoded in the unary alphabet belongs to the interval $[c_3(v^2+e), c_4 v(v+e)]$. Here $v = |V|$, $e = |E|$, and c_1, c_2, c_3, c_4 are constants independent of v and e $(0 < c_1 \le c_2,\ 0 < c_3 \le c_4)$.

The clique problem is *NP*-hard in the strong sense.

4.1. The section examines the following problems.

Problem 4.1. The jobs of a set $N = \{1, 2,..., n\}$ are processed on a single machine starting at time $d = 0$. All processing times are unit. No preemption is allowed. Each job $i \in N$ is given the due date D_i. A precedence relation with the reduction graph G is defined over set N. It is required to find a schedule s^* for processing the jobs of set N that is feasible with respect to G and minimizes the total tardiness $F(s) = \sum_{i \in N} z_i(s)$, where $z_i(s) = \max\{0, \bar{t}_i(s) - D_i\}$ and $\bar{t}_i(s)$ is the completion time of job i in schedule s.

Problem 4.2. The jobs of a set $N = \{1, 2,..., n\}$ enter the processing system consisting of two identical parallel machines at time $d = 0$. Any job $i \in N$ can be processed on any of the machines. This takes $t_i \in \{1, 2\}$ time units, and no preemption is allowed. A precedence relation is defined over set N, and G is the reduction graph of that relation. It is required to find a schedule s^* for processing the jobs of set N that is feasible with respect to G and minimizes the function $F(s)$ in the following cases:

(a) $F(s) = \bar{t}_{max}(s) = \max\{\bar{t}_i(s) \,|\, i \in N\}$;

(b) $F(s) = \sum_{i \in N} \bar{t}_i(s)$.

Problem 4.3. The jobs of a set $N = \{1, 2,..., n\}$ enter a processing system consisting of M identical parallel machines, $2 \leq M < n$, at time moment $d = 0$. All processing times are unit. No preemption is allowed. A precedence relation with the reduction graph G is defined over set N. It is required to find a schedule s^* for processing the jobs of set N that is feasible with respect to G and minimizes the function $F(s)$ in the following cases:

(a) $F(s) = \bar{t}_{max}(s)$;

(b) $F(s) = \sum_{i \in N} \bar{t}_i(s)$.

In what follows, the formulated problems are shown to be NP-hard.

4.2. The following decision problem corresponds to Problem 4.1: determine whether there exists such a schedule s^0 for processing the jobs of set N that is feasible with respect to G and such that $\sum_{i \in N} z_i(s^0) \leq y$ for a given y.

We show that the clique problem reduces to the formulated decision problem in polynomial time.

The set N of jobs is to be formed as follows. Associate each vertex j of graph Γ with the vertex job V_j, $j = 1, 2,..., v$. Associate each edge (j, k) of graph Γ with a group of v edge jobs $E^{(r)}_{j,k}$, $r = 1, 2,..., v$.

The precedence relation $\rightarrow$ is defined over the constructed set N as follows (only the reduction of the relation is presented):

$V_j \rightarrow E^{(1)}_{j,k}$, and $V_k \rightarrow E^{(1)}_{j,k}$, $(j, k) \in E$;

$E^{(r-1)}_{j,k} \rightarrow E^{(r)}_{j,k}$, $r = 2, 3,..., v$, $(j, k) \in E$.

Define

$y = (e - y_0(y_0-1)/2)(v-y_0) + (e - y_0(y_0-1)/2)(e - y_0(y_0-1)/2+1)v/2;$

$D(V_j) = v + y_0(y_0-1)v/2,\ j = 1, 2,\ldots, v;$

$D(E_{j,k}^{(r)}) = v + ve,\ (j, k) \in E,\ r = 1, 2,\ldots, v-1;$

$D(E_{j,k}^{(v)}) = y_0 + y_0(y_0-1)v/2.$

Suppose that graph Γ contains a clique with at least y_0 vertices. Then there exists a clique $\Gamma' = (V', E')$ such that $|V'| = y_0$. Consider a schedule s^0 for processing the jobs of the constructed set N in which (a) in the time interval $[0, y_0]$ exactly y_0 vertex jobs corresponding to the vertices of clique Γ' are processed; (b) in the time interval $[y_0, y_0+y_0(y_0-1)v/2]$ the jobs corresponding to the edges of Γ' (there are $y_0(y_0-1)v/2$ such jobs) are processed according to a sequence that is feasible with respect to G (G is the reduction graph of relation $\longrightarrow$); (c) starting at time $y_0+y_0(y_0-1)v/2$ the remaining vertex jobs are processed, and when these are completed, the remaining edge jobs are processed in such a way that the jobs connected by an arc in G are processed with no intermediate idle time. It is obvious that s^0 is feasible with respect to G, and $\sum_{i \in N} z_i(s^0) = (e - y_0(y_0-1)/2)(3v + ve - 2y_0 - y_0(y_0-1)v/2)/2 = y$.

Suppose now that any clique in graph Γ contains at most y_0-1 vertices. Then, in any schedule that is feasible with respect to G, at most $y_0(y_0-1)/2-1$ edge jobs $E_{j,k}^{(v)}$ can be processed in the time interval $[0, y_0+y_0(y_0-1)v/2]$. If, in some schedule s', all vertex jobs are completed in the time interval $[0, v+y_0(y_0-1)v/2]$, then $\sum_{i \in N} z_i(s') \geq \sum_{i \in N} z_i(s^0)+1 \geq y+1$. Let N_1 denote a set of all vertex jobs, and N_2 denote a set of all edge jobs $E_{j,k}^{(v)}$, $(j, k) \in E$. It is obvious that $\sum_{i \in N} z_i(s) = \sum_{i \in N_1 \cup N_2} z_i(s)$ for any schedule s. Hence, it follows that the search for a schedule s that satisfies the inequality $\sum_{i \in N} z_i(s) \leq y$ may be restricted to considering the schedules in which the jobs $E_{j,k}^{(r)}$, $r = 1, 2,\ldots, v$, (j and k are fixed) are processed immediately one after another, and the processing of a vertex job V_j starts at time $t = 0$, or immediately after the completion of some vertex job, or immediately after the completion of some job $E_{j,k}^{(v)}$. Suppose that in a schedule s'' that is feasible with respect to G a vertex job $V_{j'}$ is processed outside the time interval $[0, v+y_0(y_0-1)v/2]$ and processing starts after a job $E_{j'',k''}^{(v)}$ is completed. Transform schedule s'' into a schedule s''' in which (a) the starting time of $V_{j'}$ is v time units earlier than that in schedule s''; (b) the starting time of each of the jobs $E_{j'',k''}^{(r)}$, $r = 1, 2,\ldots, v$, is delayed by one time unit; (c) the rest of the jobs are processed as in schedule s''. Since $\bar{t}_{V_{j'}}(s'') \geq v+y_0(y_0-1)v/2+1$, it is easy to verify that

$$\sum_{i \in N} z_i(s''') = \sum_{i \in N} z_i(s'')+1-(\bar{t}_{V_{j'}}(s'')-v-y_0(y_0-1)v/2)$$

$$+ \max\{0, \bar{t}_{V_j}(s'') - 2v - y_0(y_0-1)v/2\} \le \sum_{i \in N} z_i(s'').$$

This implies that $\sum_{i \in N} z_i(s') \le \sum_{i \in N} z_i(s)$ for all feasible schedules s in which less than v vertex jobs are processed in the time interval $[0, v+y_0(y_0-1)v/2]$. Hence, when Γ does not have a clique containing at least y_0 vertices, the inequality $\sum_{i \in N} z_i(s) \ge y+1$ holds for any schedule s that is feasible with respect to G.

The implementation of the described reduction requires at most $O(ve)$ time.

Thus, Problem 4.1 is *NP*-hard in the strong sense. Remark 1.4 (see Section 1 of this chapter) implies that Problem 4.1 also remains *NP*-hard in the strong sense in the preemptive case.

Provided that $G = (N, \varnothing)$, the problem is solvable in $O(n^3)$ time (see Section 4.5 of Chapter 2).

4.3. The following decision problem corresponds to Problem 4.2(a): determine whether there exists a schedule s^0 for processing the jobs of set N that is feasible with respect to G and such that $\bar{t}_{max}(s^0) \le y$ for a given y.

We construct a polynomial reduction of the clique problem to the formulated decision problem.

The set N of jobs is to be formed as follows. For graph Γ, associate each vertex $j \in V$ with the vertex job V_j. Associate each edge $(j, k) \in E$ with the edge job $E_{j,k}$. Introduce $3v+2e$ auxiliary jobs denoted by J_r, $r = 1, 2,..., 3v+2e$.

Denote $a = 2(v+e)$ and define the precedence relation $\rightarrow$ over the constructed set N (only the reduction of the relation is presented):

$V_j \rightarrow E_{j,k}$ and $V_k \rightarrow E_{j,k}$ for all $(j, k) \in E$;

$J_{r-1} \rightarrow J_r$, $r = 2, 3,..., a$;

$J_{a+r} \rightarrow J_{2r}$, $r = 1, 2,..., y_0$;

$J_{2r} \rightarrow J_{a+r+1}$, $r = 1, 2,..., y_0-1$;

$J_{a+y_0+r} \rightarrow J_{y_0(y_0+1)+2r}$, $J_{y_0(y_0+1)+2(r-1)} \rightarrow J_{a+y_0+r}$, $r = 1, 2,..., v-y_0$.

The subgraph of the reduction graph G of the constructed precedence relation $\rightarrow$ induced by the set of all auxiliary jobs is shown in Fig. 4.1.

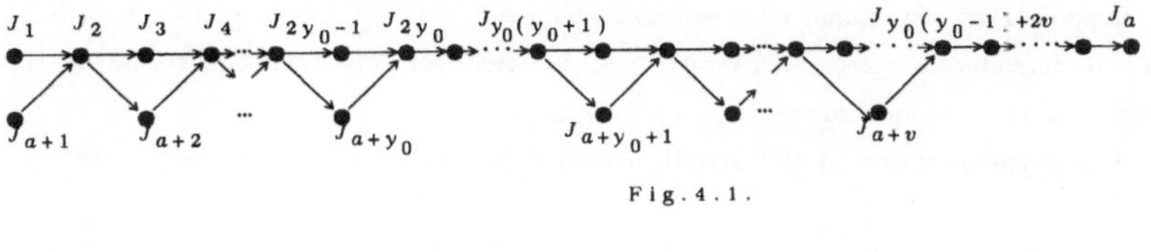

Fig. 4.1.

Define $y = 2(v+e)$; $t(V_j) = 1$, $j = 1, 2,\ldots, v$; $t(E_{j,k}) = 2$ for all $(j, k) \in E$; $t(J_r) = 1$, $r = 1, 2,\ldots, 3v+2e$.

We show that, in the constructed decision problem, a schedule s^0 that is feasible with respect to G and such that $\bar{t}_{max}(s^0) \leq y$ exists if and only if graph Γ has a clique containing at least y_0 vertices.

It is easy to verify (see Fig. 4.1) that if the relation $\bar{t}_{max}(s) \leq y$ holds for a schedule s that is feasible with respect to G, then in schedule s each job J_r, $r = 1, 2,\ldots, a$, is processed in the time interval $[r-1, r]$, each of the jobs J_{a+r}, $r = 1, 2,\ldots, y_0$, is processed in the time interval $[2r-2, 2r-1]$, and each of the jobs J_{a+y_0+r}, $r = 1, 2,\ldots, v-y_0$, is processed in the time interval $[y_0(y_0+1)+2r-2, y_0(y_0+1)+2r-1]$. Therefore, in the following, we consider only those schedules that satisfy these conditions. Without loss of generality, we may restrict our search to considering only the schedules s in which the jobs J_r, $r = 1, 2,\ldots, a$, and only those, are processed on the first machine. Since the total processing time of all vertex jobs, all edge jobs and the jobs J_{a+r}, $r = 1, 2,\ldots, v$, is equal to $2(v+e) = y$, it follows that the second machine has no idle time in the time interval $[0, a]$. Hence, if the relation $\bar{t}_{max}(s) \leq y$ holds for a schedule s that is feasible with respect to G, then exactly y_0-1 vertex jobs have to be completed by time $t = 2y_0-2$, and $v-y_0-1$ vertex jobs have to be processed in the time interval $[y_0(y_0+1)+1, y_0(y_0-1)+2v-2]$. The length of the time interval between the completion time of job J_{a+y_0} and the starting time of job J_{a+y_0+1} is equal to $y_0(y_0-1)+1$. The number $y_0(y_0-1)+1$ is odd, the number of the vertex jobs which are not yet assigned is equal to two, the processing time of each edge job is equal to two time units. Hence, in the interval in question only one of these jobs can be processed. Therefore, time left in this time interval must be used to process $y_0(y_0-1)/2$ edge jobs.

It is obvious that a schedule s in which $y_0(y_0-1)/2$ edge jobs are processed after y_0 vertex jobs are completed is feasible with respect to G if and only if graph Γ has a clique containing at least y_0 vertices.

If graph Γ contains such a clique, then a schedule s^0 that satisfies all of the above conditions and such that in s^0 one vertex job is processed on the second machine in the time interval $[y_0(y_0-1)+2v-1, y_0(y_0-1)+2v]$, and the remaining jobs are processed in the interval $[y_0(y_0-1)+2v, a]$, is feasible with respect to G and $\bar{t}_{max}(s^0) = y$.

Suppose that any clique in Γ contains at most y_0-1 vertices. In this case, at least one of the introduced necessary conditions is violated for any schedule s that is feasible with respect to G and, hence, $\bar{t}_{max}(s) \geq y+1$.

The implementation of the described reduction requires at most $O(v+e)$ time.

Thus, Problem 4.2(a) is NP-hard in the strong sense.

Note that if the condition $t_i \in \{1, 2\}$ is replaced by the condition $t_i = 1$, $i = 1, 2,\dots, n$, then Problem 4.2(a) becomes solvable in polynomial time. The running time of the corresponding algorithm is $O(n^2)$ (see Section 5.5 of Chapter 2).

4.4. The following decision problem corresponds to Problem 4.2(b): determine whether there exists a schedule s^0 for processing the jobs of set N that is feasible with respect to G and such that $\sum_{i\in N} \bar{t}_i(s^0) \le y$ for a given y.

Let the decision problem constructed in Section 4.3 be called Problem A. To prove the NP-hardness of Problem 4.2(b), we use Problem A. For Problem A, consider schedule s^0, and denote the set of jobs processed on the first and second machines by N_1 and N_2, respectively. Since $\sum_{i\in N_1} \bar{t}_i(s^0) = a(a+1)/2$ and $\sum_{i\in N_2} \bar{t}_i(s^0) \le \sum_{i\in N_1} \bar{t}_i(s^0)$, we have $\sum_{i\in N} \bar{t}_i(s^0) \le a(a+1)$.

For the decision problem corresponding to Problem 4.2(b), form the set N' of jobs by adding $a(a+1)$ auxiliary jobs J'_r, $t(J'_r) = 1$, $r = 1, 2,\dots, a(a+1)$, to set N. Extend the precedence relation defined over set N to set N' and complete its definition as follows:

$$J'_{r-1} \to J'_r,\ r = 2, 3,\dots, a(a+1);$$

$$J_a \to J'_1;\ E_{j,k} \to J'_1 \text{ for all } (j, k) \in E.$$

Let the reduction graph of the precedence relation defined over N' be denoted by G'.

Define $y' = a(a+1) + \sum_{r=1}^{a(a+1)} (a+r)$. We show that a schedule s' for processing the jobs of set N' that is feasible with respect to G' and such that $\sum_{i\in N'} \bar{t}_i(s') \le y$ exists if and only if graph Γ has a clique containing at least y_0 vertices.

Suppose that in Problem A there exists a schedule s^0 for processing the jobs of set N that is feasible with respect to G and such that $\bar{t}_{max}(s^0) \le y = a$. Construct a schedule s' for processing the jobs of set N' as follows. All vertex jobs, edge jobs and auxiliary jobs J_r, $r = 1, 2,\dots, 3v+2e$, are processed in schedule s' as in schedule s^0, while the auxiliary jobs J'_r, $r = 1, 2,\dots, a(a+1)$, are processed in the time interval $[a, a(a+2)]$ on any of the available machines in numerical order.

It is easy to check that schedule s' is feasible with respect to G' and

$$\sum_{i\in N'} \bar{t}_i(s') \le a(a+1)/2 + \sum_{r=1}^{a(a+1)} (a+r) = y'.$$

Suppose that for Problem A the relation $\bar{t}_{max}(s) \ge a+1$ is valid for any schedule s that is feasible with respect to G. It is obvious that, in this case, in any schedule s'' for processing the jobs of set N' that is feasible with respect to G', the starting time of job J_1' is not less than $a+1$, and

$$\sum_{i \in N'} \overline{t}_i(s'') \geq a(a+1)/2 + \sum_{r=1}^{a(a+1)} (a+r+1) = y' - a(a+1)/2 + a(a+1) > y'.$$

For Problem A, a schedule s^0 that is feasible with respect to G and such that $\overline{t}_{max}(s^0) \leq a$ exists if and only if graph Γ has a clique containing at least y_0 vertices. Therefore, a feasible schedule s' for processing the jobs of set N' for which $\sum_{i \in N'} \overline{t}_i(s') \leq y'$ exists if and only if graph Γ contains a required clique.

The implementation of the described reduction requires at most $O((v+e)^2)$ time.

Thus, Problem 4.2(b) is NP-hard in the strong sense.

If $G = (N, \varnothing)$, Problem 4.2(b) is solvable in polynomial time even in a more general situation where t_i are arbitrary positive numbers, the machines operate at different speeds and their number is more than two (see Section 9.3 of Chapter 2). The corresponding algorithm requires $O(n\log n)$ time.

4.5. The decision problem corresponding to Problem 4.3(a) can be formulated as follows: determine whether there exists such a schedule s^0 for processing the jobs of set N that is feasible with respect G and such that $\overline{t}_{max}(s^0) \leq y$ for a given y.

Define $M = 1 + \max\{y_0,\ v + y_0(y_0-3)/2,\ e - y_0(y_0-1)/2\}$ and $y = 3$.

Form the set N of jobs in the following way. Associate each vertex j of graph Γ with the vertex job V_j. Associate each edge (j, k) with the edge job $E_{j,k}$. Include also three groups of auxiliary jobs: $J_r^{(1)}$, $r = 1, 2,..., M-y_0$; $J_p^{(2)}$, $p = 1, 2,..., M-v-y_0(y_0-3)/2$; $J_q^{(3)}$, $q = 1, 2,..., M-e+y_0(y_0-1)/2$.

Define the precedence relation $\rightarrow$ over the constructed set N by specifying its reduction as follows:

$V_j \rightarrow E_{j,k}$ and $V_k \rightarrow E_{j,k}$ for all $(j, k) \in E$;

$J_r^{(1)} \rightarrow J_p^{(2)}$ for all r and p;

$J_p^{(2)} \rightarrow J_q^{(3)}$ for all p and q.

The definition of the number of machines M implies that one of the three groups of auxiliary jobs contains exactly one job.

If a schedule s that is feasible with respect to G satisfies the condition $\overline{t}_{max}(s) \leq 3$, then the following conditions hold:

(a) all jobs $J_r^{(1)}$, $r = 1, 2,..., M-y_0$, are processed during the time interval $[0, 1]$;

(b) all jobs $J_p^{(2)}$, $p = 1, 2,..., M-v-y_0(y_0-3)/2$, are processed during the interval

[1, 2];

(c) all jobs $J_q^{(3)}$, $q = 1, 2, \ldots, M-e+y_0(y_0-1)/2$, are processed during the interval [2, 3].

Therefore, the search for a schedules s that is feasible with respect to G and such that $\overline{t}_{max}(s) \leq 3$ can be restricted to considering the schedules for which:

(a) y_0 vertex jobs are processed during the time interval [0, 1];

(b) $e-y_0(y_0-1)/2$ edge jobs are processed during the interval [2, 3];

(c) the remaining vertex jobs (their number is $v-y_0$) and the remaining edge jobs (their number is $y_0(y_0-1)/2$) are processed during the interval [1, 2].

From considerations similar to those used in Sections 4.3 and 4.5, we derive that a schedule s^0 that satisfies the above conditions is feasible if and only if graph Γ has a clique containing at least y_0 vertices. In this case, in schedule s^0 the vertex jobs corresponding to the vertices of the clique are processed during the time interval [0, 1], and the edge jobs corresponding to the edges of the clique are processed during the time interval [1, 2].

The implementation of the described reduction requires at most $O((v+e)^2)$ time.

Thus, Problem 4.3(a) is *NP*-hard in the strong sense.

If $M = 2$, Problem 4.3(a) is solvable in $O(n^2)$ time (see Section 5.5 of Chapter 2).

4.. The following decision problem corresponds to Problem 4.3(b): determine whether there exists a schedule s^0 for processing the jobs of set N that is feasible with respect to G and such that $\sum_{i \in N} \overline{t}_i(s^0) \leq y$ for a given y.

We show that the clique problem reduces to the formulated decision problem in polynomial time. To do this, we use the decision problem described in Section 4.5. The only required change is $y = 6M$.

Suppose that in the constructed problem there exists a feasible schedule s^0 such that $\overline{t}_{max}(s^0) = 3$. Note that since each of the machines processes at most one job at a time, a schedule s with $\overline{t}_{max}(s^0) < 3$ does not exist. It is easy to calculate that $\sum_{i \in N} \overline{t}_i(s^0) = 6M$. If a schedule s is such that $\overline{t}_{max}(s) \geq 4$, then it is obvious that $\sum_{i \in N} \overline{t}_i(s) \geq 6M+1 > y$.

As shown in Section 4.5, a schedule s^0 that is feasible with respect to G and such that $\overline{t}_{max}(s^0) \leq 3$ exists if and only if graph Γ has a clique containing at least y_0 vertices. Thus the required reduction is constructed.

If $G = (N, \varnothing)$, Problem 4.3(b) is solvable in $O(n\log n)$ time. The corresponding algorithm (see Section 9.3 of Chapter 2) solves the problem (provided that $G = (N, \varnothing)$) even if the processing times are arbitrary and the machines operate at different speeds.

4.7. Denote $L_i(s) = \bar{t}_i(s) - D_i$, $z_i(s) = \max\{0, L_i(s)\}$, $u_i(s) = \operatorname{sgn}(z_i(s))$, where D_i, $i = 1, 2,..., n$, are the due dates. Define $L_{max}(s) = \max\{L_i(s) \mid i \in N\}$, $z_{max}(s) = \max\{z_i(s) \mid i \in N\}$.

Due to Remarks 1.2 and 1.3 (see Section 1 of this chapter), the facts that Problem 4.2(a) and Problem 4.3(a) are *NP*-hard in the strong sense imply the following results.

Problem 4.2 is *NP*-hard in the strong sense in the following cases:

(c) $F(s) = L_{max}(s)$;

(d) $F(s) = z_{max}(s)$; $D_i = D$, $i = 1, 2,..., n$;

(e) $F(s) = \sum_{i \in N} z_i(s)$; $D_i = D$, $i = 1, 2,..., n$;

(f) $F(s) = \sum_{i \in N} u_i(s)$; $D_i = D$, $i = 1, 2,..., n$.

Problem 4.3 is *NP*-hard in the strong sense in the following cases:

(c) $F(s) = z_{max}(s)$; $D_i = D$, $i = 1, 2,..., n$;

(d) $F(s) = \sum_{i \in N} z_i(s)$; $D_i = D$, $i = 1, 2,..., n$;

(e) $F(s) = \sum_{i \in N} u_i(s)$; $D_i = D$, $i = 1, 2,..., n$.

Remark 1.4 (see Section 1 of this chapter) implies that Problem 4.1 remains *NP*-hard in the strong sense if preemption is allowed.

Due to Remark 2.2 (see Section 2 of this chapter), the facts that Problems 4.1 and 4.3(a)-(e) are *NP*-hard in the strong sense imply that these problems remain *NP*-hard in the strong sense if the processing times are arbitrary integers are preemption is allowed only at integer times.

4.8. If the due dates D_i are assigned for jobs $i \in N$, it is often required to find a schedule s with no late jobs with respect to these due dates, i.e., a schedule such that $\bar{t}_i(s) \le D_i$ for all $i \in N$.

Remark 4.1. Let *the only difference* between decision Problems A and B be as follows. In Problem A, it is required to check the existence of a schedule s' for processing the jobs of set N such that $L_{max}(s') \le y$ (or $z_{max}(s') \le y$) for a given y, while in Problem B it is required to verify the existence of a schedule s'' for processing the jobs of the same set N that is feasible with respect to the given deadlines. Then *there exist both polynomial and pseudopolynomial reductions of Problem A to Problem B.*

In fact, if D_i, $i \in N$, are the due dates in Problem A, then take $D_i' = D_i + y$ as the deadlines D_i' in Problem B. It is easy to verify that in Problem B, a schedule for processing the jobs of set N that is feasible with respect to the deadlines D_i' exists if and only if in Problem A there exists such a schedule s' that $L_{max}(s') \le y$ (or $z_{max}(s') \le y$).

This implies that each of the problems of the existence of a schedule that is feasible with respect to the assigned deadlines corresponding to Problems 1.1(d), 1.2(a), 2.1(d), 2.4(d), 3.2, 4.2(d) and 4.3(c) is NP-hard in the strong sense except the first problem, which is NP-hard but not in the strong sense.

Let in Problems 2.1(d), 2.4(d), 3.2, 4.3(c) the processing times are not unit but arbitrary integers, and preemption is allowed only at integer times. Then, due to Remark 2.2 (see Section 2 of this chapter), the problems of the existence of a schedule that is feasible with respect to the assigned deadlines, which correspond to the problems listed above, are NP-hard in the strong sense.

5. Reducibility of the Linear Arrangement Problem

This section uses the *linear arrangement* problem as a standard problem. This problem can be formulated as follows.

A non-directed graph $\Gamma = (V, E)$ with no multiple edges or loops is given such that $|V| = v$, $|E| = e$, and a positive integer y_0 are given. The vertices of the graph are arranged at integer points in the interval $[0, v-1]$. For a given arrangement, the length of an edge $(i, j) \in E$ is defined as $|x_i - x_j|$, where x_i and x_j are the coordinates of the points at which the vertices i and j are arranged, respectively. It is required to determine whether there exists an arrangement of the graph such that the total length of the edges $\sum_{(i,j)\in E} |x_i - x_j|$ does not exceed y_0.

It is clear that a linear arrangement of graph Γ is specified by a permutation of the vertices.

The input length of the formulated problem in the binary alphabet belongs to the interval $[c_1(v+e),\ c_2(v+e)\log v]$, while that encoded in the unary alphabet belongs to the interval $[c_3(v^2+e),\ c_4 v(v^2+e)]$, where c_1, c_2, c_3, c_4 are constants independent of v and e, $0 < c_1 \le c_2$, $0 < c_3 \le c_4$.

The linear arrangement problem is NP-complete in the strong sense.

5.1. This section examines the following scheduling problems.

Problem 5.1. A single machine processes the job of a set $N = \{1, 2, \ldots, n\}$ available at time $d = 0$. The machine must operate with no idle time. The processing of each job $i \in N$

requires t_i time units, and no preemption is allowed. Each job $i \in N$ is associated with the real number (the weight) α_i. A precedence relation with the reduction graph G is defined over the set N. It is required to find a schedule s^* for processing the jobs of set N that is feasible with respect to G and minimizes the function $F(s)$ in the following cases:

(a) $F(s) = \sum_{i \in N} \alpha_i \bar{t}_i(s)$; $t_i = 1$, $\alpha_i \in \{\lambda, \lambda+1, \lambda+2\}$, $i = 1, 2,..., n$, $\lambda \in \{0, \pm 1, \pm 2,...\}$; here $\bar{t}_i(s)$ is the completion time of job i in a schedule s;

(b) $F(s) = \sum_{i \in N} \bar{t}_i(s)$, $t_i \in \{1, 2\}$, $i = 1, 2,..., n$;

(c) $F(s) = \sum_{i \in N} \bar{t}_i(s)$, $t_i \in \{0, 1\}$, $i = 1, 2,..., n$;

(d) $F(s) = \sum_{i \in N} \alpha_i \bar{t}_i(s)$, $t_i = 1$, $\alpha_i \in \{0, 1\}$, $i = 1, 2,..., n$.

We show that these problems are NP-hard. Since in each of the above problems a schedule s is specified by a permutation π of the elements of set N, along with the notation $F(s)$ and $\bar{t}_i(s)$ we write $F(\pi)$ and $\bar{t}_i(\pi)$, respectively.

5.2. The following decision problem corresponds to Problem 5.1(a): determine whether there exists a schedule s^0 for processing the jobs of set N that is feasible with respect to G and such that $\sum_{i \in N} \alpha_i \bar{t}_i(s) \le y$ for a given y.

We show that there exists a polynomial-time reduction of the linear arrangement problem to the formulated decision problem.

We start by establishing the NP-hardness of Problem 5.1 in the case $F(s) = \sum_{i \in N} \alpha_i \bar{t}_i(s)$ and $\alpha_i \in \{-1, 0, 1\}$, $i = 1, 2,..., n$.

The set N of jobs is to be formed as follows. Associate each vertex j of graph Γ with the vertex job V_j. Associate each edge (j, k) with the pair of edge jobs $E_{j,k}^{(1)}$ and $E_{j,k}^{(2)}$. Define the precedence relation $\rightarrow$ over the constructed set N: $E_{j,k}^{(1)} \rightarrow V_j$, $E_{j,k}^{(1)} \rightarrow V_k$, $V_j \rightarrow E_{j,k}^{(2)}$, $V_k \rightarrow E_{j,k}^{(2)}$ for all edges (j, k) of graph Γ.

Define $y = y_0 v^4 + (v^4 + 2e - 1)e$; $t(V_j) = v^4$, $\alpha(V_j) = 0$, $j \in V$; $t(E_{j,k}^{(1)}) = t(E_{j,k}^{(2)}) = 1$, $\alpha(E_{j,k}^{(1)}) = -1$, $\alpha(E_{j,k}^{(2)}) = 1$, $(j, k) \in E$.

Let Problem A be the following decision problem: determine whether there exists a schedule s^0 for processing the jobs of set N such that $\sum_{i \in N} \alpha_i \bar{t}_i(s^0) \le y$.

It is not difficult to check that, in Problem A, a feasible schedule is specified by a permutation of the jobs. Since it is required to verify the existence of a schedule s^0 that is feasible with respect to G and such that $\sum_{i \in N} \alpha_i \bar{t}_i(s^0) \le y$, the search can be restricted to examining only those permutations in which there are no other vertex jobs between a job $E_{j,k}^{(1)}$ and the leftmost of the jobs V_j and V_k. In fact, we have $\alpha(E_{j,k}^{(1)}) = -1$. Therefore, moving job $E_{j,k}^{(1)}$ to the right and maintaining the sequence of the other jobs will only decrease the value of the objective function. Similarly, since $\alpha(E_{j,k}^{(2)}) = 1$, the search can be restricted to considering only those permutations in which there are no other vertex jobs between a job $E_{j,k}^{(2)}$ and the rightmost of the jobs V_j and V_k.

Besides, observe that preemption, if allowed at integer times, does not extend the available possibilities, i.e., among the preemptive schedules there are no schedules with a smaller value of the objective function, as compared with the best of the non-preemptive schedules. In fact, consider a permutation π in which there is no vertex job between the vertex jobs V_{j_r} and V_{j_s}. If, in π, there is no edge job between V_{j_r} and V_{j_s}, then an interruption of the processing of V_{j_r} followed by the processing of V_{j_s} does not change the value of the objective function since $\alpha(V_j) = 0$, $j \in V$. Suppose that, in π, there is at least one edge job $E_{j,k}^{(p)}$, $p \in \{1, 2\}$, between the jobs V_{j_r} and V_{j_s}. Taking into account that only those permutations which satisfy the restrictions presented above, are considered, we derive that either $V_{j_r} \rightarrow E_{j,k}^{(p)}$ (if $p = 2$) or $E_{j,k}^{(p)} \rightarrow V_{j_s}$ (if $p = 1$) must be satisfied. Therefore, if the processing of job V_{j_r} is interrupted, job $E_{j,k}^{(p)}$ may start either only later, as compared with the sequence π, if $p = 2$, or only earlier if $p = 1$. In any case, the value of the objective function may only increase. If, in π, there are some other vertex jobs between V_{j_r} and V_{j_s}, then an interruption of the processing of V_{j_r} followed by the processing of V_{j_s} does not decrease the objective function value either.

Thus, consider the problem of finding a processing sequence for the jobs of set N. Suppose that a permutation π is feasible with respect to graph G of the reduction of the precedence relation defined over N. In π, fix the sequence π' formed by all vertex jobs. The sequence π' specifies the sequence π'' of the numbers of the vertices of graph Γ. Let $\pi''(j)$ denote the position at which j is located in permutation π''. If the jobs are processed according to the sequence π, the completion time of job $E_{j,k}^{(p)}$ is denoted by $\bar{t}(E_{j,k}^{(p)}(\pi))$, $p \in \{1, 2\}$.

It is easy to verify that for the permutations of the elements of set N that satisfy all restrictions introduced above, the following inequalities

$$v^4 \min\{\pi''(j), \pi''(k)\} - v^4 + 1 \leq \bar{t}(E_{j,k}^{(1)}(\pi)) \leq v^4 \min\{\pi''(j), \pi''(k)\} - v^4 + 2e - 1,$$
$$v^4 \max\{\pi''(j), \pi''(k)\} + 2 \leq \bar{t}(E_{j,k}^{(2)}(\pi)) \leq v^4 \max\{\pi''(j), \pi''(k)\} + 2e$$

hold.

Hence, we have

$$v^4\ (\max\{\pi''(j), \pi''(k)\} - \min\{\pi''(j), \pi''(k)\}) + v^4 - 2e + 3$$

$$\leq \bar{t}(E_{j,k}^{(2)}(\pi)) - \bar{t}(E_{j,k}^{(1)}(\pi)) \leq v^4 \ (\max\{\pi''(j),\ \pi''(k)\}$$
$$-\ \min\{\pi''(j),\ \pi''(k)\}) + v^4 + 2e - 1,$$

which is equivalent to

$$v^4 |\pi(j) - \pi(k)| + v^4 - 2e + 3 \leq \bar{t}(E_{j,k}^{(2)}(\pi)) - \bar{t}(E_{j,k}^{(1)}(\pi)) \leq v^4 |\pi(j) - \pi(k)| + v^4 + 2e - 1.$$

Since $\alpha(V_j) = 0$, $\alpha(E_{j,k}^{(1)}) = -1$, $\alpha(E_{j,k}^{(2)}) = 1$ we have

$$\sum_{i \in N} \alpha_i \bar{t}_i(\pi) = \sum_{(j,k) \in E} (\bar{t}(E_{j,k}^{(2)}(\pi)) - \bar{t}(E_{j,k}^{(1)}(\pi)))$$

and, hence,

$$v^4 \sum_{(j,k) \in E} |\pi''(j) - \pi''(k)| + (v^4 - 2e + 3)e \leq \sum_{i \in N} \alpha_i \bar{t}_i(\pi)$$
$$\leq v^4 \sum_{(j,k) \in E} |\pi''(j) - \pi''(k)| + (v^4 + 2e - 1)e.$$

For graph Γ, let there exist a linear arrangement (i.e. a permutation π'') for which the total length of the edges does not exceed y_0. Then, for any permutation π of the elements of N that is feasible with respect to G and satisfies the additional restrictions we have

$$\sum_{i \in N} \alpha_i \bar{t}_i(\pi) \leq v^4 y_0 + (v^4 + 2e - 1)e = y.$$

If for any linear arrangement of the vertices of graph Γ the total length of the edges is greater than y_0, then, for any permutation π of the elements of N that is feasible with respect to G, we have due to $v^4 > 4e^2$ that

$$\sum_{i \in N} \alpha_i \bar{t}_i(\pi) \geq v^4(y_0 + 1) + v^4 e - 2e^2 + 3e > y.$$

Thus, Problem A has a solution if and only if the linear arrangement problem has a solution. Note that the implementation of the described reduction requires $O(v+e)$ time.

Let us construct a polynomial transformation of Problem A to Problem 5.1(a), replacing each vertex job V_j by a group of v^4 jobs $V_j^{(q)}$, $q = 1, 2, \ldots, v^4$, and setting $t(V_j^{(q)}) = 1$, $\alpha(V_j^{(q)}) = 0$, $q = 1, 2, \ldots, v^4$, $j \in V$; $V_j^{(q)} \rightarrow V_j^{(q+1)}$, $q = 1, 2, \ldots, v^4 - 1$.

Due to the above remark on the preemptive schedules, it suffices to consider the schedules in which the jobs $V_j^{(1)}, V_j^{(2)}, \ldots, V_j^{(v^4)}$ are processed immediately one after another. Since $\alpha(V_j^{(q)}) = 0$, we conclude that the constructed problem (let us denote it by B) is equivalent to Problem A.

In Problem B, let us denote the set of jobs by N'. Increase all α_i, $i \in N'$, by the constant $\lambda' = 1 + \lambda$ and examine how this changes the value of $\sum_{i \in N'} \alpha_i \bar{t}_i(\pi)$. Here π is a permutation of the elements of N'. Since $|N'| = v^5+2e$ and all processing times are unit, we have (assuming $\alpha' = \alpha_i+\lambda'$)

$$\sum_{i\in N'} \alpha_i'\bar{t}_i(\pi) = \sum_{i\in N'} (\alpha_i+\lambda')\bar{t}_i(\pi) = \sum_{i\in N'} \alpha_i\bar{t}_i(\pi)+\lambda'(v^5+2e+1)(v^5+2e)/2.$$

Defining $y = v^4y_0+(v^4+2e-1)+\lambda'(v^5+2e+1)(v^5+2e)/2$, we obtain the problem equivalent to Problem B.

The implementation of the described reduction of the linear arrangement problem to Problem 5.1(a) requires $O(v^5)$ time.

Thus, Problem 5.1(a) is NP-hard in the strong sense.

Note that if graph G is series-parallel, Problem 5.1(a) is solvable in $O(n\log n)$ time (see Section 4 of Chapter 3). Chapter 3 describes some other polynomially solvable special cases of Problem 5.1(a).

5.3. The following decision problem corresponds to Problem 5.1(b): determine whether there exists a schedule s^0 for processing the jobs of set N that is feasible with respect to G and such that $\sum_{i\in N} \bar{t}_i(s^0) \leq y$ for a given y.

We show that the decision problem corresponding to Problem 5.1(a) with $\lambda = 2$ reduces to the formulated decision problem in polynomial time.

In Problem 5.1(a), replace the condition $t_i = 1$, $i = 1, 2,..., n$, by the condition $t_i = 4$, $i = 1, 2,..., n$. Such a replacement results in the problem equivalent to the initial one if we choose $y' = 4y_0$ (here y_0 is the constant in the initial problem) and consider the problem of the existence of a schedule that is feasible with respect to G and such that the value of the objective function does not exceed y'. It is easy to verify that, for this problem, preemption, if allowed at integer times, does not extend the available possibilities.

Let us make the following transformations in Problem 5.1(a) (assuming $t_i = 4$). Replace each job i such that $\alpha_i = 4$ (let the number of such jobs be r_1), by a group of four jobs $i^{(1)}$, $i^{(2)}$, $i^{(3)}$, $i^{(4)}$, assuming $\alpha_{i^{(q)}} = 1$, $t_{i^{(q)}} = 1$, $q = 1, 2, 3, 4$; $i^{(q)} \rightarrow i^{(q+1)}$, $q = 1, 2, 3$. Replace each job j such that $\alpha_j = 3$ (let the number of such jobs be r_2), by a group of three jobs $j^{(1)}$, $j^{(2)}$, $j^{(3)}$, assuming $\alpha_{j^{(q)}} = 1$, $q = 1, 2, 3$; $t_{j^{(1)}} = 2$, $t_{j^{(2)}} = t_{j^{(3)}} = 1$, $j^{(q)} \rightarrow j^{(q+1)}$, $q = 1, 2$. Replace each job k such that $\alpha_k = 2$ (let the number of such jobs be r_3), by a pair of jobs $k^{(1)}$ and $k^{(2)}$, setting $\alpha_{k^{(1)}} = \alpha_{k^{(2)}} = 1$, $t_{k^{(1)}} = t_{k^{(2)}} = 2$; $k^{(1)} \rightarrow k^{(2)}$. Besides, if in the original problem the relation $i \rightarrow j$ holds, then replace it by the relation $i^{(q_1)} \rightarrow j^{(q_2)}$ for all the jobs of the groups by which the jobs i and j have been replaced.

Due to the above remark on preemptive schedules, the search can be restricted to considering those schedules for processing the jobs of the constructed set N' in which the jobs of each new group are processed immediately one after another.

Let s be a schedule in Problem 5.1(a) (with $t_i = 4$) and s' be the corresponding schedule for processing the jobs of set N' (each job of set N is replaced by the corresponding group). It is easy to verify that

$$\sum_{i \in N} \alpha_i \bar{t}_i(s) = \sum_{j \in N'} \bar{t}_j(s') + 6r_1 + 3r_2 + 2r_3.$$

Define $y = y' - 6r_1 - 3r_2 - 2r_3$. It is obvious that a schedule s^0 for processing the jobs of set N' that is feasible with respect to the precedence relation defined over N' and such that $\sum_{i \in N'} \bar{t}_j(s^0) \le y$ exists if and only if in Problem 5.1(a) with $\lambda = 2$ and $t_i = 4$, there exists a schedule s for processing the jobs of set N that is feasible with respect to G and such that $\sum_{i \in N} \alpha_i \bar{t}_i(s) \le y' = 4y_0$.

The implementation of the described reduction requires $O(n)$ time.

Thus, Problem 5.1(b) is NP-hard in the strong sense.

5.4. A proof of the NP-hardness of Problem 5.1(c) can be done in a similar way using the following transformations. Define $\lambda = 1$. Replace each job $i \in N$ such that $\alpha_i = 3$ by a group of three jobs $i^{(1)}$, $i^{(2)}$, $i^{(3)}$, assuming $\alpha_{i^{(q)}} = 1$, $q = 1, 2, 3$; $t_{i^{(1)}} = 1$, $t_{i^{(2)}} = t_{i^{(3)}} = 0$, $i^{(q)} \rightarrow i^{(q+1)}$, $q = 1, 2$.

Replace each job j such that $\alpha_j = 2$ by a pair $j^{(1)}$, $j^{(2)}$, assuming $\alpha_{j^{(1)}} = \alpha_{j^{(2)}} = 1$, $t_{j^{(1)}} = 1$, $t_{j^{(2)}} = 0$, $j^{(1)} \rightarrow j^{(2)}$.

The NP-hardness in the strong sense of Problem 5.1(d) follows from the NP-hardness in the strong sense of Problem 5.1(b) and Lemma 2.1 (see Section 2 of this chapter).

5.5. Remark 1.2 (see Section 1 of this chapter) and the fact that Problems 5.1(a)-(d) are NP-hard in the strong sense imply that Problem 5.1 is also NP-hard in the strong sense in the following cases:

(e) $F(s) = \sum_{i \in N} \alpha_i z_i(s)$; $t_i = 1$, $D_i = D$, $\alpha_i \in \{\lambda, \lambda+1, \lambda+2\}$, $i = 1, 2, \ldots, n$, $\lambda \in \{0, \pm 1, \pm 2, \ldots\}$;

(f) $F(s) = \sum_{i \in N} z_i(s)$; $t_i \in \{1, 2\}$, $D_i = D$, $i = 1, 2, \ldots, n$;

(g) $F(s) = \sum_{i \in N} z_i(s)$; $t_i \in \{0, 1\}$, $D_i = D$, $i = 1, 2, \ldots, n$;

(h) $F(s) = \sum_{i \in N} \alpha_i z_i(s)$; $t_i = 1$, $D_i = D$, $\alpha_i \in \{0, 1\}$, $i = 1, 2, \ldots, n$.

Due to Remark 2.2 (see Section 2 of this chapter), the facts that Problem 5.1(a),

5.1(d), 5.1(e), 5.1(h) are *NP*-hard in the strong sense imply that these problems are *NP*-hard in the strong sense if the processing times are not unit but arbitrary integers, provided that preemption is allowed only at integer times. Similarly, Problems 5.1(c) and 5.1(g) are *NP*-hard in the strong sense, so it follows that these problems remain *NP*-hard in the strong sense if the condition $t_i \in \{0, 1\}$, $i = 1, 2,..., n$, is replaced by the condition: t_i are arbitrary integers, provided that preemption is allowed only at integer times.

Remark 1.4 (see Section 1 of this chapter) implies that Problems 5.1(a)–(h) are also *NP*-hard in the strong sense in the preemptive case.

6. Bibliographic Notes

The partition problem, the vertex covering problem and the clique problem have been proved to be *NP*-complete by Karp [74]. A proof of the *NP*-completeness of clique problem was earlier outlined by Cook [82]. The *NP*-completeness of the 3-partition problem has been established by Garey and Johnson [271, 56]. The same authors together with Stockmeyer [277] have proved the linear arrangement problem to be *NP*-complete.

In this section, we use a special notation for describing scheduling problems. The five-field notation $\alpha_1|\alpha_2|\alpha_3|\alpha_4|\alpha_5$ corresponds to the description of the problems given in Table I.2 of Introduction, where the fields α_1, α_2, α_3, α_4, α_5 correspond to the first five columns Table I.2, respectively. For example, the description $1|t_i;\ d_i|\,|\,|\bar{t}_i \le D_i$ corresponds to the first line of Table I.2.

Problems 1.1(a) $(2|t_i;\ d_i = 0|\,|\,|\bar{t}_{max})$ and 1.1(c) $(2|t_i;\ d_i = 0|\,|\,|\sum\alpha_i\bar{t}_i)$ are proved to be *NP*-hard by Livshits and Rublinetsky [100], see also [220]. The *NP*-hardness of Problem 1.1(b) $(2|t_i;\ d_i = 0|\,|\,|\bar{t}_{max}\sum\bar{t}_i)$ is established by Lenstra [345]. The *NP*-hardness of Problem 1.2(a) $(1|t_i;\ d_i|\,|\,|L_{max})$ is proved by Brucker et al. [217]. The problem of finding a schedule that is feasible with respect to the deadlines corresponding to Problem 1.2(a) is *NP*-hard in the strong sense (the 3-partition problem reduces to the latter problem [56]). Hence, Problems 1.2(a), 1.2(c) $(1|t_i;\ d_i|\,|\,|z_{max})$ and 1.2(d) $(1|t_i;\ d_i|\,|\,|\sum u_i)$ are also *NP*-hard in the strong sense. The *NP*-hardness of Problem 1.2(b) $(1|t_i;\ d_i = 0|\,|\,|\sum\alpha_i u_i)$ is independently proved in [74] and [100]. Note that earlier Lawler and Moore [342] established that Problem 1.2(b) was equivalent to the well-known knapsack problem (which is *NP*-hard [74]). The proof of the *NP*-hardness (in the ordinary sense) of Problem 1.3 $(M|t_i;\ d_i = 0|\,|M = M(N, D)|M;\ \bar{t}_i \le D)$ belongs to Sahni [392]. In fact, Problem

1.3 is NP-hard in the strong sense. To construct the reduction of the 3-partition problem to that problem, it suffices to define $n = 3n_0$; $t_i = \gamma_i$, $i = 1, 2,..., n$; $D = \delta$; $y = n_0$.

Problem 2.1(a) ($3|t_i = 1;\ d_i = 0|Rs(1)||\bar{t}_{max}$) is shown to be NP-hard in the strong sense by Garey and Johnson [271]. Blazewicz [209] asserts that Problem 2.1(b) ($3|t_i = 1;\ d_i = 0|Rs(1)||\Sigma\bar{t}_i$) is NP-hard (a proof is not given). Problem 2.2(a) ($1|t_i;\ d_i|||\Sigma\bar{t}_i$) is proved to be NP-hard in the strong sense by Lenstra et al. [349], the same complexity status for Problem 2.2(b) ($1|t_i;\ d_i|Pr||\Sigma\alpha_i\bar{t}_i$) is determined by Labetoulle et al. [324]. Lenstra and Rinnooy Kan [348] prove Problems 2.2(e) ($1|t_i = 1;\ d_i = 0|C||\Sigma\alpha_i z_i$), 2.2(g) ($1|t_i = 1;\ d_i = 0|C||\Sigma\alpha_i\bar{t}_i;\ \bar{t}_i \le D_i$), and 2.2(h) ($1|t_i = 1;\ d_i = 0|C||\Sigma u_i$) to be NP-hard in the strong sense, and assert that Problem 2.2(c) ($1|t_i = 1;\ d_i|C||\Sigma\alpha_i\bar{t}_i$) is of the same complexity (a proof is not presented). Problem 2.2(d) ($1|t_i;\ d_i = 0|||\Sigma\alpha_i z_i$) is shown to be NP-hard in the strong sense by Lawler [335] and Lenstra et al. [349]. The NP-hardness in the strong sense of Problem 2.2(f) ($1|t_i;\ d_i = 0|||\Sigma\alpha_i\bar{t}_i;\ \bar{t}_i \le D_i$) is established in [349]. Note that earlier these problems were proved to be NP-hard (in the ordinary sense) in [100] and [217], respectively. Problem 2.3 ($2|t_i;\ d_i = 0|\mathcal{T}^-||\Sigma\bar{t}_i$) is shown to be NP-hard in the strong sense by Sethi [405]. Blazewicz et al. [211] prove Problems 2.4(a) ($2|t_i = 1;\ d_i = 0|C;\ Rs(1)|r_i \in \{0, 1\}|\bar{t}_{max}$) and 2.4(b) ($2|t_i = 1;\ d_i = 0|C;\ Rs(1)|r_i \in \{0, 1\}|\Sigma\bar{t}_i$) to be NP-hard in the strong sense. That survey paper examines the complexity of other scheduling problems under resource constraints. Let the problem $3|t_i = 1;\ d_i = 0|Rs(q)|R_k = 1,\ r_{ik} \in \{0, 1\}|F(s)$ be called Problem A. Here we are given q types of resources, and R_k is the total amount of kth resource available at any time, $R_k = 1$, $k = 1, 2,..., q$. At any time of its processing, a job $i \in N = \{1, 2,..., n\}$ consumes $r_{ik} \in \{0, 1\}$ units of the kth resource. It is required to find a resource-feasible schedule s for processing the jobs of set N that minimizes the function $F(s)$, assuming that (a) $F(s) = \bar{t}_{max}(s)$ or (b) $F(s) = \sum_{i \in N} \bar{t}_i(s)$. Problem B differs from Problem A only in that here the processing system consists of two uniform parallel machines and the processing time t_{iH} of job $i \in N$ on machine H is equal to a_H (i.e., Problem B is $2|t_{iH} = a_H;\ d_i = 0|Rs(q)|R_k = 1,\ r_{ik} \in \{0, 1\}|F(s)$). Remarks 1.2 and 1.3 (see Section 1) imply that Problems A and B are also NP-hard in the strong sense in the following cases: (c) $F(s) = L_{max}(s)$, (d) $F(s) = z_{max}(s)$, (e) $F(s) = \sum_{i \in N} z_i(s)$, and (f) $F(s) = \sum_{i \in N} u_i(s)$. In all these cases it is assumed that $D_i = D$, $i = 1, 2,..., n$. Remark 2.2 (see Section 2) implies that Problems A(a)-(f) remain NP-hard in the strong sense if the processing times are not unit but arbitrary integers, provided that preemption is allowed only at integer times. It follows from Remark 4.1 (see Section 4) that the problems of finding a schedule that is feasible with respect to the deadlines corresponding to

Problems A(d) and B(d) are also NP-hard in the strong sense.

The first proof of the NP-hardness in the strong sense of Problem 3.1 ($2|t_i$; $d_i = 0|\mathcal{T}^+||\Sigma\bar{t}_i$) was given by Sethi [405]; Section 3.2 presents a simpler proof. The first attempt to prove that Problem 3.2 ($M|t_i = 1$; $d_i = 0|\mathcal{T}^+||L_{max}$) is NP-hard in the strong sense was made by Brucker et al. [216]. The proof given in Section 3.3 is based on the scheme proposed in [216]. As mentioned in Section 4.8, the NP-hardness in the strong sense of Problem 3.2 implies that the corresponding problem of the existence of a schedule that is feasible with respect to the deadlines is NP-hard in the strong sense as well. Transform the latter problem in the following way. Replace the condition $d_i = 0$, $i = 1, 2,..., n$, by the condition $d_i \geq 0$, $i = 1, 2,..., n$ (the jobs do not enter the processing system simultaneously); replace the condition "each connected component of G is outtree" condition by the condition "each connected component of G is intree", and introduce the condition $D_i = D$, $i = 1, 2,..., n$. The obtained problem is, in fact, $M|t_i = 1$; $d_i|\mathcal{T}^-|D_i = D|\bar{t}_i \leq D_i$. Let us call it Problem H. It is easy to verify that Problem H is equivalent to the original one (see [216]) and, hence, is NP-hard in the strong sense. Thus, the problem of finding a time-optimal schedule under the conditions of Problem H is also NP-hard in the strong sense. Due to Remarks 1.2 and 1.3, the NP-hardness in the strong sense of Problem H for $F(s) = \bar{t}_{max}(s)$ implies that it is also NP-hard in the strong sense in the following cases: $F(s) = L_{max}(s)$; $F(s) = z_{max}(s)$, $F(s) = \sum_{i \in N} z_i(s)$, $F(s) = \sum_{i \in N} u_i(s)$. In all the cases $D_i = D$, $i = 1, 2,..., n$. Remark 2.2 implies that the above problems remain NP-hard in the strong sense if the processing times are not unit but arbitrary integers, provided that preemption is allowed only at integer times.

Problems 4.1 ($1|t_i = 1$; $d_i = 0|G||\Sigma z_i$) and 4.2(b) ($2|t_i \in \{1, 2\}$; $d_i = 0|G||\Sigma\bar{t}_i$) are proved to be NP-hard in the strong sense by Lenstra [345]. The same complexity status of Problem 4.3(b) ($M|t_i = 1$; $d_i = 0|G||\Sigma\bar{t}_i$) is established in [217] by Brucker et al. The proof of the NP-hardness in the strong sense of Problems 4.2(a) ($2|t_i \in \{1, 2\}$; $d_i = 0|G||\bar{t}_{max}$) and 4.3(a) ($M|t_i = 1$; $d_i = 0|G||\bar{t}_{max}$) is due to Ullman [425, 426]. Ullman [427] also shows that Problem 4.3(a) remains to be NP-hard in the strong sense if preemption is allowed. Remarks 1.2 and 1.3 imply that this problem ($M|t_i = 1$; $d_i = 0|Pr$; $G||\bar{t}_{max}$) remains NP-hard in the strong sense if the objective function $\bar{t}_{max}(s)$ is replaced by any of the following: $z_{max}(s)$, $L_{max}(s)$, $\sum_{i \in N} z_i(s)$, or $\sum_{i \in N} u_i(s)$ (in all cases $D_i = D$, $i = 1, 2,..., n$). Due to Remark 4.1, the fact that the problem $M|t_i = 1$; $d_i = 0|Pr$; $G||z_{max}$ is NP-hard in the strong sense implies that the problem $M|t_i = 1$; $d_i = 0|Pr$; $G|D_i = D|\bar{t}_i \leq D_i$ has the same complexity status.

Lawler [336] proves Problem 5.1(a) ($1|t_i = 1;\ d_i = 0|G|\alpha_i \in \{\lambda, \lambda+1, \lambda+2\}, \lambda \in \{0, \pm 1, \pm 2,...\}|\sum\alpha_i\bar{t}_i$) to be NP-hard in the strong sense. The ideas developed in [336] are the basis for the proof of the NP-hardness in the strong sense of Problems 5.1(b) ($1|t_i \in \{1, 2\};\ d_i = 0|G||\sum\bar{t}_i$) and 5.1(c) ($1|t_i \in \{0, 1\};\ d_i = 0|G||\sum\bar{t}_i$) presented in Sections 5.3 and 5.4, respectively. The NP-hardness in the strong sense of Problem 5.1(d) is proved by Lenstra and Rinnooy Kan in [346].

Lawler [112*] shows that the following problem $1|d_i = 0||D_i' \geq D_i|\sum u_i,\ \bar{t}_i \leq D_i'$ is NP-hard. Here the goal is to minimize the number of late jobs (with respect to their due dates D_i) under the condition that all jobs have to meet given the deadlines D_i' such that $D_i' \geq D_i$.

Du and Leung [41*] prove the NP-hardness of the problem $1|d_i|Pr||\sum\bar{t}_i, \bar{t}_i \leq D_i$. This problem involves minimizing the total flow time on a single machine provided that preemption is allowed, the jobs are not simultaneously available, and all jobs have to meet the given deadlines D_i.

A pseudopolynomial-time algorithm for Problem 2.2(d) for $\alpha_i = 1$, $i = 1, 2,..., n$, ($1|d_i = 0|||\sum z_i$) is due Lawler [335]. Du and Leung [44*] prove this problem to be NP-hard in the ordinary sense, thus answering a question which was open for more than 15 years. Yuan [161*] shows that Problem 2.2(d) with $D_i = D$, $i = 1, 2,..., n$, ($1|d_i = 0||D_i = D|\sum\alpha_i z_i$) is also NP-hard in the ordinary sense. Due to Remark 1.4 (see Section 1), all these problems remain NP-hard if preemption is allowed ($1|d_i = 0|Pr||\sum z_i$ and $1|d_i = 0|Pr|D_i = D|\sum\alpha_i z_i$).

Leung and Young [122*] have improved results by Lenstra and Rinnooy Kan [348, 345]. They have shown that the problem $1|t_i = 1; d_i = 0|C||\sum z_i$ is NP-hard in the strong sense. This problem corresponds to Problem 2.2(e) if $\alpha_i = 1$, $i = 1, 2,..., n$, as well as to Problem 4.1 in the case of chain-like precedence constraints. As follows from Remark 1.4, this problem in the preemptive case ($1|t_i = 1; d_i = 0|Pr;\ C||\sum z_i$) remains NP-hard in the strong sense. In turn, Remark 2.2 implies that the latter problem remains NP-hard in the strong sense if the processing times are arbitrary integers, provided that preemption is allowed only at integer times ($1|[t_i]; d_i = 0|[Pr];\ C||\sum z_i$).

As mentioned above, Problem 1.1(a) ($2|t_i; d_i = 0|||\bar{t}_{max}$) is NP-hard only in the ordinary sense. Du et al. [47*] have shown that this problem under chain-like precedence constraints ($2|t_i;\ d_i = 0|C||\bar{t}_{max}$) becomes NP-hard in the strong sense. Remarks 1.2 and 1.3 imply that this problem remains NP-hard in the strong sense if the objective function $\bar{t}_{max}(s)$ is replaced by any of the following: $z_{max}(s)$, $L_{max}(s)$, $\sum_{i \in N} z_i(s)$, or $\sum_{i \in N} u_i(s)$ (in all cases $D_i = D$, $i = 1, 2,..., n$). Due to Remark 4.1, the fact that the problem $2|t_i;\ d_i = 0|C||z_{max}$ is *NP*-hard in the strong sense implies that the problem $2|t_i;\ d_i = 0|C|D_i = D|\bar{t}_i \le D_i$ has the same complexity status. The same applies to the problem $M|t_i;\ d_i = 0|||\bar{t}_{max}$ (and some related problems). The *NP*-hardness in the strong sense of the latter problem is established by Garey and Johnson [275].

The problems to minimize the makespan on two unrelated machines under tree-like precedence constraints ion the preemptive case ($2|t_{iH};\ d_i = 0|Pr;\ \mathcal{T}^-||\bar{t}_{max}$ and $2|t_{iH};\ d_i = 0|Pr;\ \mathcal{T}^+||\bar{t}_{max}$) are *NP*-hard in the strong sense. It is an unpublished result by Lawler (see [115*]). Remark 1.2 implies that both these problems remain *NP*-hard in the strong sense if the objective function $\bar{t}_{max}(s)$ is replaced by either $z_{max}(s)$ or $L_{max}(s)$ for $D_i = D$, $i = 1, 2, \ldots, n$. Remark 4.1 implies that the problems $2|t_{iH};\ d_i = 0|Pr;\ \mathcal{T}^-|D_i = D|\bar{t}_i \le D_i$ and $2|t_{iH};\ d_i = 0|Pr;\ \mathcal{T}^+|D_i = D|\bar{t}_i \le D_i$ are also *NP*-hard in the strong sense.

Du et al. [47*] have improved a result by Sethi [405]. They have proved that Problem 2.3 ($2|t_i;\ d_i = 0|\mathcal{T}^-||\sum\bar{t}_i$) and Problem 3.1 ($2|t_i;\ d_i = 0|\mathcal{T}^+||\sum\bar{t}_i$) are *NP*-hard in the strong sense not only for intrees and outtrees but even for chains ($2|t_i;\ d_i = 0|C||\sum\bar{t}_i$). They have proved also that both Problem 2.3 and Problem 3.1 remain *NP*-hard in the strong sense if preemption is allowed ($2|t_i;\ d_i = 0|Pr;\ \mathcal{T}^-||\sum\bar{t}_i$ and $2|t_i;\ d_i = 0|Pr;\ \mathcal{T}^+||\sum\bar{t}_i$). It follows from Remark 1.2 that both latter problems remain *NP*-hard in the strong sense if the objective is replaced by $\sum_{i \in N} z_i$ for $D_i = D$, $i = 1, 2, \ldots, n$.

The problem of minimizing the total flow time on two identical machines provided that preemption is allowed and the jobs are not available simultaneously ($2|t_i;\ d_i|Pr||\sum\bar{t}_i$) is *NP*-hard. This result is due to Du et al. [46*]. As shown in [45*] by Du et al., this problem remains *NP*-hard if the objective function $\sum_{i \in N}\bar{t}_i(s)$ is replaced by $\sum_{i \in N} u_i(s)$ ($2|t_i;\ d_i|Pr||\sum u_i$). As follows from Remark 1.2 (See Section 1), the problem also remains *NP*-hard if the objective is $\sum_{i \in N} z_i$, provided that $D_i = D$, $i = 1, 2, \ldots, n$.

Garey et al. [58*] have improved a result by Ullman [425, 426]. They have shown that Problem 4.3(a) remains *NP*-hard in the strong sense even for tree-like precedence constraints ($M|t_i = 1;\ d_i = 0|\mathcal{T}||\bar{t}_{max}$). Note that the latter problem can be solved in polynomial time for any fixed M. Problem 4.3(a) also remains *NP*-hard in the strong sense for intrees and outtrees if $t_i \in \{1, t\}$, $i = 1, 2, \ldots, n$, ($M|t_i \in \{1, t\};\ d_i = 0|\mathcal{T}^-||\bar{t}_{max}$ and $M|t_i \in \{1, t\};\ d_i = 0|\mathcal{T}^+||\bar{t}_{max}$). This result is due to Du and Leung [42*]. Both latter problems are *NP*-hard in the ordinary sense if $M = 2$ and $t_i \in \{t^p\colon p \ge 0\}$ for any integer $t > 1$ ($2|t_i \in \{t^p\colon p \ge 0\}$, $t > 1;\ d_i = 0|\mathcal{T}^-||\bar{t}_{max}$ and $2|t_i \in \{t^p\colon p \ge 0\}$, $p > 1;\ d_i = 0|\mathcal{T}^+||\bar{t}_{max}$) [42*]. Due to Remarks 1.2 and 1.3, all problems mentioned in this paragraph remain NP-hard (either in the strong or in the ordinary sense, respectively) if the objective $\bar{t}_{max}(s)$ is replaced by one of the functions $z_{max}(s)$, $L_{max}(s)$, $\sum_{i \in N} z_i(s)$ or $\sum_{i \in N} u_i(s)$ (in all cases $D_i = D$, $i = 1, 2,..., n$). Due to Remark 2.2, the first of mentioned problems and the problems obtained from it by the described replacements are NP-hard in the strong sense if job processing times are arbitrary integers, provided that preemption is allowed only at integer times ($M|[t_i]$; $d_i = 0|[Pr]$; $\mathcal{T}||\bar{t}_{max}$, $M|[t_i]$; $d_i = 0|[Pr]$; $\mathcal{T}|D_i = D|L_{max}$, $M|[t_i]$; $d_i = 0|[Pr]$; $\mathcal{T}|D_i = D|z_{max}$, $M|[t_i]$; $d_i = 0|[Pr]$; $\mathcal{T}|D_i = D|\sum z_i$ and $M|[t_i]$; $d_i = 0|[Pr]$; $\mathcal{T}|D_i = D|\sum u_i$). Due to Remark 4.1, all problems mentioned in this paragraph with the objective $z_{max}(s)$ remain NP-hard (in the strong or in the ordinary sense, respectively) if formulated as the problems of finding a schedule s that feasible with respect to the deadlines ($\bar{t}_i(s) \le D_i$).

Lenstra et al. [120*] establish the NP-hardness in the strong sense of the problem of minimizing the makespan on M unrelated machines if t_{iH} may have only two values, i.e., if $t_{iH} \in \{t, t'\}$ where $t < t'$, $2t \ne t'$, and all jobs are simultaneously available ($M|t_{iH} \in \{t, t'\}$, $t < t'$, $2t \ne t'$; $d_i = 0|||\bar{t}_{max}$. Note that this problem is polynomially solvable if either $t_{iH} = t$ or $t_{iH} \in \{1, 2\}$, $i = 1, 2,..., n$, $H = 1, 2,..., M$ [120*].

Lawler [114*] shows that minimizing the number of late jobs when scheduling n independent jobs on M identical machines with preemption ($M|t_i$; $d_i = 0|Pr||\sum u_i$) is an NP-hard problem. It should be noted that for any fixed M this problem can be solved in pseudopolynomial time [337]. If the machines are unrelated and the jobs have different release times, then this problem ($M|t_{iH}$; $d_i|Pr||\sum u_i$) is NP-hard in the strong sense. The latter result is obtained in [45*] by Du et al.

Problem 4.3(b) ($M|t_i = 1$; $d_i = 0|G||\sum \bar{t}_i$) is solvable in polynomial time if $G = (N, \varnothing)$ even if the jobs have different processing times [294, 219, 220]. However the latter problem is NP-hard if the speeds of machines decrease over time ($M|t_i$; $d_i = 0||$*machine speed* $\downarrow|\sum \bar{t}_i$). This result is due to Meilijson and Tamir [128*]. If the speeds increase then the problem is solvable in $O(n\log n)$ time [128*]. It follows from Remark 1.2 that the problem is NP-hard if the objective $\sum_{i \in N} \bar{t}_i(s)$ is replaced by $\sum_{i \in N} z_i(s)$, provided that $D_i = D$, $i = 1, 2,..., n$.

Potts and Van Wassenhove [141*] consider the single-machine problem to minimize the so-called late work $\sum_{i \in N} \min\{t_i, z_i(s)\}$. They show that this problem ($1|t_i$; $d_i = 0|$ $||\sum \min\{t_i, z_i\}$) is NP-hard in the ordinary sense. This problem is pseudopolynomially solvable (see [77*] by Hariri et al.).

The problem of preemptive scheduling jobs with equal processing times and different release dates on M identical machines to minimize the weighted flow time ($M|t_i = t$; $d_i|Pr||\Sigma\alpha_i\bar{t}_i$) is proved to be NP-hard by Leung and Young [123]. It follows from Remark 1.2 that this problem remains NP-hard if the objective function $\sum_{i\in N}\alpha_i\bar{t}_i(s)$ is replaced by $\sum_{i\in N}\alpha_i z_i(s)$, provided that $D_i = D$, $i = 1, 2,\ldots, n$, ($M|t_i = t$; $d_i|Pr|D_i = D|\Sigma\alpha_i z_i$).

Sin and Cheng [151*] established the NP-hardiness (in the ordinary sense) of the following problem. Independent and simultaneously available jobs have to be scheduled on M identical parallel machines without preemption. The objective is to minimize $\sum_{j=1}^{M}\{(\sum_{i\in N_j}\alpha_i)\sum_{i\in N_j}t_i\}$, where N_j is the set of jobs assigned to machine j.

The single-machine scheduling problem to minimize $\sum_{i\in N}\alpha_i|\bar{t}_i(s)-D_i|$ has been studied by several authors. Garey et al. [59*] show this problem to be NP-hard if $\alpha_i = 1$. $i =1, 2,\ldots, n$, ($1|d_i = 0|||\Sigma|\bar{t}_i - D_i|$). If the weights α_i are different but the jobs have a common due date, i.e., $D_i = D$, $i = 1, 2,\ldots, n$, ($1|d_i = 0||D_i = D|\Sigma\alpha_i|\bar{t}_i - D_i|$) the problem is proved to be NP-hard by Hall and Posner [74*, 75*]. In the latter case, the problem is pseudopolynomially solvable and it is polynomially solvable if either $t_i = t$ or $t_i = \alpha_i$, $i = 1, 2,\ldots, n$, (see [84*] by Hoogeven and van de Velde). The case in which $\alpha_i = 1$ and $D_i = D$, $i =1, 2,\ldots, n$, is more complicated. Let the jobs be numbered in such a way that $t_{i+1} \geq t_i$ and define $\tau = t_n + t_{n-2} + t_{n-4} + \ldots$. Then for $D < \tau$ this problem ($1|d_i = 0||D_i = D < \tau|\Sigma|\bar{t}_i - D_i|$) is NP-hard (it is, of course, pseudopolynomially solvable). This result has been independently obtained by Hall et al. [72*, 73*] and by Hoogeven and van de Velde [84*]. If $D \geq \tau$ then the problem is solvable in $O(n\log n)$ time (see [87*] by Kanet, [7*] by Baker and Scudder).

Kubiak [105*] has proved the NP-hardness of the single-machine scheduling problem to minimize the completion time variance $\sum_{i\in N}\left[\bar{t}_i(s) - \frac{1}{n}\sum_{i\in N}\bar{t}_i(s)\right]^2$ for simultaneously available jobs ($1|t_i$; $d_i = 0|||\Sigma(\bar{t}_i - \frac{1}{n}\Sigma\bar{t}_i)^2$). This problem is pseudopolynomially solvable (see [34*] by De et al.).

Chand and Schneeberger [14*] have established the NP-hardness in the strong sense of the single-machine scheduling problem to minimize $\sum_{i\in N}(D_i - \bar{t}_i(s))$ for simultaneously available jobs, provided that $\bar{t}_i(s) \leq D_i$. Moreover, this problem is NP-hard in the strong sense if the condition $D_i \leq \sum_{i\in N}t_i$, $i = 1, 2,\ldots, n$, is imposed or, equivalently, the machine is not allowed to be idle ($1|t_i$; $d_i = 0||D_i \leq \Sigma t_i|\Sigma(D_i - \bar{t}_i)$, $\bar{t}_i \leq D_i$).

The statements presented in Sections 1.8 and 2.2 are based on the ideas expressed in [349] and [348], respectively.

Many enumerative methods have been developed for solving NP-hard scheduling problems.

A general formalism of the optimization methods based upon the idea of successive design, analysis and selection of variants is developed by Mikhalevich, Ermol'ev, Shkurba, Shor et al. [112–118]. The dynamic programming method is detailed in monograph [12] by Bellman. A significant development of the constructive approach was made by Moiseev and his colleagues [119–122]. Some general schemes for solving discrete optimization problems are proposed by Zhuravlev [62–64], Cherenin [170], Khachaturov [165], Emelichev and Komlik [60], Sergienko et al. [144], Levin and Tanaev [89].

Formalizations and theoretical justifications of the branch-and-bound method have been presented by Romanovsky [134], Ibaraki [299, 300], Kise [315], Köhler and Steiglitz [319], Mitten [360], Roy [391], Tang and Wong [421], Baker [193] and some others. The surveys by Korbut et al. [80], Balas and Guignard [200], Lawler and Wood [344] contain extensive bibliographies on these issues; see also [303].

Different modifications of a branch-and-bound algorithm for minimizing the sum of (weighted) job completion times have been designed by Chandra [229], Bianco and Ricciardelli [206] ($M = 1$, $d_i \geq 0$, $i = 1, 2,..., n$); Potts [378] ($M = 1$, $d_i = 0$, $i = 1, 2,..., n$, precedence constraints); Elmaghraby and Park [253], Baker and Merten [196], Barnes and Brennan [203] ($M > 1$, $d_i = 0$, $i = 1, 2,..., n$); Bansal [202] ($M = 1$, $d_i = 0$, additional condition: $\overline{t}_i(s) \leq D_i$, $i = 1, 2,..., n$); see also [75, 103, 169, 171, 255, 437, 439]. In [4, 9, 73, 142, 201, 213, 215, 223, 230, 245, 257] applications of the branch-and-bound method to finding time-optimal schedules are discussed. A number of problems for single-stage processing systems are considered in [3, 6, 8, 23, 72, 101, 141, 145, 183, 191, 205, 214, 247, 292, 438].

Computational dynamic programming schemes for solving scheduling problems for single-stage systems are described in [123, 156, 198, 222, 225, 320, 357, 402, 436].

Graph-theoretical interpretations of scheduling problems and corresponding enumerative methods have been developed by Sotskov [146–148], Grabovsky [287], Fernandez and Lang [259], Fung [265], Köhler [318], Zak [66].

A number of situations in which the quality of schedules essentially depends on organizing setup and transportation operations lead to a necessity of considering the so-called traveling salesman problem and its various generalizations. The traveling salesman problem is NP-hard (see, for example, [375]). The first branch-and-bound method for solving this problem is due to Little et al. [102] (by the way, it is in this paper that the method has got its present name). Bellman [13] describes a dynamic programming approach to the traveling salesman problem. Different aspects of the traveling salesman problem are discussed in [303]; a list of about 600 references is presented there.

Problems of minimizing the (weighted) maximum lateness are considered by McMahon and Florian [355], Dessouky and Larson [246], Carlier [227], Potts [378]; those of minimizing the maximum tardiness are studied by Baker and Su [199], Tilquin [422]; those of minimizing the total tardiness are examined by Schild and Fredman [398], Emmons [254], Baker and Martin [195], Fisher [261], Root [385], Peterson [376], Shwimer [406], Srinivasan [418] et al.

Many heuristic approaches and approximation methods have been developed for solving *NP*-hard scheduling problems.

Extensive experimental studies of comparisons of the efficiency of a number of heuristic procedures for finding non-preemptive schedules that minimize the maximum lateness have been made by Davis and Walters [242] (10 procedures, 1560 test problems; $M = 1$, $n = 5, 10, 15, 20, 25, 30$), by Larson and Dessouky [330] (11 procedures, 1200 test problems; $M = 1$, $n = 20$), and by De and Morton [244] (one procedure, 9900 test problems; $M = 2, 4, 8$, $n = 10, 20, 30$, $d_i = 0$, $i = 1, 2, \dots, n$).

Information on approximation methods with worst-case bounds on their performance guarantees can be found in Appendix and in Table I.3 of Introduction to this book.

Problems of minimizing the total job processing cost for single-stage systems are considered in [104, 132, 226, 373, 380, 383, 411, 423, 432]. Systems with "availability windows" are studied in [16, 132, 262, 367].

APPENDIX
APPROXIMATION ALGORITHMS

This Appendix presents a review of approximation algorithms with established worst-case performance guarantees, not included into the Russian edition of the book. As a rule, polynomial-time algorithms are discussed.

Given a (scheduling) problem, an algorithm Φ is called an *approximation* algorithm if for any instance of the problem it finds a feasible schedule s^0. Schedule s^0 is called *approximate*. Let F^0 and F^* denote the values of the objective function $F(s)$ for an approximate (s^0) and an optimal (s^*) schedules, respectively. To estimate the quality of an approximate schedule s^0 either $|F^0 - F^*|$ or $\Delta = |F^0 - F^*|/|F^*|$ performance guarantee is used. An algorithm Φ is called an *ε-approximation* algorithm if for any $\varepsilon > 0$ and an arbitrary problem instance it finds such a schedule s^0 that $\Delta \leq \varepsilon$. An ε-approximation algorithm is called *fully polynomial* if its running time is a polynomial in both $1/\varepsilon$ and the problem instance length under the binary encoding.

We use the following notation:

$t_{max} = \max\{t_i \mid i \in N\}$;

$t_\Sigma = \sum_{i \in N} t_i$ (in the case of a single machine or parallel identical machines).

T_H - the total processing time of the jobs assigned to machine H;

τ - the running time of an algorithm.

All presented running times depending on M hold under the assumption that M is fixed.

As in Section 6 of Chapter 4, we use the five-field notation $\alpha_1|\alpha_2|\alpha_3|\alpha_4|\alpha_5$ to describe a scheduling problem. The fields α_1, α_2, α_3, α_4, α_5 correspond to the five first columns of Table I.3 of Introduction. For brevity, some problem discussed below are given numbers.

One of the most popular approaches to developing approximation scheduling algorithms for a (multi-processor) single-stage system is the *list scheduling* technique which is as follows. The jobs are scanned according to a certain sequence (the list). A job is chosen from the list according to some rule and is assigned to a specific machine. Then this job is either deleted from the list or marked.

The ways of forming the list, the rules of choosing and assigning a job may vary. As a rule, the list is a job sequence formed in non-increasing or non-decreasing order of one of job parameters, e.g., the processing times, the due dates, and etc. The identical job sequence 1, 2,..., n may also be taken as a list. The lists of this type are called *random*.

The job to be assigned is usually the first one in the list among those ready for processing. The job is ready if its release date allows it to be processed, if all its predecessors are completed (in the case of precedence constraints) and so on. The first available machine is usually taken to process the chosen job. In list scheduling algorithms presented below, it is assumed that this rule for selecting the machine is applied, unless stated otherwise.

In the following, instead of using the expression "the list scheduling algorithm where the list is the job sequence sorted in non-increasing (or non-decreasing) order of the parameter a_i" we write "LSA($a_i\downarrow$)" or "LSA ($a_i\uparrow$)" respectively. For the random list, we write "LSA(R)".

A.1. Problem A.1 ($M\,|\,t_i;d_i = 0\,|\,|M = M(N, D)\,|\,M;\ \overline{t}_i \le D$) is to find the smallest number of machines sufficient for completing all jobs of set $N = \{1, 2,..., n\}$ by the deadline D. This problem is also known as the *bin-packing* problem.

Given a schedule, the value $D-T_H$ is called the time reserve of machine H. Two list scheduling algorithms with a random list are offered by Garey et al. [270]. According to the first one, the next job in the list is assigned to the machine with the smallest number and sufficient time reserve. The second algorithm assigns the next job to the machine with the smallest but sufficient time reserve. Both algorithms yield $\Delta \le 7/10+2/F^*$ and $\tau = O(n\log M^0)$ where M^0 is the number of machines found by the corresponding algorithm. The bound on Δ is independently obtained by Garey et al. [269] and Sahni [393]. The first algorithm is also shown to provide $\Delta \le 7/10+1/F^*$ (see [269]). Note that $F^* \ge \lceil t_\Sigma/D\rceil$ and [269] gives instances of the problem such that $\Delta > 7/10-8/F^*$.

Algorithms which differ from the above ones only in the way of list constructing are studied by Johnson et al. [308]. These algorithms are LSA($t_i\downarrow$). Both algorithms provide $\Delta \le 2/9+4/F^*$ and require $\tau = O(n\log n)$. The first of them is often used as an auxiliary algorithm for solving some other problems. We denote that algorithm by Φ_1. These and related algorithms are also considered in [307] by Johnson.

A.2. Let the problem $M|t_i;\ d_i = 0|\ |\ |\overline{t}_{max}$ be called Problem A.2. This is the problem of minimizing the makespan on M identical parallel machines for simultaneously available jobs. One of the earliest papers on worst-case analysis of approximation algorithms is [288] by Graham, and is devoted to Problem A.2. The bound $\Delta \le 1-1/M$ is proved for LSA(R), and this bound is *tight*, i.e., there are instances of Problem A.2 such that $\Delta = 1-1/M$.

Let Φ_2 be LSA $(t_i\downarrow)$ for solving Problem A.2. This algorithm is studied by Graham [289]. The guarantee $\Delta \le 1/3-1/3M$ is proved, and this bound is tight. Algorithm Φ_2 runs in $\tau = O(n\log n)$ time and guarantees $F^0-F^* \le (1-1/M)t_{max}$. The following a posteriori guarantees are determined for Φ_2 by Coffman and Sethi [236, 237]. Suppose a schedule s^0 is found by Φ_2 and $\overline{t}_{max}(s^0) = \overline{t}_{i'}$ for some job i'. Let H be a machine which processes job i' in schedule s^0 and λ be the total number of jobs assigned to H. Then $F^0-F^* \le t_{i'}(1-1/M)$ and $\Delta \le (\lambda-1)/\lambda-1/(\lambda M)$ for $\lambda \ge 3$. It is claimed in [236, 237] that $F^0 = F^*$ if $\lambda \in \{1, 2\}$. However, Chen [15*] has recently proved that in fact for $\lambda = 2$ the bound $\Delta \le 1/3-1/(3(M-1))$ holds, and this bound is tight. It is shown by Bakenrot [5*] that $\Delta \longrightarrow 0$ as $n \longrightarrow \infty$. Algorithm Φ_2 is proved to have the following property (see [40*] by Dobson). Let v be such an integer that $t_{max} < F^*/v$. Then $\Delta \le 1/(v+2)$ and, moreover, $\Delta \le \min\{1/(v+2), (M-1)/M(v+1)\}$ if $v \ge 2$. Note that $\Delta \le 1-1/M$ for LSA$(t_i\uparrow)$ (see [238] by Coffman and Sethi).

Algorithm Φ_2 can also be applied to Problem A.3 $(M|t_i;d_i|\ |\ |\overline{t}_{max})$, in which the jobs are not simultaneously available. In this case, Φ_2 provides $F^0-F^* \le (2-1/M)t_{max}$ (see [96] by Livshitz). For LSA$(d_i\uparrow)$ the bounds $F^0-F^* < (2-1/M)t_{max}$ and $\Delta < \min\{(2M-1)/M, (2M-1)t_{max}/t_\Sigma\}$ hold (see [71*] by Gusfield).

The following algorithm, further denoted by Φ_3, is designed by Coffman et al. [233] for solving Problem A.2. It works like this. Let D_1 and D_2 be such numbers that $D_1 \le F^* \le D_2$. Given an arbitrary $D \in [D_1, D_2]$ solve Problem A.1 $(M|t_i;\ d_i = 0|\ |M = M(N, D)|M;\ \overline{t}_i \le D)$ using algorithm Φ_1. If M^0 is the resulting number of machines and $M \ge M^0$, an approximate solution of the original problem is obtained. Otherwise, the value of D should be increased. The next value of $D \in [D_1, D_2]$ may be chosen by binary search. After k steps described, the algorithm generates a schedule such that $\Delta \le \rho+1/2^k$. Here $\rho = 1/7$ if $M = 2$, $\rho = 2/3$ if $M = 3$, $\rho = 3/17$ if $M \in \{4, 5, 6, 7\}$ and $\rho = 11/50$ if $M \ge 8$. Friesen [51*] improves the latter value of ρ: $\rho = 0.2$ and obtains a lower bound $\Delta \ge 2/11$. Algorithm Φ_3 has $\tau = O(n\log n+kn\log M)$.

Friesen and Langston [54*] modify algorithm Φ_3 and obtain $2/11 \le \Delta \le 11/61+1/2^k$ and

$\tau = O(n\log n + kn\log M)$. It should be noted that here a constant substituted by "O" is "much greater" than that for algorithm Φ_3.

Hochbaum and Shmoys [82*] use the idea of algorithm Φ_3 to provide another approximation algorithm for Problem A.2. They replace algorithm Φ_1 in the scheme of algorithm Φ_3 by the so-called dual approximation algorithm. Given a deadline D and a set of jobs to be scheduled on parallel identical machines, a ρ-dual approximation algorithm ($\rho > 1$) produces a schedule that uses minimal number of machines but for some machines H it is allowed that $D \leq T_H \leq \rho D$. The resulting algorithm [82*] yields $\Delta \leq 1/k + 1/2^k$ and $\tau = O((kn)^{k^2})$. Leung [121*] has reduced the running time of that algorithm to $O((kn)^{k\log k})$. For $k = 5$ and $k = 6$, Hochbaum and Shmoys have refined their approach to obtain the algorithm with $\tau = O(n\log n)$ and $\tau = O(n(M^4 + \log n))$ respectively.

A rather unusual $O(n)$ algorithm for solving Problem A.2 for $M = 2$ is developed by Kellerer and Kotov [89*]. The algorithm constructs a list where 9 first jobs have the largest processing times and $t_1 \geq t_2 \geq \ldots \geq t_9$. The remaining jobs follow them in an arbitrary order. The jobs are assigned to the machines in the following way. The current job from the list is assigned to the machine with the largest current value of T_H if the total workload of this machine after such an assignment does not exceed $12\theta/11$, where $\theta = 0.5t_{\Sigma}$. Otherwise, this job is assigned to the machine with the smallest current workload. The stopping criteria are as follows. If for a machine H its current value T_H belongs to the interval $[10\theta/11,\ 12\theta/11]$, then all remaining jobs are assigned to the other machine. If the current job has been assigned to the machine with the smallest current workload and the new workload of this machine is greater than $12\theta/11$, all the remaining jobs are assigned to the other machine. This procedure of assigning the jobs runs three times, and three schedules are constructed. For the first run, the job with the largest processing time t_1 is assigned to machine 1 and then all remaining jobs are distributed in accordance with the above procedure. For the second and third runs, the jobs with the processing times t_1, t_5, t_6 and t_1, t_4, respectively, are assigned to machine 1. The best of these three schedules is chosen as an approximate solution. It is shown in [89*] that $\Delta \leq 1/11$.

A.3. Problem A.4 ($M\,|\,t_i;\ d_i = 0\,|\,G\,|\,|\,\overline{t}_{max}$) differs from Problem A.2 by imposing precedence constraints. Most of list scheduling algorithms for solving that problem are based on the concept of "height" h_i of a vertex i in the reduction graph G. The height h_i of a vertex i is equal to the length of the longest path from i to a leaf (a terminal vertex) of G (i.e., to a vertex with no successors). Here the length of a path is the sum of t_j where j

runs over all path vertices.

For an arbitrary graph G, LSA(R) is shown to provide $\Delta \leq 1-1/M$, and this bound is tight. The algorithm LSA($h_i\downarrow$) also yields $\Delta \leq 1-1/M$. Both these algorithms run in $\tau = O(n^2)$ time (see [288] by Graham).

If each connected component of G is an intree (the problem $M|t_i;\ d_i = 0|\mathcal{T}^-||\overline{t}_{max}$) then algorithm LSA($h_i\downarrow$) guarantees $F^0-F^* \leq (1-1/M)t_{max}$ and requires $\tau = O(n\log n)$ (see [96] by Livshitz and [313] by Kaufman). For this case Kunde [106*] gives another bound: $\Delta \leq 1-2/(M+1)$. This bound also holds (see [106*]) if each connected component of G is an outtree (the problem $M|t_i;\ d_i = 0|\mathcal{T}^+||\overline{t}_{max}$). If each connected component of G is a chain (the problem $M|t_i;\ d_i = 0|C||\overline{t}_{max}$) then $\Delta \leq 2/3$ [106*].

If $t_i = 1$, $i = 1, 2,..., n$, and G is an arbitrary graph (the problem ($M|t_i = 1;$ $d_i = 0|G||\overline{t}_{max}$), let us call it Problem A.5), then LSA($h_i\downarrow$) runs in $\tau = O(n^2)$ time, while $\Delta \leq 1/3$ for $M = 2$ and $\Delta \leq 1-1/(M-1)$ for $M > 2$, and this bound is tight at least for $M = 3$ (see [231] by Chen and Liu). Lam and Sethi [328] analyze the performance of the $O(n^2)$ algorithm by Coffman and Graham [234] (see Sections 5.4-5.6 of Chapter 2) applied to Problem 5. They show that $\Delta \leq 1-2/M$, and this bound is tight. It follows from the proof of the NP-hardness of Problem A.5 (see Section 4.5 of Chapter 4) that, unless $\mathcal{P} = \mathcal{NP}$, there exists no polynomial-time approximation algorithm with $\Delta < 1/3$. It is obvious that the same applies to Problem A.4.

For Problem $M|t_i \in \{1, t\};\ d_i = 0|G||\overline{t}_{max}$ (here t is a part of the problem input) Goyal [69*] proposes a generalization of the algorithm by Coffman and Graham [234] (that algorithm solves Problem A.5 for $M = 2$ exactly and runs in $\tau = O(n^2)$ time). This generalized algorithm yields $\Delta \leq 1/3$ if $t = 2$ and $\Delta \leq 1/2-1/2t$ if $t \geq 3$.

For Problem A.6 ($M|t_i;\ d_i = 0|Pr;\ G||\overline{t}_{max}$) in which, unlike in Problem A.4, preemption is allowed, Lam and Sethi [328] analyze the performance of the $O(n^2)$ algorithm by Muntz and Coffman [369, 370] (see Section 6.5-6.6 of Chapter 2) and show that $\Delta \leq 1-2/M$, and this bound is tight.

A.4. Let us consider Problem A.7 ($M|t_i a_H;\ d_i = 0|||\overline{t}_{max}$) of minimizing the makespan on uniform machines. Cho and Sahni [232] study LSA(R) in which the next job is assigned to the first available machine. They prove that a generated schedule guarantees $\Delta \leq (\sqrt{5}-1)/2$ if $M = 2$ and $\Delta \leq \sqrt{2M-2}/2$ if $M > 2$. The latter bound is tight for $M \leq 6$, but in general the worst known example gives $\Delta = \lfloor(\log(3M-1)+1)/2\rfloor - 1$. Liu and Liu [124*] show that $\Delta \leq a_{max}/a_{min} - 1/(a_{min}\sum_{H=1}^{M}(a_H)^{-1})$. If $a_H = 1$, $H = 1, 2,..., M-1$, $a_M < 1$ (Problem A.8) then $\Delta \leq (\sqrt{5}-1)/2$ for $M = 2$ and $\Delta \leq 2-4/(M+1)$ for $M > 2$ (see [232]). In all cases

$\tau = O(n\log M)$.

Jaffe [85*] generalizes the technique of [124*] and shows that a good result may be obtained for Problem A.7 when using not all of the machines but only l fastest of them. If $a_1 \le a_2 \le \ldots \le a_M$ then $F^0/F^* \le A_M/A_l + a_l/a_1 + 1/a_1 A_l$ where $A_k = \sum_{H=1}^{k} (a_H)^{-1}$ and F^0 is the value of $\bar{t}_{max}(s^0)$ for a schedule s^0 obtained by LSA(R) applied to l fastest machines. By minimizing the ratio F^0/F^* over l, Jaffe derives an algorithm for which $\Delta \le \sqrt{M} - 1 + O(M^{1/4})$. This bound is tight up to a constant factor.

For LSA($t_i\!\downarrow$) with the next job to be assigned as described above, the following guarantees for Problem A.7 are determined by Gonzalez et al. [283]: $\Delta \le (\sqrt{17}-3)/4$ if $M = 2$ and $\Delta \le 1 - 2/(M+1)$ if $M > 2$, and for Problem A.8: $\Delta \le (\sqrt{17}-3)/4$ if $M = 2$ and $\Delta \le 1/2 - 1/(2M)$ if $M > 2$. Besides, Morrison [133*] proves that $\Delta \le \max\{1, a_{max}/(2a_{min}) - 1\}$ for Problem A.7. The same algorithm is studied by Liu and Liu [351] and the following guarantees are obtained for Problem A.9 $(M|t_i a_H;\ d_i = 0||a_H = 1,\ H \ne M|\bar{t}_{max})$: $\Delta \le (Ma_M + 1 - 3a_M)/2a_M$ if $a_M < 1/2$, $\Delta \le (2Ma_M + 1 - 4a_M)/(2a_M + 1)$ if $1/2 \le a_M \le 1$ and $\Delta \le 1/a_M + 1/(Ma_M + 1 - a_M)$ if $a_M > 1$. In all cases $\tau = O(n\log n)$.

For LSA($t_i\!\downarrow$) with the next job to be assigned to that machine which would complete its processing earlier, the bounds $\Delta \le 1 - 2/(M+1)$ and $\Delta \le 7/12$ for Problem A.7 are obtained by Gonzalez et al. [283] and by Dobson [40*], respectively. The instances of Problem A.7 are known such that $\Delta = 13/25$ for this algorithm. For Problem A.8, $\Delta \le (\sqrt{17}-3)/4$ if $M = 2$ and $1/3 \le \Delta \le 1/2 - 1/(2M)$ if $M > 2$ [283]. In both cases $\tau = O(n\log n)$.

Algorithm Φ_3 can also be applied for finding an approximate solution of Problem A.7. If the machines are numbered in non-increasing order of a_H, that algorithm yields $\Delta \le (\sqrt{17}-3)/4 + 1/2^k$ if $M = 3$ (see [110*] by Kunde and Steppat), $\Delta \le 1/2 - 1/(2M) + 1/2^k$ if $M \in \{4, 5\}$ (see [107*, 108*] by Kunde and [153*] by Steppat) and $\Delta \le 2/5 + 1/2^k$ if $M \ge 6$ (see [53*] by Friesen and Langston). An example is provided in [53*] such that $\Delta = 0.341$. For Problem A.8, it follows from [108*] by Kunde and [111*] by Langston and Liuwe that $\Delta \le (\sqrt{6}-2)/2 + 1/2^k$ if $M = 2$ [108*] and $\Delta \le (\sqrt{17}-3)/4 + 1/2^k$ if $M \ge 3$ [108*, 111*]. If the machines are numbered in order of non-decreasing a_H, algorithm Φ_3 yields $\Delta \le 1 - 1/M + 1/2^k$ for Problem A.7, while for Problem A.8 the bounds $\Delta \le (\sqrt{17}-3)/4 + 1/2^k$ if $M = 2$ and $1/3 \le 1/(2M) + \sqrt{8M^2 - 8M + 1} \le \Delta \le \sqrt{2} - 1 + 1/2^k$ if $M > 2$ hold [110*]. There exist such examples (see [110*]) that $\Delta = 1/2$ for Problem A.7. For the third case of Problem A.9 $(a_M > 1)$, Kunde et al. [109*] modify algorithm Φ_3 and obtain $\Delta \le (\sqrt{17}-3)/4 + 1/2^k$.

For Problem A.7 Hochbaum and Shmoys [83*] provide an $O(n\log n + M)$ algorithm with Δ arbitrarily close to 1/2.

A.5. Let the problem $M|t_ia_H;\ d_i = 0|G||\bar{t}_{max}$ be called Problem A.10. It differs from Problem A.7 by imposing precedence constraints. It follows from [352] by Liu and Liu that LSA(R) yields $\Delta \le 1-1/\sum_{H=1}^{M}(a_H)^{-1}$ and $\tau = O(n^2)$.

For Problem A.10 with $M = 2$ and $t_i = 1$ $(t_{iH} = a_H)$, $i = 1, 2,..., n$, Gabow [55*] considers the algorithm which schedules the jobs as if the both machines were identical (the problem $2|t_i = 1;\ d_i = 0|G||\bar{t}_{max}$ is polynomially solvable; see Section 5.4 of Chapter 2). He establishes that $\Delta \le 1-\min\{a_1, a_2\}/\max\{a_1, a_2\}$.

For Problem A.11 $(M|t_ia_H;\ d_i = 0|Pr;\ G||\bar{t}_{max})$, which is a version of Problem A.10 when preemption is allowed, Horvath et al. [297] prove that the algorithm by Muntz and Coffman mentioned in Section A.3 guarantees $\Delta \le \sqrt{3M/2}-1$ (here $\tau = O(n^2)$). The bound on Δ is known to be tight up to a constant factor. The algorithm for solving Problem A.11 proposed by Jaffe [306] provides $\Delta \le \sqrt{M}-1/2$. There are examples in [306] for which the bound $\sqrt{M-1}-1$ is approached arbitrarily close. Röck and Schmidt [143*] offer the algorithm for solving the same problem and prove that $\Delta \le \sum_{H=1}^{(M-1)/2}\max\{a_1/a_{2H-1}, a_2/a_{2H}\}+a_1/a_M-1$ if M is odd, and $\Delta \le \sum_{H=1}^{M/2}\max\{a_1/a_{2H-1}, a_2/a_{2H}\}-1$ if M is even (here $a_1 \le a_2 \le ... \le a_M$). This algorithm ignores $M-2$ machines and schedules the jobs on two remaining machines. Note that the problem is solvable in $O(n^2)$ time for $M = 2$ (see [297, 328]). The same guarantees hold for the problem $M|t_{iH} = a_H;\ d_i = 0|Rs(1)||\bar{t}_{max}$ (see 143*]). The latter problem is solvable in $O(n\log n)$ time if $M = 2$ (see [211]).

A.6. Davis and Jaffe [241] consider Problem A.12 $(M|t_{iH};\ d_i = 0|||\bar{t}_{max})$ to minimize the makespan on unrelated machines. They describe approximation algorithms with the following parameters: $\Delta \le \sqrt{6M}+\sqrt{3}/\sqrt{8M}$ and $\tau = O(Mn\log n)$, $\Delta \le 2\sqrt{M}-1$ and $\tau = O(Mn\log n)$, $\Delta \le \sqrt{2M}+1/\sqrt{8M}$ and $\tau = O(M^M+Mn\log n)$. For the first of these algorithms they provide examples such that $\Delta = (2M-\sqrt{M}-2-2\delta)/(\sqrt{M}+\delta)$ for any sufficiently small $\delta > 0$. For the other two algorithms there are examples such that $\Delta = \sqrt{M}-1$.

An algorithm with $\Delta \le M-1$ and $\tau = O(Mn)$ for solving Problem A.12 is described by Ibarra and Kim [302]. Here the next job and the machine for its processing are chosen so as to minimize the makespan for the current partial schedule. Spinrad [152*] has studied this algorithm and has constructed the examples of Problem A.12 such that $\Delta = 1+(\lambda-1)(M-2)/\lambda$ for any $\lambda \ge 1$ (this implies that for those examples the value of Δ can be made very close to $M-1$).

Potts [138*] suggests reducing Problem A.12 to an integer linear programming problem.

Then a partial schedule is constructed by relaxing the integrity of variables. This partial schedule is used to obtain the final schedule by enumerating all the possible variants of assigning the unscheduled jobs. Note that there are no more than M such jobs to be assigned. In this case τ depends on n polynomially and on M exponentially while $\Delta \leq (\sqrt{5}-1)/2$ if $M = 2$ and $\Delta \leq 1$ if $M > 2$. For $M = 2$ the algorithm based on the same approach gives $\Delta \leq 1/2$ and $\tau = O(n)$ [138*]. Lenstra et al. [120*] improve the algorithm from [138*]. Their algorithm runs in polynomial time and satisfies $\Delta < 1$. They also show that checking whether there exists a feasible schedule with $\bar{t}_{max}(s) \leq 2$ is an $\mathcal{NP}$-complete problem. This implies that there is no polynomial-time algorithm with $\Delta < 1/2$ unless $\mathcal{P} = \mathcal{NP}$.

For Problem A.12 with $M = 2$ another algorithm is suggested by Ibarra and Kim [302]. It provides $\Delta \leq (\sqrt{5}-1)/2$ and runs in $\tau = O(n\log n)$ time.

If precedence constraints are imposed (the problem $M|t_{iH};\ d_i = 0|G||\bar{t}_{max}$) the algorithm with $\Delta \leq M-1$ and $\tau = O(Mn+n^2)$ is described in [241] by Davis and Jaffe. The algorithm assigns an available job to a machine that provides the smallest processing time for the job. The bound on Δ is tight.

A.7. For Problem A.13 ($M|t_i;\ d_i = 0|Rs(q)||\bar{t}_{max}$), Garey and Graham [268] show that LSA(R) yields $\Delta \leq \min\{(M-1)/2,\ q+1-(2q+1)/M\}$. The algorithm runs in $\tau = O(n\log n)$ time. There is an example such that $\Delta = (M-1)/2-(M-q-1)/(2k)$ where k can be arbitrarily large.

If $q = 1$ and $t_i = 1$, $i = 1, 2,\ldots, n$, in Problem A.13, then LSA(R) yields $\Delta < 17/10-12/5M+2/F^*$ and $\tau = O(n\log n)$, while LSA($r_{i1}\downarrow$) gives $\Delta < 1-2/M+1/F^*$ and requires $\tau = O(n\log n)$ (see [321] by Krause et al.). For LSA(R) and LSA($r_{i1}\downarrow$), there are examples in [321] such that $\Delta \geq 17/10-\lceil 37/(10M)\rceil$ and $\Delta \geq 1-2/M+2/(MF^*)-1/F^*$, respectively.

The problem $M|t_i = 1;\ d_i = 0|Rs(q)||\bar{t}_{max}$ is known to be solvable in $O(qn^2+n^{5/2})$ time if $M = 2$ [272]. For this problem, an algorithm with $\Delta \leq \lceil M/2\rceil$ based on the idea of scheduling jobs on two machines only is given in [143*] by Röck and Schmidt. Recall that the same idea is discussed in Section A.5. A similar approach can be also applied to the problem $M|t_i = 1;\ d_i = 0|Rs(q)||\sum\bar{t}_i$ yielding the same performance guarantee.

A.8. For the problem $M|t_i;\ d_i = 0|G;\ Rs(q)||\bar{t}_{max}$, LSA(R) gives $\Delta \leq M-1$ and requires $\tau = O(n^2)$ (see [267, 268] by Garey and Graham). This guarantee cannot be improved for $q = 1$ because there is an example of the problem $M|t_i;\ d_i = 0|G;\ Rs(1)||\bar{t}_{max}$ such that $\Delta = (M-1)/(1+M\delta)$ where δ can be arbitrarily small.

Problem A.14 which differs from $M|t_i = 1;\ d_i = 0|G;\ Rs(q)||\bar{t}_{max}$ in that here each job needs at most one resource type (this may be described by the condition $\max\{r_{ij} | 1 \le j \le q\} = \sum_{j=1}^{q} r_{ij}$) is considered by Leung [350]. For this problem LSA(R) finds a schedule with $\Delta \le \min\{M-1,\ q+1-(q+1)/M\}$, and this bound is tight.

Problem A.15, which differs from $M|t_i;\ d_i = 0|G;\ Rs(1)||\overline{t}_{max}$ in that here the resource amount is distributed over machines in advance, is considered in [310] by Kafura and Shen. LSA(R) is shown to provide $\Delta \le M-1$. If $G = (N, \varnothing)$, then $\Delta \le \log M$ for each of LSA(R), LSA($t_i\uparrow$), and LSA($r_{i1}\uparrow$), while $\Delta \le 1-1/M$ for LSA($r_{i1}\downarrow$). If for the latter algorithm those jobs with equal r_{i1} are sorted in non-increasing order of t_i then $\Delta \le 1/4$ for $M = 2$ and $\Delta \le 1-1/(M-1)$ for $M > 2$. All the bounds given in [310] are tight.

A.9. In Problem A.16 ($M|t_i = 1;\ d_i = 0|G;\ Rs(q)|M \ge n|\overline{t}_{max}$) the jobs with unit processing times are to be scheduled on parallel identical machines to minimize the makespan under precedence and resource constraints. Here, at any time job i needs r_{ij} units of resource j, and the total amount of resource j available at a time does not exceed $R_j > 0$, $j \in \{1, 2,..., q\}$. It is assumed that there are sufficiently many machines available to process any number of jobs simultaneously. For this problem LSA(R) yields $\Delta \le q(1+F^*)/2$ and $\tau = O(n^2)$ (see [269] by Garey et al.). This bound is tight; [269] provides an instance of Problem A.16 such that $\Delta \ge q(1+F^*)/2-\delta$ for any $\delta > 0$. If LSA($h_i\downarrow$) is applied (here h_i is the height of a vertex i in graph G, see Section A.3) then an essentially better guarantee $\Delta \le 17q/10$ holds [269] and this is again tight. The same bound holds for LSA($\max\{r_{ij} | 1 \le j \le q\}\downarrow$), and there is an example where $\Delta \ge \lambda q-\delta-1$ for any sufficiently small $\delta > 0$ and $1.69 < \lambda < 1.7$ (see [269]).

For a version of Problem A.16 in which each job needs at most one of q resource types (i.e., $\max\{r_{ij} | 1 \le j \le q\} = \sum_{j=1}^{q} r_{ij}$), LSA($R$) gives $\Delta \le q$, and this bound is tight (see [350] by Leung).

If $G = (N, \varnothing)$ in Problem A.16, LSA(R) provides $\Delta \le q-3/10+5/2F^*$ and requires $\tau = O(n\log M)$, while LSA($\max\{r_{ij} | 1 \le j \le q\}\downarrow$) yields $\Delta \le q-2/3$ and $\tau = O(nq+n\log n)$ [269]. For these algorithms, [269] presents such examples that for any $\delta > 0$, $\Delta \ge q-3/10-\delta$ and $\Delta \ge q-(q^2+1)/(q^2+q)-\delta$, respectively. If both $G = (N, \varnothing)$ and $q = 1$, then $\Delta < 7/10+1/F^*$ for LSA(R) [269]. For the latter case, Krause et al. [321] propose an algorithm with τ no greater than $O(n^2)$ (the authors do not estimate the running time) such that $\Delta < 1/3+1/F^*$; they provide an example such that Δ tends to 1/3 as M tends to infinity.

For the problem $M|t_i;\ d_i = 0|Rs(q)|M \ge n|\overline{t}_{max}$, LSA($R$) yields $\Delta \le q$ and $\tau = O(n\log M)$. This result is due to Garey and Graham [268]. An example is given in [268] such that

$\Delta = (q - q/\nu)/(1 + q/\nu)$ where $n = q(\nu+1)+1$.

A.10. To solve Problem A.17 ($1|t_i;\ d_i|\,|\,|L_{max}$), Schrage proposes the following algorithm (1971, an unpublished result, see [379] by Potts). The job sequence according to which the jobs are processed is formed in the following way. Let k ($k \le n$) first jobs in the sequence be determined and $\bar{t}$ be the completion time of the last job among them. Then the unscheduled jobs such that $d_i \le \bar{t}$ are analyzed, if any, and the job with the minimum value of D_i is chosen (in case of a tie for D_i the job with the maximal value of t_i is taken). If for all unscheduled jobs $d_i > \bar{t}$ holds then the job with the minimum value of d_i is taken. The chosen job is placed at the $(k+1)$th position of the current sequence. This algorithm is shown to guarantee $(F^0 - F^*)/(F^* + D_{max}) < 1 - 2t_{min}/t_\Sigma$ (see [317] by Kise et al.). Potts [379] offers a procedure of l steps ($l \le n-1$) for solving Problem A.17. While moving to the next step, the original problem is transformed in a certain manner. In each step, the sequence of jobs is constructed by Schrage's algorithm. It follows from the above relation and [379] that $(F^0 - F^*)/(F^* + D_{max}) < \min\{1/2,\ t_{max}/t_\Sigma,\ 1 - 2t_{min}/t_\Sigma\}$ for the procedure by Potts. For this procedure $\tau = O(n^2 \log n)$. Hall and Shmoys [76*] modify the algorithm from [379] and obtain $(F^0 - F^*)/(F^* + D_{max}) < 1/3$.

A.11. For Problem A.18 ($M|t_i;\ d_i|\,|\,|L_{max}$), LSA($D_i\uparrow$) gives $F^0 - F^* \le (2M-1)t_{max}/M$ while $\tau = O(n \log n)$ (see [71*] by Gusfield).

If $d_i = 0$, $i = 1, 2, \ldots, n$, then $(F^0 - F^*)/(F^* + D_{max}) \le 1 - 1/M$ for the same algorithm (see [126*] by Masuda et al.). In [126*], the following LSA($t_i\downarrow$) is described to solve this case of Problem A.18. The jobs are distributed over the machines according to the list and those jobs assigned to the same machine are arranged in non-decreasing order of D_i. It is shown that $(F^0 - F^*)/(F^* + D_{max}) \le \min\{4/3 - 1/(3M) - Mt_{min}/t_\Sigma,\ 1/3 - 1/(3M) - M(D_{max} - D_{min})/t_\Sigma\}$ and $\tau = O(n \log n)$.

If $t_i = t$, $i = 1, 2, \ldots, n$, in Problem A.18, then $F^0 - F^* < t$ for LSA($D_i\uparrow$) [71*], although the latter problem is known to be solvable in $O(n^3 \log^2 n)$ time [413].

A.12. For the problem $M|t_i;\ d_i = 0|\,|\,|\Sigma\alpha_i\bar{t}_i$, Eastman et al. [249] prove that LSA($\alpha_i/t_i\downarrow$) yields $\Delta \le (M-1)/2M$ and $\tau = O(n \log n)$. Kawaguchi and Kyan [88*] improve this bound by showing that $\Delta \le (\sqrt{2} - 1)/2$.

Coffman and Labetoulle [235] show that LSA($t_i\uparrow$) for the problem $M|t_i;\ d_i = 0|\,|\,|\Sigma\bar{t}_i/\bar{t}_{max}$ guarantees $\Delta \le (M-1)/(M+1)$ while $\tau = O(n \log n)$.

The problem $M|t_i;\ d_i = 0|\,|\,|\Sigma T_H^2$ is studied by Chandra and Wong [228]. They prove that

LSA($t_i\downarrow$) yields $\Delta \leq 1/24$ and $\tau = O(n\log n)$. On the other hand, there are examples for which $\Delta \geq 1/36 - 1/36M$.

A.13. In this section we concentrate on ε-approximation algorithms (fully polynomial, as a rule).

Fully polynomial algorithms were first developed for the Boolean knapsack problem and certain scheduling problems by Babat [3*, 4*], Ibarra and Kim [301], Sahni [393]. Later, the necessary and sufficient conditions of the existence of those algorithms were established for the problems of maximizing additive functions over so-called independence systems (by Korte and Schrader [92*]) and for the class of problems including practically all combinatorial optimization problems (by Paz and Moran [136*]). Rather general techniques of designing ε-approximation algorithms for combinatorial problems have been developed by Kovalyov and Shafransky [101*].

The problems $1|t_i;\ d_i = 0|||\sum\alpha_i(1-u_i) \to \max$ and $1|t_i;\ d_i = 0|\mathcal{T};\ D_i = D||\sum\alpha_i(1-u_i) \to \max$ are considered by Sahni [394] and Gens [60*], respectively, and ε-approximation algorithms with $\tau = O(n^2/\varepsilon)$ are offered.

The ε-approximation algorithms for the problems $1|t_i;\ d_i = 0|||\sum\alpha_i u_i$ and $1|t_i;\ d_i||d_i < d_j \Longrightarrow D_i \leq D_j|\sum\alpha_i u_i$ are presented by Gens and Levner [33, 61*] and by Kovalyov and Shafransky [100*, 101*], respectively. Both algorithms require $\tau = O(n^2\log n + n^2/\varepsilon)$.

Hall and Shmoys [76*] propose two ε-approximation algorithms for Problem A.17 ($1|t_i;\ d_i|||L_{max}$) (here $(F^0 - F^*)/(F^* + D_{max}) \leq \varepsilon$). The algorithms are not fully polynomial, the running times are are $O(n(1/\varepsilon)^{16/\varepsilon^2+8/\varepsilon} + n\log n)$ and $O(2^{4/\varepsilon}(n/\varepsilon)^{3+4/\varepsilon})$, respectively.

Lawler [113*] proposes an ε-approximation algorithm for the problem $1|t_i;\ d_i = 0|||\sum z_i$ with $\tau = O(n^7/\varepsilon)$. Kovalyov [98] gives an improved algorithm with $\tau = O(n^6/\varepsilon + n^6\log n)$.

The algorithm with $\tau = O(n^3\log n + n^3/\varepsilon)$ for the problem $1|t_i;\ d_i = 0|||\sum\alpha_i\min\{t_i,\ z_i\}$ is proposed by Kovalyov et al. [99*]. For the case $\alpha_i = 1$, $i = 1, 2,\ldots, n$, Potts and Van Wassenhove [139*] give the algorithm with $\tau = O(n^2/\varepsilon)$.

The problems $2|t_i;\ d_i = 0|||\bar{t}_{max}$ and $2|t_i a_H;\ d_i = 0|||\bar{t}_{max}$ formulated, however, in somewhat different terms, are considered in [33, 93] by Gens and Levner. The algorithms developed there require $\tau = O(\min\{n/\varepsilon,\ n + 1/\varepsilon^2\})$ and $\tau = O(\min\{n/\varepsilon,\ n + 1/\varepsilon^3\})$, respectively.

Sahni [393] offers the algorithm with $\tau = O(n^{2M-1}/\varepsilon^{M-1})$ for solving Problem A.2 ($M|t_i;\ d_i = 0|||\bar{t}_{max}$). This result has been improved by Kovalyov [93*] who has designed the algorithm with $\tau = O(n^M/\varepsilon^{M-1})$ for solving the problem $M|t_i;\ d_i = 0|||\bar{t}_{max};\ T_M \leq D$.

Kovalyov [93*] offers the algorithm for solving the problem $M|t_i;\ d_i = 0|||L_{max}$ with $(F^0 - F^*)/(F^* - D_{max}) \leq \varepsilon$ and $\tau = O((n + \log(1/\varepsilon))n^M/\varepsilon^{M-1})$. The same paper describes an

ε-approximation algorithm for the problem $M|t_i;\ d_i = 0|\,|\,|\bar{t}_{max}\Sigma\alpha_i\bar{t}_i$ with $\tau = O(n^M/\varepsilon^M)$.

For the problem $2|t_i;\ d_i = 0|\,|\,|\Sigma\alpha_i\bar{t}_i$ the algorithm designed by Sahni [393] requires $\tau = O(n^2/\varepsilon)$.

The algorithms for the problem $2|t_i;\ d_i = 0|\,|D_i = D|\Sigma z_i$ developed by Kovalyov [95*, 100*] and for the problem $2|t_i;\ d_i = 0|\,|\,|L_{max}$ by Kovalyov and Shafransky [100*, 101*]) yield $(F^0 - F^*)/(F^* + D_{max}) \leq \varepsilon$ and run in $\tau = O(n^3/\varepsilon)$ and $\tau = O((n/\varepsilon)(n+\log(1/\varepsilon))$ time, respectively.

An algorithm with $\tau = O(n^M/\varepsilon^M)$ for the problem $M|t_i;\ d_i = 0|\,|\,|\Sigma T_H^2$ is constructed by Kovalyov [93*].

A schedule s^0 is treated as an ε-approximate solution of problem $M|t_i;\ d_i = 0|\,|\,|\bar{t}_i \leq D_i$ if $(\bar{t}_i(s^0) - D_i)/D_i \leq \varepsilon,\ i = 1, 2, \ldots, n$. To find such a schedule the algorithm with $\tau = O(n^M/\varepsilon^{M-1})$ is offered by Kovalyov in [93*, 100*] while for the case $M = 2$, the algorithm with $\tau = O(n/\varepsilon)$ is developed (see [100*, 101*] by Kovalyov and Shafransky).

For Problem A.7 ($M|t_ia_H;\ d_i = 0|\,|\,|\bar{t}_{max}$) Horowitz and Sahni [296] give the algorithm with $\tau = O(n^{2M}/\varepsilon^{M-1})$. The algorithm with $\tau = O(Mn^{3+10/\varepsilon^2})$ for this problem is due to Hochbaum and Shmoys [83*].

The algorithm described in [296] by Horowitz and Sahni for the problem $M|t_ia_H;\ d_i = 0|\,|\,|\Sigma\alpha_i\bar{t}_i$ requires $\tau = O(n^{2M-2}/\varepsilon^{M-1})$. The algorithm offered by Kovalyov [93*] for the problem $M|t_i;\ d_i = 0|\,|\,|\Sigma\alpha_i\bar{t}_i$ runs in $\tau = O(n^M/\varepsilon^M)$ time. It should be mentioned that earlier Sahni [393] developed an ε-approximation algorithm for that problem with $\tau = O(n^{2M-1}/\varepsilon^{M-1})$.

Lawler and Martel [116*] derive a fully polynomial algorithm for the problem $2|t_ia_H;\ d_i = 0|Pr|\,|\Sigma\alpha_iu_i$.

The algorithm for the problem $2|t_{iH};\ d_i = 0|\,|\,|\bar{t}_{max}$ with $\tau = O(n^2/\varepsilon)$ is offered by Sahni [393].

De et al. [34*] propose an $O(n^3/\varepsilon)$ algorithm for the problem $1|t_i;\ d_i = 0|\,|\,|\Sigma(\bar{t}_i - \frac{1}{n}\Sigma\bar{t}_i)^2$.

REFERENCES

1. Ageenko N.I., Begun S.M., Vizing V.G., Zarovnyj V.P., Strokov V.I. The operational planning problem for equipment maintenance. *Ekonomika i Matematicheskie Metody*, 1976, **12**, N 2, p. 396–399 (in Russian).
2. Adel'son-Vel'sky G.M., Dinits E.A., Karzanov A.V. *Algorithms for flow problems*. Moscow, Nauka Publishers, 1975, 119 pp. (in Russian).
3. Andon F.I., Kuksa A.I., Polyachenko B.E. On optimal planning for computer processing of interconnected problems. *Kibernetika* (Kiev), 1980, N 3, p. 51–53 (in Russian).
4. Antipov V.I. Solving a production scheduling problem by method of states comparison. In: Systems of Resources Distribution on Graphs, Moscow, 1970, p. 7–24 (in Russian).
5. Arlazarov V.L., Dinits E.A., Kronrod M.A., Faradzhev I.A. On fast constructing the transitive closure of a graph. *Doklady Akademii Nauk SSSR*, 1970, **194**, N 3, p. 487–488 (in Russian).
6. Aronovich A.B. On a production scheduling problem. *Ekonomika i Matematicheskie Metody*, 1968, **4**, N 3, p. 401–406 (in Russian).
7. Aho A.V., Hopcroft J.E., Ullman J.D. *The Design and Analysis of Computer Algorithms*. Reading, Addison-Wesley, 1976.
8. Akhpatelov E.A., Cherenin V.P. Optimal sequence for a single machine. *Zhurnal Vychislitel'noj Matemematiki i Matematicheskoj Fiziki*, 1977, **17**, N 2, p. 328–338 (in Russian).
9. Barsky A.B. An automatic distribution of tasks for two identical processors.

Izvestiya Akademii Nauk SSSR. Tekhnicheskaya Kibernetika, 1968, N 4, p. 32–38 (in Russian).

10. Barsky A.B. Two problems of optimization of the usage of heterogeneous computer systems. *Izvestiya Akademii Nauk SSSR. Tekhnicheskaya Kibernetika*, 1971, N 4, p. 119–125 (in Russian).
11. Bekishev C.A. The justification of T.S. Hu algorithm. In: Mathematical Analysis and its Applications. Vol.5., Rostov-na-Donu, 1974, p. 120–126 (in Russian)
12. Bellman R. *Dynamic Programming*. Princeton, Princeton University Press, 1957.
13. Bellman R. Dynamic programming treatment of the traveling salesman problem. *J. Assoc. Comput. Mach.*, 1962, **9**, N 1, p. 61–63.
14. Bellman R. *Introduction to Matrix Analysis*. McGraw-Hill Book Company, 1960.
15. Berge C. *Teorie des graphes et ses applications*. Paris, 1958.
16. Bilyuba V.F., Belosludtsev N.M. Study on the planning of processing one type of job flows on a group of machines. *Izvestiya Vysshikh Uchebnykh Zavedenij. Priborostroenie*, 1974, **17**, N 4, p. 46–50 (in Russian).
17. Bondarenko A.T., Sapatyj P.S. An algorithm for assigning independent jobs to parallel processors. *Izvestiya Akademii Nauk SSSR. Tekhnicheskaya Kibernetika*, N 4, p. 101–103 (in Russian).
18. Burdyuk V.Y. The regular g-orderings and Smith's functions. *Kibernetika* (Kiev), 1975, N 2, p. 17–25 (in Russian).
19. Burdyuk V.Y., Reva V.N. A method for optimizing functions over permutations under constraints. *Kibernetika* (Kiev), 1980, N 1, p. 99–103 (in Russian).
20. Burdyuk V.Y., Shkurba V.V. Scheduling theory. Problems and solution methods. *Kibernetika* (Kiev), 1971, N 1, p. 89–102 (in Russian).
21. Burdyuk T.A. Intervals of priorities in the penalty minimization problem. *Kibernetika* (Kiev), 1973, N 5, p. 106–110 (in Russian).
22. Burdyuk T.A. On extrema of the Elmaghraby problem. *Vestsi Akademii Navuk BSSR. Ser. fizika – matematychnykh navuk*, 1974, N 2, p. 120–124 (in Russian).
23. Burdyuk T.A. The problem of minimizing the penalties related to workpieces delays in production. In: Improvement of Economical Effectiveness of Production. Dnepropetrovsk, 1974, 1, p. 157–171 (in Russian).
24. Burkov V.N., Lovetsky S.E. Solution methods for extremal combinatorial problems (a survey). *Izvestiya Akademii Nauk SSSR. Tekhnicheskaya Kibernetika*, 1968, N 4, p. 82–93 (in Russian).
25. Burkov V.N., Lovetsky S.E. Methods for solving extremal problems of combinatorial

type (a survey). *Avtomatika i Telemekhanika*, 1968, N 11, p. 68-93 (in Russian).

26. Burkov V.N., Sokolov V.B. Optimal allocation of data arrays in magnetic tape storage for the case of two-directed search. *Avtomatika i Telemekhanika*, 1969, N 4, p. 107-117. (in Russian).
27. Vizgunov N.P. Some approaches to solving problems for parallel machines in production scheduling. In: Analysis and Modeling of Economical Processes, Gorky, 1978, p. 15-27 (in Russian).
28. Vizing V.G. Scheduling with target times. *Cybernetics*, 1981, **17**, p. 143-152.
29. Vizing V.G. Minimization of the maximum delay in servicing systems with interruption. *U.S.S.R. Computational Mathematics and Mathematical Physics*, 1982, **22**, N 3, p. 227-233.
30. Vizing V.G. Optimal choice of job execution intensities for a convex function of penalties for intensity. *Kibernetika* (Kiev), 1982, N 3, p. 125-127 (in Russian).
31. Vizing V.G., Klipker I.A. A polynomial algorithm for solving the Garey-Johnson scheduling problem. *Soobshcheniya Akademii Nauk Gruzinskoj SSR*, 1981, **102**, N 1, p. 29-32 (in Russian).
32. Vizing V.G., Komzakova L.N., Tarchenko A.V. An algorithm for selecting the execution intensity of jobs in a schedule. *Cybernetics*, 1982, **17**, p. 646-649.
33. Gens G.V., Levner E.V. Efficient approximation algorithms for combinatorial problems. Moscow, Central Institute of Economics and Mathematics of Academy of Sciences of the USSR, Preprint, 1981, 66 pp. (in Russian).
34. Gens G.V., Levner E.V. Approximate algorithms for certain universal problems in scheduling theory. *Engineering Cybernetics*, 1978, **16**, N 6, p. 31-36.
35. Gens G.V., Levner E.V. Discrete optimization problems and efficient approximate algorithms. *Engineering Cybernetics*, 1979, **17**, N 6, p. 1-11.
36. Gik E.Y., Levner E.V. Solving some single-machine production scheduling problems. In: Topics of Mathematical Theory of Control Systems and Its Application in Metallurgy, Moscow, 1974, p. 22-25 (in Russian).
37. Gimadi E.K., Glebov N.I., Perepelitsa V.A. Studies in scheduling theory. *Upravlyaemye Sistemy*, 1974, **12**, p. 3-10 (in Russian).
38. Golovkin B.A. Methods and means for parallel data processing. *J. Sov. Math.*, 1981, **17**, p. 1876-1934.
39. Golovkin B.A. Classification of dispatching methods for multiprocessor and multimachine computing systems. *Upravlyayushchie Sistemy i Mashiny*, 1982, N 3, p. 3-11 (in Russian).

40. Gol'dgaber E.M. Problem of minimizing the completion time for the job set given by a tree. *Kibernetika* (Kiev), 1977, N 2, p. 102–107 (in Russian).

41. Gordon V.S. The deterministic single-stage scheduling with preemption. In: Computers in Engineering, Minsk, 1973, June, p. 30–38 (in Russian).

42. Gordon V.S. On optimal schedules with preemption. *Vestsi Akademii Navuk BSSR. Ser. fizika – matematychnykh navuk*, 1974, N 5, p. 129–130 (in Russian).

43. Gordon V.S. Deterministic scheduling with a minimax optimality criterion and partially ordered jobs. *Avtomatizatsiya Tekhnicheskoj Podgotovki Proizvodstva*, 1977, N 4, p. 70–75 (in Russian).

44. Gordon V.S. Some properties of series-parallel graphs. *Vestsi Akademii Navuk BSSR. Ser. fizika – matematychnykh navuk*, 1981, N 1, p. 18–23 (in Russian).

45. Gordon V.S., Tanaev V.S. Single-machine deterministic scheduling with step functions of penalties. In: Computers in Engineering, Minsk, 1971, September, p. 3–8 (in Russian).

46. Gordon V.S., Tanaev V.S. Preemptions in deterministic systems with parallel machines and different release dates of jobs. In: Optimization of Systems of Collecting, Transfer and Processing of Analogous and Discrete Data in Local Information Computing Systems. Materials of the 1st Joint Soviet–Bulgarian Seminar, Institute Engineering Cybernetics of Academy of Sciences of BSSR – Institute of Engineering Cybernetics of Bulgarian Academy of Sciences, 1973, Minsk, 1973, p. 36–50 (in Russian).

47. Gordon V.S., Tanaev V.S. Due dates in single-stage deterministic scheduling. In: Optimization of Systems of Collecting, Transfer and Processing of Analogous and Discrete Data in Local Information Computing Systems. Materials of the 1st Joint Soviet–Bulgarian Seminar, Institute Engineering Cybernetics of Academy of Sciences of BSSR – Institute of Engineering Cybernetics of Bulgarian Academy of Sciences, 1973, Minsk, 1973, p. 54–58 (in Russian).

48. Gordon V.S., Tanaev V.S. Single-machine deterministic scheduling with tree-like ordered jobs and exponential penalty functions. In: Computers in Engineering, Minsk, 1973, June, p. 3–10 (in Russian).

49. Gordon V.S., Tanaev V.S. On minimax problems of scheduling theory for a single machine. *Vestsi Akademii Navuk BSSR. Ser. fizika – matematychnykh navuk*, 1983, N 3, p. 3–9. (in Russian).

50. Gordon V.S., Chegolina E.P. The single-stage problem of scheduling theory with tree-like ordered jobs. *Vestsi Akademii Navuk BSSR. Ser. fizika – matematychnykh*

navuk, 1977, N 4, p. 36–40 (in Russian).

51. Gordon V.S., Shafransky Y.M. On a class of scheduling theory problems with partially ordered jobs. In: Proceedings of the 4th All-Union Conference on Theoretical Cybernetics Problems, 1977, Novosibirsk, 1977, p. 101–103 (in Russian).
52. Gordon V.S., Shafransky Y.M. Optimal ordering with series-parallel precedence constraints. *Doklady Akademii Nauk BSSR*, 1978, **22**, N 3, p. 244–247 (in Russian).
53. Gordon V.S., Shafransky Y.M. The decomposition approach to minimizing functions over a set of permutations of partially ordered elements. In: Proceedings of the 5th All-Union conference on Complex Systems Control, Alma-Ata, 1978, p. 51–56 (in Russian).
54. Gordon V.S., Shafransky Y.M. On optimal ordering with series-parallel precedence constraints. *Vestsi Akademii Navuk BSSR. Ser. fizika – matematychnykh navuk*, 1978, N 5, p. 135 (in Russian).
55. Gordon V.S., Shafransky Y.M. The question of minimizing functions over a set of permutations of partially ordered elements. *Vestsi Akademii Navuk BSSR. Ser. fizika – matematychnykh navuk*, 1979, N 2, p. 122–124 (in Russian).
56. Garey M.R., Johnson D.S. *Computers and Intractability: A Guide to the Theory of NP-Completeness*. Freeman, San Francisco, 1979.
57. Dinits E.A. On the solving of two assignment problems. In: Study on Discrete Optimization, Moscow, Nauka Publishers, 1976, p. 333–348 (in Russian).
58. Dinits E.A., Kronrod M.A. An algorithm for solving the assignment problem. *Doklady Akademii Nauk SSSR*, 1969, **189**, N 1, p. 23–25 (in Russian).
59. Johnson S.M. Optimal two- and three-stage production schedules with setup times included. *Naval Res. Log. Quart.*, 1954, **1**, p. 61–68.
60. Emelichev V.A., Komlik V.I. *The Method for Constructing a Sequence of Plans for Solving Discrete Optimization Problems*. Moscow, Nauka Publishers, 1981, 208 pp.
61. Emelichev V.A., Suprunenko D.A., Tanaev V.S. On research of Byelorussian mathematicians in discrete optimization. *Izvestiya Akademii Nauk SSSR. Tekhnicheskaya Kibernetika*, 1982, N 6, p. 25–45 (in Russian).
62. Zhuravlev Y.I. Complexity estimation of local algorithms for some extremal problems over finite sets. *Doklady Akademii Nauk SSSR*, 1964, **158**, N 5, p. 1018-1021 (in Russian).
63. Zhuravlev Y.I. Complexity estimations for algorithms of constructing the minimal disjunctive normal forms for logic algebra functions. *Discretnyj Analiz* (Novosibirsk), 1964, **3**, p. 41–77 (in Russian).

64. Zhuravlev Y.I. Local algorithms for calculating information. I, II. *Kibernetika* (Kiev), 1965, N 1, p. 12–19; 1966, N 2, p. 1–11 (in Russian).

65. Zhurzhenko S.L. Problems of scheduling for single-stage production (a survey). In: Modeling of Economical Processes, Vol. 2, Moscow, Moscow State University, 1968, (in Russian).

66. Zak Y.A. The determination of a sequence for executing independent jobs on parallel machines. *Izvestiya Akademii Nauk SSSR. Tekhnicheskaya Kibernetika*, 1969, N 2, p. 15–20 (in Russian).

67. Zinder Y.A. Asymptotic estimations for a scheduling theory problem. In: Problems of Designing Automated Systems of Production Control, Moscow-Gorky, 1974, p. 134–138 (in Russian).

68. Zinder Y.A. The priority solvability of a class of sequencing problems. In: Problems of Creating Automated Systems of Production Control, Kiev, 1976, p. 52–57 (in Russian).

69. Zinder Y.A. On algorithms for solving some sequencing problems. In: Algorithms and Programs, Vol. 1, Gorky, 1977, p. 114–123 (in Russian).

70. Zinder Y.A. Minimizing the maximal penalty in deterministic scheduling system with identical machines. In: System Analysis of Industrial Production, Kiev, 1978, p. 56–63 (in Russian).

71. Zinder Y.A., Podchasova T.P. Minimizing ordered criteria in a single-stage deterministic scheduling system. In: Automated Systems for Control and Data Processing, Kiev, 1976, p. 20–25 (in Russian).

72. Irikov V.A. Certain sequencing problems. *Izvestiya Akademii Nauk SSSR. Tekhnicheskaya Kibernetika*, 1970, N 4, p. 38–42 (in Russian).

73. Kantsedal S.A. Algorithm to speed up the search for solutions in a network scheduling problem. *Automated Remote Control*, 1982, **43**, p. 485–490.

74. Karp R.M. Reducibility among combinatorial problems. In: Complexity of computer computations, R.E. Miller, J.W. Thatcher (eds.), New York, Plenum Press, 1972, p. 85–104.

75. Kitik M.G. On "reducing" branching for a single-machine problem. *Kibernetika* (Kiev), 1972, N 1, p. 145–147 (in Russian).

76. Kladov G.K., Livshits E.M. On a scheduling problem to minimize the total penalty. *Kibernetika* (Kiev), 1968, N 6, p. 99–100 (in Russian).

77. Knuth D.E. *The Art of Computer Programming. Vol. 3. Sorting and Searching.* Reading, Addison-Wesley, 1973.

78. Conway R.W., Maxwell W.L., Miller L.W. *Theory of Scheduling*. Reading, Addison-Wesley, 1967.

79. Kopylov G.N. An exact solution of a single-machine problem. *Vestnik Yaroslavskogo Universiteta*, 1975, **9**, p. 52-60 (in Russian).

80. Korbut A.A., Sigal I.H., Finkel'shtein Y.Y. Branch and bound method (survey of theory, algorithms, programs and applications). *Math. Operationsforsch. Statist. Ser. Optimization*, 1977, **8**, N 2, p. 253-280 (in Russian).

81. Kostrikin A.I. *Introduction to Algebra*. New York-Heidelberg-Berlin, Springer-Verlag, 1982, 575 pp.

82. Cook S.A. The complexity of theorem-proving procedures. In: Proceedings of the 3rd Annual ACM Symposium on Theory of Computation, 1971, p. 151-158.

83. Lapin V.V. On an algorithm for scheduling of a set of jobs. *Kibernetika* (Kiev), 1976, N 2, p. 90-94 (in Russian).

84. Lapko A.A. On optimization of priority scheduling systems. *Vestsi Akademii Navuk BSSR. Ser. fizika-matematychnykh navuk*, 1977, N 2, p. 117-120 (in Russian).

85. Lapko A.A. Scheduling deterministic job flows to meet due dates. *Vestsi Akademii Navuk BSSR. Ser. fizika-matematychnykh navuk*, 1980, N 5, p. 130 (in Russian).

86. Lapko A.A. The optimal sequencing of job flows in systems with due dates. *Theory and Methods of Automated Design*, Minsk, 1980, **4**, p. 23-24 (in Russian).

87. Lebedinskaya N.B. Minimizing the maximum lateness in the preemptive case. *Zapiski Nauchnukh Seminarov Leningradskogo Otdeleniya Matematicheskogo Instituta AN SSSR*, 1978, **80**, p. 117-124 (in Russian).

88. Lebedinskaya N.B. Minimizing the maximum penalty in the preemptive case. *Zapiski Nauchnukh Seminarov Leningradskogo Otdeleniya Matematicheskogo Instituta AN SSSR*, 1980, **102**, p. 61-67 (in Russian).

89. Levin G.M., Tanaev V.S. *Decomposition Methods for Optimizing Design Decisions*. Minsk, Nauka i Tekhnika, 1978, 240 pp. (in Russian).

90. Levin M.S. Effective solution of certain problems of theory of scheduling of nets. *Cybernetics*, 1980, **16**, p. 148-154.

91. Levner E.V. The network approach to scheduling theory problems. In: Study in Discrete Mathematics. Moscow, Nauka Publishers, 1973, p. 135-150 (in Russian).

92. Levner E.V. Scheduling theory in economic systems (some mathematical topics). Moscow, Central Institute of Economics and Mathematics of Academy of Sciences of the USSR, Preprint, 1977, 53 pp. (in Russian).

93. Levner E.V., Gens G.V. Discrete optimization problems and efficient approximation

algorithms. Central Institute of Economics and Mathematics of Academy of Sciences of the USSR, Preprint, 1978. 55 pp. (in Russian).

94. Levner E.V., Gens G.V. An analysis of the computational complexity of approximation algorithms for some discrete optimization problems. *Matematicheskie Metody Resheniya Ekonomicheskikh Zadach*, 1980, **9**, p. 97–106 (in Russian).
95. Leont'ev V.K. Discrete extremal problems. *J. Sov. Math.*, 1981, **15**, p. 101–139.
96. Livshits E.M. Studies of some algorithms for optimization of network models. *Ekonomika i Matematicheskie Metody*, 1968, **4**, N 5, p.768–775 (in Russian).
97. Livshits E.M. The sequence of operations for complex machine component manufacturing. *Avtomatika i Telemekhanika*, 1968, N 11, p. 94–95 (in Russian).
98. Livshits E.M. Minimizing the maximum penalty in a single-machine problem. In: Proceedings of 1st Winter School on Mathematical Programming in Drogobych, 1968, Moscow, 1969, p. 474–475 (in Russian).
99. Livshits E.M. The position indicator in minimization problems over permutations. *Kibernetika* (Kiev), 1975, N 3, p. 80–90 (in Russian).
100. Livshits E.M., Rublinetsky Y.I. On comparative complexity of some discrete optimization problems. *Computational Mathematics and Computers*, Kharkov, 1972, **3**, p. 78–85 (in Russian).
101. Linsky V.S., Kornev M.D. Optimal scheduling for parallel processors. *Izvestiya Akademii Nauk SSSR. Tekhnicheskaya Kibernetika*, 1972, N 3, p. 160–167 (in Russian).
102. Little J.D.C., Murty K.G., Sweeney D.W., Karel C. An algorithm for the traveling salesman problem. *Oper. Res.*, 1963, **11**, p. 972–989.
103. Lokot T.V. On a problem of scheduling theory. *Sbornik Trudov Moskovskogo Inzhenerno – Stroitel'nogo Instituta imeni V.V.Kujbysheva*, 1979, **173**, p. 156–160 (in Russian).
104. Lominadze N.N., Shartava Z.K. Algorithm for sequential improvement of a job schedule for parallel machines. In: Transactions of Problem Laboratory of Automatics and Computers of Georgian Polytechnical Institute, 1972, N 3, p. 124–129 (in Russian).
105. Maksimenkov A.V. Scheduling with restriction on resource use rate. *Cybernetics*, 1980, **15**, p. 876–881.
106. Mal'tsev A.I. *Algorithms and Recursive Functions*. Moscow, Nauka Publishers, 1965, 392 pp. (in Russian).
107. Manusevich O.Z., Reva V.N. An algorithm for optimizing functions over the

permutations of a partially ordered set. In: Study on Modern Problems of Summation and Approximation of Functions and Their Applications, Dnepropetrovsk, 1979, p. 95–97 (in Russian).

108. Mel'nikov O.I., Shafransky Y.M. Parametric problem in scheduling theory. *Cybernetics*, 1980, **15**, p. 352–357.

109. Metel'sky N.N. On extremal values of a linear form over certain sets of permutations. *Vestsi Akademii Navuk BSSR. Ser. fizika – matematychnykh navuk*, 1972, N 5, p. 5–10 (in Russian).

110. Mironosetsky H.B. *Economic and Mathematical Methods of Production Scheduling*. Novosibirsk, Nauka Publishers, 1974, 140 pp. (in Russian).

111. Misyura E.B. Optimal scheduling for a single machine to minimize the total cost. *Vestnik Kievskogo Politekhnicheskogo Instituta, Tekhnicheskaya Kibernetika*, 1981, N 5, p. 58–62 (in Russian).

112. Mikhalevich V.S. Sequential optimization algorithms and their application. I, II. *Kibernetika* (Kiev), 1965, N 1, p. 45–66; 1965, N 2, p. 85–89 (in Russian).

113. Mikhalevich V.S., Volkovich V.L., Voloshin A.F., Pozdnyakov Y.M. Algorithms for sequential analysis and elimination in discrete optimization problems. *Cybernetics*, 1981, **16**, p. 389–399.

114. Mikhalevich V.S., Ermol'ev Y.M., Shkurba V.V., Shor N.Z. Complex systems and solving extremal problems. *Kibernetika* (Kiev), 1967, N 5, p. 29–37 (in Russian).

115. Mikhalevich V.S., Kuksa A.I. *Sequential Optimization Methods in Discrete Network Problems of Optimal Resource Distribution*. Moscow, Nauka Publishers, 1983, 208 pp. (in Russian).

116. Mikhalevich V.S., Shkurba V.V. Sequential schemes of optimization in job sequencing problems. *Kibernetika* (Kiev), 1966, N 2, p. 34–40.

117. Mikhalevich V.S., Shor N.Z. Numerical solution of multivariant problems by the method of sequential variant analysis. In: Scientific and Methodical Materials of Seminar on Economics and Mathematics of Laboratory of Methods in Economics and Mathematics of Academy of Sciences of the USSR, Vol. 1, Moscow, Computer Center of Academy of Sciences of the USSR, 1962, p. 15–41 (in Russian).

118. Mikhalevich V.S., Shor N.Z., Galustova L.A. et al. *Computational Methods for Choosing Optimal Design Decisions*, Kiev, Naukova Dumka, 1977, 178 pp. (in Russian).

119. Moiseev N.N. *Numerical Methods in Theory of Optimal Systems*. Moscow, Nauka Publishers, 1971, 424 pp. (in Russian).

120. Moiseev N.N. *Elements of Theory of Optimal Systems*. Moscow, Nauka Publishers, 1975, 526 pp. (in Russian).

121. Moiseev N.N. *Mathematics Carries Out an Experiment*. Moscow, Nauka Publishers, 1979, 224 pp. (in Russian).

122. Moiseev N.N., Ivanilov Y.P., Stolyarova E.M. *Methods of Optimization*. Moscow, Nauka Publishers, 1978, 352 pp. (in Russian).

123. Nikitin A.V. The application of dynamic programming for solving some sequencing problems. *Trudy Moskovskogo Energeticheskogo Instituta*, 1969, **68**, p. 165–169 (in Russian).

124. Ovsyankin B.P. On a problem of organization of data processing in multiprocessor computing systems. *Zhurnal Vychislitel'noj Matemematiki i Matematicheskoj Fiziki*, 1983, **23**, N 5, p. 1262–1266 (in Russian).

125. Panajoti B.N., P'yanzina L.Y., Chebakov V.A. Minimizing the number of preemptions in a multiprocessor schedule. *Izvestiya Akademii Nauk SSSR. Tekhnicheskaya Kibernetika*, 1971, N 4, p. 103–110 (in Russian).

126. Paramonov F.I. *Mathematical Methods for Computing Multi–Item Flows*. Moscow, Mashinostroenie, 1964, 264 pp. (in Russian).

127. Podchasova T.P., Portugal V.M., Tatarov V.A., Shkurba V.V. *Heuristic Methods for Production Scheduling*. Kiev: Tekhnika, 1980. 140 pp. (in Russian).

128. Reva V.N. On an algorithm for optimizing a function over the permutations of a partially ordered set. In: Actual Problems of Computers and Programming, Dnepropetrovsk, 1979, p. 92–95 (in Russian).

129. Reva V.N. On optimizing functions over the permutations of a partially ordered set. In: Proceedings of the 5th All-Union Conference on Theoretical Cybernetics Problems, Novosibirsk, 1980, p. 91–92 (in Russian).

130. Revchuk I.N. Conditions for the existence of a feasible schedule for a system of M machines with the availability windows. *Vestsi Akademii Navuk BSSR. Ser. fizika – matematychnykh navuk*, 1977, N 4, p. 123 (in Russian).

131. Revchuk I.N. On a class of scheduling problems with nonrigid availability windows for machines. *Theory and Methods of Automated Design*, Minsk, 1979, **2**, p. 19–28 (in Russian).

132. Revchuk I.N. On certain classes of single-stage production problems with the nonrigid availability windows for the machine. *Theory and Methods of Automated Design*, Minsk, 1979, **4**, p. 16–22 (in Russian).

133. Reingold E.M., Nievergelt J., Deo N. *Combinatorial Algorithms. Theory and Practice.*

Englewood Cliffs, Prentice-Hall, 1977.

134. Romanovsky I.V. Numerical methods of discrete programming. In: Proceedings of the 4th Winter School on Mathematical Programming and Related Topics, Vol. 5, Moscow, 1972, p. 76-128 (in Russian).

135. Sannikova A.K. On optimizing functions over permutations subject to element grouping constraints. *Vestsi Akademii Navuk BSSR. Ser. fizika-matematychnykh navuk*, 1982, N 1, p. 116-117 (in Russian).

136. Sannikova A.K., Tanaev V.S. On a class of extremal permutation problems. *Doklady Akademii Nauk BSSR*, 1979, **23**, N 9, p. 784-786 (in Russian).

137. Sarvanov V.I. Approximate solution to the problem of minimizing a linear form over the set of cyclic permutations. *Vestsi Akademii Navuk BSSR. Ser. fizika-matematychnykh navuk*, 1977, N 6, p. 5-10 (in Russian).

138. Sarvanov V.I. On optimization over permutations. *Vestsi Akademii Navuk BSSR. Ser. fizika-matematychnykh navuk*, 1979, N 4, p. 9-11 (in Russian).

139. Sarvanov V.I. On the complexity of minimizing a linear form on a set of cyclic permutations. *Sov. Math. Doklady*, 1980, **22**, p. 118-120.

140. Safonova T.E. On minimizing the number of elements of jobs which do not match their directive intervals. *Zapiski Nauchnukh Seminarov Leningradskogo Otdeleniya Matematicheskogo Instituta AN SSSR*, 1979, **90**, p. 186-193 (in Russian).

141. Safronenko V.A. Optimal schedule for a machine of the assembly line. *Kibernetika* (Kiev), 1973, N 5, p. 70-75 (in Russian).

142. Segal V.M. On a heuristic algorithm for reducing idle time of complex systems. *Trudy Moskovskogo Aviatsionnogo Instituta*, 1972, **230**, p. 55-60 (in Russian).

143. Semenov A.I., Portugal V.M. *Problems of Small-Scale Production Scheduling.* Moscow, Nauka Publishers, 1972, 183 pp. (in Russian).

144. Sergienko I.V., Lebedeva T.T., Roschin V.A. *Approximation Methods for Solving Discrete Optimization Problems.* Kiev, Naukova Dumka, 1980, 274 pp. (in Russian).

145. Sovetov B.Y., Tishkin A.I. On solving a class of scheduling theory problems. *Izvestiya Vysshikh Uchebnykh Zavedenij. Priborostroenie*, 1976, **19**, N 12, p. 56-60 (in Russian).

146. Sotskov Y.N. An optimal schedule for a set of jobs given by a mixed graph. *Vestsi Akademii Navuk BSSR. Ser. fizika-matematychnykh navuk*, 1977, N 4, p. 133 (in Russian).

147. Sotskov Y.N. Scheduling on mixed graphs with bounded maximal penalty. *Vestsi Akademii Navuk BSSR. Ser. fizika-matematychnykh navuk*, 1980, N 2, p. 37-41 (in

Russian).

148. Sotskov Y.N. Finding an optimal schedule for interconnected jobs and sequential machines. In: Methods and Programs for Solving Extremal Problems, Minsk, 1981, p. 28–34.

149. Suprunenko D.A. On values of a linear form over a set of permutations. *Kibernetika* (Kiev), 1968, N 2, p. 59–63 (in Russian).

150. Suprunenko D.A., Ajzenshtat V.S., Lepeshinsky N.A. Extremal values of functions over sets of permutations. In: Proceedings of the 1st All-Union Confetence on Operations Research, Minsk, 1972, p. 61–64 (in Russian).

151. Suprunenko D.A., Metel'sky N.N. The assignment problem and minimizing the sum of linear forms over the symmetric group. *Kibernetika* (Kiev), 1973, N 3, p. 64–68 (in Russian).

152 Tanaev V.S. On scheduling theory. *Doklady Akademii Nauk BSSR*, 1964, **8**, N 12, p. 792–794 (in Russian).

153. Tanaev V.S. Some objective functions of a single-stage production. *Doklady Akademii Nauk BSSR*, 1965, **9**, N 1, p. 11–14. (in Russian).

154. Tanaev V.S. On a class of scheduling theory problems. In: Production Control. Proceedings of the 3rd All-Union Conference on Automated Control (Engineering Cybernetics), Moscow, Nauka Publishers, 1967, p. 211–214 (in Russian).

155. Tanaev V.S. Preemptions in deterministic scheduling systems with parallel identical machines. *Vestsi Akademii Navuk BSSR. Ser. fizika – matematychnykh navuk*, 1973, N 6, p. 44–48 (in Russian).

156. Tanaev V.S. On optimal partition of a finite set into subsets. *Doklady Akademii Nauk BSSR*, 1979, **23**, p. 26–28 (in Russian).

157. Tanaev V.S., Gordon V.S. On scheduling to minimize the weighted number of late jobs. *Vestsi Akademii Navuk BSSR. Ser. fizika – matematychnykh navuk*, 1983, N 6, p. 3–9 (in Russian).

158. Tanaev V.S., Shkurba V.V. *Introduction to Scheduling Theory*. Moscow, Nauka Publishers, 1975, 256 pp. (in Russian).

159. Tuzikov A.V. On a class of problems of vector optimization over permutations. In: Algorithms and Programs for Solving Extremal Problems and Related Topics, Minsk, 1982, p. 62–70 (in Russian).

160. Tuzikov A.V., Shafransky Y.M. On a problem of lexicographic minimization over a set of permutations. *Vestsi Akademii Navuk BSSR. Ser. fizika – matematychnykh navuk*, 1983, N 6, p. 115 (in Russian).

161. Feigin L.I. Optimal scheduling for a single machine under incomplete information. *Kibernetika* (Kiev), 1971, N 4, p. 149-151 (in Russian).

162. Finkel'shtein Y.Y. *Approximation Methods and Applied Problems of Discrete Programming*. Moscow, Nauka Publishers, 1976, 264 pp. (in Russian).

163. Harary F. *Graph Theory*. Reading, Addison-Wesley, 1969.

164. Hardy G.H., Littlewood J.E., Polya G. *Inequalities*. London, Cambridge University Press, 1934.

165. Khachaturov V.R. Approximate combinatorial method and some of its applications. *Zhurnal Vychislitel'noj Matemematiki i Matematicheskoj Fiziki*, 1974, **14**, N 6, p. 1464–1487 (in Russian).

166. Khachiyan L.G. Polynomial algorithm in linear programming. *Soviet Math. Doklady*, 1979, **20**, p. 191-194.

167. Khenkin V.E. On permutation functions admitting a solving rule. In: Problems of Discrete Information Analysis, Part 2, Novosibirsk, 1976, p. 106–116 (in Russian).

168 Hu T.C. Parallel sequencing and assembly line problems. *Oper. Res.*, 1961, **9**, p. 841–848.

169. Tsallagova O.N. On a problem of scheduling. *Matematicheskie Metody Resheniya Ekonomicheskikh Zadach*, 1974, **5**, p. 130-134 (in Russian).

170. Cherenin V.P. Solving some combinatorial problems of optimal planning by methods of sequential calculations. In: Scientific and Methodical Materials of Seminar on Economics and Mathematics of Laboratory of Methods in Economics and Mathematics of Academy of Sciences of the USSR, Moscow, Computer Center of Academy of Sciences of the USSR, 1962, **2**, 44 pp. (in Russian).

171. Shapiro A.D. Some asymptotic properties of a scheduling problem. *Ekonomika i Matematicheskie Metody*, 1975, **11**, N 6, p. 1214-1216 (in Russian).

172. Shafransky Y.M. On optimal sequencing for deterministic systems with tree-like partial order. *Vestsi Akademii Navuk BSSR. Ser. fizika – matematychnykh navuk*, 1978, N 2, p. 120 (in Russian).

173. Shafransky Y.M. Optimization for deterministic scheduling systems with tree-like partial order. *Vestsi Akademii Navuk BSSR. Ser. fizika – matematychnykh navuk*, 1978, N 2, p. 119 (in Russian).

174. Shafransky Y.M. On a problem of minimizing functions over a set of permutations of partially ordered elements. I, II. *Vestsi Akademii Navuk BSSR. Ser. fizika – matematychnykh navuk*, 1980, N 5, p. 132; 1982, N 1, p. 113 (in Russian).

175. Shafransky Y.M. On a certain property of priority-generating functions. *Vestsi*

Akademii Navuk BSSR. Ser. fizika-matematychnykh navuk, 1981, N 6, p. 15-18 (in Russian).

176. Shafransky Y.M. On an algorithm for finding the minimum of priority-generating functions over special sets of permutations. I, II. *Vestsi Akademii Navuk BSSR. Ser. fizika-matematychnykh navuk*,Nauk, 1982, N 3, p. 38-42; 1983, N 1, p. 15-20 (in Russian).
177. Shafransky Y.M., Andreev G.V. Minimizing priority-generating functions over a set of permutations of partially ordered elements. In: Methods and Programs for Solving Extremal Problems. Minsk, 1981, p. 13-27 (in Russian).
178. Shafransky Y.M., Yanova O.V. Recognition and decomposition of series-parallel graphs. In: Algorithms and Programs for Solving Optimization Problems, Minsk, 1980, p. 17-24 (in Russian).
179. Shakhbazyan K.V. The ordering of the structure set of jobs minimizing the total cost. *Zapiski Nauchnukh Seminarov Leningradskogo Otdeleniya Matematicheskogo Instituta AN SSSR*, 1979, **90**, p. 229-264 (in Russian).
180. Shakhbazyan K.V. On scheduling problems of the $n|1||\sum C_i(t)$ type. *Zapiski Nauchnukh Seminarov Leningradskogo Otdeleniya Matematicheskogo Instituta AN SSSR*, 1980, **102**, p. 147-155 (in Russian).
181. Shakhbazyan K.V., Lebedinskaya N.B. Minimization of the total cost in the problem of parallel ordering of independent tasks. *Sov. Math. Doklady*, 1977, **18**, p. 1503-1506.
182. Shakhbazyan K.V., Lebedinskaya N.B. Effective optimization methods for single-machine scheduling (survey). *J. Sov. Math.*, 1984, **24**, p. 133-148.
183. Shakhbazyan K.V., Tushkina T.A. The branch and bound method for parallel ordering problem. *Zapiski Nauchnukh Seminarov Leningradskogo Otdeleniya Matematicheskogo Instituta AN SSSR*, 1973, **35**, p. 146-155 (in Russian).
184. Shkurba V.V. The queuing intervals in sequencing problems. *Kibernetika* (Kiev), 1970, N 2, p. 77-79 (in Russian).
185. Shkurba V.V., Podchasova T.P., Pshichuk A.N., Tur L.P. *Production Scheduling Problems and Methods for Their Solution*. Kiev, Naukova Dumka, 1966, 156 pp. (in Russian).
186. Shreider Y.A. *Equality, Similarity, Order*. Moscow, Nauka Publishers, 1971, 256 pp. (in Russian).
187. Abdel-Wahab H.M., Kameda T. Scheduling to minimize maximum cumulative cost subject to series-parallel precedence constraints. *Oper. Res.*, 1978, **26**, N 1,

p.141-158.

188. Adolphson D.L. Single machine job sequencing with precedence constraints. *SIAM J. Comput.*, 1977, **6**, N 1, p. 40-54.

189. Adolphson D.L., Hu T.C. Optimal linear ordering. *SIAM J. Appl. Math.* 1973, **25**, N 3, p. 403-423.

190. Adrabinski A., Wodecki M. An algorithm for solving the machine sequencing problem with parallel machines. *Zast. Matem.*, 1979, **16**, N 3, p. 513-541.

191. Agin N. Optimum seeking with branch and bound. *Manag. Sci.*, 1966, **13**, N 4, p. B175-B185.

192. Ashour S. Sequencing theory. *Lect. Notes Econ. Math. Syst.*, 1972, **69**, 133 pp.

193 Baker K.R. *Introduction to Sequencing and Scheduling.* New York, Wiley, 1974, 305 pp.

194. Baker K.R., Lawler E.L., Lenstra J.K., Rinnooy Kan A.H.G. Preemptive scheduling of a single machine to minimize maximum cost subject to release dates and precedence constraints. *Oper. Res.*, 1983, **31**, N2, p. 381-386.

195. Baker K.R., Martin J.B. An experimental comparison of solution algorithms for single-machine tardiness problem. *Naval Res. Log. Quart.*, 1974, **21**, N 1, p. 187-199.

196. Baker K.R., Merten A.G. Scheduling with parallel processors and linear delay costs. *Naval Res. Log. Quart.*, 1973, **20**, N 4, p. 793-804.

197. Baker K.R., Nuttle H.L.W. Sequencing independent jobs with a single resource. *Naval Res. Log. Quart.*, 1980, **27**, N 3, p. 499-510.

198. Baker K.R., Schrage L.E. Finding an optimal sequence by dynamic programming: an extension to precedence-related tasks. *Oper. Res.*, 1978, **26**, N 1, p. 111-120.

199. Baker K.R., Su Z.S. Sequencing with due-dates and early start times to minimize maximum tardiness. *Naval Res. Log. Quart.*, **21**, N 1, p. 171-176.

200. Balas E., Guignard M. Report of the session on branch and bound/implicit enumeration. *Ann. Discr. Math.*, 1979, **5**, p. 185-191.

201. Bank B. "Branch and Bound" - Algorithmen fur zwei Reihenfolgeprobleme. *Math. Operationsforsch. Statist.*, 1970, **1**, N 3, p. 217-228.

202. Bansal S.P. Single machine scheduling to minimize weighted sum of completion times with secondary criterion - a branch and bound approach. *Eur. J. Oper. Res.*, 1980, **5**, N 3, p. 177-181.

203. Barnes J.W., Brennan J.J. An improved algorithm for scheduling jobs on identical machines. *AIIE Trans.*, 1977, **9**, N 1, p. 25-31.

204. Bellman R. Mathematical aspects of scheduling theory. *SIAM J.*, 1956, **4**, N 3, p. 168-205.

205. Bianco L., Nicoletti B., Ricciardelli S. An algorithm for optimal sequencing of aircraft in the near terminal area. *Lect. Notes Inform. Sci.*, 1978, **7**, p. 443-453.

206. Bianco L., Ricciardelli S. On scheduling with ready times to minimize total weighted completion time. In: Contr. Sci. and Technol. Progr. Soc. Proc. 8th Trienni. World Congr. Int. Fed. Autom. Contr., Kyoto, 24-28 Aug., 1981, Vol. 1. Oxford e.a., 1982, p.501-505.

207. Blau R.A. *N*-job, one machine sequencing problems under uncertainty. *Manag. Sci.*, 1973, **20**, N 1, p. 101-109.

208 Blazewicz J. Deadline scheduling of tasks - a survey. *Found. Contr. Engin.*, 1976, **1**, N 4, p. 203-216.

209. Blazewicz J. Complexity of computer scheduling algorithms under resource constraints. *Found. Contr. Engin.*, **3**, N 2, p. 51-57.

210. Blazewicz J. Deadline scheduling of tasks with ready times and resource constraints. *Inform. Process. Lett.*, 1979, **8**, N 2, p. 60-63.

211. Blazewicz J., Lenstra J.K., Rinnooy Kan A.H.G. Scheduling subject to resource constraints: classification and complexity. *Discr. Appl. Math.*, 1983, **5**, N 1. p. 11-24.

212. Blazewicz J., Weglarz J. Deterministyczne problemy szeregowania zadan na procesorach systemow komputerowych z uwzglednieniem dodatkowych zasobow. *Arch. automat. telemechan.*, 1978, **23**, N 4, p. 485-509.

213. Bozoki G., Richard J.P. A branch-and-bound algorithm for the continuous-process job-shop scheduling problem. *AIIE Trans.*, 1970, **2**, N 3, p. 246-252.

214. Bratley P., Florian M., Robillard P. Scheduling with earliest, start and due date constraints. *Naval Res. Log. Quart.*, 1971, **18**, N 3, p. 511-519.

215. Bratley P., Florian M., Robillard P. Scheduling with earliest and due date constraints on multiple machines. *Naval Res Log. Quart.*, 1975, **22**, N 1, p. 165-173.

216. Brucker P., Garey M.R., Johnson D.S. Scheduling equal-length tasks under treelike precedence constraints to minimize maximum lateness. *Math. Oper. Res.*, 1977, **2**, N 3, p. 275-284.

217. Brucker P., Lenstra J.K., Rinnooy Kan A.H.G. Complexity of machine scheduling problems. Report BW 43, Mathematische Centrum, Amsterdam, 1975, 29 pp.

218. Bruno J. Mean weighted flow-time criterion. In: Computer and Job-Shop Scheduling

Theory, Coffman E.G., Jr. (ed.), New York, Wiley, 1976, p. 101-137.

219. Bruno J., Coffman E.G., Jr., Sethi R. Algorithms for minimizing mean flow time. In: Inform. Process. 74, Amsterdam-London, 1974, p. 504-510.

220. Bruno J., Coffman E.G., Jr., Sethi R. Scheduling independent tasks to reduce mean finishing time. *Comm. ACM*, 1974, **17**, N 7, p. 382-387.

221. Bruno J., Hofri M. On scheduling chains of jobs on one processor with limited preemption. *SIAM J. Comput.*, 1975, **4**, N 4, p. 478-490.

222. Bruno J., Sethi R. Task sequencing in a batch environment with setup times. In: Modelling and Performance Eval. Comput. Syst., Amsterdam, 1977, p. 81-88.

223. Bulfin R.L., Parker R.G. On a two facility scheduling problem with sequence dependent processing time. *AIIE Trans.*, 1976, **8**, N 2, p. 202-209.

224. Burns R.N. Scheduling to minimize the weighted sum of completion times with secondary criteria. *Naval Res. Log. Quart.*, 1976, **23**, N 1, p. 125-129.

225. Burns R.N., Steiner G. Single machine scheduling with series-parallel precedence constraints. *Oper. Res.*, 1981, **29**, N 6, p. 1195-1207.

226. Buzacott J.A., Dutta S.K. Sequencing many jobs on a multipurpose facility. *Naval Res. Log. Quart.*, 1971, 18, N 1, p. 75-82.

227. Carlier J. The one-machine sequencing problem. *Eur. J. Oper. Res.*, 1982, **11**, N 1, p. 42-47.

228. Chandra A.K., Wong C.K. Worst-case analysis of a placement algorithm related to storage allocation. *SIAM J. Comput.*, 1975, **4**, N 3, p. 249-263.

229. Chandra R. On $n/1/\bar{F}$ dynamic deterministic problems. *Naval Res. Log. Quart.*, 1979, **26**, N 3, p. 537-544.

230. Charlton J.M., Death C.C. A method of solution for general machine-scheduling problems. *Oper. Res.*, 1970, **18**, N 4, p. 689-707.

231. Chen N.F., Liu C.L. On a class of scheduling algorithms for multiprocessors computing systems. *Lect. Notes Comput. Sci.*, 1975, **24**, p. 1-16.

232. Cho Y., Sahni S. Bounds for list schedules on uniform processors. *SIAM J. Comput.*, 1980, 9, N 1, p. 91-103.

233. Coffman E.G., Jr., Garey M.R., Johnson D.S. An application of bin-packing to multiprocessor scheduling. *SIAM J. Comput.*, 1978, **7**, N 1, p. 1-17.

234. Coffman E.G., Jr., Graham R.L. Optimal scheduling for two-processor systems. *Acta Inf.*, 1972, **1**, N 3, p. 200-213.

235. Coffman E.G., Jr., Labetoulle J. Scheduling to minimize mean number in system. In: Proc. 9th Haw. Int. Conf. Syst. Sci., Honolulu, 1976, p. 119-121.

236. Coffman E.G., Jr., Sethi R. A generalized bound on LPT sequencing. In: Int. Symp. Comput. Perform. Model., Meas. and Eval., Cambridge, 1976, New York, 1976, p. 306–310.

237. Coffman E.G., Jr., Sethi R. A generalized bound on LPT sequencing. *Rev. franc. automat. inform. rech. oper.*, 1976, **10**, N 5, p. 17–25.

238. Coffman E.G., Jr., Sethi R. Algorithms minimizing mean flow time: schedule-length properties. *Acta Inf.*, 1976, **6**, p. 1–14.

239. *Computer and Job–Shop Scheduling Theory*. Coffman E.G., Jr. (ed.), New York, Wiley, 1976, 299 pp.

240. Davida G.L., Linton D.J. A new algorithm for the scheduling of tree structured tasks. In: 1976 Conf. Inform. Sci. and Syst., Baltimore, 1976, p. 543–548.

241. Davis E., Jaffe J.M. Algorithms for scheduling tasks on unrelated processors. *J. Assoc. Comput. Mach.*, 1981, **28**, N 4, p. 721–736.

242. Davis K.R., Walters J.E. Addressing the N/1 scheduling problem – a heuristic approach. *Comput. Oper. Res.*, 1977, **4**, N 2, p. 89–100.

243. Day J., Hottenstein M.P. Review of scheduling research. *Naval Res. Log. Quart.*, 1970, **17**, N 1, p. 11–39.

244. De P., Morton T.E. Scheduling to minimize maximum lateness on unequal parallel processors. *Comput. Oper. Res.*, 1982, **9**, N 3, p. 221–232.

245. Dessouky M.I., Deogun J.S. Sequencing jobs with unequal ready times to minimize mean flow time. *SIAM J. Comput.*, 1981, **10**, N 1, p. 192–202.

246. Dessouky M.I., Larson R.R. Symmetry and optimality properties of the single machine problem. *AIIE Trans.*, 1978, **10**, N 2, p. 170–175.

247. Dessouky M.I., Margenthaler C.R. The one-machine sequencing problem with early starts and due dates. *AIIE Trans.*, 1972, **4**, N 3, p. 214–222.

248. Duffin R.J. Topology of series-parallel network. *J. Math. Anal. and Appl.*, 1965, **10**, N 2, p. 303–318.

249. Eastman W.L., Even S., Isaacs I.M. Bounds for the optimal scheduling of n jobs on m processors. *Manag. Sci.*, 1964, **11**, N 2, p. 268–279.

250. Eilon S., Chowdhury I.G. Minimizing waiting time variance in the single machine problem. *Manag. Sci.*, 1977, **23**, N 6, p. 567–575.

251. Elmaghraby S.E. The sequencing of related jobs. *Naval Res. Log. Quart.*, 1968, **15**, N 1, p. 23–32.

252. Elmaghraby S.E. The one machine sequencing problem with delay costs. *J. Industr. Eng.*, 1968, **19**, N 2, p. 105–108.

253. Elmaghraby S.E., Park S.H. Scheduling jobs on a number of identical machines. *AIIE Trans.*, 1974, **6**, N 1, p. 1–13.

254. Emmons H. One-machine sequencing to minimize certain functions of job tardiness. *Oper. Res.*, 1969, **17**, N 4, p. 701–715.

255. Emmons H. One machine sequencing to minimize mean flow time with minimum number tardy. *Naval Res. Log. Quart.*, 1975, **22**, N 3, p. 585–592.

256. Emmons H. A note on a scheduling problem with dual criteria. *Naval Res. Log. Quart.*, 1975, **22**, N 3, p. 615–616.

257. Erschler J., Fontan G., Merce C., Roubellat F. Applying new dominance concepts to job scheduling optimization. *Eur. J. Oper. Res.*, 1982, **11**, N 1, p. 60–66.

258. Erchler J., Roubellat F., Vernhes J.P. Characterizing the set of feasible sequences for n jobs to be carried out on a single machine. *Eur. J. Oper. Res.*, 1980, **4**, N 3, p. 189–194.

259. Fernandez E.B., Lang T. Scheduling as graph transformation. *IBM J. Res. Develop.*, 1976, **20**, N 6, p. 551–559.

260. Fischer M.J., Meyer A.R. Boolean matrix multiplication and transitive closure. In: Conf. Rec. 12th Annu. IEEE Symp. Switch. and Automata Theory, New York, 1971, p. 129–131.

261. Fisher M.L. A dual algorithm for the one-machine scheduling problem. *Math. Program.*, 1976, **11**, N 3, p. 229–251.

262. Fisher M.L., Jaikumar R. An algorithm for the space-shuttle scheduling problem. *Oper. Res.*, 1978, **26**, N 1, p. 166–182.

263. Fujii M. Erratum: Optimal sequencing of two equivalent processors. *SIAM J. Appl. Math.*, 1971, **20**, N 1, p. 141.

264. Fujii M., Kasami T., Ninomija K. Optimal sequencing of two equivalent processors. *SIAM J. Appl. Math.*, 1969, **17**, N 4, p. 784–789.

265. Fung K.T. On a restrictive scheduling scheme. *INFOR. Can. J. Oper. Res. Inform. Process.*, 1978, **16**, N 3, p. 288–293.

266. Garey M.R. Optimal task sequencing with precedence constraints. *Discr. Math.*, 1973, **4**, N 1, p. 37–56.

267. Garey M.R., Graham R.L. Bounds on scheduling with limited resources. *Oper. Syst. Rev.*, 1973, **7**, N 4, p. 104–111.

268. Garey M.R., Graham R.L. Bounds for multiprocessor scheduling with resource constraints. *SIAM J. Comput.*, 1975, **4**, N 2, p. 187–200.

269. Garey M.R., Graham R.L., Johnson D.S., Yao A.C.C. Resource constrained scheduling

as generalized bin packing. *J. Combin. Theory (A)*, 1976, **21**, N 3, p. 257–298.

270. Garey M.R., Graham R.L., Ullman J.D. Worst-case analysis of memory allocation algorithms. In: 4th Annu. ACM Symp. Theory Comput., Denver, 1972, p. 143–150.

271. Garey M.R., Johnson D.S. Complexity results for multi-processor scheduling under resource constraints. *SIAM J. Comput.*, 1975, **4**, N 4, p. 397–411.

272. Garey M.R., Johnson D.S. Scheduling tasks with nonuniform deadlines on two processors. *J. Assoc. Comput. Mach.*, 1976, **23**, N 3, p. 461–467.

273. Garey M.R., Johnson D.S. Approximation algorithms for combinatorial problems: an annotated bibliography. In: Algorithms and Complexity. New Dir. and Recent Results. Proc. Symp. Carnegie-Mellon Univ., 1976, New York, 1976, p. 41–52.

274. Garey M.R., Johnson D.S. Two-processor scheduling with start-times and deadlines. *SIAM J. Comput.*, 1977, **6**, N 3, p. 416–426.

275. Garey M.R., Johnson D.S. "Strong" *NP*-completeness results: motivation, examples, and implications. *J. Assoc. Comput. Mach.*, 1978, **25**, N 3, p. 499–508.

276. Garey M.R., Johnson D.S., Simons B.B., Tarjan R.E. Scheduling unite-time tasks with arbitrary release times and deadlines. *SIAM J. Comput.*, 1981, **10**, N 2, p. 256–269.

277. Garey M.R., Johnson D.S., Stockmeyer L. Some simplified graph problems. *Theor. Comput. Sci.*, 1976, **1**, N 3, p. 237–267.

278. Gens G.V., Levner E.V. Computational complexity of approximation algorithms for combinatorial problems. In: Proc. 8th Symp. on Math. Foundation of Comput. Sci., Berlin, 1979, p. 292–300.

279. Glazebrook K.D. On single-machine sequencing with order constraints. *Naval Res. Log. Quart.*, 1980, **27**, N 1, p. 123–130.

280. Glazebrook K.D., Gittins J.C. On single-machine scheduling with precedence relations and linear or discounted costs. *Oper. Res.*, 1981, **29**, N 1, p. 161–173.

281 Gonzalez M.J.,Jr. Deterministic processor scheduling. *Comput. Surv.*, 1977, **9**, N3, p. 173–204.

282. Gonzalez T.F. Optimal mean finish time preemptive schedules. Technical Report 220, Computer Science Department, Pennsylvania State University. 1977.

283. Gonzalez T.F., Ibarra O.H., Sahni S. Bounds for LPT schedules on uniform processors. *SIAM J. Comput.*, 1977, **6**, N 1, p. 155–166.

284. Gonzalez T.F., Johnson D.B. A new algorithm for preemptive scheduling of trees. *J. Assoc. Comput. Mach.*, 1980, **27**, N 2, p. 287–312.

285. Gonzalez T.F., Sahni S. Open shop scheduling to minimize finish time. *J. Assoc. Comput. Mach.*, 1976, **23**, N 4, p. 665–679.

286. Gonzalez T.F., Sahni S. Preemptive scheduling of uniform processor systems. *J. Assoc. Comput. Mach.*, 1978, **25**, N 1, p. 92-101.

287. Grabowski J. Formulation and solution of sequencing problem with parallel machines. *Lect. Notes Contr. Inform. Sci.*, 1978, **7**, p. 400-410.

288. Graham R.L. Bounds for certain multiprocessing anomalies. *Bell Syst. Tech. J.*, 1966, **45**, N 9, p. 1563-1581.

289. Graham R.L. Bounds on multiprocessing timing anomalies. *SIAM J. Appl. Math.*, 1969, **17**, N 2, p. 416-429.

290. Graham R.L., Lawler E.L., Lenstra J.K., Rinnooy Kan A.H.G. Optimization and approximation in deterministic sequencing and scheduling: a survey. *Ann. Discr. Math.*, 1979, **5**, p. 287-326.

291. Heck H., Roberts S. A note on the extension of a result on scheduling with secondary criteria. *Naval Res. Log. Quart.*, 1972, **19**, N 2, p. 403-405.

292. Hempel L. Einege Bemerkungen zum Branch-and-Bound Prinzip und seiner Anwendung sur Losung von Reihenfolgeproblemen. In: 21. Int. Wiss. Kolloq. Techn. Hochsh., Ilmenau, 1976, Ht. 3, p. 131-134.

293. Horn W.A. Single-machine job sequencing with treelike precedence ordering and linear delay penalties. *SIAM J. Appl. Math.*, 1972, **23**, N 2, p. 189-202.

294. Horn W.A. Minimizing average flow time with parallel machines. *Oper. Res.*, 1973, **21**, N 3, p. 846-847.

295. Horn W.A. Some simple scheduling algorithms. *Naval Res. Log. Quart.*, 1974, **21**, N 1, p. 177-185.

296. Horowitz E., Sahni S. Exact and approximate algorithms for scheduling nonidentical processors. *J. Assoc. Comput. Mach.*, 1976, **23**, N 2, p. 317-326.

297. Horvath E.C., Lam S., Sethi R. A level algorithm for preemptive scheduling. *J. Assoc. Comput. Mach.*, 1977, **24**, N 1, p. 32-43.

298. Hsu N.C. Elementary proof of Hu's theorem on isotone mappings. *Proc. Amer. Math. Soc.*, 1966, **17**, N 1, p. 111-114.

299. Ibaraki T. On the computational efficiency of branch-and-bound algorithms. *J. Oper. Res. Soc. Jap.*, 1977, **20**, N 1, p. 16-35.

300. Ibaraki T. Branch-and-bound procedure and state-space representation of combinatorial optimization problems. *Inform. and Contr.*, 1978, **36**, N 1, p. 1-27.

301. Ibarra O.H., Kim C.E. Fast approximation algorithms for the knapsack and sum of subset problems. *J. Assoc. Comput. Mach.*, 1975, **22**, N 4, p. 463-468.

302. Ibarra O.H., Kim C.E. Heuristic algorithms for scheduling independent tasks on

nonidentical processors. *J. Assoc. Comput. Mach.*, 1977, **24**, N 2, p. 280–289.

303. Integer programming and related areas. A classified bibliography. *Lect. Notes Econ. Math. Syst.*, 1976, **128**; 1978, **160**; 1982, **197**.

304. Jackson J.R. Scheduling a production line to minimize maximum tardiness. Res. Report 43, Manag. Sci. Res. Project, UCLA, 1955.

305. Jaeschke G. Vicinal sequencing problems. *Oper. Res.*, 1972, **20**, N 5, p. 984–992.

306. Jaffe J.M. An analysis of preemptive multiprocessor job scheduling. *Math. Oper. Res.*, 1980, **5**, N 3, p. 415–421.

307. Johnson D.S. Fast algorithms for bin packing. *J. Comput. Syst. Sci.*, 1974, **8**, N 3, p. 272–314.

308. Johnson D.S., Demers A., Ullman J.D., Garey M.R., Graham R.L. Worst-case performance bounds for simple one-dimensional packing algorithms. *SIAM J. Comput.*, 1974, **3**, N 4, p. 299–325.

309. Johnson S.M. Discussion: sequencing n-jobs on two machines with arbitrary time lags. *Manag. Sci.*, 1959, **5**, N 3, p. 299–303.

310. Kafura D.G., Shen V.Y. Task scheduling on a multiprocessor system with independent memories. *SIAM J. Comput.*, 1977, **6**, N 1, p. 167–187.

311. Kanet J.J. Minimizing the average deviation of job completion times about a common due date. *Naval Res. Log. Quart.*, 1981, **28**, N 4, p. 643–651.

312. Kao E.P.C. Multiple objective decision theoretic approach to one-machine scheduling problems. *Comput. Oper. Res.*, 1980, **7**, N 3, p. 251–260.

313. Kaufman M.T. An almost-optimal algorithm for the assembly line scheduling problem. *IEEE Trans. Comput.*, 1974, **23**, N 11, p. 1169–1174.

314. King J.R., Spachis A.S. Scheduling: bibliography and review. *Int. J. Physical Distribution and Materials Manag.*, 1980, **10**, N3, p. 100–132.

315. Kise H. Scheduling and branch-and-bound method. *Syst. and Contr.*, 1973, **17**, N 5, p. 275–281.

316. Kise H., Ibaraki T., Mine H. A solvable case of the one-machine scheduling problem with ready and due-times. *Oper. Res.*, 1978, **26**, N 1, p. 121–126.

317. Kise H., Ibaraki T., Mine H. Performance analysis of six approximation algorithms for the one-machine maximum lateness scheduling problem with ready times. *J. Oper. Res. Soc. Jap.*, 1979, **22**, N 3, p. 205–224.

318. Köhler W.H. A preliminary evaluation of the critical path method for scheduling tasks on multiprocessor systems. *IEEE Trans. Comput.*, 1975, **24**, N 12, p. 1235–1238.

319. Köhler W.H., Steiglitz K. Characterization and theoretical comparison of branch-and-bound algorithms for permutation problems. *J. Assoc. Comput. Mach.*, 1974, **21**, N 1, p. 140-156.

320. Kolnikova Z. Dve metody riesenia problemov usporiadania s danymi podmienkami pre predchadzanie. *Ekon. mat. obz.*, 1970, **6**, N 1, p. 56-72.

321. Krause K.L., Shen V.Y., Schwetman H.D. Analysis of several task-scheduling algorithms for a model of multiprogramming computer systems. *J. Assoc. Comput. Mach.*, 1975, **22**, N 4, p. 522-550.

322. Kurisu T. Two-machine scheduling under required precedence among jobs. *J. Oper. Res. Soc. Jap.*, 1976, **19**, N 1, p. 1-13.

323. Kurisu T. Two-machine scheduling under arbitrary precedence constraints. *J. Oper. Res. Soc. Jap.*, 1977, **20**, N 2, p. 113-131.

324. Labetoulle J., Lawler E.L., Lenstra J.K., Rinnooy Kan A.H.G. Preemptive scheduling of uniform machines subject to release dates. Preprint BW 99, Mathematische Centrum, Amsterdam, 1979, 19 pp.

325. Lageweg B.J., Lawler E.L., Lenstra J.K., Rinnooy Kan A.H.G. Computer aided complexity classification of deterministic scheduling problems. Preprint BW 138, Mathematische Centrum, Amsterdam, 1981. 20 pp.

326. Lageweg B.J., Lenstra J.K., Rinnooy Kan A.H.G. Minimizing maximum lateness on one machine: computational experience and some applications. *Statist. Neerlandica*, 1976, **30**, N 1. p. 25-41.

327. Lakshminarayan S., Lakshmanan R., Papineau R.L., Rochette R. Optimal single-machine scheduling with earliness and tardiness penalties. *Oper. Res.*, 1978, **26**, N 6, p. 1079-1082.

328. Lam S., Sethi R. Worst case analysis of two scheduling algorithms. *SIAM J. Comput.*, 1977, **6**, N 3, p. 518-536.

329. Lang T., Fernandez E.B. Scheduling of unit-length independent tasks with execution constraints. *Inf. Process. Lett.*, 1976,**4**, N 4, p. 95-98.

330. Larson R.E., Dessouky M.I. Heuristic procedures for the single machine problem to minimize maximum lateness. *AIIE Trans.*, 1978, **10**, N 2, p. 176-183.

331. Lawler E.L. On scheduling problems with deferral costs. *Manag. Sci.*, 1964, **11**, N 2, p. 280-288.

332. Lawler E.L. Optimal sequencing of a single machine subject to precedence constraints. *Manag. Sci.*, 1973, **19**, N 5, p. 544-546.

333. Lawler E.L. Optimal sequencing of jobs subject to series-parallel precedence

constraints. Preprint BW 54, Mathematische Centrum, Amsterdam, 1975, 15 pp.

334. Lawler E.L. Sequencing to minimize the weighted number of tardy jobs. *Rev. franc. automat., inform., rech. oper.*, 1976, **10**, N 5. p. 27–33.

335. Lawler E.L. A "pseudopolynomial" algorithm for sequencing jobs to minimize total tardiness. *Ann. Discr. Math.*, 1977, **1**, p. 331–342.

336. Lawler E.L. Sequencing jobs to minimize total weighted completion time subject to precedence constraints. *Ann. Discr. Math.*, 1978, **2**, p. 75–90.

337. Lawler E.L. Preemptive scheduling of uniform parallel machines to minimize the weighted number of late jobs. Preprint BW 105, Mathematische Centrum, Amsterdam, 1979, 20 p.

338. Lawler E.L. Preemptive scheduling of precedence-constrained jobs on parallel machines. In: Deterministic and Stochastic Scheduling. Proc. NATO Adv. Study and Res. Inst. Theor. Approaches Scheduling Probl., Dordrecht, 1982, p. 101–123.

339. Lawler E.L., Labetoulle J. On preemptive scheduling of unrelated parallel processors by linear programming. *J. Assoc. Comput. Mach.*, 1978, **25**, N 4, p. 612–619.

340. Lawler E.L., Lenstra J.K. Machine scheduling with precedence constraints. Preprint BW 148, Mathematische Centrum, Amsterdam, 1981, 18 pp.

341. Lawler E.L., Lenstra J.K., Rinnooy Kan A.H.G. Recent developments in deterministic sequencing and scheduling: a survey. In: Deterministic and Stochastic Scheduling. Proc. NATO Adv. Study and Res. Inst. Theor. Approaches Scheduling Probl., Dordrecht, 1982, p. 35–73.

342. Lawler E.L., Moore J.M. A functional equation and its application to resource allocation and sequencing problems. *Manag. Sci.*, 1969, **16**, N 1, p. 77–84.

343. Lawler E.L., Sivazlian B.D. Minimization of time-varying costs in single-machine scheduling. *Oper. Res.*, 1978, **26**, N 4, p. 563–569.

344. Lawler E.L., Wood D.E. Branch-and-bound methods: a survey. *Oper. Res.*, 1966, **14**, N 4, p. 699–719.

345. Lenstra J.K. *Sequencing by Enumerative Methods.* Amsterdam, Math. Center Tracts, **69**, 1977, 202 pp.

346. Lenstra J.K., Rinnooy Kan A.H.G. Complexity of scheduling under precedence constraints. *Oper. Res.* 1978, **26**, N 1, p. 22–35.

347. Lenstra J.K., Rinnooy Kan A.H.G. An introduction to multiprocessor scheduling. Preprint BW 121, Mathematische Centrum, Amsterdam, 1980, 18 pp.

348. Lenstra J.K., Rinnooy Kan A.H.G. Complexity results for scheduling chains on a

single machine. *Eur. J. Oper. Res.*, 1980, **4**, N 4, p. 270–275.

349. Lenstra G.K., Rinnooy Kan A.H.G., Brucker P. Complexity of machine scheduling problems. *Ann. Discr. Math.*, 1977, **1**, p. 343-362.

350. Leung J.Y.T. Bounds on list scheduling of UET tasks with restricted resource constraints. *Inf. Process. Lett.*, 1979, **9**, N 4, p. 167–170.

351. Liu J.W.S., Liu C.L. Bounds on scheduling algorithms for heterogeneous computing systems. In: Proc. 1974 IFIP Congress. Amsterdam, 1974, p. 349-353.

352. Liu J.W.S., Liu C.L. Performance analysis of heterogeneous multiprocessor computing systems. In: Comput. Archit. and Networks, Amsterdam–Oxford, 1974, p. 331–343.

353. Martel C. Preemptive scheduling with release times, deadlines and due times. *J. Assoc. Comput. Mach.*, 1982, **29**, N 3, p. 812–829.

354. Maxwell W.L. On sequencing n jobs on one machine to minimize the number of late jobs. *Manag. Sci.*, 1970, **16**, N 5, p. 295–297.

355. McMahon G., Florian M. On scheduling with ready time and due dates to minimize maximum lateness. *Oper. Res.*, 1975, **23**, N 3, p. 475–482.

356. McNaughton R. Scheduling with deadlines and loss functions. *Manag. Sci.*, 1956, **6**, N 1, p. 1–12.

357. Mehta S., Chandrasekaran R., Emmons H. Order-preserving allocation of jobs to two machines. *Naval Res. Log. Quart.*, 1974, **21**, N 2, p. 361–364.

358. Merten A.G., Muller M.E. Variance minimization in single machine sequencing problems. *Manag. Sci.*, 1972, **18**, N 9, p. 518–528.

359. Mitten L.G. Sequencing n jobs on two machines with arbitrary time lags. *Manag. Sci.*, 1959, **5**, N 3, p. 293–298.

360. Mitten L.G. Branch-and-bound methods: general formulation and properties. *Oper. Res.* 1970, **18**, N 1, p. 24–34.

361. Miyazaki S. One machine scheduling problem with dual criteria. *J. Oper. Res. Soc. Jap.*, 1981, **24**, N 1, p. 37–50.

362. Monma C.L. The two-machine maximum flow time problem with series-parallel precedence constraints: an algorithm and extensions. *Oper. Res.* 1979, **27**, N 4, p. 792–798.

363. Monma C.L. Sequencing to minimize the maximum job cost. *Oper. Res.*, 1980, **28**, N 4, p. 942–951.

364. Monma C.L. Sequencing with general precedence constraints. *Discr. Appl. Math.*, 1981, **3**, N 2, p. 137–150.

365. Monma C.L., Sidney J.B. Sequencing with series-parallel precedence constraints.

Math. Oper. Res., 1979, **4**, N 3, p. 215-224.

366. Moore J.M. An n job, one machine sequencing algorithm for minimizing the number of late jobs. *Manag. Sci.*, 1968 , **15**, N 1, p. 102-109.

367. Moore J.M. An algorithm for a single machine scheduling problem with sequence dependent setup times and scheduling windows. *AIIE Trans.*, 1975, **7**, N 1, p. 35-41.

368. Müller-Merbach H. *Optimale Reinfolgen.* (Ökon. und Unternehmensforsch., 15.), Berlin-Heidelberg-New York, Springer, 1970, **10**, 225 p.

369. Muntz R.R., Coffman E.G., Jr. Optimal preemptive scheduling on two-processor systems. *IEEE Trans. Comput.*, 1969, **18**, N 11, p. 1014-1020.

370. Muntz R.R., Coffman E.G., Jr. Preemptive scheduling of real-time tasks on multiprocessor systems. *J. Assoc. Comput. Mach.*, 1970, **17**, N 2, p. 324-338.

371. Nabeshima I. Sequencing on two machines with start lag and stop lang. *J. Oper. Res. Soc. Jap.*, 1963, **5**, N 3, p. 97-101.

372. Nakamura N., Yoshida T., Hitomi K. Group production scheduling for minimum total tardiness. Part 1. *AIIE Trans.*, 1978, **10**, N 2, p. 157-162.

373. Nèmeti L. On the scheduling problem in the case of several machines of the same type. *Mathematica* (RSR), 1971, **13**, N 2, p. 251-261.

374. Panwalkar S.S., Iskander W. A survey of scheduling rules. *Oper. Res.*, 1977, **25**, N 1, p. 45-61.

375. Papadimitriou C.H. The euclidian traveling salesman problem is *NP*-complete. *Theor. Comput. Sci.*, 1977, **4**, N 3, p. 237-244.

376. Petersen C.C. Solving sequencing problems through reordering operations. *AIIE Trans.*, 1973, **5**, N 1, p. 68-73.

377. Potrzebowski H. Planowanie kalendarzowe (przeglad zagadnien). In: Pr. Inst. cybern. stosowan. PAN, 1973, N 13, 82 pp.

378. Potts C.N. An algorithm for the single machine sequencing problem with precedence constraints. In: Comb. Optimiz. 2, Proc. CO'79 Conf., Amsterdam, 1980, p. 78-87.

379. Potts C.N. Analysis of a heuristic for one machine sequencing with release dates and delivery times. *Oper. Res.*, 1980, **28**, N 6, p. 1436-1441.

380. Rajaraman M.K. A parallel sequencing algorithm for minimizing total cost. *Naval Res. Log. Quart.*, 1977, **24**, N 3, p. 473-481.

381. Rau J.G. Minimizing a function of permutations of n integers. *Oper. Res.*, 1971, **19**, N 1, p. 237-240.

382. Rinnooy Kan A.H.G. The machine scheduling problem. Preprint BW 27, Mathematische Centrum, Amsterdam, 1973. 153 pp.

383. Rinnooy Kan A.H.G., Lageweg B.J., Lenstra J.K. Minimizing total costs in one-machine scheduling. In: Combinatorial Programm.: Meth. and Appl., Dordrecht-Boston, 1975, p. 343-350.

384. Rinnooy Kan A.H.G. *Machine Scheduling Problems: Classification, Complexity and Computations*. The Hague, Hijhoff, 1976.

385. Root J.G. Scheduling with deadlines and loss functions on k parallel machines. *Manag. Sci.*, 1965, **11**, N 3, p. 460-475.

386. Rothkopf M.H. Scheduling independent tasks on parallel processors. *Manag. Sci.*, 1966, **12**, N 5, p. 437-447.

387. Rothkopf M.H. Scheduling with random service time. *Manag. Sci.*, 1966, **12**, N 9, p. 707-713.

388. Rothkopf M.H. A note on allocating jobs to two machines. *Naval Res. Log. Quart.*, 1975, **22**, N 4, p. 829-830.

389. Rowicki A. On optimal preemptive scheduling of independent tasks for multiprocessor systems. *Bull. Acad. pol. sci. Ser. sci. math., astron. et phys.*, 1978, **26**, N 7, p. 643-650.

390. Rowicki A. On the optimal preemptive scheduling for multiprocessor system. *Bull. Acad. pol. sci. Ser. sci. math., astron. et. phys.*, 1978, **26**, N 7, p. 651-660.

391. Roy B. Procedure d'exploration par separation et evaluation. *Rev. franc. inform. rech. oper.*, 1969, **3**, N V-1, p. 61-90.

392. Sahni S. Computationally related problems. *SIAM J. Comput.*, 1974, **3**, N 4, p. 262-279.

393. Sahni S. Algorithms for scheduling independent tasks. *J. Assoc. Comput. Mach.*, 1976, **23**, N 1, p. 116-127.

394. Sahni S. General techniques for combinatorial approximation. *Oper. Res.*, 1977, **25**, N 6, p. 920-936.

395 Sahni S. Preemptive scheduling with due-dates. *Oper. Res.*, 1979, **27**, N 5, p. 925-934.

396. Sahni S., Cho Y. Nearly on line scheduling of a uniform processor system with release times, *SIAM J. Comput.*, 1979, **8**, N 2, p. 275-278.

397. Sahni S., Cho Y. Scheduling independent tasks with due times on a uniform processor system. *J. Assoc. Comput. Mach.*, 1980, **27**, N 3, p. 550-563.

398. Schild A., Fredman I.J. On scheduling tasks with associated linear loss functions. *Manag. Sci.*, 1961, **7**, N 3, p. 280-285.

399. Schild A., Fredman I.J. Scheduling tasks with deadlines and non-linear functions.

Manag. Sci., 1962, **8**, N 1, p. 73-81.

400. Schindler S. Scheduling general monitor systems. In: Proc. 9th Haw. Int. Conf. Syst. Sci., Honolulu, 1976. p. 122-124.

401. Schrage L. Minimizing the time-in-system variance for a job-set. *Manag. Sci.*, 1975, **21**, N 5, p. 540-543.

402. Schrage L., Baker K.R. Dynamic programming solution of sequencing problems with precedence constraints. *Oper. Res.*, 1978, **26**, N 3, p. 444-449.

403 Sethi R. Scheduling graphs on two processors. *SIAM J. Comput.*, 1976, **5**, N 1, p. 73-82.

404. Sethi R. Algorithms for minimal-length schedules. In: Computer and Job-Shop Scheduling Theory, Coffman E.G., Jr. (ed.), New York, Wiley, 1976, p. 51-99.

405. Sethi R. On the complexity of mean flow time scheduling. *Math. Oper. Res.*, 1977, **2**, N 4, p. 320-330.

406 Shwimer J. On the *N*-job, one-machine, sequencing-independent scheduling problem with tardiness penalties. A branch-bound solution. *Manag. Sci.*, 1972, **18**, N 6, p. 301-313.

407. Sidney J.B. An extension of Moore's due date algorithm. *Lect. Notes Econ. Math. Syst.*, 1973. **86**, p. 393-398.

408. Sidney J.B. Decomposition algorithms for single-machine sequencing with precedence relations and deferral costs. *Oper. Res.*, 1975, **23**, N 2, p. 283-298.

409. Sidney J.B. Optimal single machine scheduling with earliness and tardiness penalties. *Oper. Res.*, 1977, **25**, N 1, p. 62-69.

410. Sidney J.B. The two-machine maximum flow time problem with series parallel precedence relations. *Oper. Res.*, 1979, **27**, N 4, p. 782-791.

411. Sielken R.L., Jr. Sequencing with setup costs by zero-one mixed integer linear programming. *AIIE Trans.*, 1976, **8**, N 3, p. 369-371.

412. Simons B. A fast algorithm for single processor scheduling. In: 19th Annu. Symp. Found. Comput. Sci., New York, 1978, p. 246-252.

413. Simons B. A fast algorithm for multiprocessor scheduling. In: 21-st Annu. Symp. Found. Comput. Sci. 1980, New York, 1980, p. 50-53.

414. Simons B. On scheduling with release times and deadlines. In: Deterministic and Stochastic Scheduling. Proc. NATO Adv. Study and Res. Inst. Theor. Approaches Scheduling Probl., Dordrecht, 1982, p. 75-88.

415. Slowinski R. Scheduling preemptable tasks on unrelated processors with additional resources to minimize schedule length. *Lect. Notes Comput. Sci.*, 1978, **65**,

p. 536–547.

416. Slowinski R. L'ordonnancement des tâches préemptives sur les processeurs independants on présence de ressources supplémentairs. *RAIRO Inf.*, 1981, **15**, N 2, p. 155–166.

417. Smith W.E. Various optimizers for single stage production. *Naval Res. Log. Quart.*, 1956, **3**, N 1-2, p. 59–66.

418. Srinivasan V. A hybrid algorithm for the one machine sequencing problem to minimize total tardiness. *Naval Res. Log. Quart.*, 1971, **18**, N 3, p. 317–327.

419. Su Z.S., Sevcik K.C. A combinatorial approach to dynamic scheduling problems. *Oper. Res.*, 1978, **26**, N 5, p. 836–844.

420 Szwarc W. O pewnum zagadnieniu kolejnosci. Cz I. *Prz. statyst.*, 1962, **9**, N 4, p. 367–382.

421. Tang D.T., Wong C.K. A modified branch-and-bound strategy. *Inf. Process. Lett.*, 1973, **2**, N 3, p. 65–69.

422. Tilquin C. On scheduling with earliest and due dates on a group of identical machines. *Naval Res. Log. Quart.*, 1975, **22**, N 4, p. 777–785.

423. Townsend W. A branch-and-bound method for sequencing problems with linear and exponential penalty functions. *Oper. Res. Quart.*, 1977, **28**, N 1ii, p. 191–200.

424. Tremolieres R. Scheduling jobs of equal durations with tardiness costs and resource limitations. *J. Oper. Res. Soc.*, 1978, **29**, N 3, p. 229–233.

425. Ullman J.D. Polynomial complete scheduling problems. *Oper. Syst. Rev.*, 1973, **7**, N 4, p. 96–101.

426. Ullman J.D. *NP*-complete scheduling problems. *J. Comput. Syst. Sci.*, 1975, 10, N 3, p. 384–393.

427. Ullman J.D. Complexity of sequencing problems. In: Computer and Job-Shop Scheduling Theory, Coffman E.G., Jr. (ed.), New York, Wiley, 1976, p. 139–164.

428 Usn S.J.B. Scheduling to minimize the number of late jobs when set-up and processing times are uncertain. *Manag. Sci.*, 1973, **19**, N 11, p. 1283–1288.

429. Valdes J., Tarjan R.E., Lawler E.L. The recognition of series parallel digraphs. In: Proc. 11th Annu. ACM Symp. Theory Comput., New York, 1979, p. 1–12.

430. Valdes J., Tarjan R.E., Lawler E.L. The recognition of series parallel digraphs. *SIAM J. Comput.*, 1982. **11**, N2, p. 298–313.

431. Van Wassenhove L.N., Baker K.R. A bicriterion approach to time/cost trade-offs in sequencing. *Eur. J. Oper. Res.*, 1982, **11**, N 1, p. 48–54.

432. Van Wassenhove L.N., Gelders L. Four solution techniques for a general one machine

scheduling problem. A comparative study. *Eur. J. Oper. Res.*, 1978, **2**, N 4, p. 281-290.

433. Van Wassenhove L.N., Gelders L. Solving a bicriterion scheduling problems. *Eur. J. Oper. Res.*, 1980, **4**, N 1, p. 42-48.

434. Vickson R.G. Two single machine sequencing problems involving controllable job processing times. *AIIE Trans.*, 1980, **12**, N 3, p. 258-262.

435. Vickson R.G. Choosing the job sequence and processing times to minimize total processing plus flow cost on a single machine. *Oper. Res.*, 1980, **28**, N 5, p. 1155-1167.

436. Weeda P.J. A dynamic programming formulation for the one machine sequencing problem. *Eur. J. Oper. Res.*, 1978, **2**, N 4, p. 298-300.

437. Weiss H.J. A greedy heuristic for single machine sequencing with precedence constraints. *Manag. Sci.*, 1981, **27**, N 10, p. 1209-1216.

438. Zaloom V.A. Research abstract on a feasibility approach to optimal schedules. *Lect. Notes Econ. Math. Syst.*, 1973, **86**, p. 433-437.

439. Zaloom V.A., Vatz D. A note on the optimal scheduling of two parallel processors. *Naval Res. Log. Quart.*, 1975, **22**, N 4, p. 823-827.

ADDITIONAL REFERENCES

1*. Achugbue J.O., Chin F.Y. Bounds on schedules for independent tasks with similar execution times. *J. Assoc. Comput. Mach.*, 1981, **28**, p. 81-99.

2*. Albers S., Brucker P. The complexity of one-machine batching problems. Report 131, Department of Mathematics Computer Science, Osnabruck University, Osnabruck, 1990.

3*. Babat L.G. Linear functions on N-dimensional unit cube. *Doklady Akademii Nauk SSSR*, 1975, **222**, N 4, 761-762 (in Russian).

4*. Babat L.G. Approximate evaluation of a linear function on vertices of N-dimensional unit cube. In: Studies in Discrete Optimization. Moscow, Nauka Publishers, 1976, p. 156-169 (in Russian).

5*. Bakenrot V.Y. On the efficiency of an algorithm for constructing a schedule. *Kibernetika* (Kiev), 1980, N 1, p. 140-143 (in Russian).

6*. Baker K.R., Scudder G.D. On the assignment of optimal due dates. *J. Oper. Res. Soc.*, 1989, **40**, p. 93-95.

7*. Baker K.R., Scudder G.D. Sequencing with earliness and tardiness penalties: a review. *Oper. Res.*, 1990, **38**, N 1, p. 22-36.

8*. Baranova E.V., Gordon V.S. Optimal single machine scheduling with a subset of jobs scheduled on time. In: Algorithms and Programs for Solving Optimization Problems. Minsk, 1983, p. 67-76 (in Russian).

9*. Bellman R., Esogbue A., Nabeshima I. *Mathematical Aspects of Scheduling and Applications*. 1982, Oxford, Pergamon Press, 329 pp.

10*. Blazewicz J. Selected topics in scheduling theory. *Ann. Discr. Math.*, 1987, **31**, p. 1-60.

11*. Blazewicz J., Dror M., Weglarz J. Mathematical programming formulations for machine scheduling: a survey. *Eur. J. Oper. Res.*, 1991, **51**, p. 283–300.

12*. Blazewicz J., Finke G., Haupt R., Schmidt G. New trends in machine scheduling. *Eur. J. Oper. Res.*, 1988, **37**, p. 303–317.

13*. Blazewicz J., Lenstra J.K., Rinnooy Kan A.H.G. Scheduling subject to resource constraints: classification and complexity. *Discr. Appl. Math.*, 1983, **5**, N 1, p. 11–24.

14*. Chand S., Schneeberger H. Single machine scheduling to minimize weighted earliness subject to no tardy jobs. *Eur. J. Oper. Res.*, 1988, **34**, p. 221–230.

15*. Chen B. A note on LPT scheduling. *Oper. Res. Lett.*, 1993, **14**, p. 139–142.

16*. Cheng T.C.E. Optimal due-date determination and sequencing of n jobs on a single machine. *J. Oper. Res. Soc.*, 1984, **35**, p. 433–437.

17*. Cheng T.C.E. Optimal TWK-power due-date determination and sequencing. *Int. J. Syst. Sci.*, 1987, **18**, p. 1–7.

18*. Cheng T.C.E. Determination of optimal total-work-content due-dates for a single machine sequencing problem. *Eng. Optim.*, 1988, **14**, N 2, p. 121–125.

19*. Cheng T.C.E. Optimal constant due-date assignment and sequencing. *Int. J. Syst. Sci.*, 1988, **19**, p. 1351–1354.

20*. Cheng T.C.E. Optimal due-date assignment and sequencing in a single machine shop. *Appl. Math. Lett.*, 1989, **2**, N 1, p. 21–24.

21*. Cheng T.C.E. Optimal assignment of slack due-dates and sequencing in a single machine shop. *Appl. Math. Lett.*, 1989, **2**, p. 333–335.

22*. Cheng T.C.E. Optimal assignment of total-work-content due-dates and sequencing in a single-machine shop. *J. Oper. Res. Soc.*, 1991, **42**, p. 177–181.

23*. Cheng T.C.E., Gordon V.S. On optimal assignment of processing-time-plus-wait due-dates for preemptive single machine scheduling. Preprint N 28, Institute of Engineering Cybernetics, Byelorussian Academy of Sciences, Minsk, 1991, 16 pp.

24*. Cheng T.C.E., Gordon V.S. On batch delivery scheduling on a single machine. Preprint N 9, Institute of Engineering Cybernetics, Belarussian Academy of Sciences, Minsk, 1992, 15 pp.

25*. Cheng T.C.E., Gordon V.S. On optimal assignment of due-dates for preemptive single-machine scheduling. *Int. J. Math. Comput. Modelling*, to appear.

26*. Cheng T.C.E., Gordon V.S., Kovalyov M.Y. Batch delivery single machine scheduling as a parallel machine scheduling. Preprint, Institute of Engineering Cybernetics, Belarussian Academy of Sciences, Minsk, 1993, 14 pp.

27*. Cheng T.C.E., Gupta M.C. Survey of scheduling research involving the due-date determination decisions. *Eur. J. Oper. Res.*, 1989, **38**, p. 156-166.

28*. Cheng T.C.E., Kahlbacher H.G. Scheduling with delivery and earliness penalties. Working Paper, University of Manitoba, 1990, 9 pp.

29*. Cheng T.C.E., Li S. Some observations and extensions of the optimal TWK-power due-date determination and sequencing problem. *Comput. Math. Appl.*, 1989, **17**, p. 1103-1107.

30*. Coffman E.G., Yannakakis M., Magazine M.J., Santos C. Batch sizing and sequencing on a single machine. Report, AT&T Laboratories, Murray Hill, 1988.

31*. Coffman E.G., Nozari A., Yannakakis M. Optimal scheduling of products with two subassemblies on a single machine. *Oper. Res.*, 1989, **37**, N 3, p. 426-436.

32*. Cook S. Towards a complexity theory of synchronous parallel computation. *Enseign. Math.*, 1981, **2**, N 27, p. 99-124.

33*. Davis E., Jaffe J.M. Algorithms for scheduling tasks on unrelated processors. *J. Assoc. Comput. Mach.*, 1981, **28**, N 4, p. 721-736.

34*. De P., Ghosh J.B., Wells C.E. On the minimization of completion time variance with a bi-criteria extension. Working paper, University of Dayton, Dayton, Ohio, 1989.

35*. Dekel E., Sahni S. Parallel scheduling algorithms. In: Proc. 1981 Int. Conf. Parallel Process, New York, 1981, p. 350-351.

36*. Dekel E., Sahni S. Binary trees and parallel scheduling algorithms. *IEEE Trans. Comput.*, 1983, **32**, N3, p. 307-315.

37*. Dekel E., Sahni S. Parallel scheduling algorithms. *Oper. Res.*, 1983, **31**, N 1, p. 24-49.

38*. Dempster M.A.H., Lenstra J.K., Rinnooy Kan A.H.G. (eds.) *Deterministic and Stochastic Scheduling*, Dordrecht, Reidel, 1982.

39*. Dessouky M.I., Lageweg B.J., Lenstra J.K., van de Velde S.L. Scheduling identical jobs on uniform parallel machines. *Statist. Neerland.*, 1990, **44**, p. 115-123.

40*. Dobson G. Scheduling independent tasks on uniform processors. *SIAM J. Comput.*, 1984, **13**, p. 705-716.

41*. Du J., Leung J.Y.-T. Minimizing mean flow time with release time and deadline constraints. Technical Report, Computer Science Program, University of Texas, Dallas, 1988.

42*. Du J., Leung J.Y.-T. Scheduling tree-structured tasks with restricted execution times. *Inform. Process. Lett.*, 1988, **28**, p. 183-188.

43*. Du J., Leung J. Y, T. Scheduling tree-structured tasks on two processors to minimize schedule length. *SIAM J. Discrete Math.*, 1989, **2**, p. 176-196.

44*. Du J., Leung J.Y.-T. Minimizing total tardiness on one machine is NP-hard. *Math. Oper. Res.*, 1990, **15**, p. 483-495.

45*. Du J., Leung J.Y.-T., Wong C.S. Minimizing the number of late jobs with release time constraints. Technical report, Computer Science Program, University of Texas, Dallas, 1989.

46*. Du J., Leung J.Y.-T., Young G.H. Minimizing mean flow time with release time constraint. Technical report, Computer Science Program, University of Texas, Dallas, 1988.

47*. Du J., Leung J.Y.-T., Young G.H. Scheduling chain-structured tasks to minimize makespan and mean flow time. *Inform. and Comput.*, 1991, **92**, p. 219-236.

48*. Federgruen A., Groenevelt H. Preemptive scheduling of uniform machines by ordinary network flow techniques. *Manag. Sci.*, 1986, **32**, N 3, p. 341-349.

49*. Fischetti M., Martello S. Worst-case analysis of the differencing method for the partition problem. *Math. Programming*, 1987, **37**, p. 117-120.

50*. Frederickson G.N. Scheduling unit-time tasks with integer release times and deadlines. *Inform. Process. Lett.*, 1983, **16**, p. 171-173.

51*. Friesen D.K. Tighter bounds for the multifit processor scheduling algorithm. *SIAM J. Comput.*, 1984, **13**, p. 170-181.

52*. Friesen D.K. Tighter bounds for LPT scheduling on uniform processors. *SIAM J. Comput.*, 1987, **16**, p. 554-560.

53*. Friesen D.K., Langston M.A. Bounds for multifit scheduling on uniform processors. *SIAM J. Comput.*, 1983, **12**, p. 60-70.

54*. Friesen D.K., Langston M.A. Evaluation of a MULTIFIT-based scheduling algorithm. *J. Algorithms*, 1986, **7**, p. 35-59.

55*. Gabow H.N. Scheduling UET systems on two uniform processors and length two pipelines. SIAM J. Comput., 1988, **17**, p. 810-829.

56*. Gabow H.N., Tarjan R.E. A linear time algorithm for a special case of disjoint set union. *J. Comput. System Sci.*, 1985, **30**, p. 209-221.

57*. Garey M.R., Johnson D.S., Simons B.B., Tarjan R.E. Scheduling unit-time tasks with arbitrary release times and deadlines. *SIAM J. Comput.*, 1981, **10**, p. 256-269.

58*. Garey M.R., Johnson D.S., Tarjan R.E., Yannakakis M. Scheduling opposing forests. *SIAM J. Alg. Discr. Meth.* 1983, **4**, p. 72-93.

59*. Garey M.R., Tarjan R.E., Wilfong G.T. One-processor scheduling with symmetric earliness and tardiness penalties. *Math. Oper. Res.*, 1988, **13**, N2, 330-348.

60*. Gens G.V. Problems of resources allocation in hierarchical systems. *Izvestiya Akademii Nauk SSSR. Technicheskaya Kibernetika*, 1984, N 6, p. 22-28 (in Russian).

61*. Gens G.V., Levner E.V. Fast approximation algorithm for job sequencing with deadlines. *Discr. Appl. Math.*, 1981, **3**, p. 313-318.

62*. Gordon V.S. Parallel algorithms for minmax scheduling problem. Preprint, Institute of Engineering Cybernetics, Byelorussian Academy of Sciences, Minsk, 1987, 20 pp. (in Russian).

63*. Gordon V.S. A parallel algorithm for minimizing the maximal cost in a single machine scheduling. *Izvestiya Akademii Nauk SSSR. Technicheskaya Kibernetika*, 1989, N 3, p. 181-186 (in Russian).

64*. Gordon V.S. On an optimal assignment of slack due-dates and scheduling in a single-machine shop. Preprint N 7, Institute of Engineering Cybernetics, Byelorussian Academy of Sciences, Minsk, 1991, 8 p.

65*. Gordon V.S. A single machine scheduling problem to minimize the penalty related to the variable due dates. *Avtomatika i Telemekhanika*, 1992, N 2, p. 105-112 (in Russian).

66*. Gordon V.S. Parallel algorithms for scheduling problems. *Avtomatika i Telemekhanika*, 1992, N 5, p. 97-106 (in Russian).

67*. Gordon V.S. A note on optimal assignment of slack due-dates and scheduling in a single-machine shop. *Eur. J. Oper. Res.*, 1993, to appear.

68*. Gordon V.S., Baranova E.V. On a single machine scheduling problem to minimize the total cost. *Vestsi Akademii Navuk BSSR. Ser. fizika-matematychnykh navuk*, 1984, p. 113 (in Russian).

69*. Goyal D.K. Non-preemptive scheduling of unequal execution time tasks on two identical processors. Technical report CS-77-039, Computer Science Department, Washington State University, Pullman, 1977.

70*. Gupta S.K., Kyparisis J. Single machine scheduling research. *OMEGA*, 1987, **15**, p. 207-227.

71*. Gusfield D. Bounds for naive multiple machine scheduling with release times and deadlines. *J. Algorithms*, 1984, **5**, N 1, p. 1-6.

72*. Hall N.G., Kubiak W., Sethi S.P. Deviation of completion times about a restrictive common due date. Working Paper 89-19, College of Business, The Ohio State University, Columbus.

73*. Hall N.G., Kubiak W., Sethi S.P. Earliness-tardiness scheduling problems, II: Deviation of completion times about a restrictive common due date. *Oper. Res.*,

1991, **39**, N 5, 847–856.

74*. Hall N.G., Posner M.E. Weighted deviation of completion times about a common due date. Working Paper 89-15, College of Business, The Ohio State University, Columbus, 1989.

75*. Hall N.G., Posner M.E. Earliness-tardiness scheduling problems, I: weighted deviation of completion times about a common due date. *Oper. Res.*, 1991, **39**, N 5, p. 836–846.

76*. Hall L.A., Shmoys D.B. Jackson's rule for one-machine scheduling: Making a good heuristic better. *Math. Oper. Res.*, 1992, **17**, p. 22–35.

77*. Hariri A.M.A., Potts C.N., Van Wassenhove L.N. Single machine scheduling to minimize total weighted late work. Report 9012/A, Econometric Institute, Erasmus University Rotterdam, 1990.

78*. Helmbold D., Mayr E.W. Two processor scheduling is in *NC*. *SIAM J. Comput.*, 1987, **16**, p. 747–759.

79*. Helmbold D., Mayr E.W. Applications of parallel scheduling algorithms to families of perfect graphs. *Comput. Suppl.*, 1990, **7**, p. 93–107.

80*. Herrbach L.A., Leung J.Y.-T. Preemptive scheduling of equal length jobs on two machines to minimize mean flow time. *Oper. Res.*, 1990, **38**, N 3, p. 487–494.

81*. Hochbaum D.S., Shamir R. Strongly polynomial algorithms for the high multiplicity scheduling problem. *Oper. Res.*, 1991, **39**, N 4, p. 648–653.

82*. Hochbaum D.S., Shmoys D.B. Using dual approximation algorithm for scheduling problems: theoretical and practical results. *J. Assoc. Comput. Mach.*, 1987, **34**, p. 144–162.

83*. Hochbaum D.S., Shmoys D.B. A polynomial approximation scheme for machine scheduling on uniform processors: using the dual approximation approach. *SIAM J. Comput.*, 1988, **17**, p. 539–551.

84*. Hoogeven J.A., van de Velde S.L. Scheduling around a small common due date. Report BS-8914, CWI, Amsterdam, 1989.

85*. Jaffe J.M. Efficient scheduling of tasks without full use of processor resources. *Theor. Comput. Sci.*, 1980, **12**, p. 1–17.

86*. Johnson D.S. The *NP*-completeness column: an ongoing guide: seventh edition. *J. Algorithms*, 1983, **4**, p. 189–203.

87*. Kanet J.J. Minimizing the average deviation of job completion times about a common due date. *Naval Res. Logist. Quart.*, 1981, **28**, p. 643–651.

88*. Kawaguchi T., Kyan S. Worst case bound of an LRF schedule for the mean weighted flow-time problem. *SIAM J. Comput.*, 1986, **15**, p. 1119-1129.

89*. Kellerer H., Kotov V. V. A linear compound algorithm for the partitioning problem. *Inform. Proc. Lett.* (to appear).

90*. Kindervater G.A.P., Lenstra J.K. Parallel algorithms in combinatorial optimization: An annotated bibliography. Report BW 189/83, CWI, Amsterdam, 1983, 22 pp.

91*. Kindervater G.A.P., Lenstra J.K. Parallel computing in combinatorial optimization. *Ann. Oper. Res.*, 1988, **14**, p. 245-289.

92*. Korte B., Schrader R. On the existence of fast approximation schemes. In: Nonlinear Programm. 4. Proc. 4th Symp., Madison, 1980, New York, 1981, p. 415-437.

93*. Kovalyov M.Y. A decomposition approach to the construction of ε-approximation algorithms. In: Mathematical and Software Support of Computer Systems, Minsk, 1985, p. 3-8 (in Russian).

94*. Kovalyov M.Y. An efficient ε-approximate algorithm for the problem of total completion time minimizing in two-stage system consisting of parallel machines. *Vestsi Akademii Navuk BSSR. Ser. fizika – matematychnykh navuk*, 1985, N 3, p. 119 (in Russian).

95*. Kovalyov M.Y. An approximate solution to the problem of total tardiness. *Vestsi Akademii Navuk BSSR. Ser. fizika – matematychnykh navuk*, 1985, N 5, p. 110 (in Russian).

96*. Kovalyov M.Y. Minimizing the total weighted number of late jobs on a single machine. *Zhurnal Vychislitel'noj Matematiki i Matematicheskoj Fiziki*, 1991, **31**, N 11, p. 1731-1739 (in Russian).

97*. Kovalyov M.Y. Non-preemptive parallel machine scheduling with sequence independent set-ups to minimize maximum completion time. Preprint N16, Institute of Engineering Cybernetics, Byelorussian Academy of Sciences, Minsk, 1991.

98*. Kovalyov M.Y. Binary search procedure for improving bounds of objective function optimal value. *Izvestiya Akademii Nauk Belarusi. Ser. Fiziko – matematicheskich Nauk*, 1993, N 1, 92-97.

99*. Kovalyov M.Y., Potts C.N., Van Wassenhove L.N. A fully polynomial approximation scheme for scheduling a single machine to minimize total weighted late work. *Math. Oper. Res.* (to appear).

100*. Kovalyov M.Y., Shafransky Y.M. Construction of ε-approximation algorithms for solving some scheduling problems. In: Procediings of the VIII All-Union Seminar-Conference: Control in Large Systems, Alma-Ata, 1983, p. 25-26 (in Russian).

101*. Kovalyov M.Y., Shafransky Y.M. The construction of ε-approximate algorithms for the optimization of functions in successively constructed sets. *U.S.S.R. Comput. Math. Math. Phys.*, 1986, **26**, N 4, p. 30-38.

102*. Kovalyov M.Y., Shafransky Y.M., Strusevich V.A., Tanaev V.S., Tuzikov A.V. Approximation scheduling algorithms: a survey. *Optimization*, **20**, N 6, p. 859-878.

103*. Kovalyov M.Y., Tuzikov A.V. Construction of an ε-approximation of Pareto set for some bicriteria problems. In: Mathematical Questions of Computer Aided Design and Testing, Minsk, 1986, p. 126-130 (in Russian).

104*. Kovalyov M.Y., Tuzikov A.V. Minimizing maximum penalty on a single machine under group technology and precedence constraints. Preprint N 35, Institute of Engineering Cybernetics, Byelorussian Academy of Sciences, Minsk, 1991.

105*. Kubiak W. Completion time variance minimization on single machine is difficult. Technical Report at the Memorial University of Newfoundland, 1992.

106*. Kunde M. Nonpreemptive LP-scheduling on homogeneous multiprocessor systems. *SIAM J. Comput.*, 1981, **10**, 151-173.

107*. Kunde M. A multifit algorithm for uniform multiprocessor scheduling. *Lect. Notes Comput. Sci.* 1982, **145**, p. 175-185.

108*. Kunde M. Bounds for multifit scheduling algorithms on uniform multiprocessor systems. Bericht 8203, Institut für Informatik und Praktische Mathematik, Kiel, 1982.

109*. Kunde M., Langston M.A., Liu J.-M. On a special case of uniform processor scheduling. *J. Algorithms*, 1988, **9**, N 2, p. 287-296.

110*. Kunde M., Steppat H. First fit decreasing scheduling on uniform multiprocessors. *Discr. Appl. Math.*, 1985, **10**, N 2, p. 165-177.

111*. Langston M.A., Liu J.M. On special case of uniform processor scheduling. TR CS-93-107, Washington State University, Pullman, 1983.

112*. Lawler E.L. Scheduling a single machine to minimize the number of late jobs. Preprint, Computer Science Division, University of California, Berkeley. 1982.

113*. Lawler E.L. A fully polynomial approximation scheme for the total tardiness problem. *Oper. Res. Lett.*, 1982, **1**, p. 207-208.

114*. Lawler E.L. Recent results in the theory of machine scheduling. In: Mathematical Programming: the State of the Art - Bonn 1982, Grötschel M., Korte B. (eds.), Berlin, Springer, 1983, p. 202-234.

115*. Lawler E.L., Lenstra J.K., Rinnooy Kan A.H.G., Shmoys D.B. Sequencing and scheduling: algorithms and complexity. In: Handbook in Operations Research and Management Science, Vol. 4, Logistics of Production and Inventory, Graves S.C.,

Rinnooy Kan A.H.G., Zipkin P.H. (eds.), Amsterdam, North-Holland, 1993, p. 445-522.

116*. Lawler E.L., Martel C.U. Preemptive scheduling of two uniform machines to minimize the number of late jobs. *Oper. Res.*, 1989, **37**, p. 314-318.

117*. Lenstra J.K., Rinnooy Kan A.H.G. Scheduling theory since 1981: An annotated bibliography. Report 8324/O, Erasmus University, Rotterdam, 1983, 27 pp.

118*. Lenstra J.K., Rinnooy Kan A.H.G. New directions in scheduling theory. *Oper. Res. Lett.*, 1984, **2**, N 6, p. 255-259.

119*. Lenstra J.K., Rinnooy Kan A.H.G. Sequencing and scheduling. In: Combinatorial Optimization: Annotated bibliographies, O'hEigeartaigh M., Lenstra J.K., Rinnooy Kan A.H.G. (eds.), Chichester, Wiley, 1985, p. 164-189.

120*. Lenstra J.K., Shmoys D.B., Tardos É. Approximation algorithms for scheduling unrelated parallel machines. *Math. Programming*, 1990, **46**, p. 259-271.

121*. Leung J.Y.-T., Young G.H. Minimizing total tardiness on a single machine with precedence constraints. Technical report, Computer Science Program, University of Texas, Dallas, 1989.

122*. Leung J.Y.-T., Young G.H. Minimizing Total Tardiness on a Single Machine with Precedence Constraints, Technical Report, Computer Science Program, University of Texas, Dallas, 1989.

123*. Leung J.Y.-T., Young G.H. Preemptive scheduling to minimize mean weighted flow time. *Inform. Process. Lett.*, 1990, **34**, 47-50.

124*. Liu J.W.S., Liu C.L. Bounds on scheduling algorithms for heterogeneous computing systems. Technical Report UIUCDCS-R-74-632, Department of Computer Science, University of Illinois at Urbana-Champaign, 1974, 68 pp.

125*. Martel C.U. A parallel algorithm for preemptive scheduling of uniform machines. *J. Parallel Distr. Comput.*, 1988, **5**, p. 700-715.

126*. Masuda T., Ishii H., Nishida T. Some bounds on approximation algorithms for $n/m/1/L_{max}$ and $n/2/F/L_{max}$ scheduling problems. *J. Oper. Res. Soc. Japan*, 1983, **26**, N 3, p. 212-224.

127*. McCormick S.T., Pinedo M.L. Scheduling n independent jobs on m uniform machines with both flow time and makespan objectives: A parametric analysis. Preprint, Department of Industrial Engineering and Operations Research, Columbia University, New York, 1989.

128*. Meilijson I., Tamir A. Minimizing flow time on parallel identical processors with variable unit processing time. *Oper. Res.*, 1984, **32**, p. 440-446.

129*. Mohring R.H. Computationally tractable classes of ordered sets. In: Algorithms and

Order, Rival I. (ed.), Dordrecht, Kluwer Academic Publishers, 1989, p. 105–193.

130*. Monma C.L. Linear-time algorithms for scheduling on parallel processors. *Oper. Res.*, 1982, **30**, N 1, p. 116–124.

131*. Monma C.L., Potts C.N. On the complexity of scheduling with batch setup times. *Oper. Res.*, 1989, **37**, N 5, p. 798–804.

133*. Morrison J.F. A note on LPT scheduling. *Oper. Res. Lett.*, 1988, **7**, p. 77–79.

134*. Nakajima K., Leung J. Y,T., Hakimi S.L. Optimal two processor scheduling of tree precedence constrained tasks with two execution times. *Performance Evaluation*, 1981, **1**, p. 320–330.

135*. Panwalkar S.S., Smith M.L., Seidmann A. Common due-date assignment to minimize total penalty for the one machine sequencing problem, *Oper. Res.*, 1982, **30**, p. 391–399.

136*. Paz A., Moran S. Non deterministic polynomial optimization problems and their approximations. *Theor. Comput. Sci.*, 1981, **15**, N 3, p. 251–277.

137*. Potts C.N. Analysis of a heuristic for one machine sequencing with release dates and delivery times. *Oper. Res.*, 1980, **28**, p. 1436–1441.

138*. Potts C.N. Analysis of a linear programming heuristic for scheduling unrelated parallel machines. *Discr. Appl. Math.*, 1985, **10**, p. 155–164.

139*. Potts C.N., Van Wassenhove L.N. Approximation algorithms for scheduling a single machine to minimize total late work. Working Paper N 91/18/TM, INSEAD, Fontainebleau, France, 1991.

140*. Potts C.N., Van Wassenhove L.N. Integrating scheduling with batching and lot-sizing: a review of algorithms and complexity. *J. Oper. Res. Soc.*, 1991, **43**, N 5, p. 395–406.

141*. Potts C.N., Van Wassenhove L.N. Single machine scheduling to minimize total late work. *Oper. Res.*, ???

142*. Ribero C.C. Parallel computer models and combinatorial algorithms. *Ann. Discr. Math.*, 1987, **31**, p. 325–364.

143*. Röck H., Schmidt G. Machine aggregation heuristics in shop-scheduling. *Methods Oper. Res.*, 1983, **45**, p. 303–314.

144*. Seidmann A., Panwalkar S.S., Smith M.L. Optimal assignment of due-dates for a single processor scheduling problem. *Int. J. Prod. Res.*, 1981, **19**, p. 393–399.

145*. Seidmann A., Smith M.L. Due-date assignment for production systems. *Manag. Sci.*, 1981, **27**, p. 571–581.

146*. Shafransky Y.M. On the *NP*-hardness of some choice problems. In: Methods for Solving

Extremal Problems and Related Topics, Minsk, 1990, p. 98-103 (in Russian).

147*. Shafransky Y.M. The choice of parameters of interconnected jobs. In: Proceedings of Colloquium on Scheduling Theory and its Applications, Minsk, 1992, p. 46-52.

148*. Shafransky Y.M., Tuzikov A.V. Sequencing with ordered criteria. Preprint N 36, Institute of Engineering Cybernetics of Academy of Sciences of BSSR, Minsk, 1991, 37 pp.

151*. Sin C.C.S., Cheng T.C.E. On the *NP*-completeness of the $n/m/parallel/\sum_{i\in M}\{\sum w_j \sum t_j\}$ scheduling problem. *Appl. Math. Lett.*, 1989, **2**, N 4, 389-390.

152*. Spinrad J. Worst-case analysis of a scheduling algorithm. *Oper. Res. Lett.*, 1985, **4**, N 1, p. 9-11.

153*. Steppat H. Packungsalgoritmen für die Ablaufplanung in uniformen Mehrprozessor-systemen, Diplomarbeit, Kiel, 1983.

154*. Tanaev V.S. Symmetric functions in scheduling theory (single machine equal release dates problem). *Vestsi Akademii Navuk BSSR. Ser. fizika – matematychnykh navuk*, 1992, N 5-6, p. 97-101 (in Russian).

155*. Tanaev V.S. Symmetric functions in scheduling theory (single machine different release dates problem). *Vestsi Akademii Navuk BSSR. Ser. fizika – matematychnykh navuk*, 1993, N 1, p. 84-87 (in Russian).

156*. Tang C.S. Scheduling batches on parallel machines with major and minor set-ups. *Eur. J. Oper. Res.*, 1990, **46**, p. 28-37.

157*. Tuzikov A.V. Minimizing a maximal penalty and solving some classes of bicriteria scheduling problems. *Doklady Akademii Nauk BSSR*, 1983, **27**, N 10, p. 907-910 (in Russian).

158*. Tuzikov A.V. On bicriteria scheduling problems subject to variation of processing times. *Zhurnal Vychislitel'noj Matememematiki i Matematicheskoj Fiziki*, 1984, N 10, p. 1585-1590 (in Russian).

159*. Tuzikov A.V. On bicriteria scheduling problems. In: Optimization, Decision Making, Microprocessor Systems, Sofia, Bulgarian Acad. Sci., 1985, p. 163-168.

160*. van de Velde S.L. A simpler and faster algorithm for optimal total-work-content-power due-date determination. *Mathematical and Computer Modelling*, 1990, **13**, p. 81-83.

161*. Yuan J. The *NP*-hardness of the single machine common due date weighted tardiness problem. *System Sci. Math. Studies*, 1992, **5**, N 4, p. 328-333.

INDEX

Adjacency matrix
of a graph 43
Algorithm
efficient 59
packing 81, 134
polynomial–time 65
pseudopolynomial-time 66
Alphabet 59
binary 64
external (of a Turing machine) 59
inner (of a Turing machine) 59
input (of a Turing machine) 59
unary 64
Arc 42
incident to vertices 42
transitive 44

Cartesian product
of sets 42
Chain 46
Class $\mathcal{NP}$ 62
Class $\mathcal{P}$ 62
Clique 288
Circuit 42
Completion time 5
Composite element 197
mixed 236
uniform 236
Composition (of graphs)
series 44
parallel 44
ω-series 243
Component (of a graph)
connected 43
decomposition 45
ω-decomposition 244
Cost function 5
Cumulative processing cost 136

Deadlock sequence
of transformations I, II 224
Due date 4
Degree of a vertex 43

Edge 42
incident to a vertex 42
Element
maximal (with respect to a pseudo-order) 46
maximal (with respect to a strict order) 46
minimal (with respect to a pseudo-order) 46
minimal (with respect to a strict order) 46
Elements
incomparable (with respect to a strict order) 47

Flow shop system 1
Forest 44
Function
concave 73
quasi-concave 74
transition (of a Turing machine) 60, 61
e-quasi-concave 74

Graph
connected 43
complete (non-directed) 42
deadlock 224
decomposable 44
directed 42
non-decomposable 47
non-directed 42
reducible 234
reduction (of a binary transitive relation) 46
series-parallel 44

Graph
transitive 44
tree-like 44
ω-series-parallel 243
Graphs
isomorphic 42

Height
of a vertex 43
weighted (of a vertex) 146
Intree 44
Isomorphism of graphs 42
Indegree
of a vertex 43

Job shop system 1

Label
of a vertex of a 2-3-tree 51
Language 62
feasible for a Turing machine 62
NP-complete 62

Length
of a path 42
of a permutation 49
of a schedule 122, 138
of a word 59
weighted (of a path) 140
Linear form 189
Loop 42

Machines
identical 2
parallel 1

Maximal cost 5
Multi-stage system 1

Number of late jobs 6

Operation
 elementary 64
 of identifying the vertices 224
 of including an arc 197
 of parallel composition of graphs 44
 of series composition of graphs 44
Order
 non-strict 41
 strict 41
Outtree 44
 balanced 50
 binary 45
Outdegree
 of a vertex in a directed graph 43

Path 42
 simple 42
Permutation
 complete 49
 feasible with respect to a graph 49
 feasible with respect to a strict order 49
 non-increasing with respect to pseudo-order 56
 non-decreasing with respect to pseudo-order 56
 of the elements of a set 49
 partial 49
Predecessor of a vertex 42
 direct 42
Preemption 2
Priority
 of a permutation 187
 of a composite element 197
Priority-generating function 187
Priority function 187
 auto-bounded 230
Problem
 clique 288
 decision 63
 exact cover by 3-sets 272
 extremal combinatorial 63
 linear arrangement 297
 partition 254
 standard 66
 vertex covering 280
 3-partition 260
 NP-complete 64
  in the strong sense 66
 NP-hard 64
  in the strong sense 66
 Procedure
  for reconstructing a graph 45
  for uniting subsets 53
  for constructing
  a balanced 2-3-tree 51
  for transforming an intree
  into an ω-chain 205
  for transforming an outtree
  into an ω-chain 205
  for deleting an element
  from a set 55
Processing time
 of a job 2
Pseudo-order 41

Quasi-order 41

Rank (level) of a vertex 43
Reducibility 62, 65, 67

Reduction
polynomial 62, 65
pseudopolynomial 67
Relation
antireflexive binary 41
antisymmetric binary 41
asymmetric binary 41
binary 42, 41
non-strict order 41
pseudo-order 41
quasi-order 41
reflexive binary 41
symmetric binary 41
total binary 41
transitive binary 41
strict order 41
Root of an outtree 44
Route 42
Running time
of a deterministic Turing machine 61
of a non-deterministic Turing machine 61

Schedule 2
feasible with respect to deadlines 4
feasible with respect to precedence constraints 4
feasible with respect to resource constraints 4
machine-sharing 138
optimal 5
time-optimal 5
Schedules
conjugate 167
Set
finite 40
linear ordered 41
non-strictly ordered 41
Set
partially ordered 41
pseudo-ordered 41
quasi-ordered 41
strictly ordered 41
totally ordered 41
Single-stage system 1
Speed of a machine 2
Stage of processing 1
State
inner (of a Turing machine) 59
of a Turing machine 60
Step
of a Turing machine 60
Subgraph 42
induced 42
Subtree
of a vertex of an intree 44
of a vertex of an outtree 44
with the root x of an intree 44
with the root x of an outtree 44
Subproblem 66
Successor of a vertex 42
direct 42

Tape
of a Turing machine 59
Tardiness 5
total 6
weighted total 6
Time complexity
of an algorithm 64
of a Turing machine 61
polynomial 65
Total cost 5

Transformation
  I 202
  II 202
  mixed (I or II) 236
  uniform (I or II) 236
Transitive closure
  of a graph 44
Tree
  decomposition (of a graph) 45
  complete decomposition (of a graph) 46
Turing machine
  deterministic 59
  non-deterministic 61

Vertex
  initial 43
  intermediate 43
  isolated 43
  of a graph 42
  operational (of a decomposition tree) 45
  terminal(a leaf) 43
Vertices
  adjacent 42

Vertex covering 280

Weighted number of late jobs 6
Weighted total flow time 6
Word
  feasible for a deterministic Turing machine 61
  feasible for a non-deterministic Turing machine 61

1-priority function 246
1-priority-generating function over a set 246

2-3-tree 50
  balanced 50
  balanced ordered 57
D-algorithm 228
h-algorithm 122
h-schedule 122
λ-schedule 123, 129, 148
w-chain 203

Also of interest: *Scheduling Theory. Multi-Stage Systems*, by V.S. Tanaev, Y.N. Sotskov and V.A. Strusevich. 1994
ISBN 0-7923-2854-X

The Contents is shown below

CONTENTS

Preface vii

Introduction 1

Chapter 1

Flow Shop 30

1. Maximal Completion Time. Two Machines 31
2. Maximal Completion Time. Three and More Machines 41
3. Maximal Completion Time with No-Wait in Process 49
4. Maximal Lateness 65
5. Total Flow Time 69
6. Ordered Matrices of Processing Times 75
7. Dominant Matrices of Processing Times 89
8. Approximation Algorithms 104
9. Bibliography and Review 113

Chapter 2

Job Shop 128

1. Optimal Processing of Two Jobs 129
2. Maximal Lateness 140
3. Maximal Completion Time. Equal Processing Times 147
4. Maximal Completion Time. Arbitrary Processing Times 157
5. Maximal Completion Time with No-Wait in Process 167
6. Bibliography and Review 173

Chapter 3
Open Shop 179
1. Maximal Completion Time. Two Machines 180
2. Maximal Completion Time. Three and More Machines 203
3. Maximal Completion Time. Preemption 212
4. Maximal Completion Time. Precedence Constraints 226
5. Due Dates 242
6. Total Flow Time. Equal Processing Times 254
7. Total Flow Time. Arbitrary Processing Times 261
8. Bibliography and Review 271

Chapter 4
Mixed Graph Problems 280
1. Network Representation of Processing Systems 281
2. Mixed Graphs 292
3. Branch-and-Bound Method 301
4. Optimization of Processing Systems 312
5. Stability of Optimal Schedules 319
6. Bibliography and Review 327

References 335

Additional References 391

Index 365

Other *Mathematics and Its Applications* titles of interest:

B.S. Razumikhin: *Physical Models and Equilibrium Methods in Programming and Economics.* 1984, 368 pp. ISBN 90-277-1644-7

N.K. Bose (ed.): *Multidimensional Systems Theory. Progress, Directions and Open Problems in Multidimensional Systems.* 1985, 280 pp. ISBN 90-277-1764-8

J. Szep and F. Forgo: *Introduction to the Theory of Games.* 1985, 412 pp. ISBN 90-277-1404-5

V. Komkov: *Variational Principles of Continuum Mechanics with Engineering Applications. Volume 1: Critical Points Theory.* 1986, 398 pp. ISBN 90-277-2157-2

V. Barbu and Th. Precupanu: *Convexity and Optimization in Banach Spaces.* 1986, 416 pp. ISBN 90-277-1761-3

M. Fliess and M. Hazewinkel (eds.): *Algebraic and Geometric Methods in Nonlinear Control Theory.* 1986, 658 pp. ISBN 90-277-2286-2

P.J.M. van Laarhoven and E.H.L. Aarts: *Simulated Annealing: Theory and Applications.* 1987, 198 pp. ISBN 90-277-2513-6

B.S. Razumikhin: *Classical Principles and Optimization Problems.* 1987, 528 pp. ISBN 90-277-2605-1

S. Rolewicz: *Functional Analysis and Control Theory. Linear Systems.* 1987, 544 pp. ISBN 90-277-2186-6

V. Komkov: *Variational Principles of Continuum Mechanics with Engineering Applications. Volume 2: Introduction to Optimal Design Theory.* 1988, 288 pp. ISBN 90-277-2639-6

A.A. Pervozvanskii and V.G. Gaitsgori: *Theory of Suboptimal Decisions. Decomposition and Aggregation.* 1988, 404 pp. *out of print,* ISBN 90-277-2401-6

J. Mockus: *Bayesian Approach to Global Optimization. Theory and Applications.* 1989, 272 pp. ISBN 0-7923-0115-3

Du Dingzhu and Hu Guoding (eds.): *Combinatorics, Computing and Complexity.* 1989, 248 pp. ISBN 0-7923-0308-3

M. Iri and K. Tanabe: *Mathematical Programming. Recent Developments and Applications.* 1989, 392 pp. ISBN 0-7923-0490-X

A.T. Fomenko: *Variational Principles in Topology. Multidimensional Minimal Surface Theory.* 1990, 388 pp. ISBN 0-7923-0230-3

A.G. Butkovskiy and Yu.I. Samoilenko: *Control of Quantum-Mechanical Processes and Systems.* 1990, 246 pp. ISBN 0-7923-0689-9

A.V. Gheorghe: *Decision Processes in Dynamic Probabilistic Systems.* 1990, 372 pp. ISBN 0-7923-0544-2

Other *Mathematics and Its Applications* titles of interest:

A.G. Butkovskiy: *Phase Portraits of Control Dynamical Systems.* 1991, 180 pp.
ISBN 0-7923-1057-8

A.A. Zhigljavsky: *Theory of the Global Random Search.* 1991, 360 pp.
ISBN 0-7923-1122-1

G. Ruhe: *Algorithmic Aspects of Flows in Networks.* 1991, 220 pp.
ISBN 0-7923-1151-5

S. Walukuwiecz: *Integer Programming.* 1991, 196 pp. ISBN 0-7923-0726-7

M. Kisielewicz: *Differential Inclusions and Optimal Control.* 1991, 320 pp.
ISBN 0-7923-0675-9

J. Klamka: *Controllability of Dynamical Systems.* 1991, 260 pp.
ISBN 0-7923-0822-0

V.N. Fomin: *Discrete Linear Control Systems.* 1991, 302 pp. ISBN 0-7923-1248-1

L. Xiao-Xin: *Absolute Stability of Nonlinear Control Systems.* 1992, 180 pp.
ISBN 0-7923-1988-5

A. Halanay and V. Rasvan: *Applications of Liapunov Methods in Stability.* 1993, 238 pp.
ISBN 0-7923-2120-0

D. den Hertog: *Interior Point Approach to Linear, Quadratic and Convex Programming.* 1994, 208 pp.
ISBN 0-7923-2734-9

V.S. Tanaev, V.S. Gordon and Y.M. Shafranksy: *Scheduling Theory. Single-Stage Systems.* 1994, 380 pp.
ISBN 0-7923-2853-1

V.S. Tanaev, Y.N. Sotskov and V.A. Strusevich: *Scheduling Theory. Multi-Stage Systems.* 1994, 412 pp.
ISBN 0-7923-2854-X

L.D. Akulenko: *Problems and Methods of Optimal Control.* 1994, 356 pp.
ISBN 0-7923-2855-8